The Biology of Seeds

Recent Research Advances

The Biology of Seeds

Recent Research Advances

Proceedings of the Seventh International Workshop on Seeds, Salamanca, Spain 2002

Edited by

G. Nicolás

University of Salamanca
Spain

K.J. Bradford

University of California
Davis, USA

D. Côme

University Pierre and Marie Curie
Paris, France

and

H.W. Pritchard

Royal Botanic Gardens Kew
Wakehurst Place
Ardingly, UK

CABI Publishing

CABI is a trading name of CAB International

CABI Head Office
Nosworthy Way
Wallingford
Oxfordshire OX10 8DE
UK

Tel: +44 (0)1491 832111
Fax: +44 (0)1491 833508
Email: cabi@cabi.org
Web site: www.cabi.org

CABI North American Office
875 Massachusetts Avenue
7th Floor
Cambridge, MA 02139
USA

Tel: +1 617 395 4056
Fax: +1 617 354 6875
Email: cabi-nao@cabi.org

A catalogue record for this book is available from the British Library, London, UK.

Library of Congress Cataloging-in-Publication Data
International Workshop on Seeds (7th : 2002 : Salamanca, Spain)
The biology of seeds : recent research advances : proceedings of the Seventh International Workshop on Seeds, Salamanca, Spain 2002 / edited by G. Nicolás ... [et al.].
p. cm.
Includes bibliographical references (p.).
ISBN 0-85199-653-1
1. Seeds—Congresses. I. Nicolás, G (Gregorio) II. Title.
QK661.156 2002
575.6'8—dc21
2002156121

ISBN-13 978-085199-653-0
ISBN-10 0-85199-653-1

First printed 2003
Transferred to print on demand 2011

Printed and bound in the UK by CPI Group (UK) Ltd, Croydon, CR0 4YY

Contents

Part III. Desiccation and Other Stress Tolerances and Conservation

Part IV. Seed Ecology

Part V. Seed Biotechnology

Closing Lecture

Contributors

Z. Abraham, *Laboratorio de Bioquímica y Biología Molecular, Departamento de Biotecnología-UPM, ETSI Agrónomos, Ciudad Universitaria s/n, 28040 Madrid, Spain.*

S.W. Adkins, *Tropical and Subtropical Weed Research Unit, University of Queensland, St Lucia Campus, Brisbane, Qld 4072, Australia.*

R. Aguilar, *Departamento de Bioquímica, Facultad de Química, Universidad Nacional Autónoma de México, 04510 México DF, Mexico.*

M.K. Ashley, *School of Biological Science, University of Manchester, UK.*

J.-N. Aubertot, *Unité d'Agronomie de Laon-Reims-Mons, rue Fernand Christ, 02007 Laon cedex, France.*

C. Aubry, *Laboratoire de Génétique, UMR PMS, UFR Sciences, 2 bd Lavoisier, 49045 Angers, France.*

C. Bailly, *Physiologie Végétale Appliquée, Université Pierre et Marie Curie, Tour 53, 1er étage, 4 place Jussieu, 75252 Paris cedex 05, France.*

J.M. Baskin and C.C. Baskin, *University of Kentucky, Lexington, KY 40506-0225, USA.*

M. Bastida, *Departamento de Genética Molecular Institut de Biologia Molecular de Barcelona, CID-CSIC, Jordi Girona 18, 08034 Barcelona, Spain.*

A. Benamar, *UMR PMS, Université d'Angers, LRPV, 16 bd Lavoisier, 49045 Angers, France.*

R.L. Benech-Arnold, *IFEVA, Facultad de Agronomía, Universidad de Buenos Aires, Av. San Martín 4453, (C1417DSE) Buenos Aires, Argentina.*

P. Berjak, *School of Life and Environmental Sciences, University of Natal, Durban, 4041 South Africa.*

J.D. Bewley, *Department of Botany, University of Guelph, Guelph, Ontario N1G 2W1, Canada.*

R. Bogatek, *Department of Plant Physiology, Warsaw Agricultural University, Rakowiecka 26–30, 02-528 Warsaw, Poland.*

J. Boiffin, *Unité d'Agronomie de Laon-Reims-Mons, rue Fernand Christ, 02007 Laon cedex, France.*

L. Borisjuk, *Institut für Pflanzengenetik und Kulturpflanzenforschung (IPK), D-06466 Gatersleben, Germany.*

R. Bottini, *Laboratorio de Fisiología Vegetal, Facultad de Ciencias Exactas, Físicas, Químicas y Naturales, Universidad Nacional de Río Cuarto, Campus Universitario, (5800) Rio Cuarto, Provincia de Córdoba, Argentina.*

I. Boubriak, *Millennium Seed Bank Project, Royal Botanic Gardens Kew, Wakehurst Place, Ardingly, West Sussex RH17 6TN, UK.*

K.J. Bradford, *Department of Vegetable Crops, One Shields Avenue, University of California, Davis, CA 95616-8631, USA.*

C.M. Bray, *School of Biological Science, University of Manchester, UK.*

J. Buitink, *INH, UMR PMS, Université d'Angers, 16 bd Lavoisier, 49045 Angers, France.*

A.P. Calvo, *Departamento de Fisiología Vegetal, Facultad de Biología, Universidad de Salamanca, Plaza de los Doctores de la Reina s/n, 37007 Salamanca, Spain.*

P. Carbonero, *Laboratorio de Bioquímica y Biología Molecular, Departamento de Biotecnología-UPM, ETSI Agrónomos, Ciudad Universitaria s/n, 28040 Madrid, Spain.*

F. Carrari, *Laboratorio de Fisiología y Biología Molecular, FCEyN, Universidad de Buenos Aires, Ciudad Universitaria, Pabellón II, (1428) Buenos Aires, Argentina.*

J.M. Casacuberta, *Departamento de Genética Molecular, IBMB-CSIC, Jordi Girona 18, 08034 Barcelona, Spain.*

F.J. Cejudo, *Instituto de Bioquímica Vegetal y Fotosíntesis, Centro de Investigaciones Científicas Isla de la Cartuja, Avda Américo Vespucio s/n, 41092-Sevilla, Spain.*

K.M. Chandler, *Laboratory of Plasma Studies, Cornell University, Ithaca, NY 14853, USA.*

A.L. Chateigner-Boutin, *Laboratoire de Génétique, UMR PMS, UFR Sciences, 2 bd Lavoisier, 49045 Angers, France.*

R. Chen, *Xishuangbanna Tropical Botanical Garden, The Chinese Academy of Science, Menglun, Mengla 666303, Yunnan, China.*

D. Côme, *Physiologie Végétale Appliquée, Université Pierre et Marie Curie, Tour 53, 1er étage, 4 place Jussieu, 75252 Paris cedex 05, France.*

F. Corbineau, *Physiologie Végétale Appliquée, Université Pierre et Marie Curie, Tour 53, 1er étage, 4 place Jussieu, 75252 Paris cedex 05, France.*

H. Demol, *Flanders Interuniversity Institute for Biotechnology and Department of Biochemistry, Gent University, Gent, Belgium.*

I. Díaz, *Laboratorio de Bioquímica y Biología Molecular, Departamento de Biotecnología-UPM, ETSI Agrónomos, Ciudad Universitaria s/n, 28040 Madrid, Spain.*

T.D. Dinkova, *Departamento de Bioquímica, Facultad de Química, Universidad Nacional Autónoma de México, 04510 México DF, México.*

F. Domínguez, *Instituto de Bioquímica Vegetal y Fotosíntesis, Centro de Investigaciones Científicas Isla de la Cartuja, Avda Américo Vespucio s/n, 41092-Sevilla, Spain.*

F. Dorsainvil, *Unité d'Argronomie Laon-Reims-Mons, rue Fernand Christ, 02007 Laon cedex, France.*

B. Downie, *Department of Horticulture, University of Kentucky, Lexington, KY 40546, USA.*

P. Dubrulle, *Unité d'Agronomie de Laon-Reims-Mons, rue Fernand Christ, 02007 Laon cedex, France.*

S.V. Dudley, *CSIRO Centre for Mediterranean Agricultural Research, Private Bag 5, Wembley, WA 6014, Australia.*

C. Dürr, *Unité d'Agronomie de Laon-Reims-Mons, rue Fernand Christ, 02007 Laon cedex, France.*

M. Dutt, *Department of Horticulture, University of Kentucky, Lexington, KY 40546, USA.*

F. Duval, *UMR PMS, Université d'Angers, LRPV, 16 bd Lavoisier, 49045 Angers, France.*

A.J. Ellery, *CSIRO Centre for Mediterranean Agricultural Research, Private Bag 5, Wembley, WA 6014 Australia. New address: CSIRO Publishing, PO Box 1139, Collingwood, Vic 3066, Australia.*

R.H. Ellis, *Department of Agriculture, The University of Reading, Earley Gate, PO Box 237, Reading RG6 6AR, UK.*

F. Engelmann, *International Plant Genetic Resources Institute, Via delle Sette Chiese 142, 00145 Rome, Italy.*

A. Ferrando, *Departamento de Genética Molecular, IBMB-CSIC, Jordi Girona 18, 08034 Barcelona, Spain.*

A. Feutry, *Physiologie Végétale Appliquée, Université Pierre et Marie Curie, Tour 53, 1er étage, 4 place Jussieu, 75252 Paris cedex 05, France.*

M.E. Foley, *USDA-Agricultural Research Service, Biosciences Research Laboratory, Fargo, ND 58105-5674, USA.*

J.R. Fu, *Xishuangbanna Tropical Botanical Garden, The Chinese Academy of Science, Menglun, Mengla 666303, Yunnan, China.*

R.S. Gallagher, *Department of Crop and Soil Science, Washington State University, Pullman, WA 99164-6420, USA.*

A. Gallais, *Station de Génétique Végétale du Moulon, INRA-UPS-INAPG, Ferme du Moulon, 91190 Gif/Yvette, France.*

K. Gallardo, *Laboratoire mixte CNRS-INRA-Bayer CropScience, 14 rue Pierre Baizet, 69263 Lyon, France.*

H. Gawrońska, *Department of Pomology and Basic Sciences in Horticulture, Warsaw Agricultural University, Nowoursynowska 166, 02-787 Warsaw, Poland.*

A.H.M. van der Geest, *Plant Research International, Wageningen University and Research Centre, PO Box 16, 6700 AA Wageningen, The Netherlands.*

R. Geneve, *Department of Horticulture, University of Kentucky, Lexington, KY 40546, USA.*

G. Glevarec, *UMR PMS, Université d'Angers, 2 bd Lavoisier, 49045 Angers, France.*

M.C. Goffinet, *Department of Horticultural Sciences, New York State Agricultural Experiment Station, Cornell University, Geneva, NY 14456, USA.*

E.A. Golovina, *Laboratory of Plant Physiology, Wageningen University, Arboretumlaan 4, 6703 BD Wageningen, The Netherlands.*

T. Gonai, *Graduate School of Agricultural Science, Tohoku University, 1-1 Amamiya, Aoba-ku, Sendai, Miyagi 981-8555, Japan.*

D. Gray, *Horticulture Research International, Wellesbourne, Warwick CV35 9EF, UK.*

E. Graziano, *Departamento de Genètica Molecular, Institut de Biologia Molecular de Barcelona, CID-CSIC, Jordi Girona 18, 08034 Barcelona, Spain.*

J. Grelet, *UMR PMS, Université d'Angers, LRPV, 16 bd Lavoisier, 49045 Angers, France.*

S.P.C. Groot, *Plant Research International, Wageningen University and Research Centre, PO Box 16, 6700 AA Wageningen, The Netherlands.*

X. Gu, *Department of Plant Sciences, North Dakota State University, Fargo, ND 58105-5674, USA.*

Z. Gumede, *School of Life and Environmental Sciences, University of Natal, Durban, 4041 South Africa.*

Y. Gutterman, *Wyler Department of Drylands Agriculture, Jacob Blaustein Institute for Desert Research, Ben-Gurion University of the Negev, Sede Boker Campus 84990, Israel.*

C. Hamano, *Faculty of Regional Environment Science, Tokyo University of Agriculture, 1-1-1 Sakuragaoka, Setagaya-ku, Tokyo 156-8502, Japan.*

D.A. Hammer, *Laboratory of Plasma Studies, Cornell University, Ithaca, NY 14853, USA.*

T. Hashiba, *Graduate School of Agricultural Science, Tohoku University, 1-1 Amamiya, Aoba-ku, Sendai, Miyagi 981-8555, Japan.*

C.L. Hedley, *John Innes Institute, Colney Lane, Norwich NR4 7UH, UK.*

H.W.M. Hilhorst, *Laboratory of Plant Physiology, Wageningen University, Arboretumlaan 4, 6703 BD Wageningen, The Netherlands.*

B. Hirel, *Unité de Nutrition Azotée des Plantes, INRA, route de St Cyr, 78026 Versailles, France.*

F.A. Hoekstra, *Laboratory of Plant Physiology, Wageningen University, Arboretumlaan 4, 6703 BD Wageningen, The Netherlands.*

X.M. Huang, *Xishuangbanna Tropical Botanical Garden, The Chinese Academy of Science, Menglun, Mengla 666303, Yunnan, China.*

I. Isabel-LaMoneda, *Laboratoria de Bioquímica y Biología Molecular, Departamento de Biotecnología-UPM, ETSI Agrónomos, Civdad Universitaria s/n, 28040 Madrid, Spain.*

N. Iusem, *Laboratorio de Fisiología y Biología Molecular, FCEyN, Universidad de Buenos Aires, Civdad Universitaria, Pabéllon II, (1428) Buenos Aires, Argentina.*

R. Jalouzot, *Laboratoire de Génétique, UMR PMS, UFR Sciences, 2 bd Lavoisier, 49045 Angers, France.*

C. Jiang, *Xishuangbanna Tropical Botanical Garden, The Chinese Academy of Science, Menglun, Mengla 666303, Yunnan, China.*

J.A. Jiménez, *Departamento de Fisiología Vegetal, Facultad de Biología, Universidad de Salamanca, Plaza de los Doctores de la Reina s/n, 37007 Salamanca, Spain.*

C. Job, *Laboratoire mixte CNRS-INRA-Bayer CropScience, 14/20 rue Pierre Baizet, 69263 Lyon, France.*

D. Job, *Laboratoire mixte CNRS-INRA-Bayer CropScience, 14/20 rue Pierre Baizet, 69263 Lyon, France.*

P. John, *School of Plant Sciences, The University of Reading, Reading RG6 6AS, UK.*

E. Jurado, *Facultad de Ciencias Forestales, UANL, AP 41 Linares, NL 67700, México.*

E. Justes, *Unité d'Agronomie Laon-Reims-Mons, rue Fernand Christ, 02007 Laon cedex, France.*

S. Kawahara, *Graduate School of Agricultural Science, Tohoku University, 1-1 Amamiya, Aoba-ku, Sendai, Miyagi 981-8555, Japan.*

S. Kester, *Department of Horticulture, University of Kentucky, Lexington, KY 40546, USA.*

J. Kioko, *School of Life and Environmental Sciences, University of Natal, Durban, 4041 South Africa.*

K.P. Kollipara, *Pioneer Hi-Bred International, Inc., 7301 NW 62nd Avenue, Johnston, IA 50131, USA.*

M.C.J.M. Konings, *Plant Research International, Wageningen University and Research Centre, PO Box 16, 6700 AA Wageningen, The Netherlands.*

J. Lamb, *School of Life and Environmental Sciences, University of Natal, Durban, 4041 South Africa.*

O. Leprince, *INH, UMR PMS, Université d'Angers, 16 bd Lavoisier, 49045 Angers, France.*

G. Leubner-Metzger, *Institut für Biologie II (Botanik), Albert-Ludwigs-Universität Freiburg, Schänzlestr. 1, D-79104 Freiburg i. Br., Germany.*

J. Leymarie, *Physiologie Végétale Appliquée, Université Pierre et Marie Curie, Tour 53, 1er étage, 4 place Jussieu, 75252 Paris cedex 05, France.*

D. Lijavetzky, *Instituto de Biotecnología, CICV, INTA Castelar, (1708) Morón, Provincia de Buenos Aires, Argentina.*

A.M. Limami, *UMR PMS, Université d'Angers, 2 bd Lavoisier, 49045 Angers, France.*

B. Llompart, *Departamento de Genética Molecular, IBMB-CSIC, Jordi Girona 18, 08034 Barcelona, Spain.*

T.W. Lyall, *Department of Agriculture, The University of Reading, Earley Gate, PO Box 237, Reading RG6 6AR, UK.*

B. Ly-Vu, *INH, UMR PMS, Université d'Angers, 16 bd Lavoisier, 49045 Angers, France.*

D. Macherel, *UMR PMS, Université d'Angers, LRPV, 16 bd Lavoisier, 49045 Angers, France.*

M.H. Macherel, *INH, PMS, Université d'Angers, 16 bd Lavoisier, 49045 Angers, France.*

M. Martínez, *Laboratorio de Bioquímica y Biología Molecular, Departamento de Biotecnología-UPM, ETSI Agrónomos, Ciudad Universitaria s/n, 28040 Madrid, Spain.*

A. Mead, *Horticulture Research International, Wellesbourne, Warwick CV35 9EF, UK.*

M.D. Mitchell, *Laboratory of Plasma Studies, Cornell University, Ithaca, NY 14853, USA.*

B. Mo, *Department of Botany, University of Guelph, Guelph, Ontario N1G 2W1, Canada.*

A. Moles, *Department of Biological Sciences, Macquarie University, 2109 NSW, Australia.*

F. Montrichard, *UMR PMS, Université d'Angers, LRPV, 16 bd Lavoisier, 49045 Angers, France.*

J. Moreno, *Departamento de Biología Celular, Facultad de Biología, Universidad de Sevilla, Avda Reina Mercedes s/n, 41012-Sevilla, Spain.*

M.C. Morère-Le Paven, *Laboratoire de Génétique, UMR PMS, UFR Sciences, 2 bd Lavoisier, 49045 Angers, France.*

E. van de Mortel, *Pioneer Hi-Bred International, Inc., 7301 NW 62nd Avenue, Johnston, IA 50131, USA.*

D. Nadella, *Department of Plant Sciences, North Dakota State University, Fargo, ND 58105-5674, USA.*

S. Naidoo, *School of Life and Environmental Sciences, University of Natal, Durban, 4041 South Africa.*

H. Nakajima, *Faculty of Regional Environmental Science, Tokyo University of Agriculture, 1-1-1 Sakuragaoka, Setagaya-ku, Tokyo 156-8502, Japan.*

S.C. Navie, *Tropical and Subtropical Weed Research Unit, University of Queensland, St Lucia Campus, Brisbane, Qld 4072, Australia.*

C. Nicolás, *Departamento de Fisiología Vegetal, Facultad de Biología, Universidad de Salamanca, Plaza de los Dcotores de la Reina s/n, 37007 Salamanca, Spain.*

G. Nicolás, *Departamento de Fisiología Vegetal, Facultad de Biología, Universidad de Salamanca, Plaza de los Doctores de la Reina s/n, 37007 Salamanca, Spain.*

J. Nijsse, *Laboratory of Plant Physiology, Wageningen University, Arboretumlaan 4, Wageningen, The Netherlands.*

H. Nonogaki, *Department of Horticulture, Oregon State University, Corvallis, OR 97331-7304, USA.*

R.L. Obendorf, *Department of Crop and Soil Sciences, 617 Bradfield Hall, Cornell University Agricultural Experiment Station, Cornell University, Ithaca, NY 14853-1901, USA.*

S. Odorcic, *Department of Crop and Soil Sciences, 617 Bradfield Hall, Cornell University Agricultural Experiment Station, Cornell University, Ithaca, NY 14853-1901, USA.*

K. Oracz, *Department of Plant Physiology, Warsaw Agricultural University, Rakowiecka 26–30, 02-528 Warsaw, Poland.*

D.J. Osborne, *Oxford Research Unit, Foxcombe Hall, Boars Hill, Oxford OX1 5HR, UK.*

H.R. Osuna-Fernández, *Facultad de Ciencias, Universidad Nacional Autónoma de México, 04510 México DF, México.*

N.W. Pammenter, *School of Life and Environmental Sciences, University of Natal, Durban, 4041 South Africa.*

M.F. Paterson, *Tropical and Subtropical Weed Research Unit, University of Queensland, St Lucia Campus, Qld 4072, Australia.*

L. Pérez-Flores, *Departamento de Ciencias de la Salud, DCBS, Universidad Autónoma Metropolitana-Iztapalapa, Av. Michoacán y Purísima, Col. Vicentina, CP 09340 México DF, México.*

N.C.B. Peters, *Institute of Arable Crops Research, Long Ashton Research Station, University of Bristol, Bristol BS41 9AF, UK.*

S.A. Pikuz, *Laboratory of Plasma Studies, Cornell University, Ithaca, NY 14853, USA.*

H.W. Pritchard, *Millennium Seed Bank Project, Royal Botanic Gardens Kew, Wakehurst Place, Ardingly, West Sussex RH17 6TN, UK.*

P. Puigdomènech, *Departamento de Genética Molecular, Institut de Biologia Molecular de Barcelona, CID-CSIC, Jordi Girona 18, 08034 Barcelona, Spain.*

M. Puype, *Flanders Interuniversity Institute for Biotechnology and Department of Biochemistry, Gent University, Gent, Belgium.*

M. Renard, *UMR PMS, Université d'Angers, LRPV, 16 bd Lavoisier, 49045 Angers, France.*

G. Richard, *Unité d'Agronomie de Laon-Reims-Mons, rue Fernand Christ, 02007 Laon cedex, France.*

P. Richomme, *SCAS, Université d'Angers, LRPV, 2 bd Lavoisier, 49045 Angers, France.*

A. Río, *Departamento Genética Molecular, IBMB-CSIC, Jordi Girona 18, 08034 Barcelona, Spain.*

R. Roca, *Departamento Genética Molecular, IBMB-CSIC, Jordi Girona 18, 08034 Barcelona, Spain.*

D. Rodríguez, *Departamento de Fisiología Vegetal, Facultad de Biología, Universidad de Salamanca, Plaza de los Doctores de la Reina s/n, 37007 Salamanca, Spain.*

M.V. Rodríguez, *IFEVA, Facultad de Agronomía, Universidad de Buenos Aires, Av. San Martín 4453, (C1417DSE) Buenos Aires, Argentina.*

H. Rolletschek, *Institut für Pflanzengenetik und Kulturpflanzenforschung (IPK), D-06466 Gatersleben, Germany.*

C. Rouillon, *UMR PMS, Université d'Angers, 2 bd Lavoisier, 49045 Angers, France.*

S. Rousseau, *Physiologie Végétale Appliquée, Université Pierre et Marie Curie, Tour 53, 1er étage, 4 place Jussieu, 75252 Paris cedex 05, France.*

I.N. Saab, *Pioneer Hi-Bred International, Inc., 7301 NW 62nd Avenue, Johnston, IA 50131, USA.*

M. de la P. Sánchez, *Departamento de Bioquímica, Facultad de Química, UNAM, Avenida Universidad y Copilco, 04510 México DF, México.*

R.A. Sánchez, *IFEVA, Facultad de Agronomía, Universidad de Buenos Aires, Av. San Martín 4453, (C1417DSE) Buenos Aires, Argentina.*

E. Sánchez de Jiménez, *Departamento de Bioquímica, Facultad de Química, Universidad Nacional Autónoma de México, 04510 México DF, México.*

S. Satoh, *Graduate School of Agricultural Science, Tohoku University, 1-1 Amaniya, Aoba-ku, Sendai, Miyagi 981-8555, Japan.*

P. Satour, *UMR PMS, Université d'Angers, LRPV, 16 bd Lavoisier, 49045 Angers, France.*

T.A. Shelkovenko, *Laboratory of Plasma Studies, Cornell University, Ithaca, NY 14853,USA.*

G. Siriwitayawan, *Department of Horticulture, University of Kentucky, Lexington, KY 40546, USA.*

R.D. Smith, *Millennium Seed Bank Project, Royal Botanic Gardens Kew, Wakehurst Place, Ardingly, West Sussex RH17 6TN, UK.*

Y. Soeda, *Sumika Agrotec Ltd, 5-1 Sobirakicho, Niihama, Ehime, Japan.*

S.Q. Song, *Xishuangbanna Tropical Botanical Garden, The Chinese Academy of Science, Menglun, Mengla 666303, Yunnan, China; and School of Life Sciences, Zhongshan University, Guangzhou 510275, China.*

P. Spoelstra, *Laboratory of Plant Physiology, Wageningen University and Research Centre, Arboretumlaan 4, 6703 BD Wageningen, The Netherlands; and Syngenta Seeds BV, Westeinde 62, 1601 BK Enkhuisen, The Netherlands.*

R. Stanelloni, *Instituto de Investigaciones Bioquímicas, Fundación Campomar Argentina, Buenos Aires, Argentina.*

V. Stiefel, *Departamento de Genética Molecular, Institut de Biologia Molecular de Barcelona, CID-CSIC, Jordi Girona 18, 08034 Barcelona, Spain.*

G. Stoopen, *Plant Research International, Wageningen University and Research Centre, PO Box 16, 6700 AA Wageningen, The Netherlands.*

K. Suzuki, *Faculty of Regional Environment Science, Tokyo University of Agriculture, 1-1-1 Sakuragaoka, Setagaya-ku, Tokyo 156-8502, Japan.*

C. Tallon, *UMR PMS, Université d'Angers, LRPV, 16 bd Lavoisier, 49045 Angers, France.*

A.G. Taylor, *Department of Horticultural Sciences, New York State Agricultural Experiment Station, Cornell University, Geneva, NY 14456, USA.*

E. Teyssier, *UMR PMS, Université d'Angers, LRPV, 16 bd Lavoisier, 49045 Angers, France.*

P.E. Toorop, *Laboratory of Plant Physiology, Wageningen University, Arboretumlaan 4, 6703 BD Wageningen, The Netherlands; and Seed Conservation Department, Royal Botanic Gardens Kew, Wakehurst Place, Ardingly, West Sussex RH17 6TN, UK.*

J. Vandekerckhove, *Flanders Interuniversity Institute for Biotechnology and Department of Biochemistry, Gent University, Gent, Belgium.*

J.M. Vázquez-Ramos, *Departamento de Bioquímica, Facultad de Química, UNAM, Avenida Universidad y Copilco, 04510 México DF, México.*

J. Vicente-Carbajosa, *Laboratorio de Bioquímica y Biología Molecular, Departamento de Biotecnología-UPM, ETSI Agrónomos, Ciudad Universitaria s/n, 28040 Madrid, Spain.*

T.L. Wang, *Department of Applied Genetics, John Innes Centre, Colney, Norwich NR4 7UH, UK.*

W.M. Waterworth, *School of Biological Science, University of Manchester, UK.*

H. Weber, *Institut für Pflanzengenetik und Kulturpflanzenforschung (IPK), D-06466 Gatersleben, Germany.*

C.E. West, *School of Biological Science, University of Manchester, UK.*

U. Wobus, *Institut für Pflanzengenetik und Kulturpflanzenforschung (IPK), D-06466 Gatersleben, Germany.*

T. Yoshioka, *Graduate School of Agricultural Science, Tohoku University, 1-1 Amamiya, Aoba-ku, Sendai, Miyagi 981-8555, Japan. New Address: School of Biological Sciences, University of Southampton, Bassett Crescent East, Southampton SO16 7PX, UK.*

Preface

The Seventh International Workshop on Seed Biology was held in Salamanca, Spain in May 2002, following the earlier workshops in this series in Jerusalem (1980), Wageningen (1985), Williamsburg (1989), Angers (1992), Reading (1995) and Merida (1999). In the tradition of each of those workshops, these proceedings contain papers presented at the Salamanca Workshop. It was not possible to include all of the 290 papers and posters presented, but the 49 papers included in this volume are representative of the exciting advances being made in all areas of seed biology, ecology and technology. Particularly promising are the contributions of younger scientists using new technologies to investigate seed biology, opening the way for continuing advancement of the field.

The Workshop was attended by more than 270 participants from 33 countries around the world. In addition to the scientific sessions, the participants enjoyed the historic and cultural delights of Salamanca and its surroundings and the excellent hospitality of the hosts and organizers of the Workshop. The success of the meeting can be attributed to the hard work and dedication of the Local Organizing Committee: Gregorio Nicolás (Chairman), Dolores Rodrìguez, Carlos Nicolás, Pilar Carbonero and Pere Puigdomenech. They were assisted in the scientific planning of the meeting by the International Organizing Committee: Michael Black (UK), Kent Bradford (USA), Daniel Côme (France), Yitzchak Gutterman (Israel), Henk Hilhorst (Holland), Ralph Obendorf (USA), Hugh Pritchard (UK), Ken Thompson (UK) and Jorge Vazquez-Ramos (Mexico).

The International Society for Seed Science (ISSS) was established following the Sixth Workshop in Merida in 1999, and now has over 200 members worldwide. The Seventh Workshop in Salamanca was the first to be held under the auspices of the ISSS with Prof. Daniel Côme as its first President. Ralph Obendorf (President) and J. Derek Bewley (President-elect) currently lead this growing professional organization for seed scientists (for more information, see www.css.cornell.edu/ISSS/isss.htm). In addition to the ISSS, the Workshop thanks the following sponsors for their assistance and support:

Fundación Antama
Germain's Technology Group
Junta de Castilla y León
Consejerìa de Educación y Cultura
Ministerio de Ciencia y Tecnologìa
Monsanto Agricultura
Sociedad Española de Fisiologìa Vegetal
Universidad de Salamanca

The ISSS will hold the Eighth International Workshop on Seed Biology in Brisbane, Australia, in 2005. Dr Steve Adkins (University of Queensland) will chair the organizing committee and information will be posted on the ISSS website as the arrangements are made. We invite all those interested in seed biology to become members of the ISSS and join us in Brisbane in 2005.

Gregorio Nicolás
Kent J. Bradford
Daniel Côme
Hugh Pritchard

1 Seeds of Hope; Seeds of Conflict

J. DEREK BEWLEY

Department of Botany, University of Guelph, Guelph, Ontario N1G 2W1, Canada

Seeds in Early History

Seeds play a dominant role in agriculture, serving as the means to propagate plants from one generation to the next, as food for humans and our domesticated animals, and as an important commodity in the global economy. The grains of cereals, which comprise 90% of all cultivated crops, contribute up to half of the global per capita energy intake. The Food and Agriculture Organization (FAO) of the United Nations provides data on the production and trade of the 19 major crops from which the mature seed is used for food, and of these 15 are from the grass (*Poaceae*) and legume (*Leguminosae*) families (Gooding *et al.*, 2000).

Agriculture originated in various parts of the world some 8000–10,000 years ago as society developed from one that was predominantly hunter-gatherer, following herds of migrating animals and gathering plants as they travelled, to a more stable agrarian-oriented existence (Heiser, 1981). The harvesting and planting of seeds was a major factor in this transition. Permanent settlements in the near-east of Asia, in the so-called Fertile Crescent, stretching north and east of the land between the Tigris and Euphrates rivers into present-day Turkey, Lebanon and Israel, coincide with areas where two wild varieties of wheat – emmer and einkorn – have been found. These are the unmistakable progenitors of modern wheat, and by crossing with other wild grasses, the goatgrasses, the fundamental genetic basis of modern breadwheats was established. These breadwheats appear surprisingly early, however, having been recorded in excavations in Iran and Syria that date back to between 5000 and 6000 BC. Ancient irrigation using channels to support the development of agriculture have been uncovered between the Tigris and Euphrates; this practice may have contributed to the eventual decline in crop production in this area with the increasing introduction of salts into the soil.

In the 'New World', archaeological evidence dates the earliest domestication of plants at least 7000–9000 years ago in the Peruvian Andes and in Mexico, especially of gourds, squashes, beans, chilli peppers, avocado and, of course, maize, which is derived from a coarse wild grass, teosinte. Evidence for agriculture in the Far East may date back as long as 12,000 years, with terraces of that age meant for rice cultivation being discovered in northern India. The domestication of rice for seed production was clearly significant in providing a stable food source (ancient Indian scripts call rice the sustainer of the human race). Rice also became central to ancient cultures, with the 6th century BC kings of Nepal being called 'Pure Rice', and in the 5th century BC the Chinese Emperor alone was allowed to sow the seed. Rice was introduced into Japan in about the 2nd century BC and here, each year, the Emperor himself still joins in the ritual planting and harvesting of rice in the small paddy field in the grounds of the Imperial Palace.

The domestication of wild species inevitably led to farmers selecting traits that were advantageous for better yields and tolerance of growing conditions, even though they had no concept of genetics or plant breeding. Desirable traits were selected for and undesirable ones selected against. Over a period of several thousand years, perennial wild rice was converted to the annual cultivated variety; other traits adverse to agricultural practices were eliminated, e.g. dispersal mechanisms (shattering) and dormancy, while some, such as increased seed nutrient storage capacity, which is related to larger seed size, were fostered.

Seeds in New Locations

The ability of seeds to be desiccated and to survive in the dry state has allowed them to be stored and distributed widely. Seed dispersal for our major crop plants has long depended on human beings. Considering that wheat and rice, originating in the Near and Far East, are now grown throughout the world, their distribution by humans has by far exceeded the dispersal capacity of their wild-type progenitors! Seeds have even been taken on space voyages to the moon and the space station, and successfully germinated and propagated on their return. The introduction of seeds into lands far from their centres of origin began along the ancient trade routes over land and sea. Many seed crops are now most successfully grown, at least from a commercial perspective, in regions of the world into which they have been introduced in relatively recent times (the last few hundred years), such as maize which is now grown most extensively in the mid-west states of the USA, wheat which is grown on the Canadian prairies, and coffee which originated in Ethiopia, became part of the economy of Arabia from about 500–300 years ago, and is now vital to that of Brazil.

The introduction of crops into new lands, either as seeds or as young plantlets, all too frequently has exacted a terrible price. Indigenous populations have been decimated, local ecology destroyed, and devastating social consequences resulted that have changed the course of history. Hobhouse (1999) in his republished book, *Seeds of Change*, provides harrowing accounts of how the establishment of six plants on foreign soils led to contemptible exploitation in pursuit of commercial gains. Some were the result of the desire to grow seed, others due to the ability to transport seed to new climes. To give but one example: the introduction of cotton into the United States led to the import of 4 million slaves from Africa to support the industry, and the civil war that led to the official cessation of slavery subsequently claimed over 750,000 victims.

Seed Research: a Brief History

In the first Workshop on Seeds held in Jerusalem in 1980, the plenary speaker was the late Michael Evenari. He presented an absorbing, scholarly account of the history of seed germination, including many references to observations printed in biblical, early Greek and early Roman documents (Evenari, 1980/81). He proposed Theophrastus (372–287 BC) as being the first all-round botanist known to us, and the first seed physiologist and ecologist. Some of the writings of Theophrastus bear restatement. On seed development he wrote that 'all plant seed has in itself a certain amount of nourishment which is produced with it at the beginning just as in the case of eggs'. On germination behaviour: 'Germination begins earlier in sunny places which have an even temperature.' On after-ripening: 'Another thing which makes a difference as to the rapidity with which seeds germinate is their age; for some herbs come up quicker from fresh seeds . . . some come up quicker from old seeds.' On storage: 'No seeds [of cultivated herbs] will keep for more than four years as to be of use for sowing,' and on priming: 'Some even presoak the seed of cucumber in milk or water to stimulate germination.'

Several quotes from Pliny the Elder (AD 27–79) also appear in the article, including one that pertains to the requirement of seeds to be dried before germination: 'It is difficult for a seed contained in a pod to get dry . . . and consequently they are dried artificially to make them fertile.' This predates our own observations on the need for drying to switch seeds from a developmental mode to a germinative mode (Kermode *et al.*, 1986) by almost two millennia. According to a quote in the publication of the writings of the American poet-naturalist, Henry D. Thoreau (1817–1862), in his book, *Faith in a Seed* (Thoreau, 1993), Pliny also seemed to regard plants that do not bear seed as being somewhat ominous, although his choice of species is questionable. To quote:

> The only ones among the trees that bear nothing whatsoever, not so much as any seed even, are the tamarisk, which is used only for making brooms, the poplar, the Atinian elm, and the alaternus. These trees are regarded as sinister (*infelices*: unhappy) and are considered inauspicious.

Research, if any, on seeds seems not to be recorded until the late 17th century, although breeding of plants for improved seed traits was continuing over the centuries. The originator of modern seed physiology, and a major founding contributor to plant physiology in general (and mineral nutrition and hydroponics in particular), was Julius von Sachs (1832–1897). He made several interesting observations on the storage contents of seeds during development and following germination, and noted that tyrosine and asparagine are formed from protein reserves and transported into the young growing seedling (Evenari, 1980/81). Further biochemically related observations followed over the next 50 years, including a comprehensive account of starch breakdown in barley by Horace Brown and Harris Morris (1890).

Gregor Mendel's studies on pea plants incorporated several seed traits, including the occurrence of wrinkled (*rugosus*) peas with a recessive mutation in the *r* locus. It is now known that the primary lesion in this locus results in *rr* embryos lacking starch branching enzyme I (SBEI) activity; hence less amylopectin is synthesized during seed development, a build-up of sucrose and water ensues, and the embryo, which contains less than its full complement of reserves, shrinks during maturation drying (Wang and Hedley, 1991). The pioneering work of Mendel (1822–1884) also has an interesting tie-in with the pioneering work of Nobel laureate Barbara McClintock (1902–1992) almost a century later, which initially, like his own studies, was largely ignored. She discovered the transposable element, or 'jumping gene', and contended that certain autonomous elements within the maize gene are able to excise and become reincorporated into DNA elsewhere, resulting in a mutant allele (McClintock, 1948). Some of the consequent phenotypes were noted in maize seeds. In the pea seed, the loss of SBEI activity is due to the insertion of a small transposon into its gene, rendering it untranslatable (Wang and Hedley, 1991).

One of the more profound discoveries in the 20th century that was initially made on seeds, but has had wider implications in many aspects of plant growth and development, is that of phytochrome. Although not the first to note the effect of light on seed germination, Lewis Flint and E.D. McAlister (1935) were the first to assign specific wavelengths to the phenomenon of inhibition of germination, using the light-sensitive lettuce seed. Almost by default they also found that certain wavelengths – those in the red region of the spectrum – did not inhibit germination. This was followed up some 20 years later with the classical studies of the Beltsville group of Harry Borthwick, Sterling Hendricks and co-workers (Borthwick *et al.*, 1952) showing the promotion and inhibition of germination of lettuce seed cv. Grand Rapids by exposure to red and far-red regions of the spectrum, respectively. This was the basis of their proposal for a photoreversible pigment, later called phytochrome, with an active and an inactive form.

Seeds and Molecular Biology

Some of the earliest progress in plant molecular biology was made using seeds as the subjects for study. This is not surprising, since during their development seeds accumulate copious amounts of storage reserves, including proteins, and it was logically contended that abundant proteins will result from abundant messages. The first seed protein mRNAs to be translated *in vitro*, as extracted polysomal complexes, were for the 7S globulin storage protein (G1 protein, phaseolin) from *Phaseolus vulgaris* (French bean) (Sun *et al.*, 1975) and the prolamin zein from maize (Larkins *et al.*, 1976, with a preliminary report appearing also in 1975). Somewhat later, both polysomes and isolated mRNA for the 7S and 11S storage proteins from soybean were translated *in vitro* (Beachy *et al.*, 1978). Phaseolin is featured predominantly in the literature over the next several years, as being the first, or among the first, in any plant to be: (i) translated from its isolated mRNA (Hall *et al.*, 1978); (ii) the template for the synthesis of cDNA (Hall *et al.*, 1980); (iii) shown to be a glycosylated protein (Matthews *et al.*, 1981); (iv) derived from a gene with introns (Sun *et al.*, 1981); and (v) expressed following *Agrobacterium*-mediated transvection to another species (Murai *et al.*, 1983). Another important paper from this era reports on work using developing soybean seeds, from which a cDNA library was made using mid-maturation-stage embryos. The expression of four developmentally related messages, for storage proteins, was followed and their increase during development and decline during maturation were clearly demonstrated (Goldberg *et al.*, 1981).

Since then, progress in the molecular biology of seeds has been phenomenal, and understanding of the molecular and cellular control mechanisms for many structural and functional changes is advancing strongly. The advent of *Arabidopsis*, and its many mutants, has allowed enormous progress in the areas of embryogenesis, pattern formation, reserve synthesis, desiccation and storability, and gibberellic acid (GA) and abscisic acid responses, to mention just a few (for more details see the special edition of *Plant Physiology*, Vol. 124, December 2000, which celebrates the completion of sequencing of the *Arabidopsis* genome). Two vitally important seed-related phenomena have still evaded elucidation at the molecular level, namely germination and dormancy, although the recent report of a gene (*RGL2*) in *Arabidopsis* seed whose up-regulation negatively affects germination and whose expression is suppressed during GA-promoted germination (Lee *et al.*, 2002) should open up an exciting new avenue for research.

Because of their important role in food production and manufacturing, genetic modifications of agronomically important seeds that result in quantitatively and qualitatively better products are under way. Seeds with engineered starch, oils and proteins have been obtained and some are ripe for commercial exploitation. The potential of seeds for use as chemical factories to synthesize pharmaceuticals, nutraceuticals, antibodies and bioplastics is also being vigorously pursued.

One genetic modification of a seed that has drawn a lot of attention is that to produce *GoldenRice*™. Vitamin A deficiency affects well over 100 million children worldwide; hundreds of thousands go blind, and up to 2 million die each year of deficiency-induced diarrhoea and respiratory diseases. Ingo Potrykus (2001) recounted how he and his collaborators conceived the idea to produce, in normal rice, β-carotene, a precursor of vitamin A, by inserting a cassette of genes for its synthesis. In this way they were able to insert a biochemical pathway that was not previously present. To allow for the improved rice to be successfully distributed to the most needy, the resultant *GoldenRice*™ has required the waiving of up to 70 intellectual property rights (patents) and the non-profiting collaboration of biotechnology companies. Currently efforts are under way to transfer the new synthetic characteristics to locally adapted varieties and ecotypes in areas of the world where *GoldenRice*™ will be most useful.

GoldenRice™ and other genetically engineered products are not without their opponents, and the anti-GMO (genetically modified organism) movement is gaining voice. This was in part galvanized by the announcement of another seed-related genetic transformation, a patent-protection technology, which has been branded as 'terminator technology' by its opponents. The plant biotechnology industry spends hundreds of millions of dollars to develop and market new traits introduced by genetic modifications. Therefore, it is seeking ways of protecting its genetically engineered crops from farmers who save seed, thus depriving the industry of revenue. One possible approach was developed by researchers at the United States Department of Agriculture in collaboration with a cotton breeding company, Delta and Pine Land Company. Plants transformed with a set of interacting genes in the technology protection system produce seeds incapable of germinating (hence the sobriquet 'terminator'). This has raised considerable concerns, especially in developing countries, because of the fear that large multinational biotechnology companies might come to control the availability of all seeds, excluding poor farmers who depend entirely for their existence upon their own saved seed.

Seeds, Biotechnology and Controversy

Perhaps like never before, the application of the new biotechnologies to plant and seed research has occasioned strong ethical and moral dilemmas. There are highly polarized viewpoints with little apparent common ground for compromise; the stalemate is turning into a public relations struggle for the hearts and minds of the public, with the researcher in the middle (Charles, 2001).

An initial subject for debate is: how revolutionary is biotechnology? Advocates claim that contemporary biotechnology is an evolution of past practices, whereas the opponents claim it is a revolutionary break with them. There are advocates who regard biotechnology as a group of enabling technologies that are so commingled with conventional techniques that distinctions between old and new are without meaning. Opponents point out, on the other hand, that altering the genetic make-up of an organism by modern technologies is different because it involves the exchange of materials from unrelated species, the pace of gene transmission is more rapid, and a greater number of species is involved. Not surprisingly, advocates anticipate that the revolutionary social implications of biotechnology will be positive, and the opponents fear that it will be negative.

The polarization of viewpoints has led to deeply entrenched positions, and three divisions have been offered by Dr David Castle, University of Guelph, in which to organize a considerable amount of the debate:

(i) science vs. religion; (ii) culture vs. big business; and (iii) non-governmental organization (NGO) politics vs. moral biotechnology. To illustrate these points, a small sample of the arguments (paraphrased or rearranged in places for brevity) is presented below.

Science vs. religion

In opposing articles published in 1998 and 2000 in the British press (*Daily Telegraph*) HRH Charles, Prince of Wales, and Richard Dawkins, Charles Simonyi Professor of the Public Understanding of Science at Oxford, expressed their different viewpoints. The Prince of Wales stated:

> I happen to believe that this sort of genetic modification takes mankind into the realms that belong to God, and to God alone. Apart from certain highly beneficial and specific medical applications, do we have the right to experiment with, and commercialize, the building blocks of life? We live in an age of rights – it seems to me that it is time that God had some rights too.

Countering this view, Dawkins retorted:

> A wheat grain is a genetically modified grass seed, just as a Pekinese is a genetically modified wolf. Playing God? We have been playing God for centuries!

Culture vs. big business

Vandana Shiva, founder of Navdanya, a movement in India for biodiversity, conservation and farmer's rights, argued in her Inaugural Hopper Lecture, at the University of Guelph in 1993:

> The expansion of monocultures has more to do with politics and power than enriching and enhancing biological production. They are not just reducing rich biological diversity, but are reducing the way diverse societies organize themselves politically, the way they produce and consume goods and services, and the diverse ways they seek knowledge and innovation.

Bob Shapiro, in 2000, then CEO of Monsanto, commented on this kind of anti-biotechnology activism:

> Those of us in industry can take comfort from such obvious Luddism. After all, we are the technical experts. We know we're right. The 'antis' obviously really don't understand science and are just as obviously pushing a hidden agenda – probably to destroy capitalism.

NGO politics vs. moral dilemmas

Lord Melchett, former Director of Greenpeace UK, was asked by the European Communities Committee on the Regulation of Genetic Modification of Agriculture:

> Your opposition to the release of GMOs, that is an absolute and definite opposition? It is not one that is dependent on further scientific research or improved procedures being developed or any satisfaction you might get with regard to the safety or otherwise in the future?

His reply was:

> It is a permanent and definite and complete opposition based on the view that there will always be major uncertainties. It is the nature of the technology, indeed it is the nature of science, that there will not be any absolute proof. No scientist will ever sit before your Lordships and claim that if they were a scientist at all.

Stung by criticism of *GoldenRice*™ from Greenpeace and others, Potrykus (2001) countered:

> As a consequence of their singular logic, the success of *GoldenRice*™ has to be prevented under all circumstances, irrespective of the damage to those for whose interest Greenpeace pretends to act. Hindering a person's access to life- or sight-saving food is criminal. In my view, their management has but one real interest; to organize media-effective actions for fund raising.

Finale

Seed research, and consequently knowledge of seeds, has come a long way and is increasing at a great rate. Not all seed researchers are involved in genetic engineering, nor do all make use of molecular technologies, but the controversies impact on wider fields of research than just these, and the credibility of life and plant sciences and scientists as a whole is being called into question. As stated elegantly in an editorial in *Nature* (14 February 2002):

> Faced with continuing attacks and threats from radical environmental groups, scientists find themselves increasingly lumped in with an odd mixture of societal targets, from fast-food restaurants to mink farmers to second-hand car dealers. No amount of academic studies proclaiming the safety of GM crops, or public relations campaigns explaining the benefits of biotechnology to a hungry world, are likely to persuade [those stridently against] otherwise. So scientists should focus instead on winning over the vastly larger number of people who may be uneasy or skeptical about GM foods, but who still have open minds. And the best way to do that is to stick with two of sciences traditional strengths: the open exchange of information, and faith in rationality.

Both are cornerstones of these Workshops on Seeds.

Acknowledgements

My thanks to Dr David Castle, Department of Philosophy, University of Guelph, for use of his manuscript 'Biotechnology Revolution?', which was presented as a seminar to the Department of Botany on 5 February 2002.

References

Beachy, R.N., Thompson, J.F. and Madison, J.T. (1978) Isolation of polyribosomes and messenger RNA active in *in vitro* synthesis of soybean seed proteins. *Plant Physiology* 61, 139–144.

Borthwick, H.A., Hendricks, S.B., Parker, M.W., Toole, E.H. and Toole, V.K. (1952) A reversible photoreaction controlling seed germination. *Proceedings of the National Academy of Sciences USA* 38, 662–666.

Brown, H.T. and Morris, G.H. (1890) Researches on the germination of some Gramineae. Part I. *Chemical Society of London* 57, 458–528.

Charles, D. (2001) *Lords of the Harvest.* Perseus Publishing, Cambridge, Massachusetts, 368 pp.

Evenari, M. (1980/81) The history of germination research and the lesson it contains for today. *Israel Journal of Botany* 29, 4–21.

Flint, L.H. and McAlister, E.D. (1935) Wave lengths of radiation in the visible spectrum inhibiting the germination of light-sensitive lettuce seed. *Smithsonian Miscellaneous Collections* 94, 1–11.

Goldberg, R.B., Hoschek, G., Ditta, G.S. and Breidenbach, R.W. (1981) Developmental regulation of cloned superabundant embryo mRNAs in soybean. *Developmental Biology* 83, 218–231.

Gooding, M.J., Murdoch, A.J. and Ellis, R.H. (2000) The value of seeds. In: Black, M. and Bewley, J.D. (eds) *Seed Technology and Its Biological Basis*. Sheffield Academic Press, Sheffield, UK, pp. 3–41.

Hall, T.C., Ma, Y., Buchbinder, B.U., Pyne, J.W., Sun, S.M. and Bliss, F.A. (1978) Messenger RNA for G1 protein of French bean seeds: cell-free translation and product characterization. *Proceedings of the National Academy of Sciences USA* 75, 3196–3200.

Hall, T.C., Sun, S.M., Buchbinder, B.U., Pyne, J.W., Bliss, F.A. and Kemp, J.D. (1980) Bean seed globulin mRNA: translation, characterization, and its use as a probe towards genetic engineering of crop plants. In: Leaver, C.J. (ed.) *Genome Organization and Expression in Plants*. Plenum Publishing, New York, pp. 259–272.

Heiser, C.B. Jr (1981) *Seeds to Civilization. The Story of Food*. W.H. Freeman and Co., San Francisco, California, 254 pp.

Hobhouse, H. (1999) *Seeds of Change. Six Plants That Transformed the World*. Macmillan Publishers, London, 381 pp.

Kermode, A.R., Bewley, J.D., Dasgupta, J. and Misra, S. (1986) The transition from seed development to germination. A key role for desiccation? *HortScience* 21, 1113–1118.

Larkins, B.A., Bracker, C.E. and Tsai, C.Y. (1976) Storage protein synthesis in maize. Isolation of zein-synthesizing polyribosomes. *Plant Physiology* 57, 740–745.

Lee, S., Cheng, H., King, K.E., Wang, W., He, Y., Hussain, A., Lo, J., Harberd, N.P. and Peng, J. (2002) Gibberellin regulates *Arabidopsis* seed germination via RGL2, a GAI/RGA-like gene whose expression is up-regulated following imbibition. *Genes and Development* 16, 646–658.

Matthews, J.A., Brown, J.W.S. and Hall, T.C. (1981) Phaseolin mRNA is translated to yield glycosylated polypeptides by *Xenopus* oocytes. *Nature* 294, 175–176.

McClintock, B. (1948) Mutable loci in maize. *Carnegie Institute of Washington Yearbook* 47, 155–169.

Murai, N., Sutton, D.W., Murray, M.G., Slightom, J.L., Merlo, D.J., Reichert, N.A., Sungupta-Gopalan, C., Stock, C.A., Barker, R.F., Kemp, J.D. and Hall, T.C. (1983)

Phaseolin gene from bean is expressed after transfer to sunflower via tumor-induced plasmid vectors. *Science* 222, 476–482.
Potrykus, I. (2001) Golden rice and beyond. *Plant Physiology* 125, 1157–1161.
Sun, S.M., Buchbinder, B.U. and Hall, T.C. (1975) Cell-free synthesis of the major storage protein of the bean, *Phaseolus vulgaris* L. *Plant Physiology* 56, 780–785.
Sun, S.M., Slightom, J.L. and Hall, T.C. (1981) Intervening sequences in a plant gene – comparison of the partial sequence of cDNA and genomic DNA of French bean phaseolin. *Nature* 289, 37–41.
Thoreau, H.D. (1993) *Faith in a Seed. The Dispersion of Seeds and Other Late Natural History Discoveries.* Island Press, Washington, DC, 283 pp.
Wang, T.L. and Hedley, C.L. (1991) Seed development in peas: knowing your three 'r's' (or four or five). *Seed Science Research* 1, 3–14.

2 Physiological and Molecular Aspects of the Control of Dormancy and Germination in Developing Sorghum Caryopses

Roberto L. Benech-Arnold[1], M. Verónica Rodríguez[1], Rodolfo A. Sánchez[1], Fernando Carrari[2], Laura Pérez-Flores[3], H. Reyna Osuna-Fernández[4], Norberto Iusem[2], Diego Lijavetzky[5], Roberto Stanelloni[6] and Ruben Bottini[7]

[1]*IFEVA, Facultad de Agronomía, Universidad de Buenos Aires, Av. San Martín 4453, (C1417DSE) Buenos Aires, Argentina;* [2]*Laboratorio de Fisiología y Biología Molecular, FCEyN, Universidad de Buenos Aires, Ciudad Universitaria, Pabellón II, (1428) Buenos Aires, Argentina;* [3]*Departamento de Ciencias de la Salud, DCBS, Universidad Autónoma Metropolitana-Iztapalapa, Av. Michoacán y Purísima, Col. Vicentina, CP 09340 México DF, México;* [4]*Facultad de Ciencias, UNAM, 04510 México DF, México;* [5]*Instituto de Biotecnología, CICV, INTA Castelar, (1708) Morón, Provincia de Buenos Aires, Argentina;* [6]*Instituto de Investigaciones Bioquímicas, Fundación Campomar Argentina, Buenos Aires, Argentina;* [7]*Laboratorio de Fisiología Vegetal, Facultad de Ciencias Exactas, Físicas, Químicas y Naturales, Universidad Nacional de Río Cuarto, Campus Universitario, (5800) Río Cuarto, Provincia de Córdoba, Argentina.*

Introduction

Preharvest sprouting is a major constraint for the production of sorghum (*Sorghum bicolor*) in the humid Pampa of Argentina. This crop matures during the end of summer and the beginning of the autumn, when environmental humidity is high; hence, susceptible genotypes are prone to sprouting. Nevertheless, variability exists for this trait: among white, tanninless sorghums, some are characterized as sprouting-susceptible and others as sprouting-resistant. Since the early 1990s, progress has been made towards the elucidation of the basis of preharvest sprouting resistance in sorghum by utilizing two genotypes with contrasting sprouting behaviour as an experi-

 The Biology of Seeds: Recent Research Advances (eds G. Nicolás, K.J. Bradford, D. Côme and H.W. Pritchard)

mental system. The Redland B2 variety is susceptible to preharvest sprouting, while the IS 9530 variety is resistant. Following the acquisition of germination capacity throughout grain development of these two varieties, it was shown that developing seeds of the sprouting-susceptible Redland B2 acquire the capacity to germinate more rapidly, i.e. 20–25 days after pollination (DAP), than grains from sprouting-resistant IS 9530 (40–45 DAP) (Steinbach *et al.*, 1995) (Fig. 2.1). However, when isolated from the entire grain and incubated in distilled water, the embryos of both varieties are able to germinate very rapidly from early stages of development (15–20 DAP). These results show that the inability to germinate during seed development is the result of dormancy imposed by the presence of the seed coat tissues (i.e. endosperm plus pericarp) that surround the embryo. This coat-imposed dormancy is the barrier preventing untimely germination.

From these results it can be concluded that sprouting resistance is related to the maintenance of a sufficient dormancy level until late stages of seed development and maturation. This should be regarded as a special case of dormancy. Indeed, developing seeds have some characteristics that are not common to mature seeds: (i) immature seeds have large amounts of endogenous hormones whose content may be barely detectable in mature ones, regardless of whether or not they are dormant; (ii) immature seeds are largely in an anabolic mode, at least until separation from the mother plant, while this is not the case in dry, quiescent or dormant mature seeds. These considerations suggest that most of the mechanisms controlling dormancy

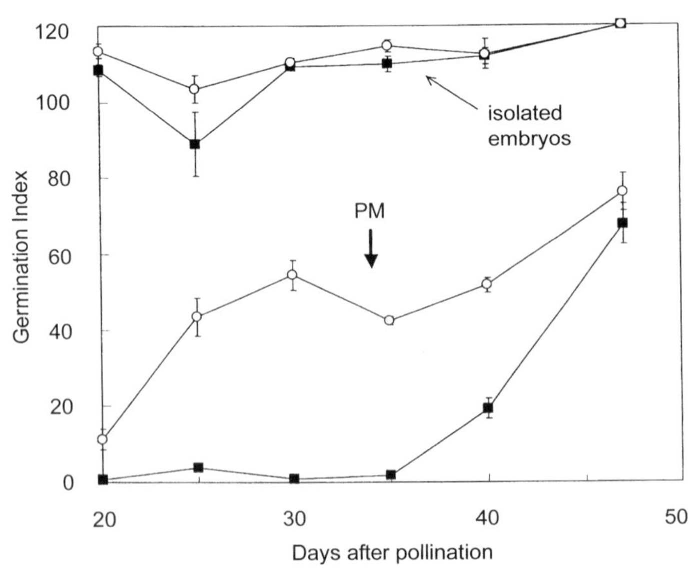

Fig. 2.1. Germination indices (see Steinbach *et al.*, 1995, for details of its construction) of developing sorghum caryopses and embryos harvested at different times after pollination and incubated at 25°C. Sprouting-susceptible variety Redland B2 (○) and sprouting-resistant variety IS 9530 (■). The arrow indicates the moment of physiological maturity (PM) (maximum dry weight). Vertical bars show SE ($n = 3$).

in developing seeds may not operate in mature ones. None the less, these two sorghum varieties constitute an attractive experimental system for studying dormancy (and, in particular, hormonal regulation of dormancy) from its imposition to its expression. The two questions addressed here are: (i) how is dormancy differentially imposed in developing grains of these two varieties; and (ii) how is dormancy differentially expressed throughout development in grains of these two varieties?

Role of the Seed Coat in the Imposition and Expression of Dormancy in Developing Grains

As mentioned before, although embryos from both Redland B2 (sprouting-susceptible) and IS 9530 (sprouting-resistant) caryopses are equally able to germinate from early stages of development if isolated from the entire grain, the presence of the seed coat allows the expression of a differential germination capacity of the immature caryopses. Therefore, regardless of the genetically imposed differential capacity of each type of embryo to overcome such restriction, it could be argued that embryo germination is differentially restricted by different types of seed coat. In other words, it could be thought that seed coats from the sprouting-resistant variety (IS 9530) have certain characteristics that make them more restrictive for embryo germination than Redland B2 seed coats.

To test for this possibility we carried out reciprocal crosses between both varieties, using each genotype as a mother. In this way, the genotype of the hybrid embryo would be the same irrespective of the direction of the cross; however, the seed coat would have, predominantly, the characteristics of the genotype that had acted as the mother (i.e. endosperms are two-thirds maternal and one-third paternal, whereas pericarps are entirely maternal). Any difference between the germination behaviour of seeds derived from one or other reciprocal cross should be attributed to a direct effect of the seed coat.

Caryopses from hybrids harvested at 22, 27 and 30 DAP presented different germination behaviour depending on the direction of the cross: caryopses from hybrids coming from crosses where Redland B2 had been used as a mother displayed higher germination indices than those coming from crosses where mother plants had been IS 9530 (Fig. 2.2). Moreover, the germination behaviour of caryopses from each hybrid resembled the germination pattern displayed by the caryopses from the pure line that had acted as its mother plant (Fig. 2.2). Unfortunately, no data were available for the Redland B2 × IS 9530 cross for 35 and 40 DAP. However, the fact that hybrid caryopses from the cross IS 9530 × Redland B2 displayed indices similar to those from the Redland B2 parent suggests that: (i) the maternal effect does not exist beyond 30 DAP and (ii) high dormancy is a recessive character.

These results suggest that, at least during early stages of development, the different dormancy levels exhibited by caryopses of each variety are, in

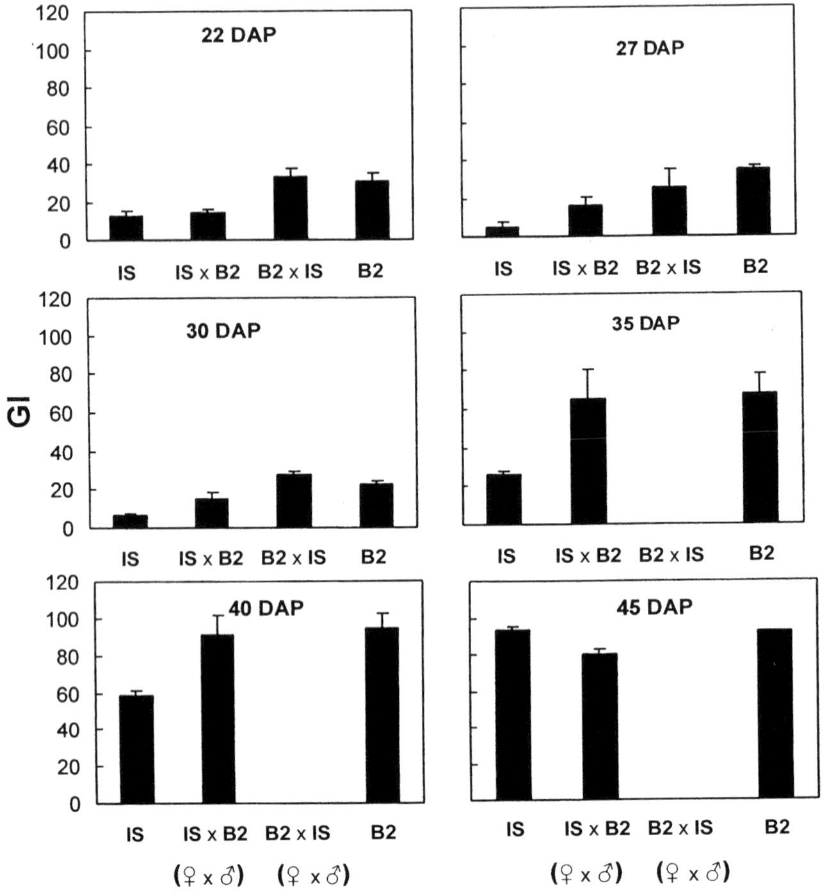

Fig. 2.2. Germination indices (GI) for caryopses of IS 9530 (IS), Redland B2 (B2) and the hybrids that resulted from reciprocal crosses (IS × B2 and B2 × IS; symbols ♀ and ♂ indicate which genotype had acted as mother and as father, respectively, in the cross) harvested at different days after pollination (DAP) and incubated at 25°C for 12 days. Data were not available for the cross B2 × IS on 35, 40 and 45 DAP. Vertical bars show SE ($n = 3$).

part, differentially imposed by some unknown characteristic of each type of seed coat.

Hormonal Regulation of Imposition and Expression of Dormancy in Developing Sorghum Caryopses: Physiological and Molecular Bases

ABA content and sensitivity

Research on the mechanisms of dormancy in the developing seeds of many species suggests a strong involvement of the phytohormone abscisic acid (ABA) (King, 1982; Fong *et al.*, 1983; Karssen *et al.*, 1983; Walker-Simmons, 1987; Black, 1991). ABA-deficient or ABA-insensitive mutants of *Arabidopsis* and maize germinate precociously (Robichaud *et al.*, 1980; Karssen *et al.*, 1983) and application of the ABA-synthesis inhibitor fluridone has been shown to reduce dormancy in developing seeds of some species (Fong *et al.*, 1983; Xu *et al.*, 1990). In sorghum, the participation of ABA in the maintenance of dormancy is evidenced by the fact that inhibition of ABA synthesis with fluridone applied at early developmental stages accelerates the termination of dormancy (Steinbach *et al.*, 1997). Despite some differences between varieties in ABA content of embryos excised from caryopses at various stages of development, it was clear that embryos from the more dormant variety, IS 9530, did not have a constantly higher ABA content than embryos from the less dormant variety, Redland B2 (Steinbach *et al.*, 1995). However, suppression of germination of Redland B2 embryos required ABA concentrations tenfold higher than those required for inhibiting germination of IS 9530 embryos (Steinbach *et al.*, 1995). This difference in embryonic sensitivity to ABA should be an important feature for both the imposition and the expression of the different patterns of dormancy release displayed by developing grains of these two varieties. For this reason, we found it necessary to explore the origin of this different sensitivity to ABA.

Although a genetic origin is quite obvious, it might also be that the presence of each type of seed coat, as described above, could be in part responsible for the different sensitivity to ABA displayed by embryos of these two varieties. If this is the case, then embryos with the same genotype but developing within different types of seed coat should exhibit different sensitivities to ABA. To test for this possibility, we used the same hybrids described above from reciprocal crosses between both sorghum varieties. We excised embryos at different stages of development and incubated them in the presence of 5 μM ABA. Embryos from hybrids harvested at 22 and 27 DAP presented different sensitivities to ABA depending on the seed coat type within which they had developed. Embryos from hybrids coming from crosses where Redland B2 had been used as the mother displayed a markedly lower ABA sensitivity than those coming from crosses where mother plants had been IS 9530 (Fig. 2.3). However, it was evident that this maternal effect does not persist beyond 27 DAP. Moreover, ABA sensitivity in embryos from the IS 9530 × Redland B2 cross was similar to that of embryos of the parental Redland B2, suggesting that high embryonic sensitivity to ABA is a recessive character.

These results show that, at least until 27 DAP, embryo sensitivity to ABA is somehow modulated by the type of seed coat within which the embryo has developed. In other words, IS 9530 seed coat type appears to make embryos more ABA sensitive than Redland B2 ones during early stages of development.

A genetic origin for this different embryo sensitivity to ABA was also evaluated. We hypothesized that the lower sensitivity to ABA that was characteristic of Redland B2 embryos was due to a physiological disruption in the function of a gene controlling sensitivity to ABA. Gene *vp1* (*Viviparous-1*) is homologous to *abi-3* from *Arabidopsis thaliana*; *abi-3* mutants are ABA insensitive and dormancy is much reduced in their seeds. *vp1* codes for a

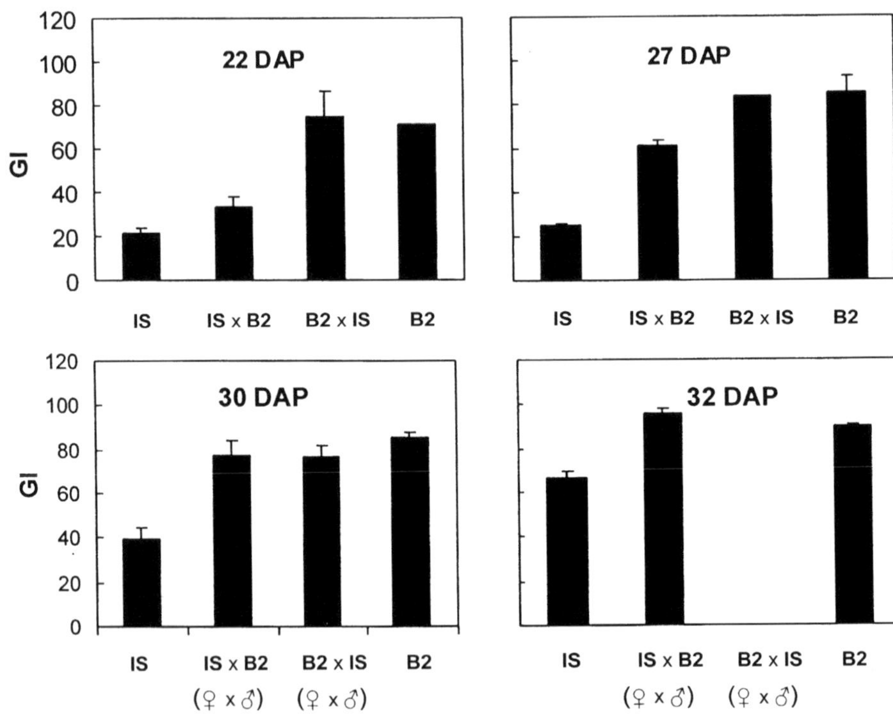

Fig. 2.3. Germination indices (GI) for embryos of IS 9530 (IS), Redland B2 (B2) and the hybrids that resulted from reciprocal crosses (IS × B2 and B2 × IS; symbols ♀ and ♂ indicate which genotype had acted as mother and as father, respectively, in the cross) harvested at different days after pollination (DAP) and incubated in presence of 5 μmol ABA at 25°C for 12 days. Data were not available for the cross B2 × IS on 32 DAP. Vertical bars show SE ($n = 3$).

transcription factor that, among other processes, controls sensitivity to ABA. Moreover, maize *vp1* mutants are ABA insensitive and, consequently, they are viviparous (a phenomenon that is phenotypically similar to preharvest sprouting) (McCarty *et al.*, 1991). Therefore, *vp1* was chosen as a candidate gene for investigating the molecular bases of the contrasting embryo sensitivity to ABA displayed by these two sorghum varieties. Cloning and sequencing *vp1* from our sorghum varieties showed that the sequences of the two alleles differed by only 2% (Carrari *et al.*, 2001). We also analysed expression of *SbVp1* during seed development in embryos from both lines and found no correlation between expression of this gene and embryo sensitivity to ABA (Carrari *et al.*, 2001). Moreover, in a genetic analysis of dormancy in sorghum, Lijavetzky *et al.* (2000) showed that the character presents continuous variation in a segregating F_2 population generated by crossing IS 9530 and Redland B2, and found two unlinked quantitative trait loci (QTLs) that, together, explained more than 80% of the observed phenotypic variance. However, *SbVp1* did not map within any of those QTLs (Carrari *et al.*, 2003). Taken together, these results suggest that *SbVp1* is not responsible for the imposition of a different timing of dormancy release through the control of embryonic sensitivity to ABA.

The possibility that gene *SbVp1* is involved in the expression of dormancy cannot be ruled out. In such a case, there should be a correlation between *SbVp1* expression during grain incubation and the level of dormancy expressed by the grains. For example, *SbVp1* expression would be maintained in seeds that will not germinate (if, say, dormancy is maintained via *SbVp1* protein) and would disappear prior to germination of seeds that will germinate (if, say, the grain was released from dormancy because the *SbVp1* protein was no longer present). To test this possibility, we analysed *SbVp1* expression in embryos excised from IS 9530 and Redland B2 (40 DAP) caryopses at different times following grain imbibition. While *SbVp1* mRNA remained at high levels after several days of incubation in dormant IS 9530 grains, mRNA abundance decreased steadily until day 3 after imbibition in the less dormant Redland B2 ones, preceding the germination of these caryopses (Fig. 2.4). We also analysed *SbVp1* expression in embryos from caryopses of both varieties coming from panicles that had been treated with fluridone at 7 DAP; these caryopses had less dormancy, due to the lower ABA content that results from application of fluridone (Steinbach *et al.*, 1997). Fluridone-treated Redland B2 caryopses germinated faster than control ones and *SbVp1* mRNA had almost disappeared at day 2 after imbibition, again preceding the initiation of germination (Fig. 2.4). These results show that *SbVp1* is differentially regulated in caryopses with different dormancy levels and suggest that *SbVp1* might be involved in the expression of dormancy rather than in the imposition of a different timing of dormancy release.

Gibberellin synthesis during incubation of grains with contrasting dormancy

The central role of gibberellins (GAs) in promoting seed germination was suggested decades ago and confirmed by the identification of GA-deficient mutants of *Arabidopsis* and tomato, seeds of which will not germinate unless exogenously supplied with GAs (Koornneef and van der Veen, 1980; Groot and Karssen, 1987). It has been proposed that endogenous GAs control germination through two processes: (i) a decrease in the mechanical resistance of the tissues surrounding the embryo (Groot and Karssen, 1987) and (ii) promotion of the growth potential of the embryo (Carpita *et al.*, 1979; Karssen *et al.*, 1989). In seeds where the tissues covering the embryo are weak or split during imbibition, this increase in embryo growth potential may be the only process required for the completion of germination (e.g. Schopfer and Plachy, 1985). The fact that inhibitors of GA biosynthesis such as paclobutrazol and tetcyclacis prevent germination (Karssen *et al.*, 1989; Nambara *et al.*, 1991) suggests the requirement for *de novo* synthesis of GAs for germination upon imbibition. On this basis, it might be hypothesized that, in addition to the existence of other constraints, dormant IS 9530 grains will not germinate because their capacity to produce GAs *de novo* is blocked. In contrast, the *de novo* synthesis of GAs should proceed normally during early stages of imbibition in non-dormant grains. The fact that dormant

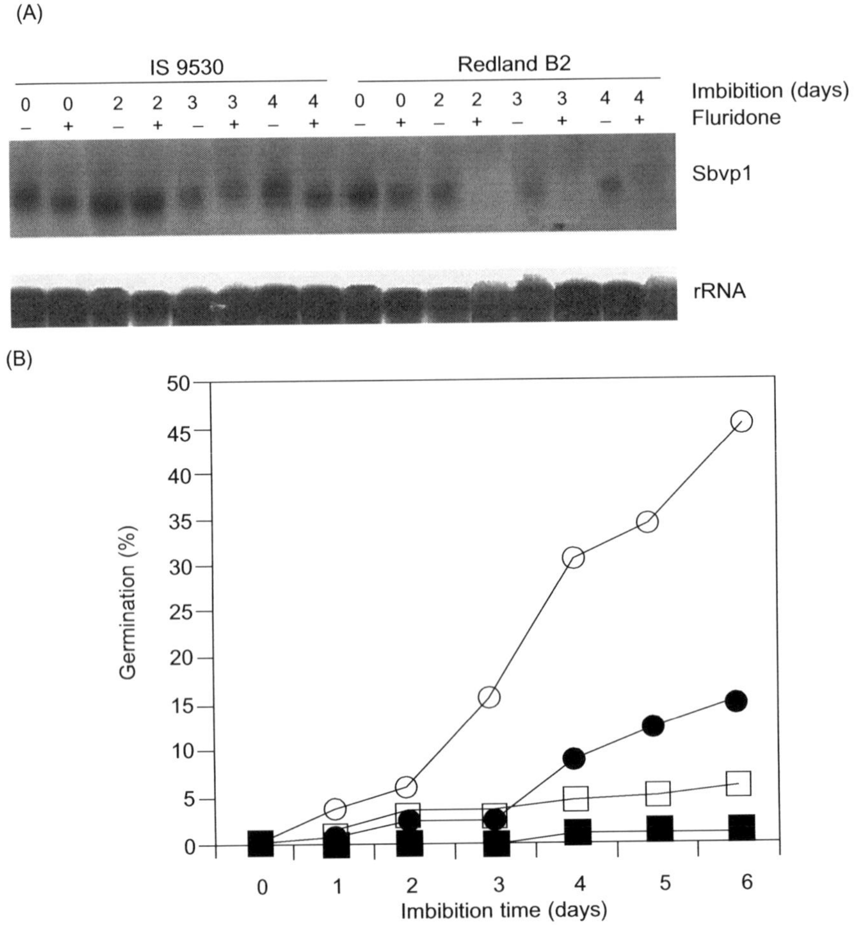

Fig. 2.4. Effect of imbibition on *SbVp1* mRNA abundance and seed germinability. Caryopses (40 DAP) were incubated in water for different lengths of time. (A) *SbVp1*-specific northern hybridization to total RNA (25 μg) isolated from embryos of fluridone-treated and control caryopses (upper panel). The same filter was hybridized to an rRNA-specific probe (lower panel) to indicate RNA loading. (B) Germination percentages of 40 DAP caryopses from control (●, ■) and fluridone-treated panicles (○, □) of IS 9530 (■, □) and Redland B2 (●, ○) varieties at different imbibition times. (From Carrari *et al.*, 2001.)

sorghum seeds can be induced to germinate through exogenously supplied GA_{4+7} supports this hypothesis (Steinbach *et al.*, 1997).

Preliminary measurements of GA content were carried out in Redland B2 and IS 9530 embryos excised from 40 DAP caryopses incubated for different periods (Fig. 2.5), following the procedure described by Gaskin *et al.* (1985). GA_3 was the most abundant GA detected in IS 9530 and Redland B2 embryos during the incubation period. The GA content was initially high in

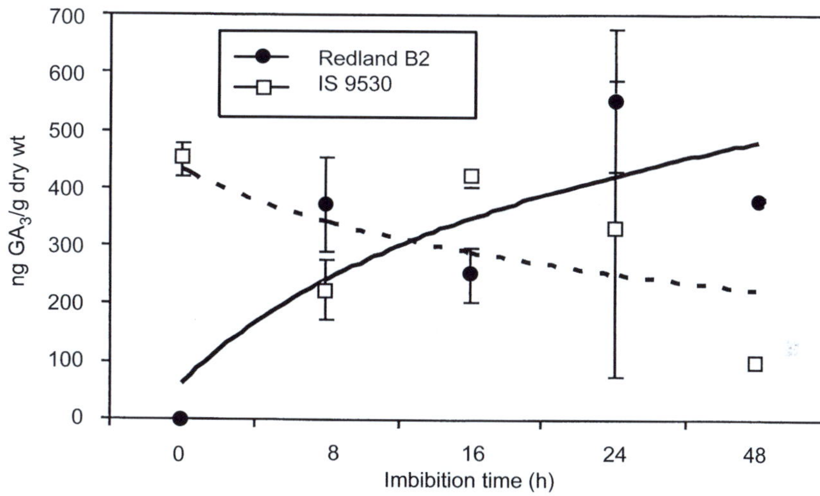

Fig. 2.5. Gibberellin content (GA_3) in IS 9530 and Redland embryos isolated from 40 DAP caryopses incubated in water at 25°C for the indicated times.

IS 9530 embryos from non-incubated caryopses, then decreased during the incubation of the whole caryopsis. Thus the pattern of GA content in IS 9530 embryos throughout the incubation period appears to be indicating breakdown or conjugation rather than synthesis (Fig. 2.5). Concomitantly, almost no germination was recorded after 12 days of incubation (data not shown). In contrast, no GAs were detected in Redland B2 embryos prior to imbibition, but they accumulated GA_3 after grain imbibition (Fig. 2.5). Forty-eight hours after imbibition (i.e. almost 48 h before radicle protrusion) Redland B2 embryos had a GA_3 content that was fourfold higher than that recorded in IS 9530 embryos. Germination of Redland B2 caryopses started 96 h after imbibition and they reached 60% germination after 12 days of incubation (data not shown). These results suggest that *de novo* synthesis of GAs proceeds upon imbibition of Redland B2 grains with low dormancy, whereas in dormant grains of IS 9530 it appears to be reduced or blocked.

Within the GA biosynthetic pathway, oxidation and elimination of C-20 catalysed by GA 20-oxidases to yield the C-19 GAs, which include the biologically active plant hormones, is thought to be a site of regulation (Lange *et al.*, 1994; Phillips *et al.*, 1995; Lange, 1998; Yamaguchi and Kamiya, 2000). If germination of dormant seeds is indeed restricted through a blockade in GA biosynthesis, a differential control of the genes encoding for GA synthesis enzymes at, for example, the transcriptional level might be expected. Genes encoding GA 20-oxidases are therefore attractive targets for expression analysis during imbibition of dormant and non-dormant grains. Therefore, we isolated a cDNA encoding GA 20-oxidase (gibberellin, 2-oxoglutarate:oxygen oxidoreductase (20-hydroxylating, oxidizing) EC 1.14.11.-) (*SbGA20-ox*) from embryos of sorghum, and carried out expression analysis in embryos excised from 40 DAP Redland B2 and IS 9530 caryopses incubated for different

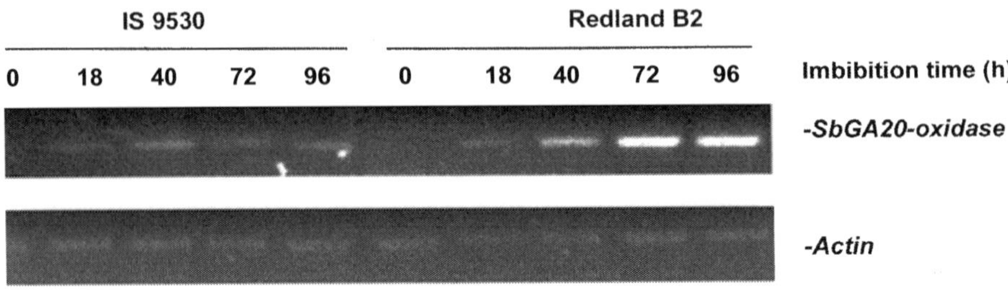

Fig. 2.6. Expression analysis of *SbGA20-ox* in Redland B2 and IS 9530 embryos isolated from caryopses (40 DAP) incubated for the indicated times. RT-PCR was performed in a linear range with 28 and 30 cycles for *SbGA20-ox* and *actin* respectively. One of three repeats is shown. (From Perez-Flores *et al.*, 2003.)

lengths of time (Perez-Flores *et al.*, 2003).

In Redland B2 embryos *SbGA20-ox* mRNA was detectable from 18 h after grain imbibition. From then onwards, the amount of the specific transcript increased steadily until 96 h after imbibition, when radicle protrusion started to be visible (Fig. 2.6). In contrast, in IS 9530 embryos the amount of *SbGA20-ox* mRNA remained barely detectable throughout the incubation period, in agreement with both the apparent absence of GA *de novo* synthesis and the high dormancy displayed by the caryopses (Fig. 2.6).

Taken together, these results show that a different dormancy level in immature sorghum grains is associated with a differential capacity of the embryo to produce GA by *de novo* synthesis upon seed imbibition and suggest that this GA synthesis is regulated at the level of expression of the GA 20-oxidase gene.

Concluding Remarks

The information presented throughout this chapter is summarized in the scheme shown in Fig. 2.7. The two sorghum lines used in this study have a contrasting pattern of exit from dormancy during grain development: Redland B2 caryopses start losing dormancy well before physiological maturity, while IS 9530 caryopses remain dormant for some time after full maturity. It should be noted that, in this chapter, 'dormancy' (a physiological condition imposed early during seed development) is distinguished from 'expression of dormancy' (the consequences of dormancy), which can be evaluated through, for example, germination capacity.

Our experiments have shown that some unknown attribute of the type of seed-coat is in part responsible for the imposition of the characteristic pattern of exit from dormancy, at least during the earlier stages of development of grains from each variety (i.e. until 30 DAP). Also, since we have shown that modifications of the endogenous levels of the growth regulators ABA (and, possibly, embryo sensitivity to ABA) and GAs during seed development (i.e. by means of the application of inhibitors of their synthesis almost immediately

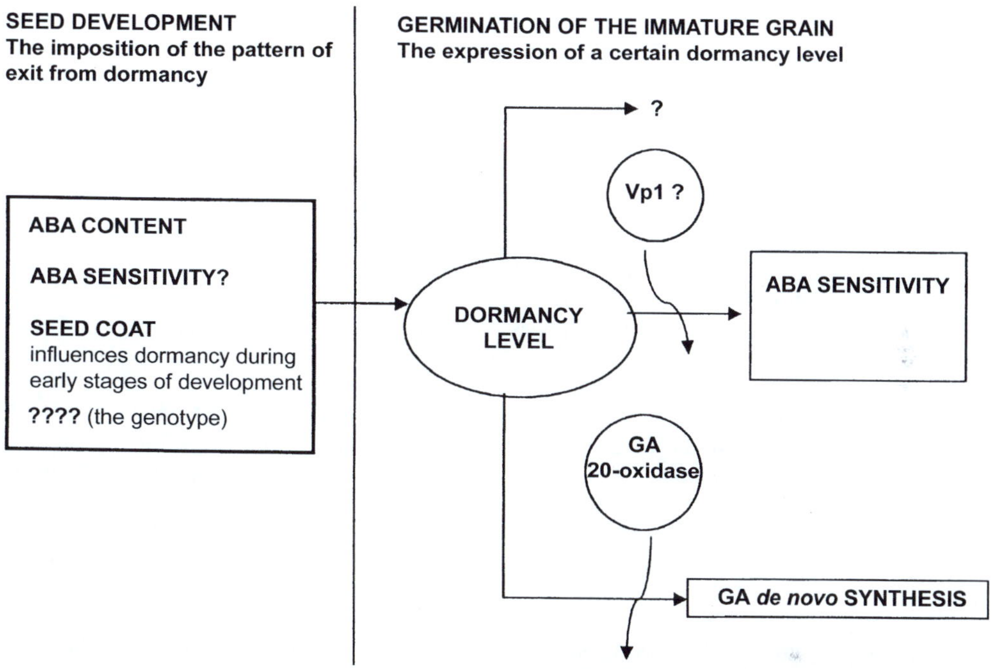

Fig. 2.7. A scheme summarizing the state of knowledge on the imposition and expression of dormancy in developing sorghum grains. On the left of the scheme are the genetic and physiological factors possibly participating in the imposition of a particular pattern of exit from dormancy for a particular genotype. This imposition is likely to occur early during seed development. The molecular and physiological mechanisms that participate in the expression of a certain dormancy level are described in the right of the panel.

after anthesis) can modulate the characteristic pattern of exit from dormancy for each line (Steinbach *et al.*, 1997), it appears that mechanisms imposing the timing of dormancy release are, in turn, under hormonal control.

If high sensitivity to ABA is characteristic of the dormant state (expression of dormancy), those processes controlling ABA sensitivity are part of the mechanism establishing dormancy, whereas the ABA-modulated processes involved in germination (e.g. embryo growth) determine the expression of dormancy. The seed coat type has been shown to modulate embryo sensitivity, at least during early stages of development (i.e. until 27 DAP). However, it is evident that differences in embryo sensitivity to ABA displayed by this pair of varieties must have a genetic origin. While the *vp1* gene has been suggested to be a major determinant of embryo sensitivity to ABA, its role (if any) in our experimental system appears to be limited to the expression of dormancy. A similar role can possibly be ascribed to the gene encoding *SbGA20-oxidase*, which is expressed to a different extent depending on a dormancy level that has been imposed during development.

In summary, future work should be directed to the better understanding of the mechanisms involved in expression of dormancy. For example,

SbGA20-oxidase function is only preliminarily characterized; moreover, it is quite possible that other enzymes of the GA biosynthetic pathway might also have a regulatory function. Most important, the imposition of dormancy is far from being understood at both a physiological and a genetic level.

References

Black, M. (1991) Involvement of ABA in the physiology of developing and mature seeds. In: Davies, W.J. (ed.) *Abscisic Acid Physiology and Biochemistry*. BIOS Scientific Publishers, Oxford, pp. 99–124.

Carpita, N.C., Ross, C.W. and Nabors, M.W. (1979) The influence of plant growth regulators on the growth of the embryonic axes of red- and far-red treated lettuce seeds. *Planta* 145, 522–516.

Carrari, F., Perez-Florez, L., Lijavetzky, D., Enciso, S., Sánchez, R.A., Benech-Arnold, R.L. and Iusem, N. (2003) Cloning and expression of a sorghum gene with homology to maize *VP1*. Its potential involvement in pre-harvest sprouting resistance. *Plant Molecular Biology* 45, 631–640.

Carrari, F., Benech-Arnold, R.L., Osuna-Fernandez, H.R., Hopp, E., Sánchez, R.A., Iusem, N.D. and Lijavetzky, D. (2003) Genetic mapping of the *Sorghum bicolor vp1* gene and its relationship with pre-harvest sprouting resistance. *Genome* 46, 253–258.

Fong, F., Smith, J.D. and Koehler, D.E. (1983) Early events in maize seed development. *Plant Physiology* 73, 899–901.

Gaskin, P., Gilmour, S.J., MacMillan, J. and Sponsel, V.M. (1985) Gibberellins in immature seeds and dark-grown shoots of *Pisum sativum*. Gibberellins identified in the tall cultivar Alaska in comparison with those in the dwarf Progress No. 9. *Planta* 163, 283–289.

Groot, S.P.C. and Karssen, C.M. (1987) Gibberellins regulate seed germination in tomato by endosperm weakening: a study with gibberellin-deficient mutants. *Planta* 171, 525–531.

Karssen, C.M., Brinkhorst-Van der Swan, D.L.C., Breekland, A.E. and Koorneef, M. (1983) Induction of dormancy during seed development by endogenous abscisic acid: studies on abscisic acid deficient genotypes of *Arabidopsis thaliana* (L.). *Planta* 157, 158–165.

Karssen, C.M., Zagorski, S., Kepczynski, J. and Groot, S.P.C. (1989) Key role for endogenous gibberellins in the control of seed germination. *Annals of Botany* 63, 71–80.

Koornneef, M. and van der Veen, J.H. (1980) Induction and analysis of gibberellin sensitive mutants in *Arabidopsis thaliana* (L.) Heynh. *Theoretical and Applied Genetics* 58, 257–263.

King, R.W. (1982) Abscisic acid in seed development. In: Kahn, A.A. (ed.) *The Physiology and Biochemistry of Seed Development, Dormancy and Germination*. Elsevier Biomedical Press, Amsterdam, pp. 157–181.

Lange, T. (1998) Molecular biology of gibberellin synthesis. *Planta* 204, 409–419.

Lange, T., Schweimer, A., Ward, D.A., Hedden, P. and Graebe, J.E. (1994) Separation and characterization of three 2-oxoglutarate-dependent dioxygenases from *Cucurbita maxima* L. endosperm involved in gibberellin biosynthesis. *Planta* 195, 98–107.

Lijavetzky, D., Martinez, C., Carrari, F. and Hopp, E. (2000) QTL analysis and mapping of preharvest sprouting resistance in *Sorghum*. *Euphytica* 112, 125–135.

McCarty, D.R., Hattori, T., Carson, C.B., Vasil, V., Lazar, M. and Vasil, I.K. (1991) The viviparous-1 developmental gene of maize encodes a novel transcriptional activator. *Cell* 66, 895–905.

Nambara, E., Akazawa, T. and McCourt, P. (1991) Effects of gibberellin biosynthesis inhibitor uniconazol on mutants of *Arabidopsis*. *Plant Physiology* 97, 736–738.

Perez-Flores, L.J., Carrari, F., Osuna-Fernández, R., Rodriquoz, M.V., Enciso, S., Stanelloni, R., Sánchez, R.A., Bottini, R. and Benech-Arnold, R.L. (2003) Expression analysis of a GA 20-oxidase in embryos from two sorghum lines with contrasting dormancy: possible participation of this gene in the hormonal control of germination. *Journal of Experimental Botany* (in press).

Phillips, A.L., Ward, D.A., Uknes, S., Appleford, N.E.J., Lange, T., Huttly, A.K., Gaskin, P., Graebe, J.A. and Hedden, P. (1995) Isolation and expression of three gibberellin 20-oxidase cDNA clones from *Arabidopsis*. *Plant Physiology* 108, 1049–1057.

Robichaud, C.S., Wong, J. and Sussex, I.M. (1980) Control of *in vitro* growth of viviparous embryo mutants of maize by abscisic acid. *Developmental Genetics* 1, 325–330.

Schopfer, P. and Plachy, C. (1985) Control of seed germination by abscisic acid. III. Effect on embryo growth potential (minimum turgor pressure) and growth coefficient (cell wall extensibility) in *Brassica napus* L. *Plant Physiology* 77, 676–686.

Steinbach, H.S., Benech-Arnold, R.L., Kristof, G., Sánchez, R.A. and Marcucci Poltri, S. (1995) Physiological basis of pre-harvest sprouting resistance in *Sorghum bicolor* (L.) Moench. ABA levels and sensitivity in developing embryos of sprouting resistant and susceptible varieties. *Journal of Experimental Botany* 45, 701–709.

Steinbach, H.S., Benech-Arnold, R.L. and Sánchez, R.A. (1997) Hormonal regulation of dormancy in developing Sorghum seeds. *Plant Physiology* 113, 149–154.

Walker-Simmons, M.K. (1987) ABA levels and sensitivity in developing wheat embryos of sprouting-resistant and -susceptible cultivars. *Plant Physiology* 84, 61–66.

Xu, N., Coulter, K.M. and Bewley, J.D. (1990) Abscisic acid and osmoticum prevent germination of developing alfalfa embryos, but only osmoticum maintains the synthesis of developmental proteins. *Planta* 182, 382–390.

Yamaguchi, S. and Kamiya, Y. (2000) Gibberellin biosynthesis: its regulation by endogenous and environmental signals. *Plant Cell Physiology* 41, 251–257.

3 *Long cell*, a Mutant from Maize Producing a Distorted Embryo and Generalized Cell Death

Enrique Graziano, Miriam Bastida, Virginia Stiefel and Pere Puigdomènech

Departamento de Genètica Molecular, Institut de Biologia Molecular de Barcelona, CID-CSIC, Jordi Girona 18, 08034 Barcelona, Spain

Introduction

Embryogenesis is the process that produces a fully functional organism from the single cell of the fertilized ovule. This process is being actively investigated in animal species, especially in *Drosophila*. In plants, much less is known about the genetic processes governing embryogenesis and a number of specific features of the plant cell indicates that some of their basic mechanisms, such as cell movement and apoptosis, may differ between plants and animals. Plant cells do not migrate as animal cells do in the embryo during the process of the formation of specific organs. Apoptosis is also different in plants. It has been shown that active programmes of cell death take part in the reaction of the plant to a pathogen attack as well as during development (Pennell and Lamb, 1997; Lam *et al.*, 1999). In some cases the structure of the dead plant cell is preserved and no system of recycling the cellular content has been described so far in plants. However, organs such as the suspensor or the endosperm disappear during the development of a plant, indicating that cell death is an active developmental process in plants. In any case, the crucial mechanisms used by plants for morphogenesis both in the embryo and in the adult plant are the precise definition of the plane of cell division and cell elongation.

Mutants having defects in embryogenesis have been collected and analysed in plants. The species on which most of the work has been done is *Arabidopsis thaliana* and genes involved in cytokinesis and hormonal control have been found in this way (Mayer *et al.*, 1993; Shevell *et al.*, 1994). Genes have been identified, such as *lec* or *pickle* (Lotan *et al.*, 1998; Ogas *et al.*, 1999), that produce somatic embryos in the adult plant. In maize, mutants with defective embryogenesis have long been known. They have been classified as *emb* (embryo specific) mutants, which produce a distorted embryo with no endosperm defects, and *dek* (defective kernel) mutants that have an effect

(eds G. Nicolás, K.J. Bradford, D. Côme and H.W. Pritchard)

in both the embryo and the endosperm. Two types of *dek* mutants can be considered: those that block embryogenesis and those that distort embryogenesis. Examples of the first type are *dek1* (Faccio and Sparvoli, 1988) and *lacchrima* (Stiefel *et al.*, 1999) and an example of the second type is *dek31* (Sheridan and Thorstenson, 1986). Initial characterization of some of these mutants at the molecular level has been reported (Scanlon and Myers, 1998; Heckel *et al.*, 1999; Stiefel *et al.*, 1999; Elster *et al.*, 2000). Besides the economic and scientific interest of the species, maize and cereals in general have features of embryogenesis that seem to indicate that some aspects of the developmental process follow specific pathways. In particular, the definition of bilateral symmetry appears after a period of cell proliferation where no specific patterns except for the formation of the suspensor and the protoderm can be observed (José-Estanyol and Puigdomènech, 1998; Ingram *et al.*, 2000). In the course of a process to obtain maize mutants by insertion of transposon *Ac* (Dellaporta and Moreno, 1994), a number of *dek* mutants were observed. One of these mutants, *lacchrima*, blocks embryo development at the transition stage of embryogenesis (Stiefel *et al.*, 1999). Another of the mutants that was obtained belongs to the second group of *dek* mutants from maize. This mutation, called *long cell* (*lc*), is described here. The mutant is lethal and recessive, and it is shown that lethality is produced by a general process of programmed cell death throughout the embryo.

Materials and Methods

Mutant lines, growing conditions and genetic analysis

The *long cell* mutant of maize (*Zea mays* L.) was produced by Dr S. Dellaporta, as previously described (Dellaporta and Moreno, 1994). All the plant lines were maintained in the greenhouse of the Departamento de Genética Molecular (IBMB-CSIC, Barcelona). Allelism tests were done according to Sheridan and Clark (1987). All the mutants used in this study except the *lc* line were obtained from the Maize Genetics Cooperative Stock Center.

Scanning electron microscopy (SEM)

Maize embryos and kernels were fixed in a mixture of ethanol, formaldehyde and acetic acid (80:3.5:5) for 30 min at room temperature. After fixation the samples were transferred to 70% ethanol and subjected to an alcoholic dehydration series. The dehydrated specimens were immersed for 10 min in successive mixtures of ethanol/isoamyl acetate (2:1, 1:1, 1:2) and finally three times in 100% isoamyl acetate. The samples were observed with a Hitachi S2300 scanning electron microscope.

Detection of cells undergoing programmed cell death (PCD)

Fixation of tissues was performed in the same conditions used for SEM. Pretreatment of tissue, incubation with proteinase K, blocking of endogenous peroxidase, cell permeabilization, labelling, signal conversion and analysis were performed according to the suppliers' protocols (In Situ Cell Death Detection, POD, Roche Molecular Biochemicals). In order to reduce background and non-specific labelling, a 1:3 dilution of converted POD in blocking solution (150 mM NaCl, 100 mM Tris-HCl pH 7.5) and a 50% reduction of the concentration of the labelling mix with the TUNEL dilution buffer (30 mM Tris-HCl pH 7.2, 140 mM sodium cacodylate and 1 mM $CoCl_2$) were used. Better results were obtained using an alternative protocol for the TUNEL method applied to difficult tissues (Roche Molecular Biochemicals). Results were viewed using bright-field microscopy and photographs were taken using Ektachrome 160T films.

Results

The *lc* mutation behaves as a Mendelian recessive trait. The *lc*/+ heterozygous plants do not show any visible abnormality. Self-pollination of the heterozygotes gives rise to 25% abnormal grains that are unable to germinate, even under *in vitro* culture conditions. The mutant grains have a reduced endosperm and a distorted embryo. Apart from the reduced size, there is no other morphological indication of endospermic abnormalities. The aleurone layer is not altered.

For a better understanding of the *lc* mutation we performed a morphological study of the embryos using optic and scanning electron microscopy (Fig. 3.1). A simple dissection of the embryos from their grains revealed that all of them had a darker colour than their wild-type siblings, probably due to an abnormal accumulation of pigments. Those pigments were hydrophilic, since the embryos were able to stain the water when submerged in it. The mutant embryos had variable shapes. Many of them were apparently amorphous, mostly an irregular ovoid (Fig. 3.1C). others had clearly distinguishable, albeit always abnormal, structures (Fig. 3.1D). In the less severe cases we could detect the presence of a scutellum, embryo axis with its corresponding tip (which may be green) and primary root, and even the suspensor (Fig. 3.1E). It is worth noting that we never found any indication of embryonic leaves, nor the presence of initial secondary roots, indicating that *lc* embryos do not reach stage 1 of maize embryogenesis.

The most characteristic feature of the *lc* phenotype is the presence of abnormal cell elongation in the affected embryos. The enlarged cells easily reach 500 μm (Fig. 3.1E and F). Those cells are enlarged in an allometric fashion by expanding only along the longitudinal axis, while keeping normal dimensions in the other directions. This indicates that the *lc* cellular phenotype is not due to a general decrease in the mechanical strength of the cell wall, since it would give rise to round cells. The enlarged cell morphology strongly supports the notion that the mutant cells are elongated by selectively changing the cell wall properties in a spatially restricted manner, indicating that the *lc* locus may be involved in the fine-tuning of the cell wall resistance that ultimately controls the cell shape.

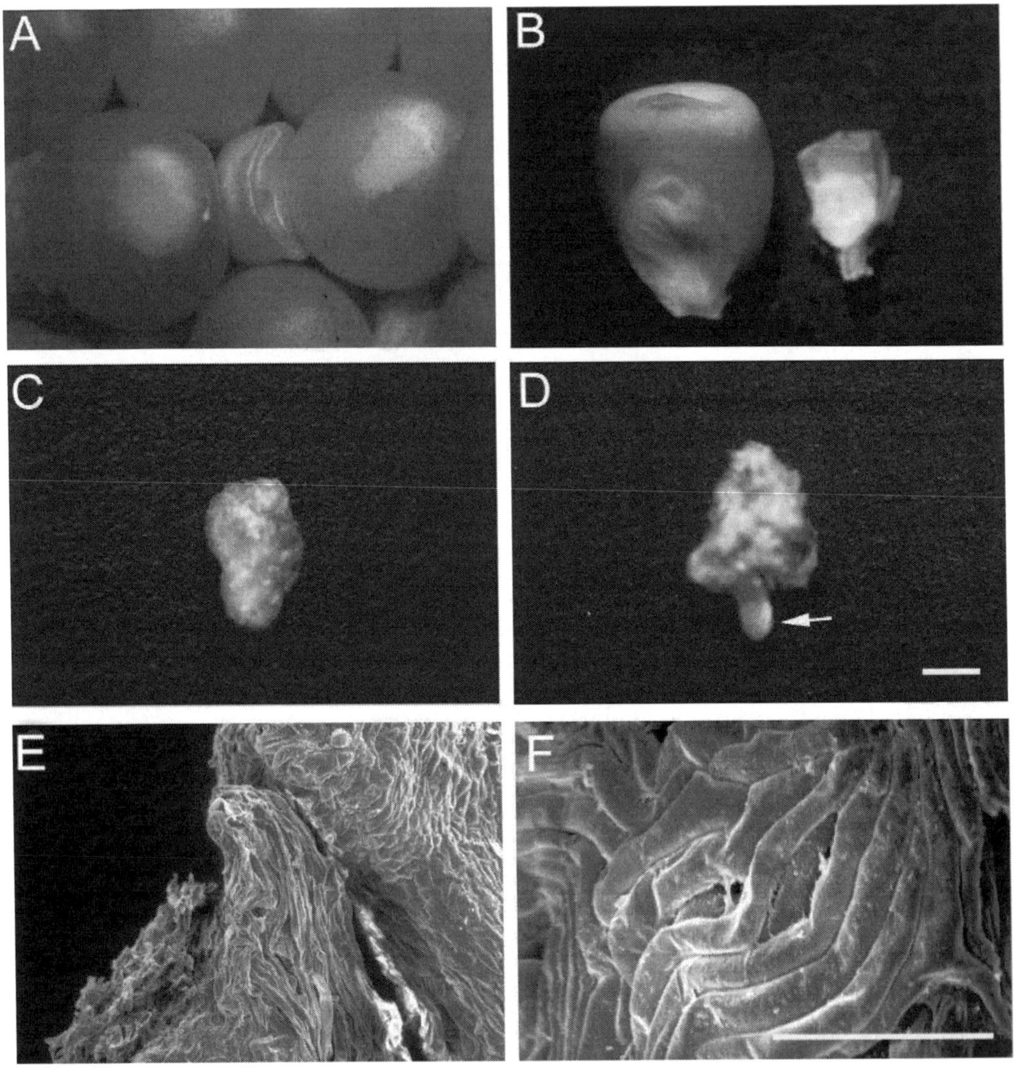

Fig. 3.1. The *lc* phenotype as observed with the binocular microscope and with scanning electron microscopy. (A) Appearance of an *lc* kernel surrounded by its wild-type siblings. (B) A normal kernel and an *lc* one. (C, D) Shapes of *lc* embryo after pericarp and endosperm removal: they look irregular with a rough surface and a yellowish colour. Although many of them have no clear structures, some of them have a visible suspensor (D, arrow). (E) Shoot apex detail of an *lc* embryo, where the shoot primordium and the scutellum are easily recognizable. This apical primordium was green. (F) A close-up image of a field of enlarged cells. All the grains and embryos shown in this figure were studied at kernel maturity. The size bar in (D) corresponds to 1.5 mm in (C) and in (D). The size bar in (F) corresponds to 300 μm in (E) and 200 μm in (F).

We inferred a possible relationship between the inability of the *lc* homozygous grains to germinate and the abnormally elongated cells. It is possible that the grains are unable to germinate due to a general failure in the cell expansion machinery in an embryo that would be otherwise viable. To check the viability of the *lc* cells in the dried seed we used agarose gel electrophoresis to examine the integrity of the genomic DNA purified from *lc* embryos. In all the nine *lc* embryos examined, the genomic DNA was degraded, while in 20 sibling embryos of wt phenotype, the genomic DNA was of high molecular weight as expected. From this result we concluded that *lc* grains are unable to germinate because of a general process of cell death. Then, from our observation of degraded DNA in *lc* embryos, we surmised that *lc* embryos may inappropriately switch on the cell death machinery. We applied the technique of TUNEL on *lc* embryos at a stage of 34 days after pollination to detect cells undergoing PCD (Fig. 3.2B, C and D). At this stage, cells located in the inner layers of the scutellum are undergoing PCD in wild-type embryos but there is no sign of this process in the embryo axis

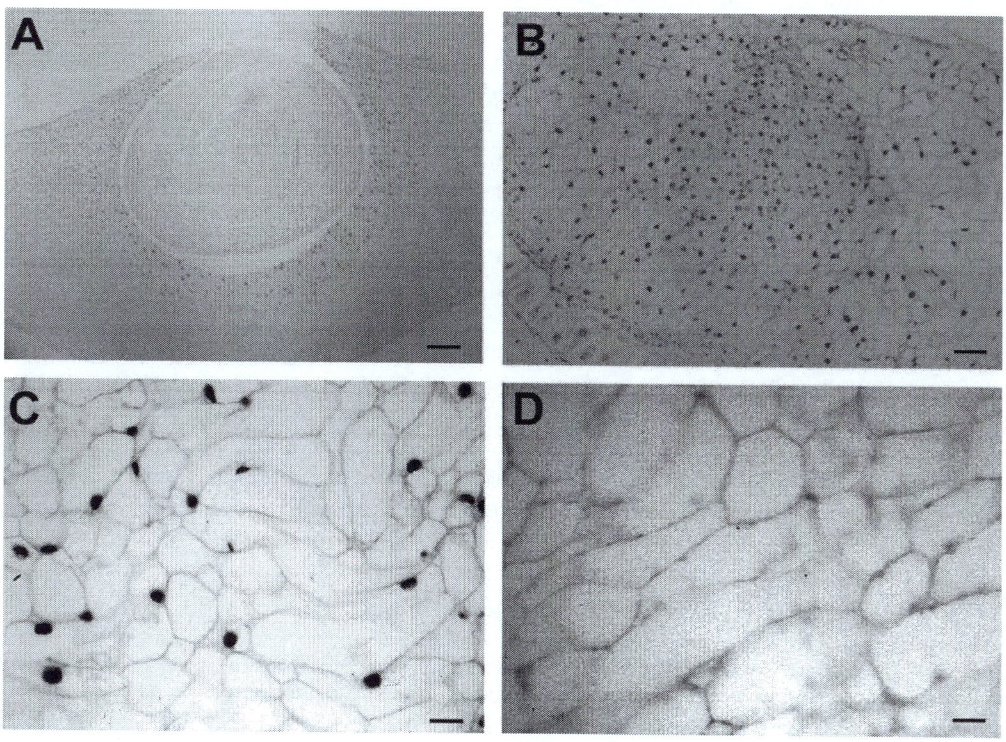

Fig. 3.2. The *lc* mutants undergo programmed cell death. Cells located near the embryo axes in wild-type embryos at 34 days after pollination undergo a process of programmed cell death revealed by TUNEL positive staining (A). At the same stage, *lc* mutant embryos exhibit a generalized breakage of nuclear DNA (B, and detail in C). The negative control (D) is obtained by performing the TUNEL reaction without the TdT enzyme on embryos at the same developmental stage. The size bar is 500 μm in (A), 200 μm in (B) and 50 μm in (C) and (D).

(Fig. 3.2A). In *lc* embryos at the same stage a broad process of cell death is suffered all over the embryo (Fig. 3.2B).

Discussion

The production of the complete embryo depends on a succession of cell divisions that occur in a well defined and ordered pattern that has been exploited in *Arabidopsis* in order to analyse the genes that control early embryogenesis in plants. This pattern may not be evident in the initial stages of maize embryogenesis (Van Lammeren, 1986). In fact the proembryo stage appears as mainly proliferative, where essentially only the suspensor and the protoderm visibly differentiate (Ingram *et al.*, 2000). Before the definition of bilateral symmetry, the expression of markers of scutellum (José-Estanyol and Puigdomènech, 1998) precedes the first indications of specific meristems and the markers that have been observed at this stage correspond to the protoderm (Sossountzov *et al.*, 1991; Ingram *et al.*, 2000). It has been shown in *Arabidopsis* that processes such as cytokinesis that take part in the control of cell division and elongation are essential for the process of embryogenesis (Jürgens, 2000).

The *lc* mutant appears to disturb the balance between elongation and division. The general appearance of the embryo also seems to be disturbed, though the differentiation of its various organs does not seem to be blocked. A number of *dek* mutants already described produce similar phenotypes (Sheridan and Neuffer, 1980), indicating that the mechanisms that define the different organs in the maize embryo are independent of the internal mechanisms that control specific planes of cell division. This fact would favour the idea that many developmental processes in plants are essentially not cell autonomous but mostly depend on general factors acting on the embryo. The phenotype of the *lc* cells in the late phases of embryogenesis is very elongated but they do not seem to have more than one nucleus. In consequence, the mechanism that appears to be perturbed by the mutation is the elongation of the cells. With current knowledge, it is difficult to say whether the *lc* gene takes part directly in the cell expansion machinery, or whether it controls it through a chain of regulative interactions.

One of the phenotypes produced by the *lc* mutation is a generalized cell death in the embryo. It has been shown that different organs of the seed undergo PCD during the formation of the seed and during its germination. This is the case for the suspensor, for the aleurone in barley (Bethke *et al.*, 1999; Fath *et al.*, 1999) and for maize or wheat endosperm and pericarp (Young *et al.*, 1997; Young and Gallie, 1999, 2000). In fact, it appears that all the tissues that will finally not contribute to the formation of the adult plant undergo PCD at specific periods of embryogenesis and germination. For this reason the *lc* mutation may act in the developing embryo in at least two ways. One may be by contributing to the control of cell elongation; if this essential process of plant morphogenesis does not occur properly, it may trigger PCD through specific checkpoints in the cell cycle. The second possibility would be that the protein encoded by the *lc* gene is essential for preserving the embryo cells fated to become part of the developing plant after germination. The *lc* gene may do this by preventing the PCD in these cells. Therefore, its mutation would result in generalized cell death; a final effect on the growing cells would be the failure of the control of cell elongation.

Acknowledgements

The work has been funded by CICYT (grant BIO97-0729) and by the European Union (Maizemb Program). M. Bastida is a recipient of a predoctoral fellowship from the Generalitat de Catalunya. The work has been carried out within the framework of the Centre de Referència de Biotecnologia de la Generalitat de Catalunya.

References

Bethke, P.C., Londsdale, J.E., Fath, A. and Jones, R.L. (1999) Hormonally regulated programmed cell death in barley aleurone cells. *Plant Cell* 11, 1033–1045.

Dellaporta, S.L. and Moreno, S.A. (1994) Gene tagging with Ac/Dc elements in maize. In: Freeling, M. and Walbot, V. (eds) *The Maize Handbook*. Springer Verlag, New York, pp. 526–233.

Elster, R., Bommert, P., Sheridan, W.F. and Werr, W. (2000) Analysis of four embryo-specific mutants in *Zea mays* reveals that incomplete radial organization of the proembryo interferes with subsequent development. *Development, Genes and Evolution* 210, 300–310.

Faccio, S. and Sparvoli, F. (1988) Cytological characterization of the embryo-lethal mutant *dek1* of maize. *Protoplasma* 144, 142–148.

Fath, A., Bethke, P.C. and Jones, R.L. (1999) Barley aleurone cell death is not apoptotic: characterization of nuclease activities and DNA degradation. *Plant Journal* 20, 305–315.

Heckel, T., Werner, K., Sheridan, W.F., Dumas, C. and Rogowsky, P. (1999) Novel phenotypes and developmental arrest in early embryo-specific mutants of maize. *Planta* 210, 1–8.

Ingram, G.C., Boisnard-Lorig, C., Dumas, C. and Rogowsky, P. (2000) Expression patterns of genes encoding HD-ZipIV homeo domain proteins define specific domains in maize embryos and meristems. *Plant Journal* 22, 401–414.

José-Estanyol, M. and Puigdomènech, P. (1998) Developmental and hormonal regulation of genes coding for proline-rich proteins in maize embryo and ovary. *Plant Physiology* 116, 485–494.

Jürgens, G. (2000) Cytokinesis: the art of partitioning. *Plant Cell* 12, 827–828.

Lam, E., Pontier, D. and del Pozo, O. (1999) Die and let live. Programmed cell death in plants. *Current Opinion in Plant Biology* 2, 502–507.

Lotan, T., Ohto, M., Yee, K.M., West, M.A.L., Lo, R., Kwong, R.W., Yamagishi, K., Fischer, R.L., Goldberg, R.B. and Harada, J.J. (1998) Arabidopsis LEAFY COTYLEDON1 is sufficient to induce embryo development in vegetative cells. *Cell* 93, 1195–1205.

Mayer, U., Büttner, G. and Jürgens, G. (1993) Apical-basal pattern formation in the Arabidopsis embryo: studies on the role of the gnom gene. *Development* 117, 149–162.

Ogas, J., Kaufmann, S., Henderson, J. and Somerville, C. (1999) PICKLE is a CHD chromatin-remodeling factor that regulates the transition from embryonic to vegetative development in Arabidopsis. *Proceedings of the National Academy of Sciences USA* 96, 13839–13844.

Pennell, R.I. and Lamb, C. (1997) Programmed cell death in plants. *Plant Cell* 9, 1157–1168.

Scanlon, M.J. and Myers, A.M. (1998) Phenotypic analysis and molecular cloning of *discolored-1* (*dsc1*), a maize gene required for early kernel development. *Plant Molecular Biology* 37, 483–493.

Sheridan, W.F. and Clark, J.K. (1987) Maize embryogeny: a promising experimental system. *Trends in Genetics* 3, 3–6.

Sheridan, W.F. and Neuffer, M. (1980) Defective kernel mutants of maize II. Morphological and embryo culture studies. *Genetics* 95, 945–960.

Sheridan, W.F. and Thorstenson, Y.R. (1986) Developmental profiles of three embryo-lethal maize mutants lacking leaf primordia: ptd*-1130, cp*-1418, bno*-747B. *Developmental Genetics* 7, 35–49.

Shevell, D.E., Leu, W., Gillmor, C.S., Xia, G., Feldmann, K.A. and Chua, N. (1994) *EMB30* is essential for normal cell division, cell expansion, and cell adhesion in Arabidopsis and encodes a protein that has similarity to Sec7. *Cell* 77, 1051–1062.

Sossountzov, L., Ruiz-Avila, L., Vignols, F., Jolliot, A., Arondel, V., Tchang, F., Grosbois, M., Guerbette, F., Miginiac, E., Delseny, M., Puigdomènech, P. and Kader, J.C. (1991) Spatial and temporal expression of a maize lipid transfer protein gene. *Plant Cell* 3, 923–933.

Stiefel, V., López Becerra, E., Roca, R., Bastida, M., Jahrmann, T., Graziano, E. and Puigdomènech, P. (1999) TM20, a new class of transmembrane proteins expressed in meristematic tissues of *Zea mays*. *Journal of Biological Chemistry* 274, 27734–27739.

Van Lammeren, A.A.M. (1986) Developmental morphology and cytology of the young maize embryo (*Zea mays* L.). *Acta Botanica Neerlandica* 35, 169–188.

Young, T.E. and Gallie, D.R. (1999) Analysis of programmed cell death in wheat endosperm reveals differences in endosperm development between cereals. *Plant Molecular Biology* 39, 915–926.

Young, T.E. and Gallie, D.R. (2000) Regulation of programmed cell death in maize endosperm by abscisic acid. *Plant Molecular Biology* 42, 397–414.

Young, T.E., Gallie, D.R. and DeMason, D.A. (1997) Ethylene-mediated programmed cell death during maize endosperm development of wild-type and *shrunken2* genotypes. *Plant Physiology* 115, 737–751.

4 The Failure of the Embryonic Epidermis Affects Seed Maturation and Embryo Growth in Pea

Ljudmilla Borisjuk[1], Trevor L. Wang[2], Hardy Rolletschek[1], Ulrich Wobus[1] and Hans Weber[1]

[1]*Institut für Pflanzengenetik und Kulturpflanzenforschung (IPK), D-06466 Gatersleben, Germany;* [2]*Department of Applied Genetics, John Innes Centre, Colney, Norwich NR4 7UH, UK*

Introduction

Transfer cell formation in growing legume embryos starts with the establishment of the protoderm during the late globular stage. Its cells divide anticlinally and form an epidermis. Epidermal cells are rectangular in shape and densely cytoplasmic whereas underlying parenchyma cells are isodiametric, highly vacuolated and divide in both anti- and periclinal planes (Borisjuk *et al.*, 1995). When morphogenesis and organogenesis of the young legume embryo has been accomplished, the abaxial epidermal cells differentiate into transfer cells characterized by ingrowths of the outer cell wall which increase the transport-active surface (Bonnemain *et al.*, 1991). Transfer cell formation in legume cotyledons establishes a new function that epidermal cells acquire at a certain stage of development and represents a regional specification to improve embryo nutrition. The control factors are unknown but metabolic signalling may be involved, possibly triggered by stimuli from neighbouring compartments. Hexose sugars induce transfer cell formation in *Vicia faba* embryos whereas high sucrose is inhibitory (Offler *et al.*, 1997; Weber *et al.*, 1997; Farley *et al.*, 2000). Transfer cells mediate sucrose uptake by H^+ co-transport. Their formation is coupled with up-regulated expression of genes encoding transporters for sucrose, hexoses, amino acids and H^+-ATPases (Patrick and Offler, 2001). In *V. faba* cotyledons increased gene expression of the sucrose transporter is accompanied by accumulation of large amounts of sucrose in the underlying tissue (Borisjuk *et al.*, 2002b).

The embryos of the pea seed mutant *E2748* are strongly reduced in growth and abort before complete maturation. The embryo appears to lack a proper epidermis or the epidermal layer adopts an altered morphology (Johnson *et al.*, 1994). Because the mutant embryo is obviously affected in cotyledonary transfer cell formation, assimilate supply could be impaired. The mutant provides a suitable model to study nutrient uptake into the

 The Biology of Seeds: Recent Research Advances (eds G. Nicolás, K.J. Bradford, D. Côme and H.W. Pritchard)

embryo and filial–maternal interactions, as well as maturation and differentiation events in the embryo.

Materials and Methods

Pea plants were grown in growth chambers under a light/dark regime of 16/8h at 16/19°C. Because the recessive *E2748* seeds were lethal, the plants were maintained as heterozygotes. The use of heterozygous plants allowed both mutant and wild-type seeds to be analysed from the same pod and thus provided suitable controls of the same age. The volume of the vacuolar liquid in the endosperm was determined by gradual sampling from embryo sacs of freshly harvested seeds at different developmental stages using a microsyringe. Tissue preparation and microscopy were performed as described by Borisjuk *et al.* (1995, 2002a). Biochemical procedures as well as *in situ* techniques were performed according to Weber *et al.* (1997).

Results

The *E2748* seed mutation affects embryo growth during the maturation phase

Fresh weight of mutant and wild-type embryos is not different up to the stage when morphogenesis and cell division have terminated and maturation started. Mutant embryos grow slowly and reach only approximately 10% of the wild-type values (Fig. 4.1A) whereas seed coat growth is nearly unchanged (not shown). The endospermal vacuoles increase similarly up to approximately 150 mg seed weight in both wild type and mutant (Fig. 4.1B) then decrease continuously in the wild type, when their place is occupied by the fast-growing embryo. Within the mutant, endospermal vacuoles further increase until the seeds reach approximately 250 mg. The endospermal vacuoles bearing mutant or wild-type embryos contain similar concentrations of sugars (Fig. 4.1C) indicating that retarded embryo growth is not due to limiting carbohydrates. Moreover, the change of the hexose to sucrose ratio within the endospermal vacuole is unaltered in the mutant and apparently independent of normal embryo growth. Mutant embryos are not able to accumulate normal levels of sucrose (Fig. 4.1D) and starch (Fig. 4.1E).

Cotyledonary epidermal cells of the mutant de-differentiate at the beginning of seed maturation

Epidermal cells in wild-type embryos of 50 mg seeds can already be distinguished from the parenchyma cells by anticlinal cell divisions and smaller size (Fig. 4.2A). Their subcellular structure, however, is similar to that of the underlying parenchyma cells. Subsequently epidermal cells become more vacuolated and tend to expand periclinally. Differences in size between epidermal and parenchyma cells now become apparent (Fig. 4.2B,C). Cell divisions occur equally and exclusively anticlinally. The outer cell wall is thicker than the inner one, indicating asymmetrical cell wall thickening (Fig. 4.3A).

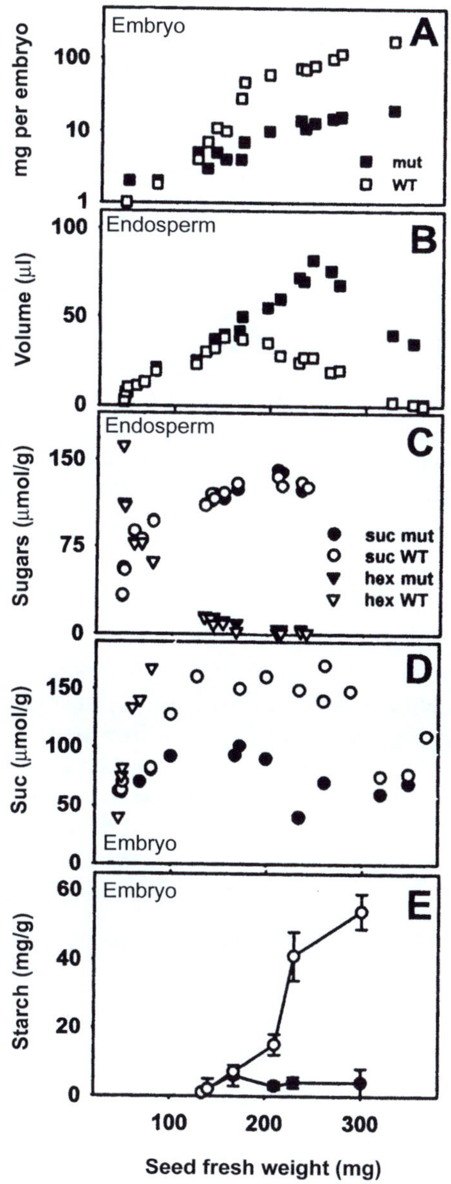

Fig. 4.1. Growth parameters of mutant *E2748* and wild-type seeds. (A) Fresh weight of embryos and (B) endospermal vacuole volumes. (C) Hexoses (sum of glucose and fructose) and sucrose within endospermal vacuoles. Concentration in embryo of (D) sucrose and (E) starch. Modified from Borisjuk *et al.* (2002a).

The outward side of the cell walls differs from the inner one by ingrowths to the cytoplasm (Fig. 4.3B) representing typical morphological features of transfer cells. The inner wall has many symplasmic connections to the parenchyma via branched-type plasmodesmata (Fig. 4.3C). Consequently,

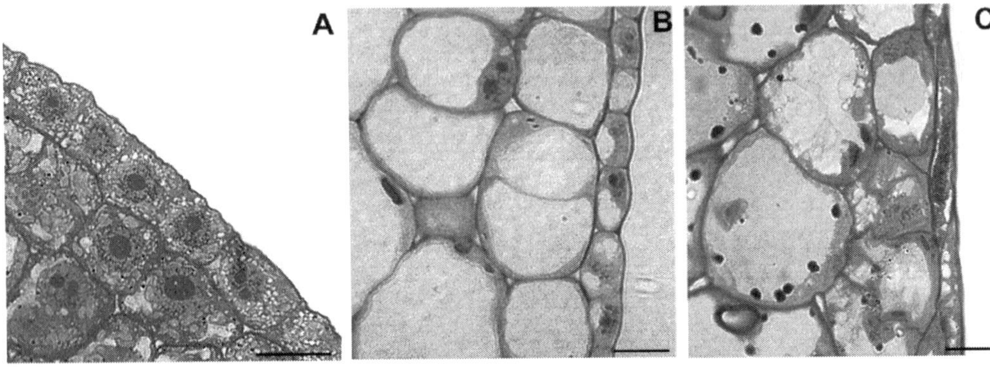

Fig. 4.2. Epidermal and transfer cell structure in wild-type pea embryos following toluidine blue staining. (A) Young embryo at late heart stage with single epidermal cell layer and underlying parenchyma cells. (B) Embryo from mid-developmental stage seeds. (C) Embryo from late-developmental stage seeds. Scale bars, (A) 10 µm, (B, C) 6 µm. Selected from Borisjuk *et al.* (2002a).

the smooth surface of wild-type embryos (260 mg seed weight) is formed by a continuous thin epidermal cell layer which covers the storage parenchyma. Unlike in the mutant (compare Fig. 4.3A with Fig. 4.3D), this layer represents a clear boundary between embryo and neighbouring tissue.

A mutant phenotype can be identified from early cotyledon stage on. Initially the mutant cotyledonary epidermis still appears predominantly like that of the wild type (Fig. 4.4.A,B). Later its morphology changes in certain

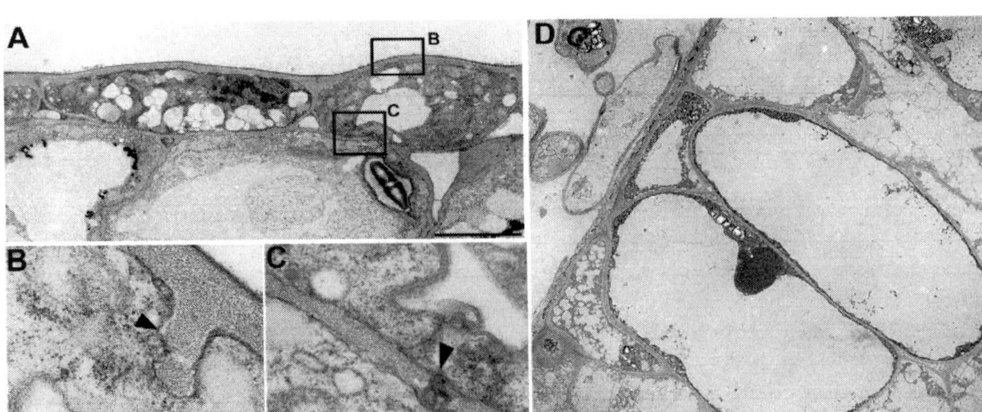

Fig. 4.3. Transfer cell ultrastructure in wild-type and mutant pea embryos. (A) Transfer layer in the wild type. (B) Outer cell wall morphology of transfer cell with cell wall ingrowth. (C) Inner cell wall morphology of epidermal cell, faced to the parenchyma with branched-type plasmodesmata. (D) Cotyledon to seed coat contact zone in the mutant. Note the large differences in the size of the outermost cells. Scale bars, (A) 4.5 µm, (B, C) 0.56 µm, (D) 40 µm. Selected from Borisjuk *et al.* (2002a).

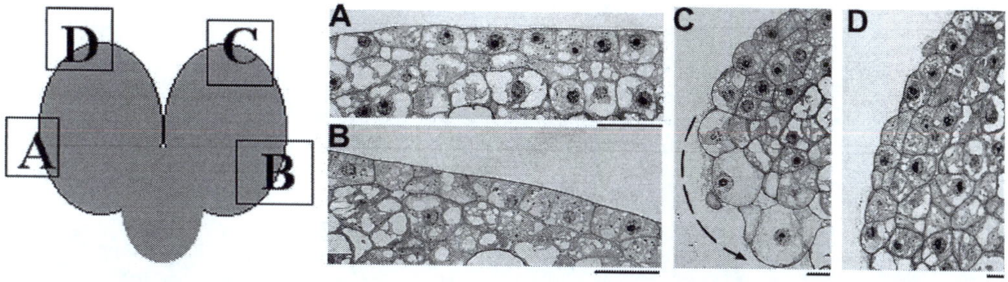

Fig. 4.4. Epidermal structure in a single *E2748* embryo during the transition stage, with toluidine blue staining. (A,B) Embryo at the early cotyledon stage having a wild-type-like epidermis. (C) Highly vacuolated modified epidermal cells with altered surfaces; the arrow shows increasing morphological aberrations. (D) Modified epidermal cell with periclinal as well as anticlinal cell divisions. Modified from Borisjuk *et al.* (2002a).

areas, most often abaxially or at the tip (Fig. 4.3C,D). Some cells become more vacuolated and enlarged, forming a highly irregular morphology with a rough surface (Fig. 4.3C). Periclinally as well as anticlinally oriented cell walls are visible (Fig. 4.4D). During further development (160 mg seeds), the surface layer of the cotyledon become unusually enlarged and vacuolated. Epidermal cells cannot be clearly distinguished from parenchyma cells, and the cotyledonary cell population becomes very heterogenous in size, degree of vacuolization and storage product accumulation (Fig. 4.3D).

The pattern of sucrose transporter gene expression is altered in mutant cotyledons

A sucrose transporter-specific probe (*VfSUT1*; Weber *et al.*, 1997) was used for *in situ* hybridizations on wild-type cotyledons of 60 mg seeds (Fig. 4.5A). SUT1-specific signals are present within the outermost layer, indicating gene expression within the epidermal cell layer (Fig. 4.5B,C). In mutant cotyledons a similar pattern occurs (not shown). In wild-type cotyledons of 200 mg seeds (Fig. 4.5D) SUT1-specific signal is present within the outer layer with higher intensity in the abaxial region where transfer cells have been formed (Fig. 4.5E). A steep gradient of label intensity is present, decreasing from the abaxial towards the interior, indicating lower expression also in the parenchyma cells. In the mutant (Fig. 4.5F) the pattern is completely irregular. Stronger label occurs within the parenchyma whereas it is much weaker in the outermost cell layer (Fig. 4.5G).

Discussion

Epidermal differentiation regulates response to external sugars

It is not known what triggers the differentiation of epidermal into transfer cells. There is some evidence that metabolic signalling is involved. During transfer cell formation in faba bean cotyledons, the hexoses to sucrose ratio

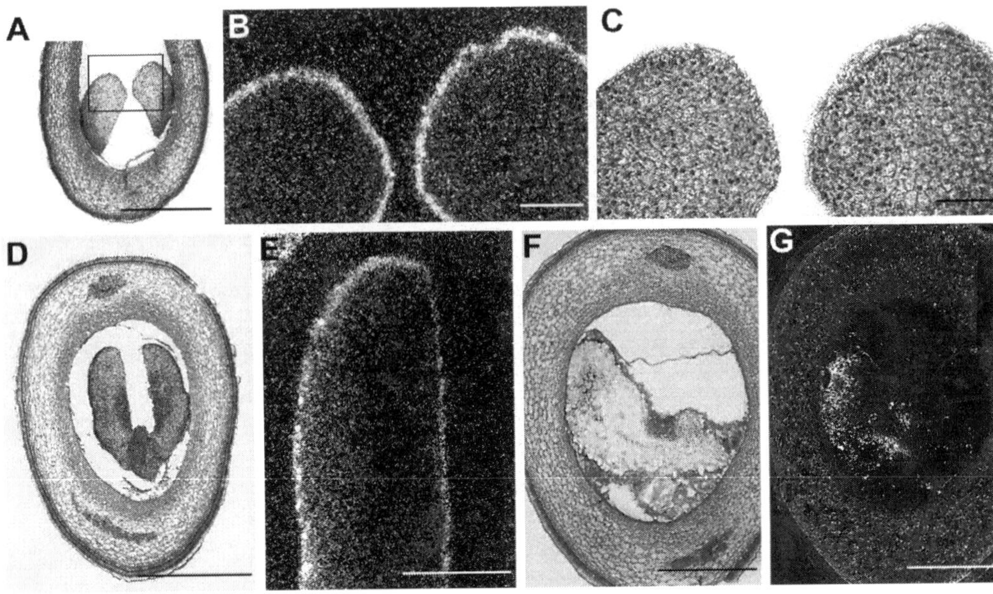

Fig. 4.5. Sucrose transporter gene expression of wild-type and mutant cotyledons. Toluidine blue staining (bright field) and after *in situ* hybridization (dark field). (A) Wild-type seed of approximately 60 mg weight; (B) its cotyledons and (C) the same section after toluidine blue staining. (D): Wild-type seed at mid-cotyledon stage. (E) Fragment of cotyledon after *in situ* hybridization. (F, G) Mutant seed: note the completely irregular pattern of the signal (G). Scale bars, (A) 0.7 mm, (B) 0.1 mm, (E) 0.4 mm, (D, F) 1.5 mm. Selected from Borisjuk *et al.* (2002a).

within the endospermal vacuole changes from 100:1 to 3:1 (Offler *et al.*, 1997). Possibly, the *E2748* gene product could play a role in perceiving or transmitting a sugar signal.

Changes in epidermal morphology occur in mutant embryos at the time when sucrose in endospermal vacuoles becomes high. Possibly the mutant with impaired epidermis cannot tolerate elevated levels of sucrose in its environment and its outermost cells react to sucrose with a kind of wound response that induces callus-like growth, thereby preventing coordinated cotyledon development. This hypothesis is underlined by *in vitro* experiments: feeding sucrose to young faba bean embryos, before they have developed transfer cells, dramatically alters their shape and initiates callus-like growth, e.g. de-differentiation. In contrast, older embryos at the mid-cotyledon stage do not respond (Weber *et al.*, 1996). Thus, the sucrose response only occurs when epidermal cells are still undifferentiated. The time when cells lose competence coincides with the time of transfer cell differentiation.

In vivo, de-differentiation into callus-like growth is prevented by the particular sugar state within the endospermal vacuole: it does not occur when sucrose is < 20–30 mM. At the time when sucrose rises, epidermal cells of wild-type embryos are already differentiated and provide a barrier preventing underlying parenchyma tissues from responding with callus-like growth. In the *E2748* mutant, epidermal differentiation is blocked and the parenchyma cells adopt callus-like growth at a time when sucrose within the endospermal vacuole increases to approximately 30 mM. It is supposed that the de-differentiation of epidermis disrupts the normal developmental pathway and ultimately leads to seed abortion (for a summarizing scheme, see Fig. 4.6).

Conclusions

The outermost cell layer of the mutant cotyledons cannot acquire transfer cell morphology but loses epidermal cell identity and does not function as a sucrose uptake system. As a consequence of lacking the transfer cell layer, the cells adopt callus-like growth at a time when sucrose increases. This de-

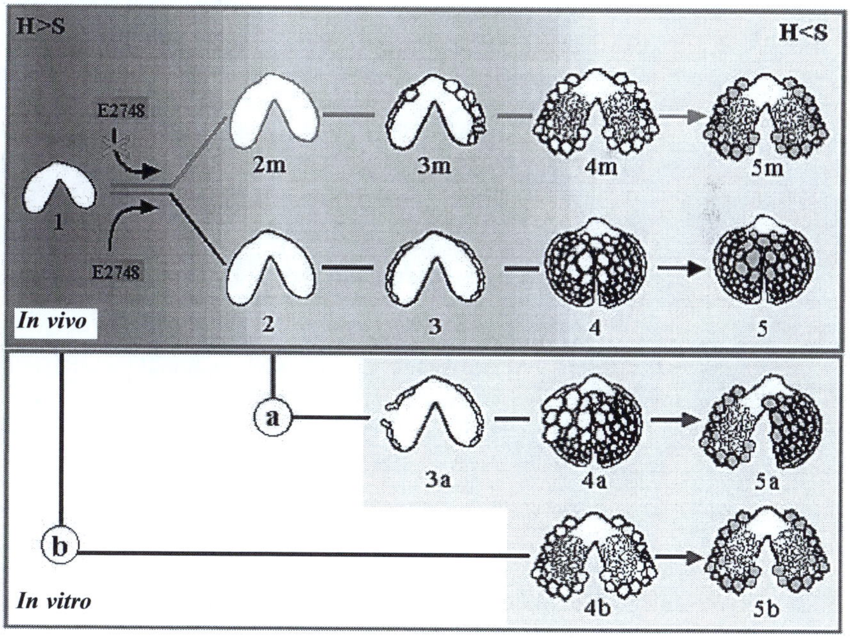

Fig. 4.6. Major histological changes in *E2748* (**1–5m**) and wild-type (**1–5**) embryos. The underlying shading indicates the sugar status from a high hexose to sucrose ratio (H > S) to a low hexose to sucrose ratio (H < S). The *E2748* mutation blocks epidermal differentiation. Primary changes in epidermal morphology occur during early stages (**2m** and **2**) when cells of outer cell layer of mutant become vacuolated (**3m–5m**) instead of forming transfer cells (**3–5**). Removal of transfer cells in wild type (*in vitro* **a** passage to **3a**) causes disturbed differentiation of parenchyma cells mimicking the mutant phenotype (**4a–5a**). Feeding sucrose to young normal embryos before transfer cells are differentiated (*in vitro* **b** passage) induces callus-like growth (**4b**) resulting in a mutant-like phenotype (**5b**). Extended from Borisjuk *et al.* (2002a).

differentiation disrupts further coordinated development and ultimately causes seed abortion. When *E2748* embryos are rescued *in vitro*, the embryo axis grows to a normal and fully differentiated pea plant in which the mutant phenotype only becomes evident in all of the embryos at seed set (Borisjuk *et al.*, in preparation). The *E2748* gene product, therefore, is required only for transfer cell formation in developing cotyledons and has apparently no other function during normal plant growth. It is concluded that the *E2748* gene controls the differentiation of the cotyledonary epidermis into transfer cells and thus is required for the regional specialization with a function in embryo nutrition (Borisjuk *et al.*, 2002a).

References

Bonnemain, J.L., Bourquin, S., Renault, S., Offler, C. and Fisher, D.G. (1991) Transfer cells: structure and physiology. In: Bonnemain, J.L., Delrot, S., Lucas, W.J. and Dainty, J. (eds) *Phloem Transport and Assimilate Compartmentation*. Quest Editions, Nantes, France, pp. 178–186.

Borisjuk, L., Weber, H., Panitz, R., Manteuffel, R. and Wobus, U. (1995) Embryogenesis of *Vicia faba*. Histodifferentiation in relation to starch and storage protein synthesis. *Journal of Plant Physiology* 147, 203–218.

Borisjuk, L., Wang, T., Rolletschek, H., Wobus, U. and Weber, H. (2002a) A pea seed mutant affected in the differentiation of the embryonic epidermis leads to deregulated seed maturation and impaired embryo growth and seed maturation. *Development* 129, 1595–1607.

Borisjuk, L., Walenta, S., Rolletschek, H., Mueller-Klieser, W., Wobus, U. and Weber, H. (2002b) Spatial analysis of plant development: sucrose imaging within *Vicia faba* cotyledons reveals specific developmental patterns. *Plant Journal* 29, 521–530.

Farley, S.J., Patrick, J.W. and Offler, C.E. (2000) Functional transfer cells differentiate in cultured cotyledons of *Vicia faba* L. seeds. *Protoplasma* 214, 102–117.

Johnson, S., Liu, C.M., Hedley, C.L. and Wang, T.L. (1994) An analysis of seed development in *P. sativum* XVIII. The isolation of mutants defective in embryo development. *Journal of Experimental Botany* 45, 1503–1511.

Offler, C.E., Liet, E. and Sutton, E.G. (1997) Transfer cell induction in cotyledons of *V. faba*. *Protoplasma* 200, 51–64.

Patrick, J.W. and Offler, C.E. (2001) Compartmentation of transport and transfer events in developing seeds. *Journal of Experimental Botany* 52, 551–564.

Weber, H., Borisjuk, L. and Wobus, U. (1996) Controlling seed development and seed size in *Vicia faba*: a role for seed coat-associated invertases and carbohydrate state. *Plant Journal* 10, 823–834.

Weber, H., Borisjuk, L., Heim, U., Sauer, N. and Wobus, U. (1997) A role for sugar transporters during seed development: molecular characterization of a hexose and a sucrose carrier in Faba bean seeds. *Plant Cell* 9, 895–908.

5 Seed Quality Development in Wrinkled-seeded Mutant Peas (*Pisum sativum* L.)

T.W. Lyall[1], R.H. Ellis[1], P. John[2], C.L. Hedley[3] and T.L. Wang[3]

[1]*Department of Agriculture, The University of Reading, Earley Gate, PO Box 237, Reading RG6 6AR, UK;* [2]*School of Plant Sciences, The University of Reading, Reading RG6 6AS, UK;* [3]*John Innes Institute, Colney Lane, Norwich NR4 7UH, UK*

Introduction

The role of maturation drying in improving seed quality has been widely discussed. Some authors suggest that seed quality is maximal at physiological maturity (i.e. the end of seed filling) (Shaw and Loomis, 1950) and subsequently declines (Harrington, 1972; TeKrony and Egli, 1997). However, improvements in seed quality after this developmental stage have been shown in many contrasting species (e.g. Demir and Ellis, 1992a,b; Hong and Ellis, 1992; Zanakis *et al.*, 1994; Sinniah *et al.*, 1998a).

Several factors appear necessary to confer desiccation tolerance. Evidence implicates the accumulation of soluble sugars, especially sucrose and raffinose family oligosaccharides (Sun and Leopold, 1993; Górecki *et al.*, 1997; Corbineau *et al.*, 2000). However, such sugars have also been detected in immature desiccation-intolerant embryos of maize and wheat (Bochicchio *et al.*, 1997; Black *et al.*, 1999). Other factors, such as heat-stable late embryogenesis abundant proteins, may be involved (Dure *et al.*, 1989; Galau *et al.*, 1991; Blackman *et al.*, 1992, 1995), but some of these have been identified in recalcitrant (desiccation-intolerant) seeds (Finch-Savage *et al.*, 1994). Hence comparing the response of desiccation tolerant and intolerant seeds or embryos to drying fails to provide conclusive evidence of a role in desiccation tolerance for either soluble sugars or heat-stable proteins.

Breaking the normal time course of seed development by denying water has been used to reduce the normal association between seed development and duration from flowering (Sinniah *et al.*, 1998b). In this way it was shown that both soluble sugars and heat-stable proteins were equally likely (or unlikely) to be involved in the development of seed quality.

(eds G. Nicolás, K.J. Bradford, D. Côme and H.W. Pritchard)

The objective of the present study is to determine whether genetic differences can help to identify the significant biochemical changes in developing seeds that confer desiccation tolerance and improve subsequent seed survival. We used a set of near-isogenic lines (NILs) of pea with lesions in the starch biosynthetic pathway inducing variable soluble sugar contents, determining temporal changes in the accumulation of soluble sugars and heat-stable proteins in three mutant lines (*RRrbrb*, *rrRbRb* and *rrrbrb*) and the wild type (*RRRbRb*).

Materials and Methods

Seed of each of four NILs of pea (*Pisum sativum* L.) backcrossed six times to the original parent (Hedley *et al*., 1994) were sown on 3 May 2000. The first day of flowering among uniformly produced plants (median day of the flowering period) was 18 June 2000; growth analyses are described as days after flowering (DAF) from this date.

Sequential harvests were made between 10 and 45 DAF. Seed moisture content (%) was determined using either the high-constant-temperature method or the two-stage method (when seed moisture content was above 17%) (ISTA, 1999a,b). The ability to germinate before and after desiccation was tested in 100 (two replicates of 50) seeds between moistened rolled paper towels at 20°C for 8 days: seedlings were evaluated for normal development (ISTA, 1999a,b). Seeds were dried to 15% moisture content at 15°C with 10% relative humidity. At each harvest, five replicates of 20 seeds were hermetically sealed in laminated aluminium foil packets and placed in liquid nitrogen for 30 min and then stored at –20°C for quantitative biochemical analyses.

Seed longevity was estimated in seeds rapidly dried to 16 ± 0.3% moisture content (at which value the four NILs have similar equilibrium relative humidities) and then hermetically stored at 40°C. Seed survival curves were fitted by probit analysis to the equation $v = K_i - p/\sigma$, where v is the probit of percentage viability after p days in storage and σ is the standard deviation of the frequency distribution of seed deaths in time (d) (Ellis and Roberts, 1980). Values for the 50% viability period, p_{50}, were calculated (the product of K_i and σ) and used as the measure of seed longevity.

Soluble sugars were quantitatively analysed in three replicates of 500 mg of pea flour according to Sinniah *et al*. (1998b), except that after the first extraction the resulting pellet was washed in 5 ml of 80% ethanol three times. Soluble sugars were analysed in a 20 µl aliquot of the supernatant, by HPLC as in Sinniah *et al*. (1998b), except that a Bio-Rad Aminex HPX-42C column (300 × 7.8 mm) was used. Profiles of heat-stable proteins were produced from two replicates (each of 1 g) of pea flour, as in Sinniah *et al*. (1998b).

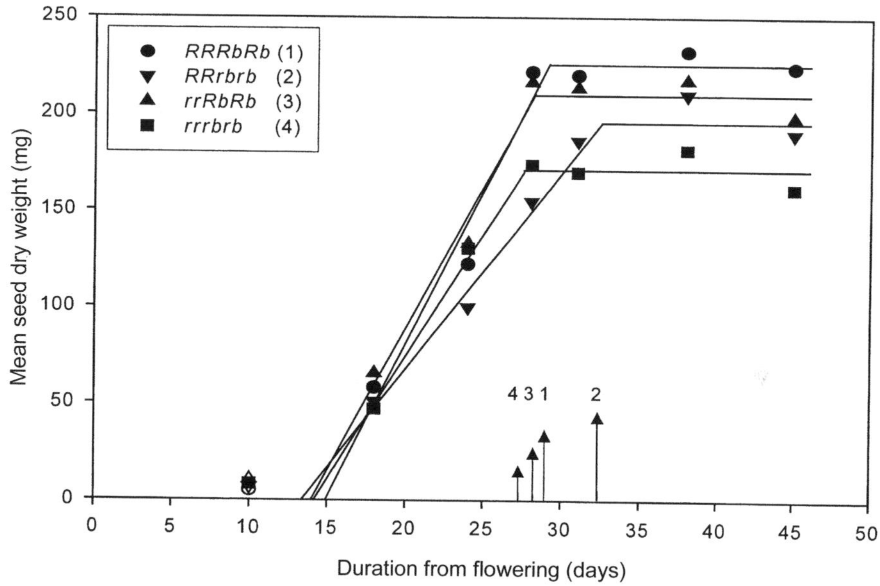

Fig. 5.1. Changes in seed dry weight of four near-isogenic lines of pea during development and maturation. Mass maturity (end of seed filling) is the point of intersection between the horizontal and diagonal lines and is shown by an arrow for each NIL.

Results

Seeds filled at a similar rate among NILs ($P > 0.25$) but final mean dry weights varied significantly among NILs ($P = 0.001$; Fig. 5.1); seeds of the wild type (*RRRbRb*) had the greatest final dry weight and those of the double mutant (*rrrbrb*) the lowest. Mass maturity was attained earliest in seeds of *rrrbrb* (27.5 days) and latest in *RRrbrb* (32.5 days).

A slow reduction in the seed moisture content of all NILs occurred between 10 and 30 DAF (Fig. 5.2) after which, at around mass maturity, a rapid decline in moisture content occurred. Throughout seed filling, *RRRbRb* had a lower moisture content than any of the mutant NILs.

The onset of the ability to germinate occurred between 10 and 18 DAF for all NILs except *rrRbRb*, which showed some germinability at 10 DAF (Fig. 5.3a). Seeds of all NILs showed good germinability (> 70%) prior to mass maturity (24 DAF), and maximal germination coincided with mass maturity in most NILs. Over the period when ability to germinate improved, seed moisture content remained between 85 and 78% in all mutant NILs, and between 82 and 68% in the wild type.

The onset of ability to germinate after rapid-enforced drying to 15% moisture content (relative humidity = 70–75%) occurred between 24 and 31 DAF (Fig. 5.3b). All NILs attained full desiccation tolerance by 38 DAF, when seed moisture content had fallen to 35–40%.

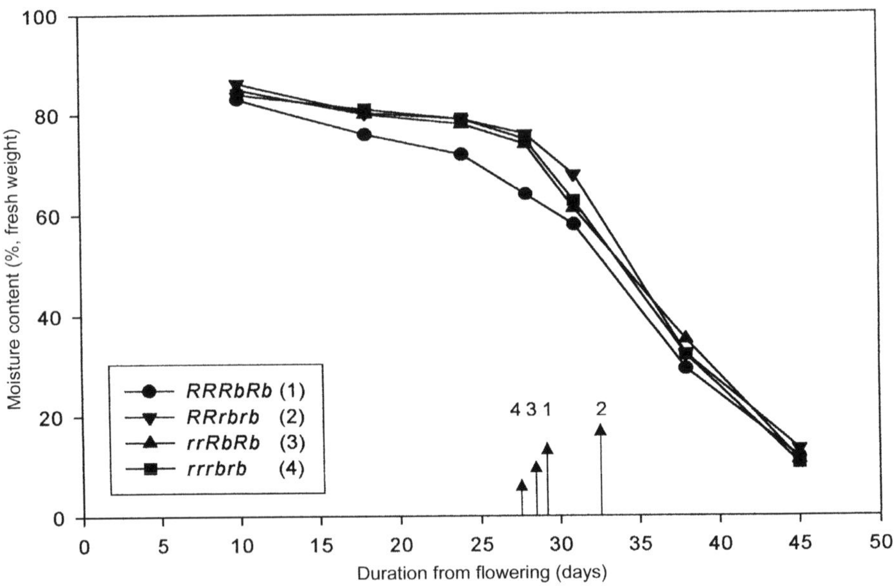

Fig. 5.2. Changes in the moisture content (%, fresh weight) of serially harvested seeds of four near-isogenic lines of pea during seed development and maturation. Mass maturity (end of seed filling) is shown by an arrow for each NIL.

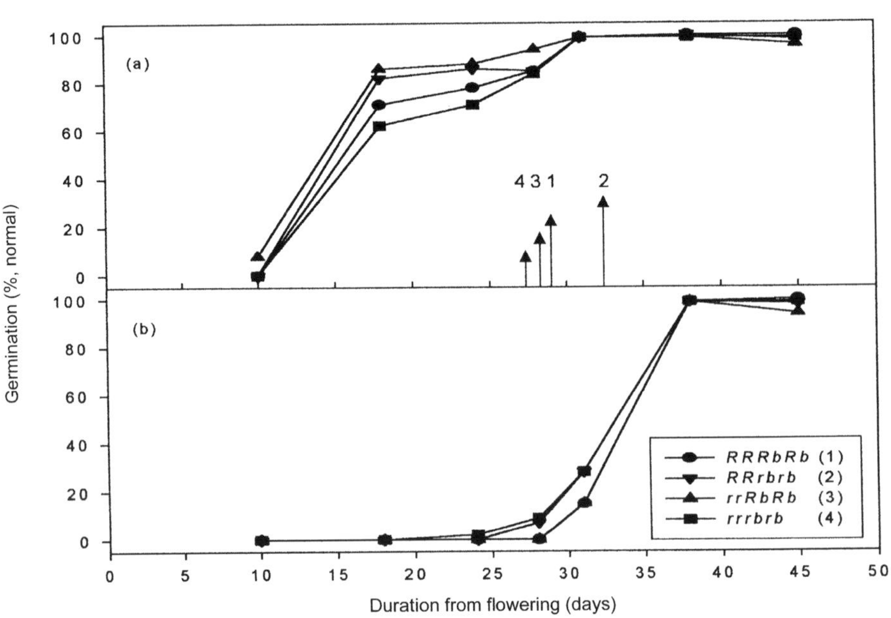

Fig. 5.3. Changes in the ability to germinate normally of (a) fresh seeds and (b) rapidly dried seeds (to 15% moisture content). Mass maturity (end of seed filling) is shown by an arrow for each NIL.

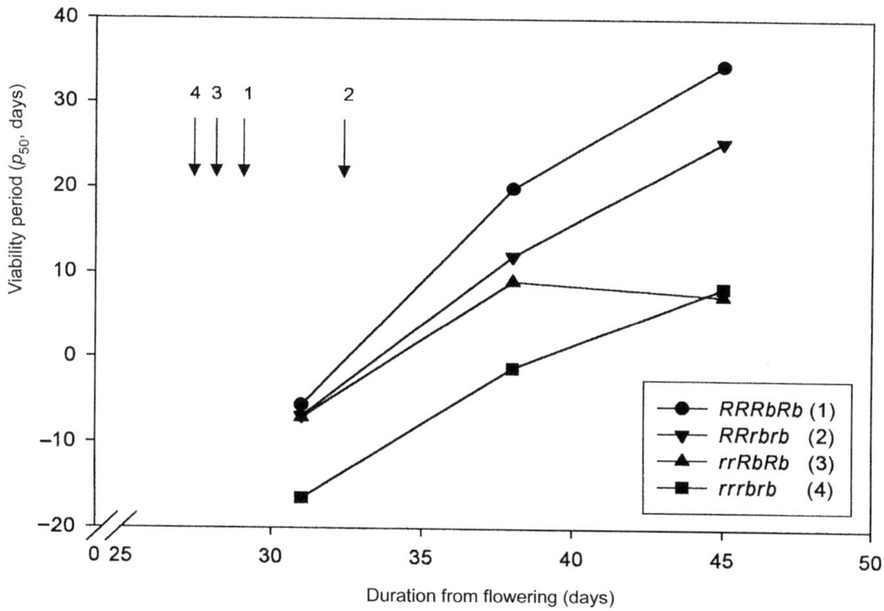

Fig. 5.4. Changes in subsequent air-dry seed longevity (p_{50}) at 40°C and 16.5 ± 0.3% moisture content during the development and maturation of pea seeds of four near-isogenic lines of pea. Mass maturity (end of seed filling) is shown by an arrow for each NIL.

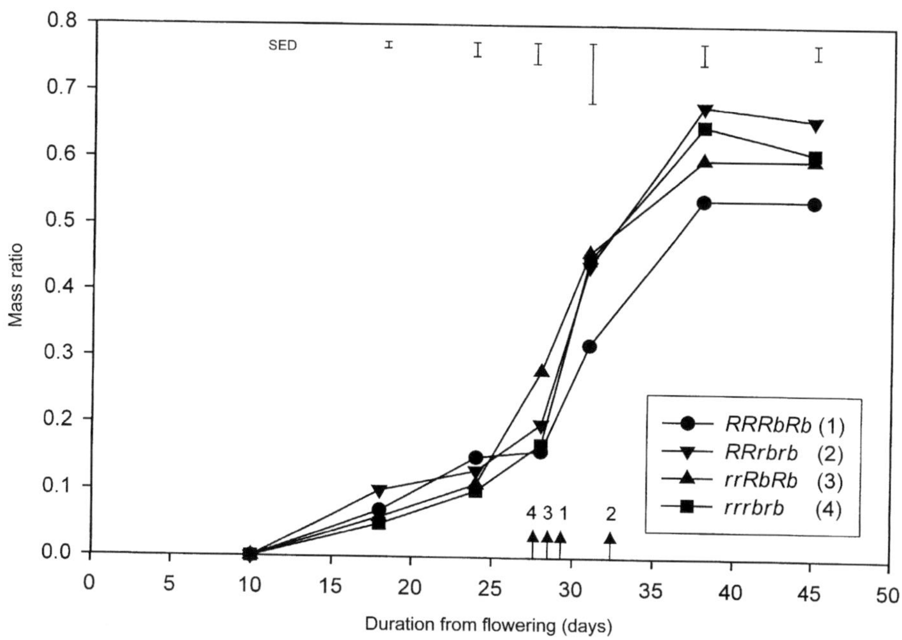

Fig. 5.5. Changes in the ratio of raffinose family oligosaccharides to total sugars during seed development and maturation among four near-isogenic lines of pea. Mass maturity (end of seed filling) is shown by an arrow for each NIL.

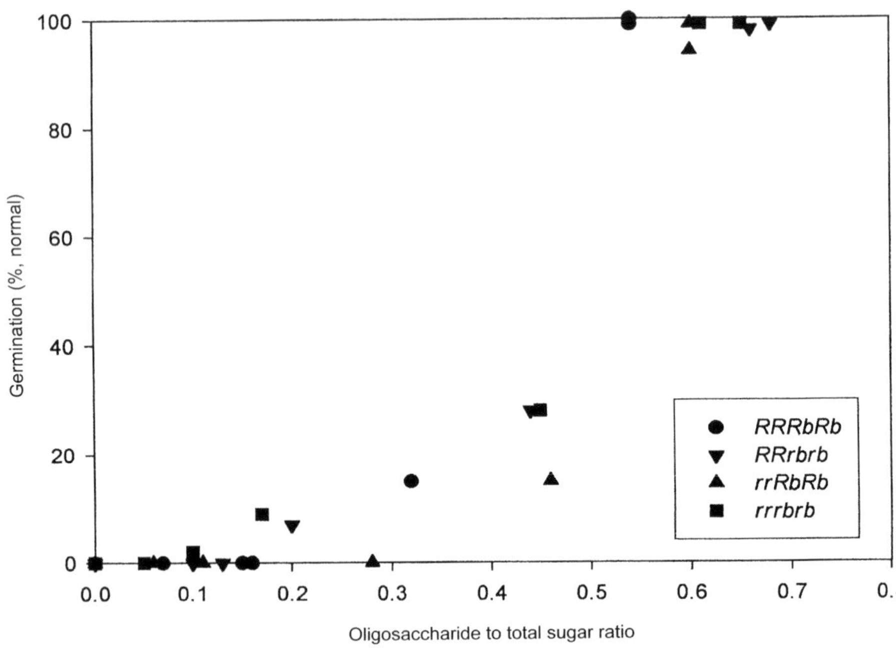

Fig. 5.6. Relations between the oligosaccharide to total sugar ratio and the ability to germinate following rapid enforced drying (to 15% moisture content) for seeds of four near-isogenic lines of pea.

Seed longevity improved after mass maturity in all NILs (Fig. 5.4). Longevity was consistently greatest in serially harvested seeds of the wild type *RRRbRb* and poorest in the double mutant *rrrbrb*.

When fresh seeds were able to germinate (> 70%), the ratio of oligosaccharide to total sugars (Fig. 5.5) had increased to values between 0.05 (*rrRbRb*) and 0.10 (*rrrbrb*). When seeds were first able to tolerate rapid desiccation, the ratio had risen to between 0.54 (*RRRbRb*) and 0.66 (*RRrbrb*) (Fig. 5.6). However, correlation analysis showed no relation between this mass ratio and either the ability to germinate following drying ($r = -0.046$) or seed longevity ($r = -0.007$).

Certain bands of heat-stable proteins were significantly and positively correlated with the ability to germinate after drying, namely those of 7.7 kDa ($r = 0.614$), 24.4 kDa ($r = 0.639$), 27.7 kDa ($r = 0.419$), 29.1 kDa ($r = 0.567$), 32.7 kDa ($r = 0.537$) and 46.9 kDa ($r = 0.800$). Generally their initial detection matched that of the onset of desiccation tolerance. Other (small) heat-stable proteins were detected which, although not significantly correlated with desiccation tolerance, accumulated concurrently with the onset of desiccation tolerance. Only one protein (11.6 kDa) was significantly associated ($r = 0.598$, $P < 0.05$) with seed longevity.

Discussion

The higher moisture contents of mutant NIL seeds before mass maturity were a direct consequence of lesions in starch branching enzyme (*r* mutant) and ADPglucose pyrophosphorylase (*rb* mutant) within the starch biosynthetic pathway, which decrease the osmotic potential of seeds due to higher sucrose contents (Wang and Hedley, 1991). Germination of fresh seeds coincided with an oligosaccharide to total sugars value of between 0.05 and 0.1. This occurred at about 24 DAF. By about 38 DAF, the ratio had risen to between 0.54 (*RRRbRb*) and 0.66 (*RRrbrb*), moisture content had declined to between 30 and 35%, and seeds had become desiccation tolerant. The coincidence of oligosaccharide accumulation with the onset of desiccation tolerance in legumes is well documented (Górecki *et al.*, 1997; Corbineau *et al.*, 2000; Bailly *et al.*, 2001). Similarly, a period of water loss conferred desiccation tolerance, which is consistent with reports elsewhere for legume seed development (Matthews, 1973; Zanakis *et al.*, 1994; Sanhewe and Ellis, 1996a).

Seed survival in air-dry storage continued to improve after seed filling in all four NILs of pea. This is consistent with results from a wide range of contrasting species (e.g. Pieta Filho and Ellis, 1991; Sanhewe and Ellis, 1996b; Sinniah *et al.*, 1998a). Maturation drying improved seed longevity in seeds of all NILs, but the presence of mutant alleles at the two *rugosus* loci studied (*r* and *rb*) reduced longevity compared with the wild type.

Although the ratio of oligosaccharides to total sugars varied considerably among the NILs during development, this did not affect variation in the onset of desiccation tolerance among NILs. Based on the correlative evidence, this suggests that the quantity of soluble sugars per se is less important than a threshold ratio in conferring desiccation tolerance. Hence our results support the suggestion (Caffrey *et al.*, 1988) that oligosaccharides could prevent the crystallization of sucrose in dry seeds. Soluble sugars may therefore favour vitrification and offer protection from dehydration (Koster and Leopold, 1988; Koster, 1991).

The coincidental appearance of heat-stable proteins and the ability to tolerate desiccation may suggest a causal role. This would support the earlier views (Blackman *et al.*, 1992; Finch-Savage *et al.*, 1994) that the accumulation of either soluble sugars or heat-stable proteins alone is insufficient to confer desiccation tolerance.

This work is ongoing. Clearly, while a causal relationship has yet to be identified, we have shown that the greater proportion of sucrose and consequential greater initial accumulation and subsequent loss of water during seed development in the double mutant compared with the wild type is negatively associated with the latter's better air-dry seed survival.

Acknowledgements

TWL thanks the University of Reading's Research Endowment Fund Trust for a postgraduate scholarship.

References

Bailly, C., Audigier, C., Ladonne, F., Wagner, M.H., Coste, F., Corbineau, F. and Côme, D. (2001) Changes in oligosaccharide content and antioxidant enzyme activities in developing bean seeds as related to acquisition of drying tolerance and seed quality. *Journal of Experimental Botany* 52, 701–708.

Black, M., Corbineau, F., Gee, H. and Côme, D. (1999) Water content, raffinose, and dehydrins in the induction of desiccation tolerance in immature wheat embryos. *Plant Physiology* 120, 463–471.

Blackman, S.A., Obendorf, R.L. and Leopold, A.C. (1992) Maturation proteins and sugar in desiccation tolerance of developing seeds. *Plant Physiology* 100, 225–230.

Blackman, S.A., Obendorf, R.L. and Leopold, A.C. (1995) Desiccation tolerance in developing soybean seeds: the role of stress proteins. *Physiologia Plantarum* 93, 630–638.

Bochicchio, A., Vernieri, P., Puliga, S., Murelli, C. and Vazzana, C. (1997) Desiccation tolerance in immature embryos of maize: sucrose, raffinose and ABA-sucrose relation. In: Ellis, R.H., Black, M., Murdoch, A.J. and Hong, T.D. (eds) *Basic and Applied Aspects of Seed Biology. Proceedings of the Fifth International Workshop on Seeds, Reading, 1995.* Kluwer Academic Publishers, Dordrecht, The Netherlands, pp. 13–22.

Caffrey, M., Fonseca, V. and Leopold, A.C. (1988) Lipid–sugar interactions: relevance to anhydrous biology. *Plant Physiology* 86, 754–758.

Corbineau, F., Picard, M.A., Fougereux, J.A., Ladonne, F. and Côme, D. (2000) Effects of dehydration conditions on desiccation tolerance of developing pea seeds as related to oligosaccharide content and cell membrane properties. *Seed Science Research* 10, 329–339.

Demir, I. and Ellis, R.H. (1992a) Changes in seed quality during seed development and maturation in tomato. *Seed Science Research* 2, 81–87.

Demir, I. and Ellis, R.H. (1992b) Development of pepper (*Capsicum annuum*) seed quality. *Annals of Applied Biology* 121, 385–399.

Dure, L., Crouch, M., Harada, J., Ho, T.-H.D., Mundy, J., Quatrano, R., Thomas, T. and Sung, Z.R. (1989) Common amino acid sequence domains among the LEA proteins of higher plants. *Plant Molecular Biology* 12, 475–486.

Ellis, R.H. and Roberts, E.H. (1980) Improved equations for the prediction of seed longevity. *Annals of Botany* 45, 13–30.

Finch-Savage, W.E., Pramanik, S.K. and Bewley, J.D. (1994) The expression of dehydrin proteins in desiccation-sensitive (recalcitrant) seeds of temperate trees. *Planta* 193, 478–485.

Galau, G.A., Jakobsen, K.S. and Hughes, D.W. (1991) The control of late dicot embryogenesis and early germination. *Physiologia Plantarum* 81, 280–288.

Górecki, R.J., Piotrowicz-Cieślak, A.I., Lahuta, L.B. and Obendorf, R.L. (1997) Soluble carbohydrates in desiccation tolerance of yellow lupin seeds during maturation and germination. *Seed Science Research* 7, 107–115.

Harrington, J.K. (1972) Seed storage and longevity. In: Kozlowski, T.T. (ed.) *Seed Biology*. Academic Press, New York, pp. 145–245.

Hedley, C.L., Lloyd, J.R., Ambrose, M.J. and Wang, T.L. (1994) An analysis of seed development in *Pisum sativum* XVII. The effect of the *rb* locus alone and in combination with *r* on the growth and development of the seed. *Annals of Botany* 74, 365–371.

Hong, T.D. and Ellis, R.H. (1992) Development of desiccation tolerance in Norway maple (*Acer platanoides* L.) seeds during maturation drying. *Seed Science Research* 2, 169–172.

ISTA (International Seed Testing Association) (1999a) International rules for seed testing. Rules 1999. *Seed Science and Technology* 27 Supplement.

ISTA (International Seed Testing Association) (1999b) International rules for seed testing. Annexes 1999. *Seed Science and Technology* 27 Supplement.

Koster, K.L. (1991) Glass formation and desiccation tolerance in seeds. *Plant Physiology* 96, 302–304.

Koster, K.L. and Leopold, A.C. (1988) Sugars and desiccation tolerance in seeds. *Plant Physiology* 88, 829–832.

Matthews, S. (1973) Changes in developing pea (*Pisum sativum*) seeds in relation to their ability to withstand desiccation. *Annals of Applied Biology* 75, 93–105.

Pieta Filho, C. and Ellis, R.H. (1991) The development of seed quality in spring barley in four environments. I. Germination and longevity. *Seed Science Research* 1, 163–177.

Sanhewe, A.J. and Ellis, R.H. (1996a) Seed development and maturation in *Phaseolus vulgaris*. I. Ability to germinate and to tolerate desiccation. *Journal of Experimental Botany* 47, 949–958.

Sanhewe, A.J. and Ellis, R.H. (1996b) Seed development and maturation in *Phaseolus vulgaris*. II. Post-harvest longevity in air-dry storage. *Journal of Experimental Botany* 47, 959–965.

Shaw, R.H. and Loomis, W.E. (1950) Bases for the prediction of corn yields. *Plant Physiology* 25, 225–244.

Sinniah, U.R., Ellis, R.H. and John, P. (1998a) Irrigation and seed quality development in rapid-cycling brassica: seed germination and longevity. *Annals of Botany* 82, 309–314.

Sinniah, U.R., Ellis, R.H. and John, P. (1998b) Irrigation and seed quality development in rapid-cycling brassica: soluble carbohydrates and heat stable proteins. *Annals of Botany* 82, 647–655.

Sun, W.Q. and Leopold, A.C. (1993) Acquisition of desiccation tolerance in soybeans. *Physiologia Plantarum* 87, 403–409.

TeKrony, D.M. and Egli, D.B (1997) Accumulation of seed vigour during development and maturation. In: Ellis, R.H., Black, M., Murdoch, A.J. and Hong, T.D. (eds) *Basic and Applied Aspects of Seed Biology. Proceedings of the Fifth International Workshop on Seeds, Reading, 1995*. Kluwer Academic Publishers, Dordrecht, The Netherlands, pp. 369–384.

Wang, T.L. and Hedley, C.L. (1991) Seed development in peas: knowing your three 'r's' (or four, or five). *Seed Science Research* 1, 3–14.

Zanakis, G.N., Ellis, R.H. and Summerfield, R.J. (1994) Seed quality in relation to seed development and maturation in three genotypes of soyabean (*Glycine max*). *Experimental Agriculture* 30, 139–156.

6 Galactosyl Cyclitol Accumulation Enhanced by Substrate Feeding of Soybean Embryos

SILVIA ODORCIC AND RALPH L. OBENDORF

Seed Biology, Department of Crop and Soil Sciences, 617 Bradfield Hall, Cornell University Agricultural Experiment Station, Cornell University, Ithaca, NY 14853-1901, USA

Introduction

Soybean (*Glycine max* (L.) Merrill) seeds accumulate soluble carbohydrates, primarily sucrose, raffinose and stachyose and lesser amounts of galactopinitol A, galactopinitol B, ciceritol and fagopyritol B1, in axis and cotyledon tissues as part of the seed maturation process (Obendorf *et al.*, 1998). By contrast, somatic embryos appear to be deficient in D-pinitol and galactopinitols, indicating a lack of biosynthesis by embryo tissues (Obendorf *et al.*, 1996; Chanprame *et al.*, 1998). Instead of free cyclitols, seeds accumulate galactosyl cyclitols (Horbowicz and Obendorf, 1994; Obendorf, 1997). *myo*-Inositol is essential to all living cells (Loewus and Murthy, 2000) and is an important component in the metabolism of galactinol, raffinose and stachyose in seeds (Peterbauer and Richter, 2001). D-Pinitol and D-*chiro*-inositol are less widely distributed than *myo*-inositol among seeds and plants (Horbowicz and Obendorf, 1994; Obendorf, 1997). *myo*-Inositol-phosphate synthase converts glucose-6-phosphate to *myo*-inositol-phosphate in embryos of *Phaseolus* seeds (Johnson and Wang, 1996). The D-pinitol biosynthetic pathway converts *myo*-inositol to D-ononitol to D-pinitol (Dittrich and Brandl, 1987). *myo*-Inositol 6-*O*-methyltransferase, the enzyme that forms D-ononitol, is located in leaves and stems (Wanek and Richter, 1997). Soybean somatic embryos transformed with *myo*-inositol 6-*O*-methyltransferase form D-ononitol but not D-pinitol (J.J. Finer and J.G. Streeter, Ohio, 2002, personal communication), indicating that soybean somatic embryos do not form D-pinitol. Soybean leaves accumulate mostly D-pinitol with small amounts of D-*chiro*-inositol, *myo*-inositol and D-ononitol (Streeter, 2001). Free D-pinitol concentration is highest in seed coats and lower in axis and cotyledons of soybean seeds (Kuo *et al.*, 1997), suggesting transport of D-pinitol from leaves to seeds. If not biosynthesized in the embryo, feeding D-pinitol and D-*chiro*-inositol to soybean embryos should markedly increase accumu-

 The Biology of Seeds: Recent Research Advances (eds G. Nicolás, K.J. Bradford, D. Côme and H.W. Pritchard)

lation of galactopinitols and fagopyritol B1. If D-pinitol and D-*chiro*-inositol are synthesized in the embryo, feeding exogenous cyclitols should have less effect on galactosyl cyclitol accumulation. Stachyose synthase forms galactopinitols from D-pinitol and galactinol (Hoch *et al.*, 1999) and galactosylononitol from D-ononitol and galactinol (Peterbauer and Richter, 1998). Peterbauer and Richter (2001) proposed that stachyose synthase also uses D-*chiro*-inositol and galactinol to form galactosyl D-*chiro*-inositol. Galactinol synthase may use D-*chiro*-inositol and UDP-galactose to form galactosyl D-*chiro*-inositol (Frydman and Neufeld, 1963). Fagopyritol B1 and galactopinitols may be biosynthesized by different enzymes. If made by the same enzyme, product accumulation may be reduced due to competition between substrates. If made by different enzymes (galactinol synthase and stachyose synthase), product accumulation should be independent of substrate competition.

The objectives of this study were: (i) to determine if feeding free cyclitols to soybean embryos enhances the accumulation of galactosyl cyclitols; and (ii) to determine if feeding D-*chiro*-inositol and D-pinitol in combination to soybean embryos reduces the accumulation of fagopyritol B1 and galactopinitols below the amounts observed after feeding D-*chiro*-inositol and D-pinitol individually.

Materials and Methods

Plant materials

Soybean (*Glycine max* (L.) Merrill) plants were grown in the greenhouse (Obendorf *et al.*, 1980, 1998) at 27°C day (14 h) and 22°C night (10 h) under natural sunlight supplemented 14 h daily with 640 μmol/m^2/s incandescent light from metal halide lamps (Sylvania 1000 watt BU). Three embryos isolated from immature seeds (250 ± 20 mg fresh weight, approximately 35 days after flowering) by removal of the seed coat and nucellus remnants were incubated in 20 ml screw-capped vials containing 3 ml of substrate (cyclitol and/or sucrose) solutions for 24 h at 25°C and 200 μmol/m^2/s fluorescent light. Embryos were blotted, placed in small plastic Petri dishes, and subjected to slow drying by daily transfer to successive lower relative humidity (RH) controlled by saturated salt solutions (Blackman *et al.*, 1992; Obendorf *et al.*, 1998): day 1, 92% RH; day 2, 87% RH; day 3, 75% RH; day 4, 54% RH; day 5, 45% RH; day 6, 32% RH; day 7, 12% RH; and remained at 12% RH days 8–14.

Embryo feeding experiments

Four replications of three embryos each (n = 12) were incubated in *myo*-inositol-sucrose concentration series: 0 mM *myo*-inositol + 100 mM sucrose; 10 mM *myo*-inositol + 90 mM sucrose; 25 mM *myo*-inositol + 75 mM sucrose; 50 mM *myo*-inositol + 50 mM sucrose; and 100 mM *myo*-inositol + 0 mM sucrose. After 24 h, embryos were removed, blotted and slow dried for 14 days. Three replications of three embryos each ($n = 9$) were incubated with 30 mM *myo*-inositol + 100 mM sucrose for 24 h, blotted and slow dried for 0, 1, 2, 3, 4 or 14 days.

Six replications of three embryos each ($n = 18$) were incubated in D-*chiro*-inositol-sucrose concentration series: 0 mM D-*chiro*-inositol + 100 mM sucrose; 10 mM D-*chiro*-inositol + 90 mM sucrose; 25 mM D-*chiro*-inositol + 75 mM sucrose; 50 mM D-*chiro*-inositol + 50 mM sucrose; and 100 mM D-*chiro*-inositol + 0 mM sucrose. After 24 h, embryos were removed, blotted and slow dried for 14 days. Three replications of three embryos each ($n = 9$) were incubated with 100 mM D-*chiro*-inositol for 24 h, blotted and slow dried for 0, 1, 2, 3, 4, or 14 days.

Three replications of three embryos each (n = 9) were incubated in D-pinitol-sucrose concentration series: 0 mM D-pinitol + 100 mM sucrose; 10 mM D-pinitol + 90 mM sucrose; 25 mM D-pinitol + 75 mM sucrose; 50 mM D-pinitol + 50 mM sucrose; and 100 mM D-pinitol + 0 mM sucrose. After 24 h, embryos were removed, blotted and slow dried for 14 days. Three replications of three embryos each (total of 9 embryos/treatment) were incubated with 100 mM D-pinitol for 24 h, blotted and slow dried for 0, 1, 2, 3, 4 or 14 days.

Three replications of three embryos each ($n = 9$) were incubated with 100 mM D-pinitol + 100 mM D-*chiro*-inositol for 24 h, blotted and slow dried for 0, 1, 2, 3, 4 or 14 days. Three replications of three embryos ($n = 9$) were incubated in sucrose concentration series: 0 mM sucrose; 25 mM sucrose; 50 mM sucrose; 75 mM sucrose; 100 mM sucrose; and 200 mM sucrose. After 24 h, embryos were removed, blotted and slow dried for 14 days. For all treatments, cotyledon tissues were separated, weighed and analysed for soluble carbohydrates.

Substrates

Sucrose and *myo*-inositol were purchased from Sigma-Aldrich (St Louis, Missouri). D-Pinitol and D-*chiro*-inositol were purchased from Industrial Research Limited (Lower Hutt , New Zealand). When needed, substrates were purified by carbon-Celite column chromatography (Whistler and Durso, 1950) before use.

Carbohydrate analysis

Soluble carbohydrates were extracted from two cotyledons for each embryo. Two cotyledons were extracted with 2.0 ml of ethanol:water (1:1, v/v) containing 300 µg of phenyl α-D-glucoside as internal standard. One axis was extracted with 1.0 ml of ethanol:water (1:1, v/v) containing 100 µg of phenyl α-D-glucoside as internal standard. Extracts were passed through a 10,000 molecular weight cut-off filter by centrifugation, and 200 µl were dried in silylation vials under nitrogen gas, derivatized with 200 µl of trimethylsilylsylimidazole:pyridine (1:1, v/v), and analysed by high resolution gas chromatography as previously described (Horbowicz and Obendorf, 1994; Obendorf *et al.*, 1998).

Results and Discussion

Concentration series experiments were adjusted to be a constant 100 mM (cyclitol + sucrose) except for the sucrose concentration series experiment.

Feeding *myo*-inositol up to 50 mM increased free *myo*-inositol slightly (D-pinitol and D-*chiro*-inositol even less) with a doubling of galactinol after feeding 25–50 mM *myo*-inositol, but had little effect on other soluble carbohydrates (Fig. 6.1A–C). Elevated *myo*-inositol concentrations decreased as galactinol increased transiently during days 2 and 3 and then galactinol decreased as raffinose and stachyose accumulated (Fig. 6.1D–F). Total *myo*-inositol decreased, indicating metabolism to other products (phytic acid, cell wall pectins). After feeding sucrose with *myo*-inositol, sucrose concentration decreased and raffinose and stachyose concentrations increased during slow drying (Fig. 6.1F).

Feeding D-*chiro*-inositol resulted in a 50-fold increase in free D-*chiro*-inositol and a sevenfold increase in fagopyritol B1, but did not increase D-pinitol, *myo*-inositol, galactopinitol A, galactopinitol B, galactinol, raffinose or stachyose concentrations (Fig. 6.2A–C). These results suggest that D-*chiro*-inositol does not serve as precursor to other cyclitols in soybean embryos, and fagopyritol B1 does not serve as an alternate galactosyl donor for the biosynthesis of raffinose and stachyose. A large increase in fagopyritol B1 between day 2 and day 4 during slow drying (Fig. 6.2E) was accompanied by a decrease in free D-*chiro*-inositol (Fig. 6.2D). A transient increase in galactinol (Fig. 6.2E) preceded the increase in raffinose and stachyose (Fig. 6.2F).

Feeding D-pinitol resulted in an eightfold increase in D-pinitol concentration, a 4.5-fold increase in galactopinitol A concentration, and a 4.2-fold increase in concentration of galactopinitol B, but D-*chiro*-inositol, *myo*-inositol, fagopyritol B1, galactinol, raffinose and stachyose concentrations were not increased (Fig. 6.3A–C). Galactopinitol A and galactopinitol B increased during the third day of slow drying, galactinol decreased during the third day, and raffinose and stachyose increased during the third and fourth days of slow drying. *myo*-Inositol decreased (Fig. 6.3D) as galactinol increased during days 2 and 3, and then galactinol decreased (Fig. 6.3E) as raffinose and stachyose accumulated (Fig. 6.3F). The larger increases in raffinose and stachyose (Fig. 6.3C,F compared with Fig. 6.2C,F) suggest that galactopinitol A may be effective as a galactosyl donor for the biosynthesis of raffinose and stachyose, as suggested by Hoch *et al.* (1999) and Peterbauer and Richter (2001). Sucrose concentration increased through day 3.

Feeding a combination of D-pinitol and D-*chiro*-inositol resulted in high concentrations of both free D-pinitol and free D-*chiro*-inositol; a decrease in free D-pinitol was accompanied by an increase in galactopinitol A and galactopinitol B, and a decrease in free D-*chiro*-inositol was accompanied by an increase in fagopyritol B1 (Fig. 6.4A,B). Biosynthesis of fagopyritol B1 and

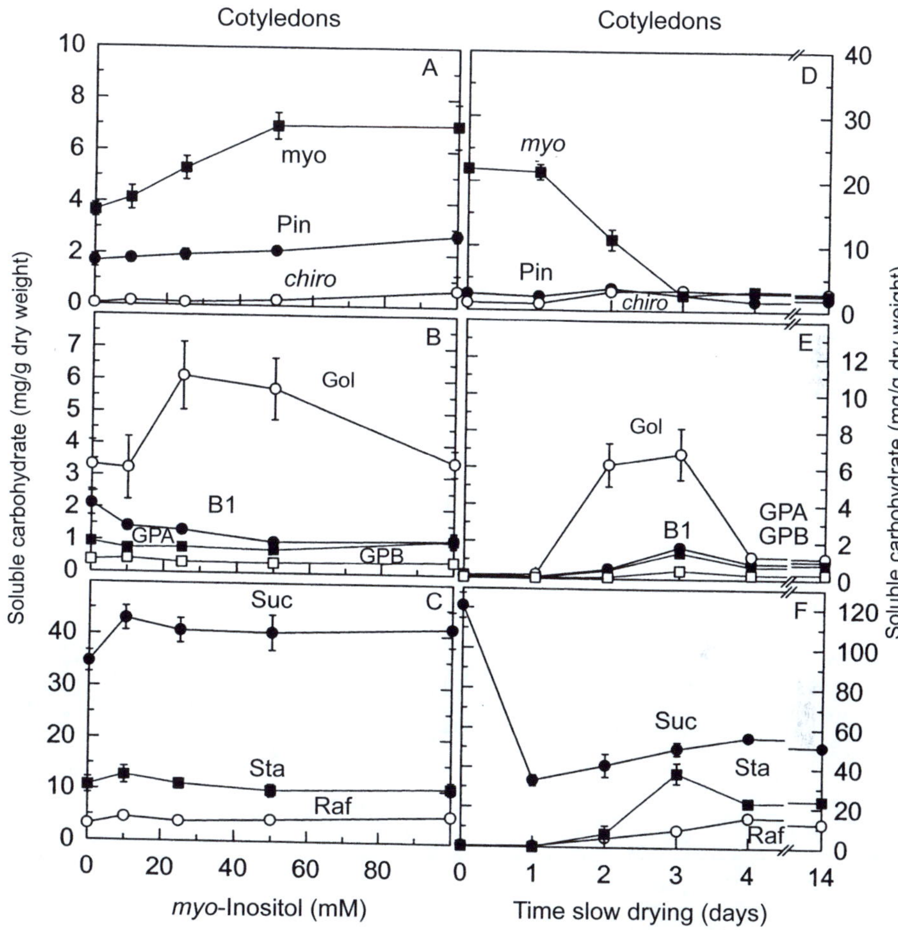

Fig. 6.1. (A–C) Accumulated soluble carbohydrates in cotyledon tissues after precocious maturation of immature soybean embryos as a function of *myo*-inositol concentration after feeding *myo*-inositol (0–100 mM) plus sucrose (100–0 mM) (100 mM total concentration) for 24 h at 25°C followed by 14 days precocious maturation in slow drying series relative humidities ($n = 12$). (D–F) Accumulated soluble carbohydrates in cotyledon tissues after precocious maturation of immature soybean embryos as a function of time of slow drying after feeding 30 mM *myo*-inositol and 100 mM sucrose for 24 h at 25°C followed by 0–14 days precocious maturation in slow drying time series relative humidities ($n = 9$). Values are mean ± SE of the mean. (A–F) *myo*-inositol (*myo*), D-pinitol (Pin), D-*chiro*-inositol (*chiro*), fagopyritol B1 (B1), galactinol (Gol), galactopinitol A (GPA), galactopinitol B (GPB), raffinose (Raf), stachyose (Sta), sucrose (Suc).

biosynthesis of galactopinitols were not in direct competition. Feeding D-pinitol and D-*chiro*-inositol in combination resulted in a 50% decrease in galactinol steady-state concentration and a 50% decrease in galactopinitol A and galactopinitol B concentrations, but only a 15% decrease in fagopyritol B1 compared with feeding D-pinitol or D-*chiro*-inositol alone. No noted decrease in stachyose and raffinose was observed (Fig. 6.4C). Galactinol con-

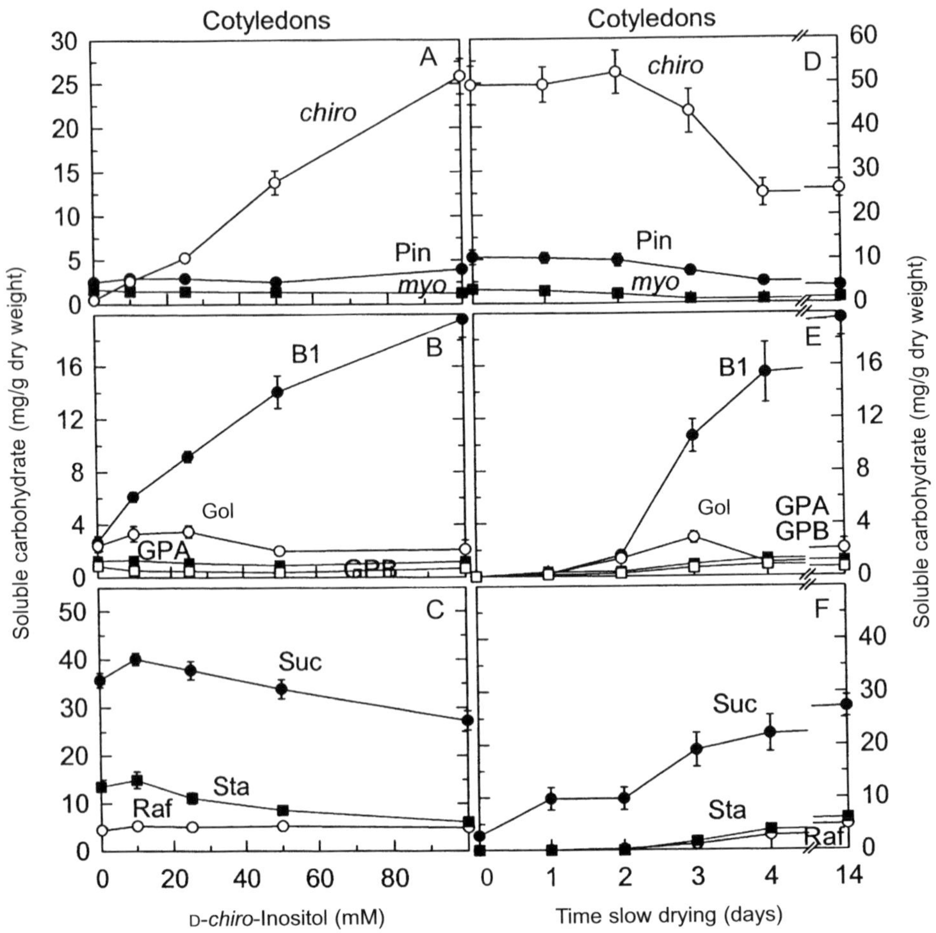

Fig. 6.2. (A–C) Accumulated soluble carbohydrates in cotyledon tissues after precocious maturation of immature soybean embryos as a function of D-*chiro*-inositol concentration after feeding D-*chiro*-inositol (0–100 mM) plus sucrose (100–0 mM) (100 mM total concentration) for 24 h at 25°C followed by 14 days precocious maturation in slow drying series relative humidities (n = 18). (D–F) Accumulated soluble carbohydrates in cotyledon tissues after precocious maturation of immature soybean embryos as a function of time of slow drying after feeding 100 mM D-*chiro*-inositol for 24 h at 25°C followed by 0–14 days precocious maturation in slow drying time series relative humidities (n = 9). Values are mean ± SE of the mean. (A–F) *myo*-inositol (*myo*), D-pinitol (Pin), D-*chiro*-inositol (*chiro*), fagopyritol B1 (B1), galactinol (Gol), galactopinitol A (GPA), galactopinitol B (GPB), raffinose (Raf), stachyose (Sta), sucrose (Suc).

centration peaked at day 2 and declined as raffinose, stachyose and galactopinitols accumulated. Accumulation of fagopyritol B1 appeared to be independent of accumulation of galactopinitols, raffinose and stachyose, suggesting that fagopyritol B1 biosynthesis is independent of galactopinitol

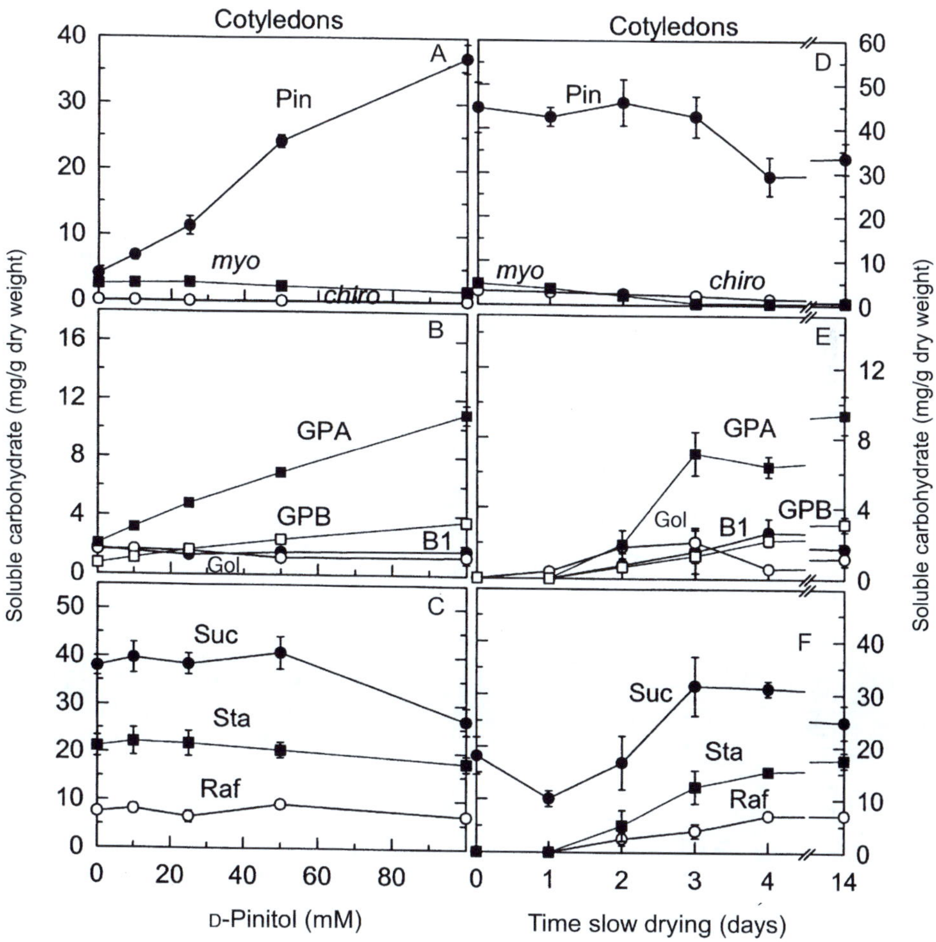

Fig. 6.3. (A–C) Accumulated soluble carbohydrates in cotyledon tissues after precocious maturation of immature soybean embryos as a function of D-pinitol concentration after feeding D-pinitol (0–100 mM) plus sucrose (100–0 mM) (100 mM total concentration) for 24 h at 25°C followed by 14 days precocious maturation in slow drying series relative humidities (n = 9). (D–F) Accumulated soluble carbohydrates in cotyledon tissues after precocious maturation of immature soybean embryos as a function of time of slow drying after feeding 100 mM D-pinitol for 24 h at 25°C followed by 0–14 days precocious maturation in slow drying time series relative humidities (n = 9). Values are mean ± SE of the mean. (A–F) *myo*-inositol (*myo*), D-pinitol (Pin), D-*chiro*-inositol (*chiro*), fagopyritol B1 (B1), galactinol (Gol), galactopinitol A (GPA), galactopinitol B (GPB), raffinose (Raf), stachyose (Sta), sucrose (Suc).

biosynthesis. Feeding a combination of 50 mM D-pinitol + 50 mM D-*chiro*-inositol resulted in patterns essentially identical to those with 100 mM (data not shown), indicating that the cyclitol substrates may be at saturating concentrations. Feeding a combination of 100 mM D-pinitol + 100 mM D-*chiro*-inositol + 100 mM sucrose resulted in patterns essentially identical to those

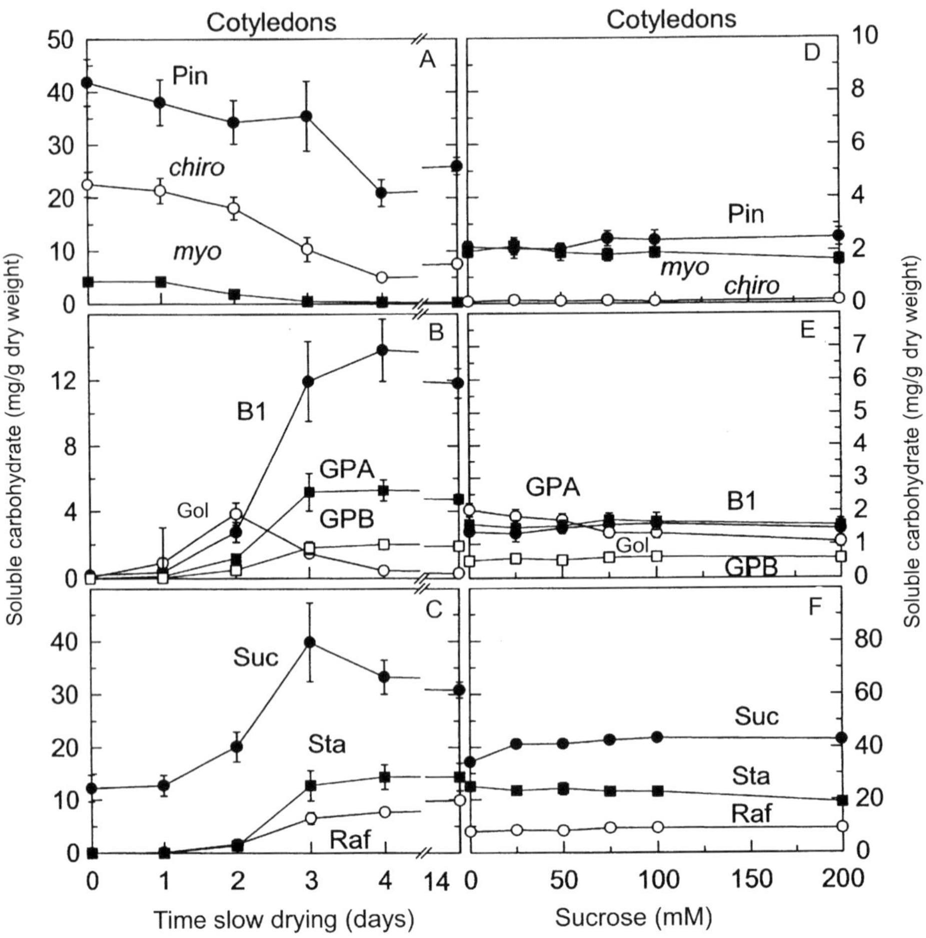

Fig. 6.4. (A–C) Accumulated soluble carbohydrates in cotyledon tissues after precocious maturation of immature soybean embryos as a function of time of slow drying after feeding 100 mM D-*chiro*-inositol + 100 mM D-pinitol for 24 h at 25°C followed by 0–14 days precocious maturation in slow drying time series relative humidities (n = 9). (D–F) Accumulated soluble carbohydrates in cotyledon tissues after precocious maturation of immature soybean embryos as a function of sucrose concentration after feeding sucrose (0–200 mM) for 24 h at 25°C followed by 14 days precocious maturation in slow drying series relative humidities (n = 9). Values are mean ± SE of the mean. (A–F) *myo*-inositol (*myo*), D-pinitol (Pin), D-*chiro*-inositol (*chiro*), fagopyritol B1 (B1), galactinol (Gol), galactopinitol A (GPA), galactopinitol B (GPB), raffinose (Raf), stachyose (Sta), sucrose (Suc).

without sucrose, except that sucrose concentrations were higher initially (data not shown). Feeding sucrose at 0 to 200 mM resulted in little change in concentrations of soluble carbohydrates (Fig. 6.4D–F), indicating that a differential in osmotic concentration during the 24 h feeding had little effect on the subsequent accumulation of soluble carbohydrates.

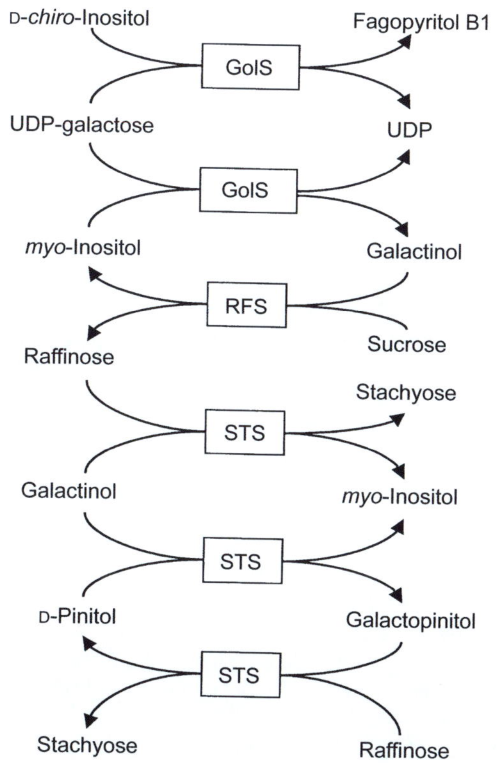

Fig. 6.5. Proposed biosynthetic pathways. Galactinol synthase (GolS), raffinose synthase (RFS), stachyose synthase (STS).

Results of these substrate feeding experiments are consistent with the interpretation that D-pinitol and D-*chiro*-inositol are transported from maternal tissues and not biosynthesized in the embryo tissues, that galactopinitols and fagopyritol B1 are biosynthesized by different pathways, that galactopinitols are biosynthesized by stachyose synthase (Fig. 6.5), that fagopyritol B1 is biosynthesized by galactinol synthase (Fig. 6.5), and that galactopinitols may serve as galactosyl donors (substituting for galactinol) for stachyose biosynthesis.

Acknowledgements

We thank Monica Johnson, Paja Sijacic, Sarah Stewart, Hirotoki Takemasa, Ahmet Tunceroglu, Alicia Rozario, Carly Gomes, Elizabeth Vassallo, Erin Cannon, Khanhvy Phan, Maria Martynovsky, Marcin Horbowicz, Melinda Lee and Regan Anderson for assistance with experiments, data analysis and substrate preparation.

References

Blackman, S.A., Obendorf, R.L. and Leopold, A.C. (1992) Maturation proteins and sugars in desiccation tolerance of developing soybean seeds. *Plant Physiology* 100, 225–230.

Chanprame, S., Kuo, T.M. and Widholm, J.M. (1998) Soluble carbohydrate content of soybean [*Glycine max* (L.) Merr.] somatic and zygotic embryos during development. *In Vitro Cellular and Developmental Biology – Plant* 34, 64–68.

Dittrich, P. and Brandl, A. (1987) Revision of the pathway of D-pinitol formation in Leguminosae. *Phytochemistry* 26, 1925–1926.

Frydman, R.B. and Neufeld, E.F. (1963) Synthesis of galactosylinositol by extracts from peas. *Biochemical and Biophysical Research Communications* 12, 121–125.

Hoch, G., Peterbauer, T. and Richter, A. (1999) Purification and characterization of stachyose synthase from lentil (*Lens culinaris*) seeds: galactopinitol and stachyose synthesis. *Archives of Biochemistry and Biophysics* 366, 75–81.

Horbowicz, M. and Obendorf, R.L. (1994) Seed desiccation tolerance and storability: dependence on flatulence-producing oligosaccharides and cyclitols – review and survey. *Seed Science Research* 4, 385–405.

Johnson, M.D. and Wang, X. (1996) Differentially expressed forms of 1-L-*myo*-inositol-1-phosphate synthase (EC 5.5.1.4) in *Phaseolus vulgaris*. *Journal of Biological Chemistry* 271, 17215–17218.

Kuo, T.M., Lowell, C.A. and Nelsen, T.C. (1997) Occurrence of pinitol in developing soybean seed tissues. *Phytochemistry* 45, 29–35.

Loewus, F.A. and Murthy, P.P.N. (2000) *myo*-Inositol metabolism in plants. *Plant Science* 150, 1–19.

Obendorf, R.L. (1997) Oligosaccharides and galactosyl cyclitols in seed desiccation tolerance (Review Update). *Seed Science Research* 7, 63–74.

Obendorf, R.L., Ashworth, E.N. and Rytko, G.T. (1980) Influence of seed maturation on germinability in soybean. *Crop Science* 20, 483–486.

Obendorf, R.L., Moon, H., Hildebrand, D.F., Torisky, R. and Collins, G.B. (1996) A comparison of pinitols in somatic and zygotic soybean embryos. *Molecular and Cellular Biology of the Soybean* 6, 40.

Obendorf, R.L., Horbowicz, M., Dickerman, A.M., Brenac, P. and Smith, M.E. (1998) Soluble oligosaccharides and galactosyl cyclitols in maturing soybean seeds *in planta* and *in vitro*. *Crop Science* 38, 78–84.

Peterbauer, T. and Richter, A. (1998) Galactosylononitol and stachyose synthesis in seeds of adzuki bean: purification and characterization of stachyose synthase. *Plant Physiology* 117, 165–172.

Peterbauer, T. and Richter, A. (2001) Biochemistry and physiology of raffinose family oligosaccharides and galactosyl cyclitols in seeds. *Seed Science Research* 11, 185–198.

Streeter, J.G. (2001) Simple partial purification of D-pinitol from soybean leaves. *Crop Science* 41, 1985–1987.

Wanek, W. and Richter, A. (1997) Biosynthesis and accumulation of D-ononitol in *Vigna umbellata* in response to drought stress. *Physiologia Plantarum* 101, 416–424.

Whistler, R.L. and Durso, D.F. (1950) Chromatographic separation of sugars on charcoal. *Journal of the American Chemical Society* 72, 677–679.

7 MARK, a Maize Atypical Receptor-like Kinase, Expressed During Embryogenesis and in the Meristems of the Adult Plant, Interacts with MIK, a New GCK/SPS1 Kinase

BLANCA LLOMPART, ADOLFO RÍO, ALEJANDRO FERRANDO, RAMÓN ROCA, VIRGÍNIA STIEFEL, PERE PUIGDOMÈNECH AND JOSEP M. CASACUBERTA

Departamento de Genética Molecular, IBMB-CSIC, Jordi Girona 18, 08034 Barcelona, Spain

Introduction

Plant embryogenesis allows the fertilized egg to be transformed into a multicellular organism, which, in contrast to the animal embryo, is not a miniature of the adult organism but a more simple juvenile structure known as the seedling. In addition to the embryonic leaves (cotyledons), stem (hypocotyl) and root (radicle), the seedling contains two sets of stem cells, the meristems, that allow the plant to develop post-embryonically most of the adult organs. The shoot apical meristem (SAM) is the source of the stem and the leaves, and it will also give rise to the secondary meristems that will produce the flowers, while the root meristem (RM) will produce the primary root. The formation of the meristems during embryogenesis and their maintenance during the plant's life are thus essential to ensure proper development. In the last few years the genetic and molecular dissection of embryogenesis of the model plant *Arabidopsis thaliana* has allowed the definition of a number of genes that participate in the control of this process. Among those, the most relevant are the *CLAVATA* (*CLV*) and *WUSCHEL* (*WUS*) genes for SAM maintenance, and the *MONOPTEROS* (*MP*) gene for RM establishment (for a review, see Jurgens, 2001).

Receptor protein kinases (RPKs) are major players in cell-to-cell communication in animal systems, and in the last few years an increasing number of receptor-like kinases (RLKs) have also been described in plants.

Although RLKs display structural similarities to animal RPKs and are supposed to function in a similar way, very little is known about their respective ligands and, in most cases, almost nothing has been described about the downstream signalling cascades that they activate. Putative ligands have only been described for SRK (Schopfer *et al.*, 1999), CLV1 (Trotochaud *et al.*, 2000), BRI1(Wang *et al.*, 2001) and FLS2 (Gomez-Gomez *et al.*, 2001) and proteins interacting with their kinase domains have only been found for RLK5 (Stone *et al.*, 1994), SRK (Bower *et al.*, 1996) and CLV1 (Trotochaud *et al.*, 1999). Moreover, the activation of the kinase domain after ligand binding has only been demonstrated for the SRK RLK (Cabrillac *et al.*, 2001).

This chapter describes an atypical receptor-like kinase from maize that we have named as MARK (after Maize Atypical Receptor-like Kinase). MARK accumulates in the membranes of highly proliferating and differentiating cells during embryogenesis and also in the meristems of the adult plant. Although MARK presents a putative intracellular domain containing the 11 conserved subdomains of protein kinases, it contains substitutions of some of the amino acids that are highly conserved among kinases. Probably because of this, MARK is not able to autophosphorylate. Yeast two-hybrid screenings performed with the putative intracellular domain of MARK allowed us to identify a protein with sequence similarities to protein kinases, which we have named as MIK (MARK Interacting Kinase). MIK is related to the GCK subfamily of MAPkinases, which in some cases have been shown to connect cell-surface receptors to intracellular MAPkinase signalling cascades (Kyriakis, 1999).

MARK Accumulates in Highly Proliferating and Differentiating Cells

In a search for immature embryo-specific cDNAs by differential screening, we obtained a partial cDNA clone displaying high sequence similarity with kinase domains of RLKs. The particularities of the predicted kinase domain deduced from this sequence (see below) prompted us to name the corresponding gene as *MARK* (from Maize Atypical Receptor Kinase).

Northern and *in situ* hybridization analysis showed that *MARK* is expressed during early and mid-embryogenesis, from 2 until 30–40 days after pollination (DAP), with a peak at 15 DAP, in embryo and endosperm (data not shown). An antibody raised against the kinase domain of MARK (MARK-KD) was used to perform immunolocalization studies. Figure 7.1 shows that MARK accumulates in the provascular tissues of the embryo proper (coleoptile, leaf primordia and the pericycle of the radicle). It also accumulates in the scutellum cells as a gradient from outside to inside, coinciding with the proliferation and differentiation activity of this organ at this stage of development. On the other hand, Fig. 7.2 shows that MARK specifically accumulates in the primary meristems of the plant, while Northern blot analyses have shown that its mRNA is not detectable in other adult plant tissues (not shown). Interestingly, MARK accumulates in most of the meristematic tissues, but it is not detected in the stem cells that are main-

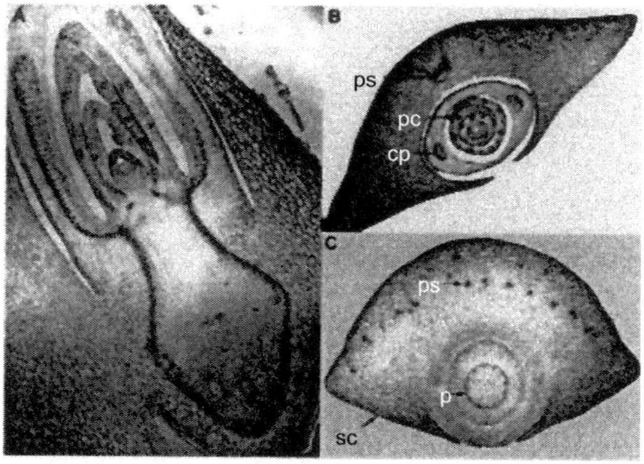

Fig. 7.1. Immunolocalization of MARK in immature maize embryos. (A) Longitudinal and transverse sections at the (B) apical or (C) root level of 20 DAP embryos are shown. The scutellum (sc), pericycle (p), coleoptilar procambium (cp), provascular cells (pc) and the provascular strands of the scutellum (ps) are indicated. Embryos were fixed and embedded in paraffin wax as described (Langdale, 1994); 8 μm sections were incubated with a polyclonal rabbit antibody raised against MARK-KD and detection was performed with Vectastain ABC kit (Vector).

tained undifferentiated and that divide at a slow rate, and which are located within the central zone of the SAM and the quiescent centre of the RM. This is particularly clear in Fig. 7.2A, where the absence of accumulation of MARK signal in the central zone of the SAM is highlighted by a strong signal accumulation in the rest of the meristematic tissues, such as the differentiated epidermal L1 layer, that surround this zone.

All these results show that MARK accumulates in cells that are proliferating and differentiating actively, and suggests a possible role for this RLK in proliferation/differentiation signalling.

MARK Presents an Atypical and Defective Kinase Domain

The partial cDNA clone obtained from the differential screening was used as a probe to screen a maize genomic library, and its sequence was used to define primers for rapid amplification of CDNA ends (RACE) experiments. A full-length cDNA clone and a genomic corresponding to the *MARK* gene were obtained, and the analysis of these sequences revealed an open reading frame (ORF) presenting a putative signal peptide, a region containing five imperfect leucine-rich repeats (LRR), a putative transmembrane domain, and a C-terminal region presenting the 11 conserved subdomains of Ser/Thr kinases. However, some of the amino acids that are invariant among kinases, considered as essential for kinase activity, are substituted in MARK-KD. In particular, a highly conserved asparagine involved in the binding of ATP in subdomain VIb is substituted for an arginine, and the DFG triplet of subdo-

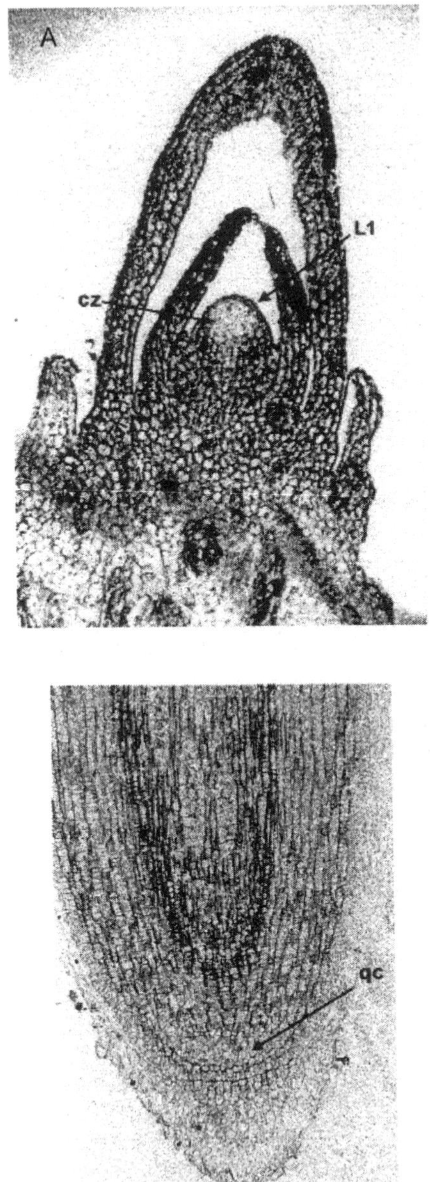

Fig. 7.2. Immunolocalization of MARK in meristems. Sections of (A) shoot and (B) root apical meristems of 7 days germinated young plantlets were processed as in Fig. 7.1. The central zone region (cz), the epidermal L1 layer (L1) and the quiescent centre (qc) are indicated.

main VII, which constitutes the activation loop, is also not conserved in MARK-KD (Fig. 7.3). Moreover, we have not been able to detect any phosphorylation activity of MARK-KD *in vitro* (not shown), which suggests that

MARK-KD is an impaired kinase domain. While only one plant RLK presenting a putative impaired kinase domain has been described (the *Arabidopsis* TMKL1 protein; Valon *et al.*, 1993), a number of expressed sequence tag and genomic sequences with similar characteristics are present in plant databases (not shown), suggesting that other atypical receptor kinases may exist in plants. In animals, a small number of atypical receptor kinases with impaired kinase domains but actively participating in signal transduction have been described (reviewed in Kroiher *et al.*, 2001).

MARK Interacts with MIK, a GCK-like Kinase

While the mechanisms by which the animal atypical receptor kinases transduce signals are not completely elucidated, it has been proposed that the impaired kinase domains may interact specifically with other proteins upon ligand binding by their receptor domain (Stein and Staros, 2000). We have thus searched for proteins that could interact with MARK-KD by performing a two-hybrid screen of a 7 DAP maize embryo cDNA library cloned in the pACT2 vector with MARK-KD fused to the Gal4 DNA binding domain.

A partial cDNA clone, DH5, presenting an ORF with high sequence similarities to the C-terminal domain of *Brassica* and *Arabidopsis* MAP4K (Leprince *et al.*, 1999) was isolated. The interaction between DH5 and MARK-KD was further analysed by pull-down experiments. Figure 7.4A shows that all the different deletions of the DH5 clone tested abolished the interaction. We next tested the role that the atypical residues in MARK-KD might have in the interaction with MIK. The substitution of the atypical amino-acids present in MARK-KD subdomains VIb and VII by those present in typical kinases does not significantly affect the interaction with MIK, suggesting that they do not play a major role in the interaction (Fig. 7.4B).

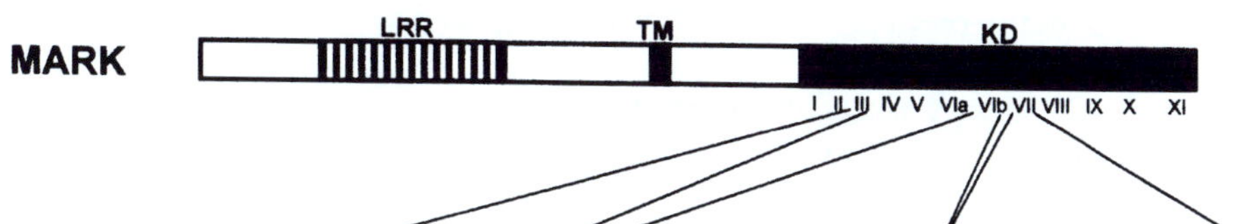

	*	*		***
CLV1	RSDHGFTAEI	LHRDVKSNNI	LLDSDFEAHV	ADFGLAKFL
SRK	QGIDEFMNEV	IHRDLKPGNI	LLDKYMIPKI	SDFGMARIF
CONS	E	D		DFG
MARK	LSEAEFRE**R**I	SHG**N**IKSSNV	LLGKSYQARV	S**EN**GLTTLV

Fig. 7.3. Scheme representing the MARK protein. The relative position of the leucine-rich repeats (LRR), the transmembrane domain (TM) and the kinase-like domain (KD) are indicated, as are the positions of the 11 subdomains conserved among kinases. The sequences of the subdomains III, VIb and VII of MARK-KD are shown compared with those of CLV1 and SRK receptor kinases, and the amino acids that are invariable among kinases are highlighted.

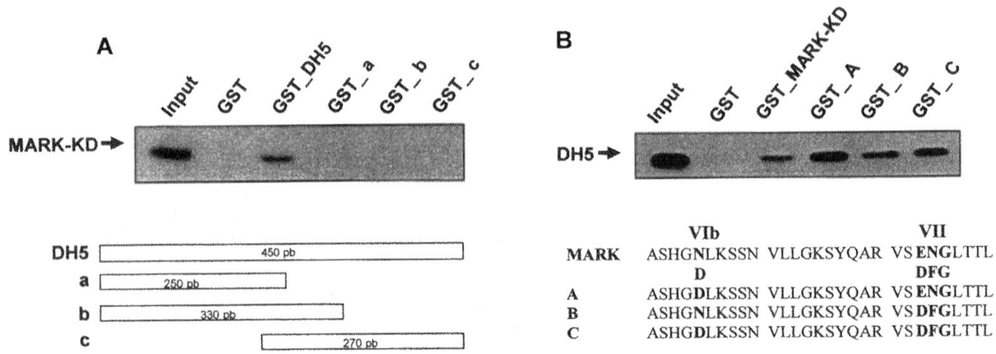

Fig. 7.4. GST pull-down experiments performed between MARK-KD and MIK. (A) The complete DH5-MIK sequence, as well as different deletions of DH5-MIK (a–c), were fused to GST and immobilized on to glutathion-S-Sepharose and were incubated with recombinant His-tagged MARK-KD, using GST as a negative control, in binding buffer (Hepes pH 7.9 20 mM, glycerol 20%, EDTA 0.2 mM, NaCl 150 mM, NP40 0.5%, PMSF 0.1 mM, DTT 0.5 mM). After extensive washing with the same buffer, proteins were eluted with Laemmli buffer, separated by SDS-PAGE and transferred to nitrocellulose. Immunodetection was performed with an anti-MARK polyclonal antibody. (B) Immobilized GST and GST fused MARK-KD, as well as three different MARK-KD mutants (A–C) were incubated with His-tagged DH5-MIK as described in (A). Immunodetection was performed with an anti-DH5-MIK polyclonal antibody.

The DH5 clone was used as a probe to screen maize genomic and cDNA libraries and the corresponding full-length and genomic clones were obtained. The deduced protein sequence, MIK, shows high similarities with the GCK subfamily of the Ste20 family of MAPkinases. Some of the Ste20-related proteins have been shown to phosphorylate MAPKKK, thus acting as MAP4K (Drogen *et al.*, 2000; Raitt *et al.*, 2000). The GCK subfamily consists of a heterogeneous group of proteins that, in most cases, are still poorly characterized. Nevertheless, some GCKs have been shown to connect cell-surface receptors to intracellular MAPK cascades (Kyriakis, 1999), suggesting that MIK could connect MARK to intracellular signalling cascades.

The expression pattern of MIK is consistent with a possible interaction with MARK, as both have a similar pattern of expression during embryogenesis, presenting a peak of expression at 15 DAP in embryo and endosperm (data not shown). Moreover, immunolocalizations of the MIK protein have shown that it accumulates preferentially in provascular cells of the coleoptile and radicle – cells that also exhibit a high accumulation of the MARK protein.

GCK proteins contain an amino-terminal kinase domain and a carboxyl-terminal regulatory domain that probably inhibits its kinase activity. The interaction shown here of the intracellular inactive kinase domain of MARK with the C-terminal regulatory domain of MIK (MIK-RD) could thus suggest a model in which, upon ligand binding, MARK-KD would interact with

MIK-RD, releasing the inhibition of the kinase activity of MIK and allowing this protein to activate intracellular signalling cascades.

Conclusion

In summary, we have described the cloning and characterization of two maize genes, *MARK* and *MIK*, that are expressed during embryogenesis and in the meristems of the adult plant and that could participate in signal transduction associated with proliferation/differentiation processes. The direct interaction of MARK, an atypical receptor kinase, with MIK, a GCK-like kinase, suggests a new mechanism by which receptors with an impaired kinase domain could participate in signal transduction.

References

Bower, M.S., Matias, D.D., Fernandes-Carvalho, E., Mazzurco, M., Gu, T., Rothstein, S.J. and Goring, D.R. (1996) Two members of the thioredoxin-h family interact with the kinase domain of a Brassica S locus receptor kinase. *Plant Cell* 8, 1641–1650.

Cabrillac, D., Cock, J.M., Dumas, C. and Gaude, T. (2001) The S-locus receptor kinase is inhibited by thioredoxins and activated by pollen coat proteins. *Nature* 410, 220–223.

Drogen, F., O'Rourke, S.M., Stucke, V.M., Jaquenoud, M., Neiman, A.M. and Peter, M. (2000) Phosphorylation of the MEKK Ste11p by the PAK-like kinase Ste20p is required for MAP kinase signaling *in vivo*. *Current Biology* 10, 630–639.

Gomez-Gomez, L., Bauer, Z. and Boller, T. (2001) Both the extracellular leucine-rich repeat domain and the kinase activity of FSL2 are required for flagellin binding and signaling in Arabidopsis. *Plant Cell* 13, 1155–1163.

Jurgens, G. (2001) Apical-basal pattern formation in Arabidopsis embryogenesis. *EMBO Journal* 20, 3609–3616.

Kroiher, M., Miller, M.A. and Steele, R.E. (2001) Deceiving appearances: signaling by 'dead' and 'fractured' receptor protein-tyrosine kinases. *Bioessays* 23, 69–76.

Kyriakis, J.M. (1999) Signalling by the germinal center kinase family of protein kinases. *Journal of Biological Chemistry* 274, 5259–5262.

Langdale, J. (1994) *In situ* hybridisation. In: Freeling, M. and Walbet, V. (eds) *The Maize Handbook*. Springer-Verlag, New York, pp. 165–180.

Leprince, A.S., Jouannic, S., Hamal, A., Kreis, M. and Henry, Y. (1999) Molecular characterisation of plant cDNAs BnMAP4Kα1 and BnMAP4α2 belonging to the GCK/SPS1 subfamily of MAP kinase kinase kinase kinase. *Biochimica et Biophysica Acta* 1444, 1–13.

Raitt, D.C., Posas, F. and Saito, H. (2000) Yeast Cdc42 GTPase and Ste20 PAK-like kinase regulate Sho1-dependent activation of the Hog1 MAPK pathway. *EMBO Journal* 19, 4623–4631.

Schopfer, C.R., Nasrallah, M.E. and Nasrallah, J.B. (1999) The male determinant of self-incompatibility in Brassica. *Science* 286, 1697–1700.

Stein, R.A. and Staros, J.V. (2000) Evolutionary analysis of the ErbB receptor and ligand families. *Journal of Molecular Evolution* 50, 397–412.

Stone, J.M., Collinge, M.A., Smith, R.D., Horn, M.A. and Walker, J.C. (1994) Interaction of a protein phosphatase with an Arabidopsis serine–threonine receptor kinase. *Science* 266, 793–795.

Trotochaud, A.E., Hao, T., Wu, G., Yang, Z. and Clark, S.E. (1999) The CLAVATA1 receptor-like kinase requires CLAVATA3 for its assembly into a signalling complex that includes KAPP and a Rho-related protein. *Plant Cell* 11, 393–406.

Trotochaud, A.E., Jeong, S. and Clark, S.E. (2000) CLAVATA3, a multimeric ligand for the CLAVATA1 receptor-kinase. *Science* 289, 613–617.

Valon, C., Smalle, J., Goodman, H.M. and Giraudat, J. (1993) Nucleotide characterization of an *Arabidopsis thaliana* gene (TMKL1) encoding a putative transmembrane protein with an unusual kinase-like domain. *Plant Molecular Biology* 23, 415–421.

Wang, Z.Y., Seto, H., Fujioka, S., Yoshida, S. and Chory, J. (2001) BRI1 is a critical component of a plasma-membrane receptor for plant steroids. *Nature* 410, 380–383.

8 Sunflower Seed Development as Related to Antioxidant Enzyme Activities

Christophe Bailly[1], Juliette Leymarie[1], Sandra Rousseau[1], Daniel Côme[1], Armand Feutry[2] and Françoise Corbineau[1]

[1]*Physiologie Végétale Appliquée, Université Pierre et Marie Curie, Tour 53, 1er étage, 4 place Jussieu, 75252 Paris cédex 05, France;* [2]*Vilmorin, route du Manoir, 49250 La Ménitré, France*

Introduction

The efficiency of reactive oxygen species (ROS) scavenging through antioxidant enzymes, namely superoxide dismutase (SOD), glutathione reductase (GR) and mainly catalase (CAT), has been shown to be closely related to sunflower seed vigour (Bailly *et al.*, 1998, 2002). Impairment or deficiency of these activities leads to accumulation of ROS and to various cellular injuries, such as lipid peroxidation, which can in turn affect seed germinability (Bailly *et al.*, 1998) and seedling emergence (Bailly *et al.*, 2002).

It is generally accepted that inability of recalcitrant seeds to withstand dehydration is, at least in part, related to an uncontrolled accumulation of ROS (Hendry *et al.*, 1992; Li and Sun, 1999; Pammenter and Berjak, 1999). Orthodox seeds would have built up various molecular and cellular events conferring dehydration tolerance during their natural desiccation on the plant (Vertucci and Farrant, 1995). However, the role of oxidative stress during desiccation and maturation drying of orthodox seeds is so far little described and evidence is lacking for a ubiquitous role of detoxifying enzymes in seed germinability and vigour.

Besides their role in the antioxidant defence system through the elimination of stress-generated hydrogen peroxide, CAT and GR, by controlling H_2O_2 homeostasis, also regulate H_2O_2-signalling pathways, which recently have emerged as playing an essential part in transduction of various cellular signals (Bowler and Fluhr, 2000; Desikan *et al.*, 2001).

The aims of the present work were to determine whether acquisition of sunflower seed vigour during seed development was related to changes in

(eds G. Nicolás, K.J. Bradford, D. Côme and H.W. Pritchard)

activities of the main detoxifying enzymes, and to study the effects of artificial drying on free-radical scavenging and lipid peroxidation.

Material and Methods

Sunflower (*Helianthus annuus* L., cv. Fructidor) plants were grown in 1999 in fields of Limagrain (Drôme, France). Flower-heads were harvested at various times from 29 to 55 days after anthesis (DAA), and were air-dried at room temperature for at least 7 days until the seed moisture content reached *c.* 4.5% fresh weight (FW) basis. Achenes were hand collected and stored at 20°C and 70% relative humidity for 5–6 months in order to break their dormancy (Corbineau *et al.*, 1990). For each developmental stage, some seeds were frozen in liquid nitrogen before drying the flower-heads and stored at –30°C until biochemical analysis.

Seed moisture content and dry weight (DW) were determined by oven-drying the seeds at 105°C for 17 h. Results are expressed as the mean moisture contents obtained with 60 g of achenes.

Germination was tested at 20°C with a 12 h photoperiod, in samples of 300 seeds (six replicates of 50 seeds) placed in boxes on a layer of filter paper soaked with deionized water, as recommended by ISTA (1993).

Seed vigour was estimated by the mean time to germination (MTG) of seeds (four replicates of 50 seeds) placed at 15 or 20°C in darkness on filter paper soaked with water or with solutions of polyethylene glycol-8000 at –0.4, –0.6 and –0.8 MPa. A seed was regarded as germinated when the radicle protruded through the envelopes (seed coat + pericarp). Germination counts were made daily for 5 days and MTG (in days) was calculated as follows:

$$\text{MTG} = [n_1 \times 1 + (n_2 - n_1) \times 2 + (n_3 - n_2) \times 3 + (n_4 - n_3) \times 4 + (n_5 - n_4) \times 5]/n_5$$

where n_1, n_2, n_3, n_4 and n_5 are the mean percentages of germinated seeds obtained at days 1, 2, 3, 4 and 5.

Controlled deterioration was carried out by equilibrating seeds at a moisture content of *c.* 8% FW and then placing them at 45°C in tightly closed boxes for 3–7 days. Germination ability of seeds thus treated was evaluated according to ISTA (1993) rules. The half-viability period (P_{50}) was determined as the time taken for germination to drop to 50%.

Superoxide dismutase (SOD, EC 1.15.11), catalase (CAT, EC 1.11.1.6) and glutathione reductase (GR, EC 1.6.4.2) were extracted and assayed as previously described (Bailly *et al.*, 1998). SOD, CAT and GR activities of each extract were measured three times, and the results, expressed per milligram of protein (specific activity), correspond to the means of the values obtained with three different extracts ± SD. Lipid peroxidation was evaluated by measuring malondialdehyde (MDA) content according to Heath and Parker (1968). Results are expressed as μmol/g dry matter of plant material and correspond to the means of five replicates ± SD.

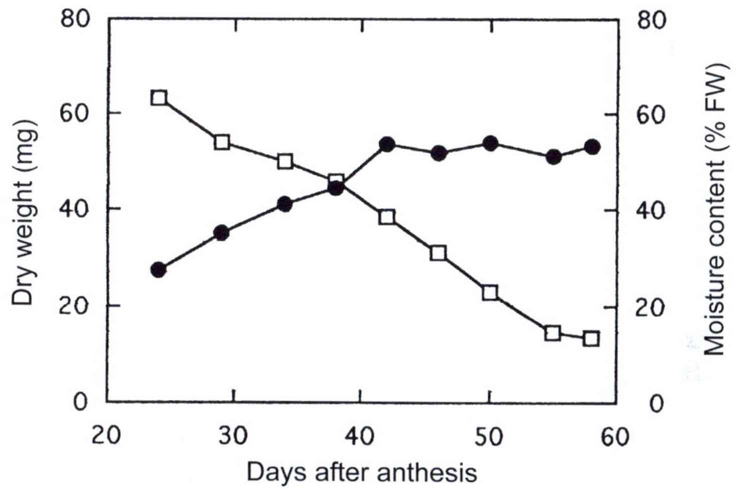

Fig. 8.1. Changes in dry weight (●) and moisture content (□) (% fresh weight) of sunflower seeds during their development.

Results and Discussion

Seed development and germination

Seed moisture content decreased regularly during seed development and was about 15% FW at full maturity, i.e. at 58 DAA (Fig. 8.1). At the end of seed filling, which occurred at *c.* 42 DAA, the mean seed moisture content was 38% FW and the mean seed dry weight was 53 mg (Fig. 8.1).

Freshly harvested seeds did not germinate, whatever their developmental stage (data not shown), because they were dormant (Corbineau *et al.*, 1990). After 5–6 months of dry storage almost all the seeds germinated easily and 90–98% of them gave normal seedlings as soon as 29 DAA (Table 8.1), i.e. at approximately 50% of the seed filling (Fig. 8.1). These results also indicate that seeds were already desiccation tolerant at 29 DAA, at least in the conditions of drying used in our experiments. Similar early acquisition of tolerance to desiccation during seed development was observed in wheat and maize embryos (Black *et al.*, 1996; Brenac *et al.*, 1997). The MTG at 15 and 20°C decreased regularly from 29 to 55 DAA, suggesting that vigour of the seeds increased during their development. The lower the water potential of the medium, the slower was the germination rate, but the sensitivity of seeds to water potential progressively decreased during their development (Table 8.1). In addition, there existed a linear positive relationship between the resistance of seeds to controlled deterioration, expressed as P_{50}, and their moisture content at harvest (Fig. 8.2). Taken together, these results show that, as in soybean (Zanakis *et al.*, 1994) and bean (Sanhewe and Ellis, 1996; Bailly *et al.*, 2001), the vigour of sunflower seeds increased progressively during their development.

Table 8.1. Germination ability, expressed as normal seedlings obtained at 20°C and mean time to germination (MTG) at 15 and 20°C on water and polyethylene glycol solutions at various water potentials, of seeds collected at various stages of development and stored dry for 5–6 months. Means of six replicates.

Days after anthesis	Normal seedlings (%)	MTG (h)				
		15°C on water	20°C on water	20°C on a PEG solution		
				–0.4 MPa	–0.6 MPa	–0.8 MPa
29	93.7	69.9	46.9	71.3	74.0	98.2
34	97.7	62.7	45.7	56.2	72.1	76.5
38	98.3	59.3	45.3	57.2	67.5	72.5
42	95.3	57.1	44.0	48.9	57.5	66.5
46	91.7	52.2	37.9	47.1	54.1	65.2
55	89.7	48.1	37.7	45.0	49.5	57.5

Enzyme activities and lipid peroxidation

SOD, CAT and GR activities and MDA content in fresh and dry seeds are shown as a function of seed water content at harvest (Fig. 8.3). SOD activity was slightly higher in dried seeds than in fresh ones whatever the developmental stage, and did not change markedly during seed development (Fig. 8.3A). CAT activity notably increased during reserve accumulation and seed desiccation *in planta* until seed moisture content dropped down to *c.* 30% FW (i.e. at 47 DAA) and then remained at the same level (Fig. 8.3B). Artificial desiccation of fresh seeds resulted in an increase in CAT activity, at least when initial seed water content was higher than 30% FW. GR activity

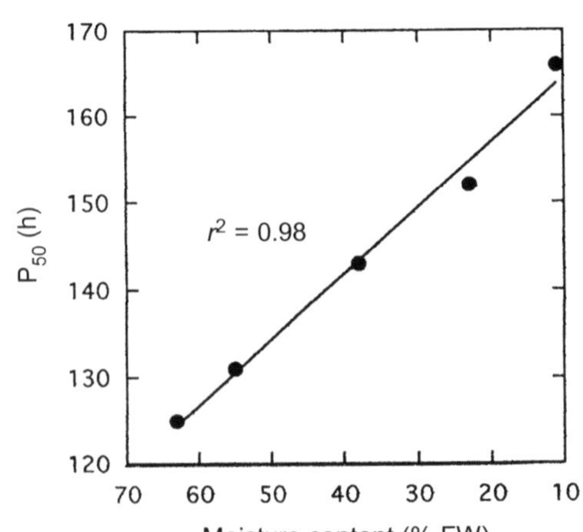

Fig. 8.2. Relationship between the moisture content of seeds at harvest and their half-viability period (P_{50}) after dry storage for 5–6 months. P_{50} calculated as indicated in Materials and Methods; r^2, correlation coefficient of linear regression.

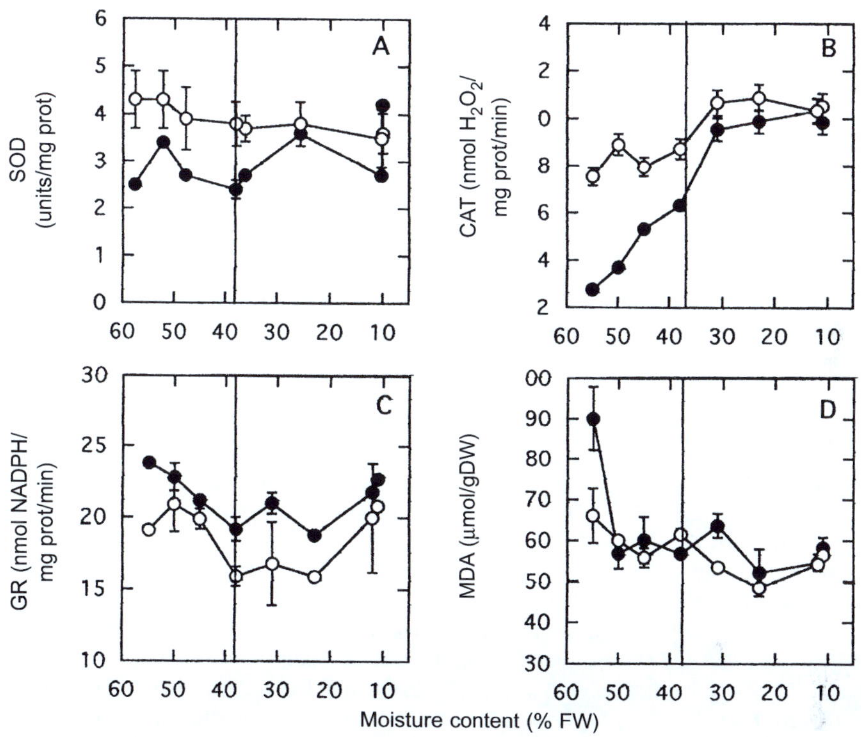

Fig. 8.3. Changes in (A) SOD, (B) CAT and (C) GR activities, and (D) in MDA content in fresh (●) and dried (○) seeds as a function of their moisture content at harvest. Vertical lines indicate the onset of mass maturity. Means of nine (enzymes) or five (MDA) measurements ± SD. Where no bars are shown, SD is less than the size of the symbols.

decreased during seed filling and then remained almost constant during further seed dehydration on the plant (Fig. 8.3C). Artificial drying induced a decrease in GR activity at all the developmental stages studied. Thus, GR responded to natural and artificial dehydration in an opposite way than CAT. CAT might be activated during drying in order to avoid H_2O_2 accumulation and related damage, as suggested by the changes in MDA content (Fig. 8.3D) which indicated the level of lipid peroxidation. Indeed, at least at the earliest stage of development studied, either natural or artificial drying induced a decrease in MDA content, concomitantly to a stimulation of CAT activity.

Conclusion

This work clearly shows that sunflower seed vigour improved regularly during seed development. A key role seems to be devoted to CAT in the acquisition of seed germination ability and possibly desiccation tolerance, whereas SOD and GR activities are probably less related to these phenomena. High quality seeds would require a high level of CAT activity, which might in turn provide indications on the seed developmental stage of fresh seeds. Its activity seems to be related to seed moisture content, suggesting that this enzyme might play a role during seed dehydration by avoiding dehydration-related oxidative stresses. Through the control of H_2O_2 homeostasis, CAT might also be involved in H_2O_2 signalling pathways. Hydrogen peroxide has been shown to control the expression of many genes (Desikan *et al.*, 2001), among which some are probably involved in the shift from a developmental to a germinative mode during seed development. In addition, a high CAT activity in mature seeds might be necessary to limit damaging peroxidative reactions during imbibition and/or during β-oxidation of fatty acids.

References

Bailly, C., Benamar, A., Corbineau, F. and Côme, D. (1998) Free radical scavenging as affected by accelerated ageing and subsequent priming in sunflower seeds. *Physiologia Plantarum* 104, 646–652.

Bailly, C., Audigier, A., Ladonne, F., Wagner, M.H., Coste, F., Corbineau, F. and Côme, D. (2001) Changes in oligosaccharide content and antioxidant enzyme activities in developing bean seeds as related to acquisition of drying tolerance and seed quality. *Journal of Experimental Botany* 52, 701–708.

Bailly, C., Bogatek-Leszczynska, R., Côme, D. and Corbineau, F. (2002) Changes in activities of antioxidant enzymes and lipoxygenase during growth of sunflower seedlings from seeds of different vigour. *Seed Science Research* 12, 47–55.

Bewley, J.D. and Black, M. (1994) *Seeds. Physiology of Development and Germination*, 2nd edn. Plenum Press, New York, 445 pp.

Black, M., Corbineau, F., Grzesik, M., Guy, P. and Côme, D. (1996) Carbohydrate metabolism in the developing and maturing wheat embryo in relation to its desiccation tolerance. *Journal of Experimental Botany* 47, 161–169.

Bowler, C. and Fluhr, R. (2000) The role of calcium and activated oxygen as signals for controlling cross-tolerance. *Trends in Plant Science* 6, 241–246.

Brenac, P., Horbowicz, M., Downer, S.M., Dickerman, A.M., Smith, M.E. and Obendorf, R.L. (1997) Raffinose accumulation related to desiccation tolerance during maize (*Zea mays* L.) seed development and maturation. *Journal of Plant Physiology* 150, 481–488.

Corbineau, F., Bagniol, S. and Côme, D. (1990) Sunflower (*Helianthus annuus* L.) seed dormancy and its regulation by ethylene. *Israel Journal of Botany* 39, 313–325.

Desikan, R., Mackerness, S.A.H., Hancock, J.T. and Neill, S.J. (2001) Regulation of the *Arabidopsis* transcriptome by oxidative stress. *Plant Physiology* 127, 159–172.

Heath, R.L. and Parker, L. (1968) Photoperoxidation in isolated chloroplasts. I. Kinetics and stochiometry of fatty acid peroxidation. *Archives of Biochemistry and Biophysics* 125, 189–198.

Hendry, G.A.F., Finch-Savage, W.E., Thorpe, P.C., Atherton, N.M., Buckland, S.H., Nilsson, K.A. and Seel, W.E. (1992) Free radical processes and loss of seed viability during desiccation in the recalcitrant species *Quercus robur* L. *New Phytologist* 122, 273–279.

ISTA (International Seed Testing Association) (1993) International rules for seed testing. Rules 1993. Annex to Chapter 5. *Seed Science and Technology* 21, supplement, 141–186.

Li, C. and Sun, W.Q. (1999) Desiccation sensitivity and activities of free radical-scavenging enzymes in recalcitrant *Theobroma cacao* seeds. *Seed Science Research* 9, 209–217.

Pammenter, N.W. and Berjak, P. (1999) A review of recalcitrant seed physiology in relation to desiccation-tolerance mechanisms. *Seed Science Research* 9, 13–37.

Sanhewe, A.J. and Ellis, R.H. (1996) Seed development and maturation in *Phaseolus vulgaris*. I. Ability to germinate and to tolerate desiccation. *Journal of Experimental Botany* 47, 949–958.

Vertucci, C.W. and Farrant, J.M. (1995) Acquisition and loss of desiccation tolerance. In: Kigel, J. and Galili, G. (eds) *Seed Development and Germination*. Marcel Dekker, New York, pp. 237–271.

Zanakis, G.N., Ellis, R.H. and Summerfield, R.J. (1994) Seed quality in relation to seed development and maturation in three genotypes of soyabean (*Glycine max*). *Experimental Agriculture* 30, 139–156.

9 GAMYB and BPBF Transcriptional Factors in the Control of Gene Expression During Development of Barley Endosperm

Isabel Díaz, Jesús Vicente-Carbajosa, Zamira Abraham, Manuel Martínez, Inés Isabel-LaMoneda and Pilar Carbonero

Laboratorio de Bioquímica y Biología Molecular, Departamento de Biotecnología-UPM, ETSI Agrónomos, Ciudad Universitaria s/n, 28040 Madrid, Spain

Introduction

During cereal seed development, the starchy endosperm cells accumulate proteins, mainly prolamines, as well as several antimicrobial peptides and inhibitors of heterologous hydrolases implicated in plant defence (Shewry *et al.*, 1995; Carbonero *et al.*, 1999). The genes encoding the storage proteins are coordinately expressed in the developing endosperm where they are under spatial and temporal transcriptional control, involving *cis*-acting motifs in their promoters and *trans*-acting transcriptional factors (TFs). A number of consensus sequences in gene promoters have been shown to be involved in conferring endosperm specificity in cereals (Albani *et al.*, 1997; Marzabal *et al.*, 1998; Mena *et al.*, 1998; Vicente-Carbajosa *et al.*, 1998; Oñate *et al.*, 1999; Carbonero *et al.*, 2000; Wu *et al.*, 2000; Díaz *et al.*, 2002). A conserved *cis*-acting motif found in most storage protein gene promoters of seeds is the endosperm-box, a bipartite motif located ~300 bp upstream of the translation initiation codon that contains two distinct binding sites: the GCN4-like motif (GLM) motif (5′-ATGAG/CTCAT-3′) that resembles the yeast GCN4 motif and the prolamine box (PB: 5′-TGTAAAG-3′). There is a third motif (5′-AACA/TA-3′), which plays a role in the expression of genes encoding glutelins in rice (Suzuki *et al.*, 1998). These three types of motifs are the target sites of different TFs.

In barley, the GLM is recognized by bZIP proteins of the Opaque2 family (BLZ1 and BLZ2) that through dimer formation, as either homo- or heterodimers (Vicente-Carbajosa *et al.*, 1998; Oñate *et al.*, 1999), *trans*-activate the expression of endosperm-specific genes such as the *Hor2* and *Itr1* genes encoding the B-hordeins and a trypsin inhibitor, respectively. The prolamine box is recognized by a TF of the DOF class, DNA binding with One Finger (BPBF), that activates transcription of a native *Hor2* promoter in co-bombarded barley endosperms through binding to the prolamine box (PB) (Mena *et al.*, 1998; Carbonero *et al.*, 2000). We describe here how the 5′-AACA/TA-3′ motif interacts with a TF of the MYB class, which was previously shown to have a role in the activation of gibberellin (GA)-induced genes upon germination in barley aleurone cells (Gubler *et al.*, 1995, 1999).

Materials and Methods

Plant material

Barley (*Hordeum vulgare*) cv. Bomi was grown in a greenhouse at 18°C under 18/6 h day/night photoperiod. Developing endosperms (10–22 days after flowering), mature embryos and 7-day-old leaves and roots were used for RNA extraction. For transient expression assays, developing endosperms collected at 18 days after flowering (DAF) were used immediately.

Northern blot analysis

RNA was isolated by the Lagrimini *et al.* (1987) procedure, electrophoresed, blotted to Hybond membranes and hybridized according to standard procedures (Sambrook *et al.*, 1989). Specific labelled probes were used to identify the *HvGAMyb*, *Pbf*, *Hor2* and *Itr1* mRNAs (Díaz *et al.*, 1995; Gubler *et al.*, 1995; Mena *et al.*, 1998).

DNA constructs and particle bombardment in developing endosperms

The effector constructs contained the cDNAs of *GAMYB* (Gubler *et al.*, 1995) or *BPBF* (Mena *et al.*, 1998) driven by the 35S promoter plus the first intron of the alcohol dehydrogenase I gene (*AdhI*). The reporter constructs included the GUS reporter gene (*uidA*) fused to the *Itr1* and *Hor2* gene promoter fragments (Díaz *et al.*, 1995; Mena *et al.*, 1998), or their mutated versions (see Fig. 9.3a), which corresponded to base changes in the specific MYB and DOF binding motifs.

Particle bombardment was carried out with a biolistic helium gun device (DuPont PDS-000, BioRad, USA). Gold particle coating and bombardment conditions were performed according to Vicente-Carbajosa *et al.* (1998).

Yeast two-hybrid system assays

Translational fusions with the GAMYB and BPBF (or derived fragments of it) to the alcohol dehydrogenase I (*AdhI*) promoter fused to the Gal4 DNA-binding and DNA-activation domains (BD- and AD fusions, respectively) were done (see Fig. 9.4a). *Saccharomyces cerevisiae* transformation was per-

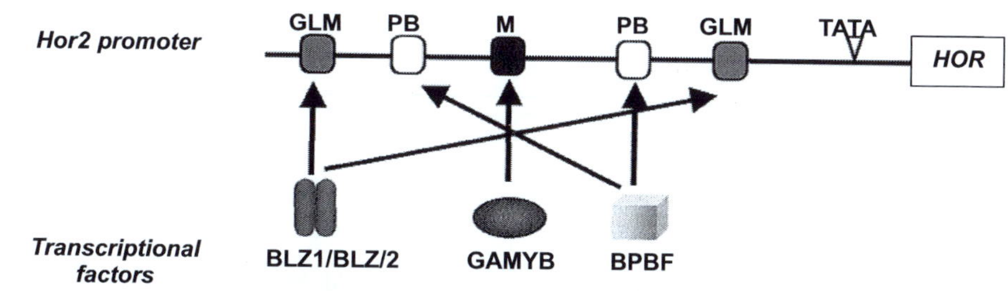

Fig. 9.1. Schematic representation of *cis*-motifs in the barley *Hor2* gene promoter and their putative interaction with transcriptional factors (TF) of the bZIP (BLZ1 and BLZ2), DOF (BPBF) and MYB (GAMYB) classes. GLM: GCN4-like motif; PB: Prolamine Box; M: MYB binding site; TATA: TATA box.

formed by the polyethyleneglycol method and transformants were screened for β-galactosidase production (*LacZ*).

Results and Discussion

Functional analyses of the promoters of several genes expressed in barley endosperm, such as those encoding B-hordeins (*Hor2*) and trypsin inhibitor CMe (*Itr1*), had shown the presence of putative bZIP, MYB and DOF binding motifs in their promoters (Fig. 9.1), which led us to investigate the possible involvement of the HvGAMYB and the BPBF protein factors in their regulation.

The HvGAMYB is a TF of the MYB class that binds specifically to the GA-responsive element (GARE) present in the promoters of α-amylases, β-glucanases and cysteine-proteases of the cathepsin B and cathepsin L types (Cercos *et al.*, 1999; Gubler *et al.*, 1999). Mutations in the GARE abolished both binding and *trans*-activation. The BPBF is a TF of the DOF class that we had previously shown to be expressed in the developing barley endosperm. The BPBF protein expressed in bacteria was able to bind specifically to the PB motif in a *Hor2* gene promoter and through direct interaction with this *cis*-element did transiently *trans*-activate transcription from the *Hor2* promoter in microprojectile-bombarded developing barley endosperms (Mena *et al.*, 1998; Carbonero *et al.*, 2000).

The expression of the encoding gene GAMYB was analysed by Northern blot and these experiments demonstrated that this gene was already present in developing endosperms at 10 DAF and could be detected throughout all stages analysed, as occurred with the DOF transcription factor BPBF (*Pbf* gene) (Fig. 9.2). The transcripts of B-hordein (*Hor2* gene) and trypsin inhibitor CMe (*Itr1* gene) were detectable at 10 DAF and peaked at 22 DAF (Fig. 9.2). The pattern of *GAMYB* accumulation was consistent with its being a regulator of the *Hor2* and *Itr1* genes, whose temporal mRNA expression overlaps with that of the Hv*GAMYB*. *In situ* hybridization studies have localized the spatial expression of the Hv*GAMYB* within the developing

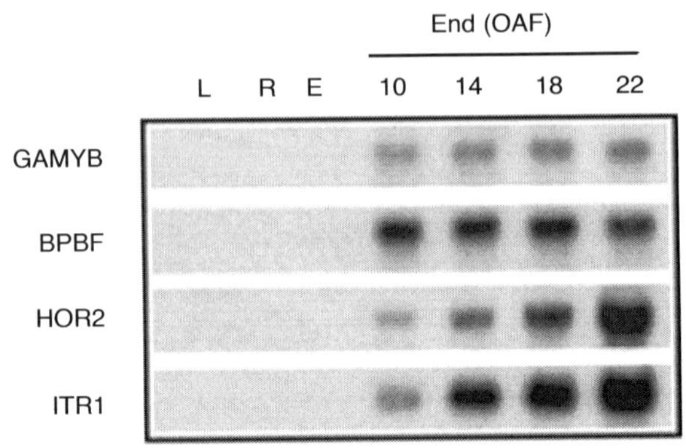

Fig. 9.2. Northern blot analysis of *GAMYB* and *BPBF* mRNA expression in different barley tissues. Total RNA (8 μg) from leaves (L), roots (R), mature embryo (E) and developing endosperm (End) at 10–22 DAF were electrophoresed and blotted. The filter was hybridized with specific probes from *GAMYB, BPBF, Hor2* and *Itr1* cDNAs.

seed in the endosperms, aleurone layers, vascular tissues and embryo (Díaz *et al.*, 2002).

The capability of *in vitro* binding of the GAMYB protein to the 5′-C/TAACT/AACC/A-3′ motif in the promoters of *Hor2* and *Itr1* genes was shown by electrophoretic mobility shift assays (EMSA), using two oligonucleotide probes whose sequences were deduced from the corresponding regions of the *Hor2* and *Itr1* gene promoters and protein extracts derived from *Escherichia coli* expressing GAMYB as a fusion protein. Retarded bands were produced only when the probes were incubated with the transformed bacterial protein extracts (data not shown). Binding specificity was demonstrated by competition titration up to 100× concentration with the same unlabelled probe, but not by the mutated probes which differed only in two base changes at the binding MYB motif. The formation of the shifted DNA-protein complexes was not observed with the mutated variants.

The functional relevance of the interaction *in vitro* was further tested *in planta* by transient expression assays in co-bombarded developing endosperms (Fig. 9.3). When a reporter gene under the control of the –343 bp or –560 bp promoter regions from the *Hor2* and *Itr1* genes, respectively, containing intact MYB and DOF binding sites (pItr1 and pBhor in Fig. 9.3A), were co-transfected with *GAMYB* or *BPBF* as effectors, a threefold to tenfold increase in GUS activity was observed (Fig. 9.3B,C). As expected, mutations in the MYB binding motif (pItr1* and pBhor** in Fig. 9.3A) that prevent *in vitro* binding by the GAMYB protein in EMSA assays abolished GUS *trans*-activation (Fig. 9.3B,C). In contrast, when the reporter construct mutated in the DOF binding site was used (pBhor* in Fig. 9.3C), GAMYB failed to *trans*-activate GUS expression, although the MYB site was intact, while the BPBF protein, as effector, was still able to increase the GUS activity over the control without effector (Fig. 9.3C), probably because there are two DOF binding sites in the promoter (see Fig. 9.1) participating in the regulation process.

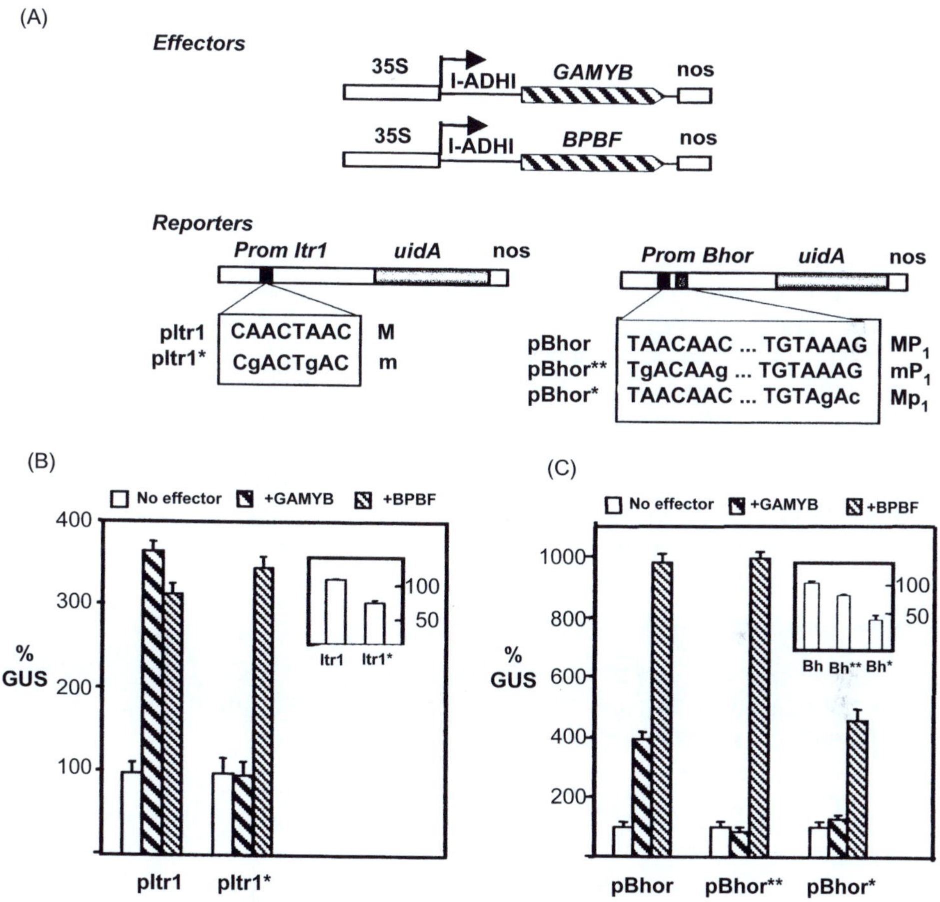

Fig. 9.3. *Trans*-activaton of the *Itr1* and *Hor2* promoters by HvGAMYB and BPBF in developing endosperm. (A) Schematic representation of the effector and reporter constructs. The effectors contained the *GAMYB* or *BPBF* cDNAs driven by the CaMV 35S promoter plus the first intron of the *AdhI* gene. The reporter constructs consisted of the GUS gene (*uidA*) under the control of the *Itr1* gene promoter (pItr1), the *Hor2* promoter (pBhor), or under the mutated promoters with the indicated base changes (pItr1*, pBhor* and pBhor**). (B), (C) Transient expression assays by co-bombardment of developing barley endosperms with the indicated combinations of reporter and effector constructs. GUS activity was expressed as percentage relative to the control plasmids without effector. Transient expression driven by the promoter mutants under study relative to their respective wild types are shown in the inserts.

Mutation at the MYB and DOF binding sites in the reporter plasmids resulted in the loss of the basal transcriptional activity of the native pBhor and pItr1 promoter constructs (inserts of Fig. 9.3B,C).

Taken together, all these results indicate that the GAMYB mediates the *trans*-activation of the *Itr1* and *Hor2* gene promoters in developing barley endosperm through binding to the MYB motif, and that an interaction with the DOF factor BPBF is necessary for the full *trans*-activation of *Hor2*. This hypothesis was checked, *in vivo*, by the yeast two-hybrid system, using the effector constructs schematically represented in Fig. 9.4A. As shown in Fig. 9.4B (lane 2), yeast cells co-transformed with the full-length cDNA fragments of both proteins activated the expression of a *Lac* reporter gene, indicating interaction between these two TFs in this system. This interaction took place between HvGAMYB and the C-terminal part of the BPBF (lane 6 in Fig. 9.4B). In addition, we observed that BPBF is a transcriptional activator in yeast and this protein factor devoid of the N-terminal DOF domain was a stronger activator than the full-length BPBF (lanes 1 and 5 in Fig. 9.4B), an observation previously reported by Kang and Singh (2000) with other DOF transcriptional factors.

These results strongly implicate the GAMYB protein from barley as a TF involved in the combinatorial regulation of genes specifically expressed in the endosperm during development. Recent data from our laboratory have demonstrated that BPBF also has a role in the control of gene expression upon germination (Mena *et al.*, 2002), which suggests that involvement of the same TFs in the processes of seed development and germination may be a more general phenomenon than previously suspected.

Acknowledgements

Financial support from the Spanish Ministerio de Ciencia y Tecnología (project BMC2000-1483) and from Comunidad Autónoma de Madrid (07G/0015/2000) is gratefully acknowledged.

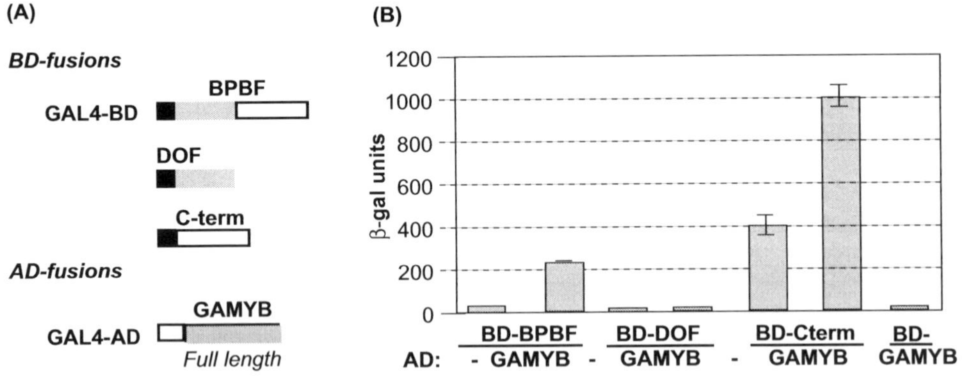

Fig. 9.4. Interaction between HvGAMYB and BPBF in the yeast *S. cerevisiae* two-hybrid system. (A) GAL4-BD and GAL4-AD: GAL4-binding and -activation domains; BPBF: full-length *BPBF* cDNA (nt 1–999); DOF: BPBF N-terminal region (nt 1–273); C-term: BPBF C-terminal region (nt 274–999); GAMYB: full-length *GAMYB* cDNA (nt 1–1719). (B) Quantitative evaluation of the β-galactosidase (*LacZ*) reporter activity in liquid medium, expressed as Miller´s units. Standard errors of the mean of four isolates were 10%. The reporter *LacZ* was driven by GAL4 truncated promoter under the control of Gal4-responsive elements.

References

Albani, D., Hammond-Kosack, M.C.U., Smith, C., Conlan, S., Colot, V., Holdsworth, M. and Bevan, M.W. (1997) The wheat transcriptional activator SPA: a seed-specific bZIP protein that recognizes the GCN-4-like motif in the bifactorial endosperm box of prolamin genes. *The Plant Cell* 9, 171–184.

Carbonero, P., Díaz, I., Vicente-Carbajosa, J., Alfonso-Rubí, J., Gaddour, K. and Lara, P. (1999) Cereal α-amylase/trypsin inhibitors and transgenic insect resistance. In: Scarasccia-Mugnozza, G.T., Porceddu, E. and Pagnotta, M.A. (eds) *Genetics and Breeding for Crop Quality and Resistance.* Kluwer Academic Publishers, Dordrecht, The Netherlands, pp. 147–158.

Carbonero, P., Vicente-Carbajosa, J., Mena, M., Oñate, L., Lara, P. and Diaz, I. (2000) bZIP and DOF transcription factors in the regulation of gene expression in barley endosperm. In: Black, M., Bradford, K.J. and Vazquez-Ramos, J. (eds) *Seed Biology.* CAB International, Wallingford, UK, pp. 27–41.

Cercos, M., Gomez-Cadenas, A. and Ho, T.H.D. (1999) Hormonal regulation of a cysteine proteinase gene, EPB-1, in barley aleurone layers: *cis-* and *trans-*acting elements involved in the co-ordinated gene expression regulated by gibberellins and abscisic acid. *The Plant Journal* 19, 107–118.

Diaz, I., Royo, J., O'Connor, A. and Carbonero, P. (1995) The promoter of the *Itr1* gene from barley confers different tissue specificity in transgenic tobacco. *Molecular General Genetics* 248, 592–298.

Diaz, I., Vicente-Carbajosa, J.V., Abraham, Z., Martínez, M., Isabel-LaMoneda, I. and Carbonero, P. (2002) The GAMYB protein from barley interacts with the DOF transcription factor BPBF and activates endosperm-specific genes during seed development. *The Plant Journal* 29, 453–464.

Gubler, F., Kalla, R., Roberts, J.K. and Jacobsen, J.V. (1995) Gibberellin-regulated expression of a *myb* gene in barley aleurone cells: evidence for Myb transactivation of a High-pI α-amylase gene promoter. *The Plant Cell* 7, 1879–1891.

Gubler, F., Raventos, D., Keys, M., Watts, R., Mundy, J. and Jacobsen, J.V. (1999) Target genes and regulatory domains of the GAMYB transcription activator in barley aleurone. *The Plant Journal* 17, 1–9.

Kang, H.-G. and Singh, K.B. (2000) Characterization of salicylic acid-responsive *Arabidopsis* Dof domain proteins: over expression of *OBO3* leads to growth defects. *The Plant Journal* 21, 329–339.

Lagrimini, L.M., Burkhart, W., Moyer, M. and Rosthein, S. (1987) Molecular cloning of complementary DNA encoding the lignin forming peroxidases from tobacco: molecular analysis and tissue-specific expression. *Proceedings of the National Academy of Sciences USA* 84, 7542–7546.

Marzabal, P., Busk, P.K., Ludevid, M.D. and Torrent, M. (1998) The bifactorial endosperm box of γ-zein gene: characterization and function of the Pb3 and GZM *cis-*acting elements. *The Plant Journal* 16, 41–52.

Mena, M., Vicente-Carbajosa, J., Schmidt, R.J. and Carbonero, P. (1998) An endosperm-specific DOF protein from barley, highly conserved in wheat, binds to and activates transcription from the prolamin-box of a native B-hordein promoter in barley endosperm. *The Plant Journal* 16, 53–62.

Mena, M., Cejudo, F.J., Isabel-Lamoneda, I. and Carbonero, P. (2002) A role for the DOF transcription factor BPBF in the regulation of gibberellin-responsive genes in barley aleurone. *Plant Physiology* 130, 111–119.

Oñate, L., Vicente-Carbajosa, J., Lara, P., Díaz, I. and Carbonero, P. (1999) Barley BLZ2, a seed-specific bZIP protein that interacts with BLZ1 *in vivo* and activates transcription from the GCN4-like motif of B-hordein promoters in barley endosperm. *Journal of Biological Chemistry* 274, 9175–9182.

Sambrook, J., Fritsch, E.F. and Maniatis, T. (1989) *Molecular Cloning: a Laboratory Manual*, 2nd edn. Cold Spring Harbor Laboratory Press, Cold Spring Harbor, New York.

Shewry, P.R., Naiser, J.A. and Tatham, A.S. (1995) Seed storage proteins: structure and biosynthesis. *The Plant Cell* 7, 945–956.

Suzuki, A., Wu, C.Y., Washida, H. and Takaiwa, F. (1998) Rice MYB protein OSMYB5 specifically binds to the AACA motif conserved among promoters of genes for storage protein glutelin. *Plant Cell Physiology* 39, 555–559.

Vicente-Carbajosa, J., Oñate, L., Lara, P., Díaz, I. and Carbonero, P. (1998) Barley BLZ1: a bZIP transcriptional activator that interacts with endosperm-specific gene promoters. *The Plant Journal* 13, 629–640.

Wu, C.-Y., Washida, H., Onodera, Y., Harada, K. and Takaiwa, F. (2000) Quantitative nature of the prolamine-box, ACGT and AACA motifs in a rice glutelin gene promoter: minimal *cis*-element requirements for endosperm-specific gene expression. *The Plant Journal* 23, 415–421.

10 Peptide Transport in the Developing Barley Grain

WANDA M. WATERWORTH, MEREWYN K. ASHLEY, CHRISTOPHER E. WEST AND CLIFFORD M. BRAY

School of Biological Sciences, University of Manchester, UK

Introduction

Peptide transporters play a major role in nitrogen mobilization during barley grain germination (reviewed by Waterworth *et al.*, 2001). Enzymatic hydrolysis of storage proteins results in the accumulation of small peptides and amino acids in the endosperm. These are translocated to the growing embryo by specific transporters localized to the plasma membrane of the scutellar epithelium in the mature imbibed cereal grain. The barley scutellar peptide transporter *HvPTR1* has been cloned and shown to be a seed-specific protein expressed early in germination (West *et al.*, 1998).

The developing grain is supplied with nutrients from vegetative tissues of the mother plant. In the absence of a direct vascular connection between maternal and filial (endosperm/embryo) tissue, it is believed that assimilates are delivered to the extracellular space (apoplast) dividing these tissues, followed by uptake into the developing grain. Developing cereal grains have no direct vascular connections with the mother plant and so short-distance transport mechanisms operate to translocate assimilates from the regions of vascular tissues to the region of reserve deposition, the endosperm. Nutrients are also required to support the growth of the developing cereal embryo. In temperate cereals such as wheat and barley, assimilates must pass through the funiculus–chalazal region and the nucellus projection before entering the starchy endosperm. Nutrients are believed to pass symplastically, via plasmodesmata, through maternal cells to the cells of the nucellus projection. Here, transfer cells redirect solutes to the apoplast (extracellular regions), possibly by active transport or passive, turgor-driven mechanisms.

The import of nitrogen into developing seeds has been poorly characterized relative to that of carbon but is thought to occur largely in the form of amino acids (reviewed by Wolswinkel, 1992; Weber *et al.*, 1998). However, little is known of the involvement of peptides in long-distance nitrogen mobilization in plants and the possible role that peptide transport may play in seed development. Evidence for a physiological role for peptide transport in *Arabidopsis* seed development has arisen from studies in which antisense repression of the peptide transporter AtPTR2-B resulted in abnormal seed development (Song *et al.*, 1996, 1997). Sopanen *et al.* (1985) also reported that isolated barley embryos from developing grain showed active peptide transport, but peptide uptake across the transfer region was not examined. Here we characterize the import of small peptides into the developing barley grain and show that HvPTR1, a peptide transporter important to reserve mobilization in the germinating barley grain, is also expressed during embryo development.

Materials and Methods

Plant material and growth conditions

Barley (*Hordeum vulgare* cv. Annabel) was grown under controlled conditions of constant humidity (70%) with 16 h light and 8 h dark cycles at 20°C. Visible light was provided by a combination of high-intensity discharge lamps (Osram, Light Source Supplies, Bishops Stortford, UK) and tungsten lamps. Plants were tagged at anthesis to determine developmental age.

[^{14}C]Gly-Sar uptake by the developing barley grain

Ears of barley at the appropriate developmental stage after anthesis were cut 3 cm below the ear and placed in 1 ml 5 mM sodium phosphate-citrate buffer (pH 5.0) containing 0.5 μCi [^{14}C]Glycyl-Sarcosine ([^{14}C]Gly-Sar) at a final concentration of 1 mM. After the appropriate time, embryo and endosperm tissue were dissected out of the barley grain and boiled for 20 min in 5 M acetic acid (1.5 ml); the extracts were added to 3.5 ml scintillation fluid and scintillation was counted.

RNA isolation and RT-PCR

RNA isolation from germinating and developing barley grain, RT-PCR and recombinant DNA procedures were performed as described in West *et al.* (1998).

SDS-PAGE and Western analysis

Membrane and total proteins were extracted from developing barley tissue as described in Waterworth *et al.* (2000). Protein concentrations were determined by the Bio-Rad protein assay (Bio-Rad Laboratories, Hemel Hempstead, UK) using bovine serum albumin as a standard. Protein samples were separated by SDS-PAGE and transferred to PVDF membrane (Bio-Rad) for 3 h at 100 V. The blots were probed with either a 1/5000 dilution of peptide antiserum raised to HvPTR1 (Waterworth *et al.*, 2000) or a 1/5000 dilution of polyclonal antiserum raised to the 7S globulin from *Avena sativa* cv. Maris Tabard (Burgess and Miflin, 1985). SDS-PAGE gels were Coomassie-blue stained for analysis of proteins.

Results

The uptake of the non-hydrolysable peptide [^{14}C]Gly-Sar into developing barley grain was investigated using grain of different ages after anthesis. Initial experiments showed that significant accumulation of [^{14}C]Gly-Sar occurred in the developing grain, indicating that peptides had passed through the transfer region of the nucellus projection. The embryo and endosperm tissues were dissected out and the radioactivity extracted to determine the destination of peptides assimilated into the developing grain. Developing embryos were too insubstantial to analyse until 10 days after anthesis (DAA) and so only data for the whole grain was obtained prior to this. The embryo showed a much larger accumulation of [^{14}C]Gly-Sar in comparison with the endosperm on the basis of fresh weight (Fig. 10.1), indicating that peptide uptake may play a more important role in the nutrition of the developing embryo. Peptide uptake into the embryo peaked between 19–23 DAA, which coincides with storage protein deposition in the barley endosperm. This is the first demonstration of peptide transport from the mother plant to the developing grain.

Dinitrophenol (DNP), an inhibitor of H^+ coupled transport, inhibited peptide transport both into the embryo and endosperm of 20 DAA developing barley grain by 70%. This level of inhibition is similar to the levels of

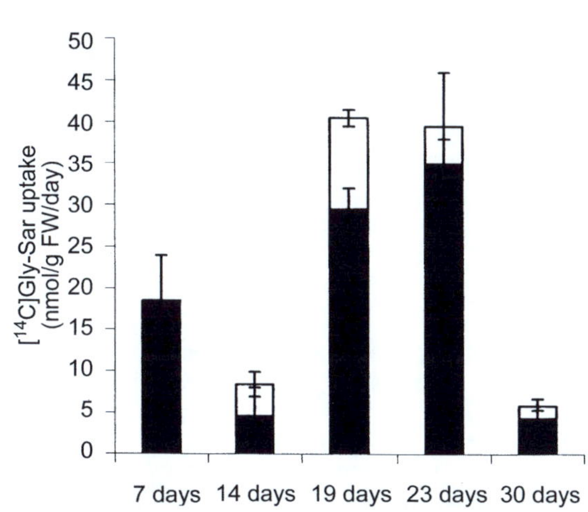

Fig. 10.1. [^{14}C]Gly-Sar uptake by the developing barley grain. Embryo (■) and endosperm (□) tissue were isolated from barley grain radiolabelled with [^{14}C]Gly-Sar for 6 h and the incorporation of radioactivity was determined. Values are the mean ± SE of two to three independent experiments.

peptide transport inhibition obtained in the germinating barley grain (Higgins and Payne, 1977) and indicates that a large component of Gly-Sar uptake into the developing grain is active and dependent on an H^+ gradient for at least one stage between feeding into the stem and translocation to embryo/endosperm tissue.

The kinetic properties of peptide transport into the developing grain were then investigated. Peptide transport into the embryo of 23 DAA barley grain displayed a Michaelis constant (K_m) of 2 mM, similar to that found in embryos isolated from germinating barley grain and indicating that a relatively low affinity uptake system is in operation. Transport displayed a pH optimum of 5.0. Active peptide transport was also found to have a comparable developmental profile and properties in the developing wheat grain.

Using degenerate primers designed to conserved regions of PTR family peptide transporters, we used an RT-PCR approach to identify peptide transporters present in the developing barley grain. *HvPTR1*, a peptide transporter localized to the scutellum of the germinating barley grain that plays an important role in nutrient reserve mobilization (West *et al.*, 1998),

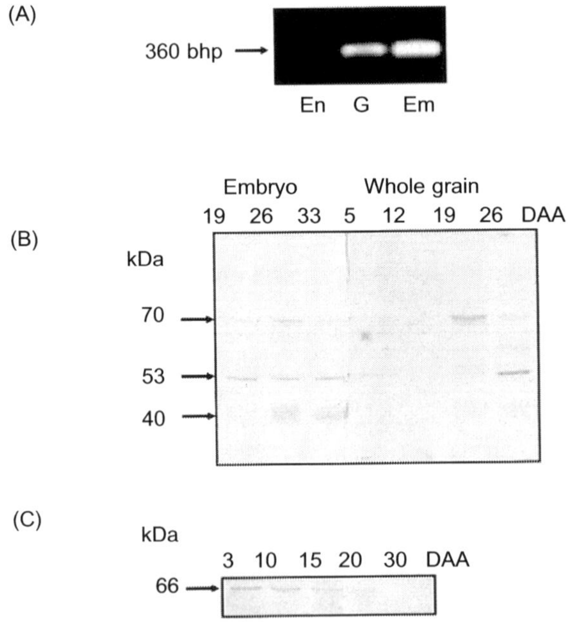

Fig. 10.2. (A) RT-PCR analysis of HvPTR1 expression in the developing barley grain. RNA isolated from the indicated tissues was reverse transcribed and amplified with degenerate primers designed to conserve regions of PTR family proteins as described in West *et al.* (1998). PCR products were separated on a 1% agarose gel. En: endosperm; Em: embryo tissue isolated from 15 DAA barley grain; G: mature barley grain germinated for 24 h. (B) Western analysis of 7S globulin expression in the developing barley embryo. (C) Western analysis of HvPTR1 expression during barley grain development.

was found to be present in 15 DAA developing grain (Fig. 10.2A). *HvPTR1* was present in embryo but not endosperm tissue, indicating that *HvPTR1* may play a role in nutrient transport during embryo development. Western analysis using peptide antiserum raised to HvPTR1 (Waterworth *et al.*, 2000) showed that the 66 kDa protein was present in the developing grain (Fig. 10.2C). HvPTR1 protein was only detectable in the barley embryo (Fig. 10.2C) and was present from as early as 3 DAA, with HvPTR1 protein levels subsequently declining between 20 and 30 DAA. Northern analysis showed that transcript levels of *HvPTR1* mRNA were much lower in the developing grain compared with the germinating barley grain. The expression of *HvPTR1* so early in grain development was unexpected since storage protein deposition, and the highest levels of peptide transport activity, do not occur until much later in grain development.

The major storage proteins of the barley embryo are globulins. Western analysis using polyclonal antiserum raised to the 7S globulin from *A. sativa* was used to analyse the accumulation of 7S globulins in the developing barley embryo (Fig. 10.2B). The antiserum recognized three groups of polypeptides of molecular mass 40, 50–55 and 66–75 kDa, which are comparable to the molecular masses of the polypeptides recognized in *A. sativa* (Burgess and Shrewry, 1986). Globulin storage protein deposition in the embryo was detected by 19 DAA, much later than the induction of HvPTR1 expression. The expression of HvPTR1 so early in barley grain development indicates that peptide transport may not be required solely for nutrient import during seed development and may in addition play an as yet undetermined role in embryo/seed development.

Discussion

Here we have presented evidence for the active transport of peptides from the mother plant into the developing barley grain. The most obvious role for peptide transport in grain development would be the mobilization of nitrogen into the developing barley grain for reserve protein deposition. The transport of nitrogen in the form of peptides is energetically more favourable than that of free amino acids (Higgins and Payne, 1980). Protein reserve deposition in the developing barley grain occurs after that of starch, from around 20 DAA onwards (Bewley and Black, 1994). The development of peptide transport in grain filling coincided with this, peaking at 19–23 DAA. These results are comparable to those of Sopanen *et al.* (1985), which demonstrated that peptide uptake in isolated developing barley embryos increased during the early and middle stages of embryo development, corresponding to 20–28 DAA in this study. However, the flux of peptides into the developing barley grain is low (reaching a maximum of 43 nmol/g/day at 19 dpa) compared with rates of peptide transport found during barley grain germination, which display a maximum transport rate of 6 μmol/g/h for Ala-Phe (West *et al.*, 1998). This may suggest that peptides are only a minor component of nitrogen import into the developing barley grain. However, the developing grain may not need such high rates of peptide transport because nitrogen import occurs over a much longer period of time than the mobilization of peptides from endosperm to embryo in cereal grain germination. Levels of small peptides in plant phloem/xylem sap and the developing barley grain need to be determined in order to confirm that peptide transport does indeed play a role in grain filling.

Peptides accumulated predominantly in the developing barley embryo, as opposed to endosperm tissue, where storage protein reserves are predominantly localized, and HvPTR1 was expressed only in developing embryo tissue. One explanation could be that peptides, once transported to the developing grain, are hydrolysed into constituent amino acids prior to translocation to the endosperm. The onward transport of amino acids to the endosperm would have been complicated by the use of the non-hydrolysable peptide Gly-Sar here. Alternatively peptides could be involved in nitrogen supply for protein synthesis and deposition in the barley embryo. The very early expression of HvPTR1 in the developing grain, before reserve protein deposition, raises the possibility that this peptide transporter may play an as yet undetermined role in seed development.

In addition to nitrogen/carbon mobilization, peptide transporters may perform a number of other functions in grain development, such as transport of sulphur in the form of the tripeptide glutathione, peptide conjugates of plant hormones or peptide signalling molecules into the developing grain. Peptide transporters might also facilitate the uptake of peptide-like bacterial toxins in plants, as peptides have been implicated as signalling molecules in host–pathogen interactions (Steiner *et al.*, 1994).

Further evidence supporting a role for peptide transport in seed development has arisen from studies of antisense *Arabidopsis* plants with reduced expression of the peptide transporter AtPTR2-B. Antisense AtPTR2-B plants showed a reduction in seed number per silique, although individual seed weights increased, and aborted seeds did not develop beyond the heart or torpedo stage (Song *et al.*, 1997). This may reflect a requirement for peptide transport during embryogenesis. Peptide transport in plants has to date been well characterized only in cereal grain germination. However, there are far more homologues of the PTR peptide transporter family in *Arabidopsis* than in other genomes sequenced to date (*Arabidopsis* Genome Initiative, 2000). Thus, although the functions of this gene family remain elusive, peptide transporters may well play a more central role in plant physiology than considered previously. Elucidating the roles of peptide transporters in seed germination and development in both dicots and cereals promises to be an exciting and challenging area of research in the next few years.

Acknowledgements

We gratefully acknowledge the UK BBSRC for financial support.

References

Arabidopsis Genome Initiative (2000) Analysis of the genome sequence of the flowering plant *Arabidopsis thaliana*. *Nature* 408, 796–815.

Bewley, J.D. and Black, M. (1994) *Seeds: Physiology of Development and Germination*, 2nd edn. Plenum Press, New York, p. 445.

Burgess, S.R. and Miflin, B.J. (1985) The localisation of oat (*Avena sativa* L.) seed globulins in protein bodies. *Journal of Experimental Botany* 36, 945–954.

Burgess, S.R. and Shrewry, P.R. (1986) Identification of homologous globulins from embryos of wheat, barley, rye and oats. *Journal of Experimental Botany* 37, 1863–1871.

Higgins, C.F. and Payne, J.W. (1977) Characterisation of active dipeptide transport by the germinating barley embryo. *Planta* 136, 71–76.

Higgins, C.F. and Payne, J.W. (1980) Transport and utilisation of amino acids and peptides by higher plants. In: Payne, J.W. (ed.) *Microorganisms and Nitrogen Sources*. Wiley, London, pp. 609–639.

Song, W., Steiner, H.-Y., Zhang, F., Stacey, G. and Becker, J.M. (1996) Cloning of a second *Arabidopsis* peptide transport gene. *Plant Physiology* 110, 171–178.

Song, W., Koh, S., Czako, M., Marton, L., Denkard, E., Becker, J.M. and Stacey, G. (1997) Antisense expression of the peptide transport gene *AtPTR2-B* delays flowering and arrests seed development in transgenic *Arabidopsis* plants. *Plant Physiology* 114, 927–935.

Sopanen, T., Sinervo, T. and Ahokas, H. (1985) Uptake of amino acids and peptides by developing barley embryos. *Physiologia Plantarum* 65, 347–351.

Steiner, H.Y., Song, W., Zhang, L., Naider, F., Becker, J.M. and Stacey, G. (1944) An *Arabidopsis* peptide transporter is a member of a new class of membrance transport proteins. *Plant Cell* 6, 1289–1299.

Waterworth, W.M., West, C.E. and Bray, C.M. (2000) The barley scutellar peptide transporter: biochemical characterisation and localisation to the plasma membrane. *Journal of Experimental Botany* 51, 1201–1209.

Waterworth, W.M., West, C.W. and Bray, C.M. (2001) The physiology and molecular biology of peptide transport in germination. *Seed Science Research* 11, 275–284.

Weber, H., Heim, U., Golombek, S., Borisjuk, L. and Wobus, U. (1998) Assimilate uptake and the regulation of seed development. *Seed Science Research* 8, 331–345.

West, C.E., Waterworth, W.M., Stephens, S.M., Smith, C.P. and Bray, C.M. (1998) Cloning and functional characterisation of a peptide transporter expressed in the scutellum of barley grain during the early stages of germination. *Plant Journal* 15, 221–229.

Wolswinkel, P. (1992) Transport of nutrients into developing seeds: a review of physiological mechanisms. *Seed Science Research* 2, 59–73.

11 Oxygen as a Control Factor in Embryogenesis of Legume Seeds

HARDY ROLLETSCHEK, LJUDMILLA BORISJUK, ULRICH WOBUS AND HANS WEBER

Institut für Pflanzengenetik und Kulturpflanzenforschung (IPK), D-06466 Gatersleben, Germany

Introduction

Low oxygen (O_2) control of plant growth has been reported at the whole plant level (Bacanamwo and Purcell, 1999) and for specialized tissues such as rhizomes (Rolletschek *et al.*, 1998) and roots (Armstrong *et al.*, 1994) that often grow in O_2-deficient environments. In addition, seed structures of legumes and other crop plants provide characteristics that promote O_2 deficiency inside the tissue. Seeds are covered with thickly cutinized cell layers that strongly restrict gas exchange (Wager, 1974). The legume pod wall has a low stomatal frequency compared with leaves, further diminishing gas exchange. Specialized systems for O_2 delivery are lacking. Low porosity results in high diffusional impedances. Moreover, embryogenic growth as well as biochemical characteristics promote O_2 shortage. The meristematic growth during the early stages of seed development is connected with high energy demand. The subsequent seed-filling phase is characterized by high synthesis rates for storage compounds (Borisjuk *et al.*, 1995). Both stages require high respiration rates, thus diminishing O_2 levels inside tissues.

Overall, these observations suggest that O_2 levels inside seeds are significantly lower than ambient levels. Indeed, several findings in the literature support this assumption, e.g. the induction of fermentation enzymes (Boyle and Yeung, 1983). Based on the decline in the adenylate energy charge, an O_2 concentration of 5–10% has been suggested for soybean seeds (Shelp *et al.*, 1995). The importance of O_2 is further supported by the observation that seed production and seed size depend on the atmospheric (external) O_2 level (Porterfield *et al.*, 1999). Therefore, it may be postulated that substantial concentration gradients of O_2 are necessary to drive diffusion into seed structures. Otherwise, diffusive influx is too low to meet O_2 demand, followed by disrupted seed development or even seed abortion.

 The Biology of Seeds: Recent Research Advances (eds G. Nicolás, K.J. Bradford, D. Côme and H.W. Pritchard)

However, all these observations give only indirect evidence for O_2-depleted zones inside seeds.

In recent years, O_2-sensitive and mechanically more stable microsensors have become available, allowing the direct measurement of O_2 levels inside plant tissues (Armstrong *et al.*, 1994; Porterfield *et al.*, 1999). By using such microsensors we were able to measure O_2 profiles across developing seeds of *Vicia faba* and *Pisum sativum* at high spatial resolution (Rolletschek *et al.*, 2002). Here we summarize our findings on O_2 distribution within seeds in relation to seed development and the environment. Analysis of data on energy status, metabolite levels and respiratory activity suggests a possible control function of O_2 deficiency in legume seed development.

Legume Embryos Develop in a Hypoxic Environment

O_2 distribution within *Vicia* and *Pisum* seeds shows a very similar pattern. Inside the seed coat the O_2 level falls strongly and reaches very low levels at its inner layers (Fig. 11.1). Entry of O_2 from the surrounding gas space into the seed is strongly restricted by the seed coat and occurs entirely through the micropylar region (Wager, 1974). On the one hand, low permeability of the seed coat restricts oxygen uptake, thereby promoting O_2 deficiency. On the other hand, low gas exchange might be helpful to minimize the loss of CO_2 coming from respiration. Indeed, high CO_2 levels within seeds (up to 11% v/v; see Wager, 1974) were found, promoting the activity of phosphoenolpyruvate carboxylase catalysing CO_2 refixation (Golombek *et al.*, 1999). Lowest O_2 levels within seeds (up to 1% saturation; note that atmospheric O_2 level of approximately 21 kPa is set to 100%) are detected within the endospermal vacuole between the seed coat and the embryo. This points to high respiratory activity of endospermal cytoplasmic strands and corresponds to earlier findings on their metabolic activity (accumulation of stor-

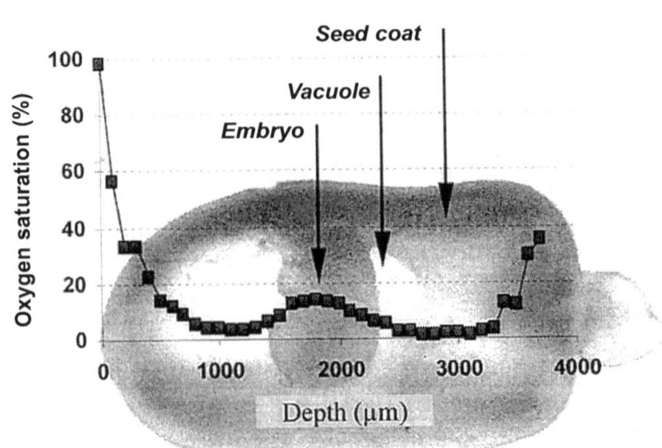

Fig. 11.1. O_2 concentration measured in light along a longitudinal transect of a *Vicia faba* seed of 100 mg fresh weight.

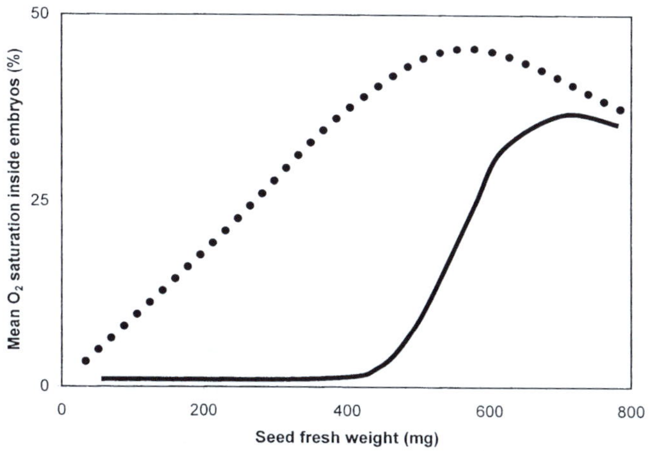

Fig. 11.2. Relationship between seed fresh weight and mean O_2 level measured inside *Vicia faba* embryos in light (•••) and in darkness (——).

age products; see Borisjuk *et al.*, 1995). Within the embryo, O_2 levels rise to higher values – at least in the light, due to photosynthesis. In darkness, the level is very low (up to 1% saturation) and depends mainly on diffusive uptake from the surrounding tissues.

Pronounced O_2 gradients within the embryo are lacking. This was unexpected because both metabolic and photosynthetic activity vary significantly within embryos (Borisjuk *et al.*, 1995; own unpublished data). Obviously, these processes are tightly coupled. Alternatively, embryonic tissue may offer relatively little diffusional impedance. This would allow efficient gas exchange and compensate for locally distinct O_2 consumption and/or production but this is unlikely, as discussed below.

Light supply and the corresponding ability for photosynthetic O_2 production significantly elevates the O_2 level within the embryo compared with its environment (endospermal vacuole or seed coat). This relationship is evident in Fig. 11.2. It is also somewhat surprising, because photosynthetic activity of legume seeds is low compared with that of leaves and was until recently regarded as of minor importance. In terms of CO_2 fixation, photosynthetic activity of seeds is insignificant (Atkins and Flinn, 1978), but in terms of O_2 supply it is obviously not. From the pattern in Fig. 11.2 some further important conclusions emerge. In light, O_2 levels within embryonic tissues rise during development, following increasing photosynthetic activity. When maturity is reached photosynthetic activity, i.e. O_2 levels, starts to decrease. This correlates well with the greening/yellowing process during embryogenesis. In darkness, O_2 concentration inside embryos remains at a very low level. In the later stages of development, O_2 levels increase to values observed under light conditions (Fig. 11.2). This increase may be caused by changes in both respiratory activity and seed coat permeability (diffusive O_2 influx).

In summary, O_2 levels inside seeds vary remarkably, with lowest levels during early seed development and in darkness. Shortage of O_2 should, therefore, have an impact on seed development and should be subject to diurnal fluctuations.

Embryo Tissue Avoids Anoxia

Although O_2 falls to very low levels in darkness, anoxic zones could never be observed within *Vicia* or *Pisum* seeds under natural conditions. Accordingly, no significant induction of fermentative pathways occurs (except for early developmental stages; unpublished data). Anoxia has to be avoided by the plant because of its detrimental effects on cell metabolism. These differ from plant response to hypoxia (Drew, 1997; Zeng *et al.*, 1999; Geigenberger *et al.*, 2000) and may completely interrupt nutrient supply for the developing embryo and biosynthetic activity (Thorne, 1982). The mechanism responsible for this avoidance is interesting, and the idea of relating it to non-symbiotic haemoglobins is attractive (Hill, 1998). These proteins are induced by low O_2 and may potentially regulate plant responses when critical O_2 pressures are reached.

Low Oxygen Effects on Seed Metabolism

As shown above, legume embryos develop in a hypoxic environment. The question arises: what is the significance for seed metabolism? The principal effects of O_2 deficiency are well described. An initial pH shift caused by ATP hydrolysis and onset of fermentation is followed by the inhibition of metabolic activity and the induction of anaerobic response genes, including fermentative and some glycolytic enzymes (Drew, 1997; Geigenberger *et al.*, 2000). The balancing of ATP demand and supply pathways is shifted towards energy-saving metabolic pathways, thereby conserving the energy charge of the cells. Can such changes be found in seed development?

Concomitant with very low O_2 levels in the early stages of seed development, adenylate energy charges as well as ATP levels are dramatically low, as shown in early embryos of *Vicia* (Rolletschek *et al.*, 2002). This suggests that O_2 deficiency limits ATP production via oxidative phosphorylation. Thus, early embryo growth may be energy limited. The lowest O_2 levels measured within embryos are still sufficient for cytochrome oxidase, due to its very high affinity for O_2 (0.013%; Drew, 1997). However, significant metabolic adaptations already occur at considerably higher O_2 concentrations (Geigenberger *et al.*, 2000). The relationship between respiratory activity of *Vicia* embryos and external O_2 concentrations is shown in Fig. 11.3. It is obvious that the respiratory flux measured as O_2 uptake by the embryo decreases almost linearly in response to falling O_2 levels. When the external O_2 level decreases, the concentration gradient decreases and, therefore, the driving force for diffusional fluxes. It seems likely that O_2 diffusibility (i.e. high impedance) limits O_2 uptake rather than demand by respiration. The extent

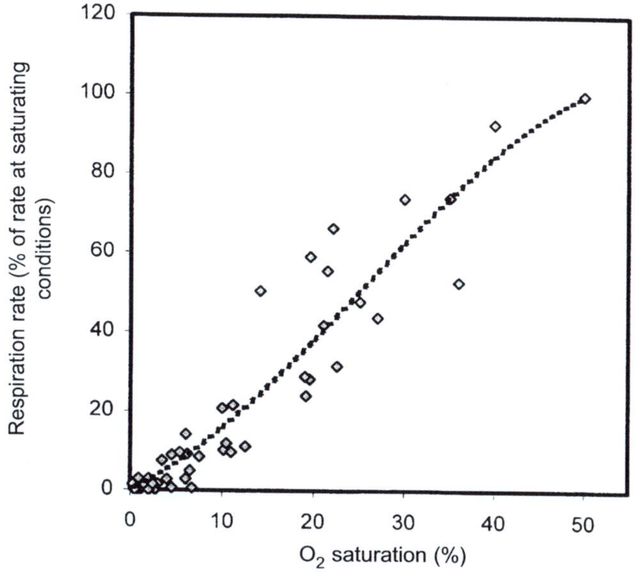

Fig. 11.3. Relationship between external O_2 levels and respiratory activity of *Vicia faba* embryos. A polynomial regression line is fitted to the data.

to which respiration is inhibited by low external O_2 is quite strong. Relating the inhibition curve of Fig. 11.3 to the *in vivo* O_2 levels measured in light and darkness (Fig. 11.2), it becomes evident that embryonic respiration is strongly limited during early stages of development and during the night. This corresponds very well to the low ATP levels during the early stages of embryo development (Rolletschek *et al.*, 2002).

The low O_2 levels found here could represent a signal that induces adaptive energy-saving metabolic responses. The repression of invertase but induction of sucrose synthase by low O_2 is considered to be part of this metabolic shift (Zeng *et al.*, 1999). Compared with invertase, sucrose synthase saves one ATP molecule and therefore represents an energy-saving mechanism. Indeed, during the switch from the pre-storage to the storage compound accumulation phase of seeds, a change from an invertase to a sucrose synthase pathway of sucrose degradation occurs (Weber *et al.*, 1995, 1996). Because the K_m value of sucrose synthase in *Vicia* seeds is high, sucrose levels increase and storage product synthesis is promoted. Sucrose synthase activity provides precursors for cell wall biosynthesis and increases osmolality (sucrose is a main osmolyte). Furthermore, low O_2 has been shown to induce cell wall proteins responsible for partial cell wall degradation. All these aspects are important prerequisites for lowering water potential followed by water uptake and cell elongation growth, all of which are observed during the main storage phase. Thus, low O_2 may provide an important signal for the regulation of seed development.

References

Armstrong, W., Strange, M.E., Cringle, S. and Beckett, P.M. (1994) Microelectrode and modelling study of oxygen distribution in roots. *Annals of Botany* 74, 287–299.

Atkins, C.A. and Flinn, A.M. (1978) Carbon dioxide fixation in the carbon economy of developing seeds of *Lupinus albus* (L.). *Plant Physiology* 62, 486–490.

Bacanamwo, M. and Purcell, L.C. (1999) Soybean dry matter and N accumulation responses to flooding stress, N sources and hypoxia. *Journal of Experimental Botany* 50, 689–696.

Borisjuk, L., Weber, H., Panitz, R., Manteuffel, R. and Wobus, U. (1995) Embryogenesis of *Vicia faba* L.: histodifferentiation in relation to starch and storage protein synthesis. *Journal of Plant Physiology* 147, 203–218.

Boyle, S.A. and Yeung, E.C. (1983) Embryogeny of *Phaseolus*: developmental pattern of lactate and alcohol dehydrogenases. *Phytochemistry* 22, 2413–2416.

Drew, M.C. (1997) Oxygen deficiency and root metabolism: injury and acclimation under hypoxia and anoxia. *Annual Reviews of Plant Physiology and Plant Molecular Biology* 48, 223–250.

Geigenberger, P., Fernie, A.R., Gibon, Y., Christ, M. and Stitt, M. (2000) Metabolic activity decreases as an adaptive response to low internal O_2 in growing potato tubers. *Biological Chemistry* 381, 723–740.

Golombek, S., Heim, U., Horstmann, C., Wobus, U. and Weber, H. (1999) Phosphoenolpyruvate carboxylase in developing seeds of *Vicia faba* L.: gene expression and metabolic regulation. *Planta* 208, 66–72.

Hill, R.D. (1998) What are hemoglobins doing in plants? *Canadian Journal of Botany* 76, 707–712.

Porterfield, D.M., Kuang, A., Smith, P.J.S., Crispi, M.L. and Musgrave, M.E. (1999) Oxygen-depleted zones inside reproductive structures of *Brassicaceae*: implications for oxygen control of seed development. *Canadian Journal of Botany* 77, 1439–1446.

Rolletschek, H., Bumiller, A., Henze, R. and Kohl, J.G. (1998) Implications of missing efflux sites on convective ventilation and amino acid metabolism in *Phragmites australis*. *New Phytologist* 140, 211–217.

Rolletschek, H., Borisjuk, L., Koschorreck, M., Wobus, U. and Weber, H. (2002) Legume embryos develop in a hypoxic environment. *Journal of Experimental Botany* 53, 1099–1107.

Shelp, B.J., Walton, C.S., Snedden, W.A., Tuin, L.G., Oresnik, I.J. and Layzell, D.B. (1995) Gaba shunt in developing soybean seeds is associated with hypoxia. *Physiologia Plantarum* 94, 219–228.

Thorne, J.H. (1982) Temperature and oxygen effects on ^{14}C-photosynthate unloading and accumulation in developing soybean seeds. *Plant Physiology* 69, 48–53.

Wager, H.G. (1974) The effect of subjecting peas to air enriched with carbon dioxide. I. The path of gaseous diffusion, the content of CO_2 and the buffering of the tissue. *Journal of Experimental Botany* 25, 330–337.

Weber, H., Borisjuk, L., Heim, U., Buchner, P. and Wobus, U. (1995) Seed coat associated invertases of fava bean control both unloading and storage functions: cloning of cDNAs and cell type-specific expression. *The Plant Cell* 7, 1835–1846.

Weber, H., Buchner, P., Borisjuk, L. and Wobus, U. (1996) Sucrose metabolism during cotyledon development of *Vicia faba* L. is controlled by the concerted action of both sucrose-phosphate synthase and sucrose synthase: expression patterns, metabolic regulation and implications for seed development. *Plant Journal* 9, 841–850.

Zeng, Y., Wu, Y., Avigne, W.T. and Koch, K.E. (1999) Rapid repression of maize invertases by low oxygen. Invertase/sucrose synthase balance, sugar signaling potential, and seedling survival. *Plant Physiology* 121, 599–608.

12 Hormonal and Molecular Events During Seed Dormancy Release and Germination

GERHARD LEUBNER-METZGER

Institut für Biologie II (Botanik), Albert-Ludwigs-Universität, Schänzlestr. 1, D-79104 Freiburg i. Br., Germany

Introduction

Seed germination of species with 'coat-imposed' dormancy is determined by the balance of forces between the growth potential of the embryo and the constraint exerted by the covering layers, e.g. testa (seed coat) and endosperm. Little is known about the key interconnected molecular processes regulating seed dormancy and germination in response to plant hormones and environmental cues. Seed dormancy can be coat-imposed and/or determined by the embryo itself and is a temporary failure or block of a viable seed to complete germination under physical conditions that normally favour the process (Hilhorst, 1995; Bewley, 1997b; Koornneef *et al.*, 2002). The testa is no hindrance during the germination of some species, such as *Brassica napus* (Schopfer and Plachy, 1984) and *Pisum sativum* (Petruzzelli *et al.*, 2000). The process of germination commences with the uptake of water by imbibition of the dry seed, followed by embryo expansive growth. This usually culminates in rupture of the covering layers and emergence of the radicle, generally considered as the completion of germination. Radicle protrusion during seed germination depends on embryo expansion, which is a growth process driven by water uptake.

Abscisic acid (ABA) is known as a positive regulator of dormancy and a negative regulator of seed germination (Hilhorst, 1995; Bewley, 1997b; Koornneef *et al.*, 2002). ABA treatment of non-endospermic, non-dormant *B. napus* seeds has no effect on the kinetics of testa rupture, but it inhibits the post-germinational extension growth of the radicle (Schopfer and Plachy, 1984). Thus, ABA does not inhibit initial imbibition of water (water uptake phases 1 and 2) needed for initial embryo extension growth. ABA inhibits the transition to the seedling growth phase (water uptake phase 3) after radicle emergence (Lopez-Molina *et al.*, 2001). In agreement with this, ABA treatment does not inhibit germination scored as initial radicle extension growth

 The Biology of Seeds: Recent Research Advances (eds G. Nicolás, K.J. Bradford, D. Côme and H.W. Pritchard)

of detipped (surgical removal of the micropylar layers covering the radicle tip) seeds of tomato (*Lycopersicon esculentum*) (e.g. Liptay and Schopfer, 1983; Groot and Karssen, 1992; Bewley, 1997a). Even 1 mM ABA does not inhibit the germination of detipped tomato seeds, whereas 100 μM ABA results in a substantial inhibition of intact seeds (Liptay and Schopfer, 1983). Detipping can also replace the requirement for treatment with gibberellin (GA), a positive regulator of dormancy release and germination, of GA-deficient *gib1* mutant seeds of tomato (Groot and Karssen, 1992). It is therefore the micropylar testa and endosperm tissue of tomato seeds, also termed the micropylar cap, that confers the primary control of germination timing.

In many species the seed envelope imposes a physical constraint to radicle protrusion and, depending on the species, testa, endosperm, perisperm, hull, glumella or the megagametophyte can confer coat-imposed dormancy (e.g. Downie and Bewley, 1996; Benech-Arnold *et al.*, 1999). The main focus of this presentation is on testa- and endosperm-imposed dormancy of dicot seeds and on the interactions among ABA, GA, ethylene and brassinosteroids (BR) in regulating the key interconnected molecular processes that determine dormancy and germination.

Testa-imposed Dormancy is Regulated by ABA–GA Interaction

In non-endospermic seeds, as well as in *Arabidopsis* (which only has a single layer of endosperm), the testa characteristics are responsible for the degree of coat-imposed dormancy (Debeaujon and Koornneef, 2000; Debeaujon *et al.*, 2000). The testa is a diploid and entirely maternal covering tissue that develops from the integuments of the ovule. The testa layers of *Arabidopsis*, cells of which die during late seed maturation, undergo considerable developmental changes that are manifested in colour, permeability and testa-imposed dormancy. A peak in ABA biosynthesis during seed development is needed for the induction of primary dormancy of *Arabidopsis* seeds (Koornneef and Karssen, 1994). Only embryonic ABA, but not maternal or exogenously applied ABA, is able to induce dormancy. ABA deficiency during seed development is associated with absence of primary dormancy of the mature seed. The formation of non-dormant seeds occurs in the ABA-deficient biosynthesis mutant *aba1* of *Arabidopsis* (Koornneef and Karssen, 1994). However, sensitivity of seeds to ABA is partially maternally controlled, and embryonic, maternal and applied ABA affect other aspects of *Arabidopsis* seed development.

According to the revised hormone-balance hypothesis for seed dormancy proposed by Karssen and Laçka (1986), ABA and GA act at different times and sites during 'seed life'. ABA induces dormancy during maturation and GA plays a key role in the promotion of germination. Seed germination of the GA-deficient biosynthesis mutant *ga1* of *Arabidopsis* depends on the addition of GA to the medium during imbibition (Koornneef and Karssen, 1994). Seed dormancy of *Arabidopsis* appears to be a quantitative trait and can be released by chilling or light treatment of imbibed seeds or by after-ripening, i.e. a period of dry storage at room temperature for several months.

Numerous *Arabidopsis* mutants that affect the development of the testa without impairing the viability of the seeds have been isolated (e.g. Debeaujon and Koornneef, 2000; Debeaujon *et al.*, 2000). In general, the alterations in the testa characteristics of these mutants caused decreased testa-imposed dormancy. Freshly harvested seeds of testa mutants germinate more readily and are more sensitive to dormancy-releasing treatments compared with wild-type seeds. Seed germination of testa mutants is more sensitive to GA treatment, and reciprocal crosses with wild type demonstrate that this effect is determined by the altered testa characteristics. Comparative studies that also include reciprocal crosses between testa and hormone-deficient mutants support the view that dormancy and germination are probably the net result of a balance between many promoting and inhibiting factors, including GA and ABA, that target the embryo and the testa. Debeaujon and Koornneef (2000) concluded that the GA requirement for dormancy release and germination is determined by: (i) ABA produced in the developing seeds and/or the state of dormancy set by ABA and (ii) ABA produced upon imbibition, especially in dormant seeds. When the restraint to radicle protrusion imposed by the seed envelopes is weakened by the testa mutations, the embryo growth potential threshold required for germination is lowered. Thus, the GA requirement for *Arabidopsis* seed germination is determined both by testa characteristics and by embryonic ABA. Therefore, testa-imposed dormancy of *Arabidopsis* appears to be regulated by an indirect ABA–GA interaction.

In endospermic seeds the contributions of both the testa and the endosperm layers to the degree of coat-imposed dormancy have to be considered (Hilhorst, 1995; Bewley, 1997a). The testa accounts for approximately 20% of the mechanical resistance during the early phase of tomato seed imbibition (Groot and Karssen, 1987) and the mechanical resistance of the testa appears to decrease only during the late phase, just prior to radicle protrusion. In agreement with this, a much thinner testa (one cell layer) of the ABA-deficient *sit^{w}* mutant compared with wild-type (four to five cell layers) tomato is correlated with faster seed germination of the mutant seeds (Hilhorst and Downie, 1995). Germination of intact *sit^{w}* seeds occurred at lower external osmotic potentials, and removal of the micropylar testa did not affect seed germination of the *sit^{w}* mutant but significantly promoted wild-type seed germination. Hilhorst and Downie (1995) concluded that, although the testa resistance is smaller compared with the endosperm resistance, it is the micropylar testa that finally controls the completion of tomato seed germination, i.e. radicle emergence. Interestingly, faster seed germination of the ABA-deficient *sit^{w}* mutant is also associated with significantly increased βGlu I expression compared with wild-type tomato (Leubner-Metzger, unpublished results). ABA-deficiency causes altered testa characteristics during seed maturation, i.e. altered testa-imposed dormancy, whereas GA positively regulates tomato germination during seed imbibition. Thus, the ABA-deficient *sit^{w}* mutant is also a testa mutant and, as in *Arabidopsis*, testa-imposed dormancy of tomato appears to be regulated by an indirect ABA–GA interaction.

Testa rupture of *Nicotiana* seeds appears also to be regulated by an indirect ABA–GA interaction. In the mature tobacco seed, three to five layers of rather thick-walled endosperm cells surround the embryo. The periphery of the endosperm is pressed against the thin testa, which consists of an outer layer of cutinized and lignified dead cells and a living inner parenchyma layer (Leubner-Metzger, 2001). Rupture of the testa and the endosperm are distinct and temporally separate events during the germination of tobacco seeds (Arcila and Mohapatra, 1983; Leubner-Metzger *et al.*, 1995; online www.leubner.ch). Testa rupture starts near the funiculus and spreads in random directions along the ridges on the testa. Channels underlying the ridges facilitate progress of testa rupture. When seeds reach the advanced testa rupture stage, the micropylar endosperm covering the radicle tip is exposed as a dome-shaped structure. Microscopic studies showed that storage reserves are degraded in the micropylar endosperm cells prior to protrusion by the radicle; and that the endospermic hole, which has a smooth outline and is always formed at the micropylar end of germinating tobacco seeds, results from tissue dissolution rather than from the pushing action of the protruding radicle (Arcila and Mohapatra, 1983; Leubner-Metzger *et al.*, 1995). Surgical removal of the micropylar testa and the endosperm tissues permits radicle growth under conditions that inhibit germination of intact seeds of tobacco (Bihlmeier, 1927; Kincaid, 1935), demonstrating that regulation of germination by the micropylar covering layers is a common characteristic of solanaceous seeds (e.g. Liptay and Schopfer, 1983; Hilhorst, 1995; Bewley, 1997a). ABA deficiency during seed development of *Nicotiana* seeds is also associated with absence of primary dormancy of the mature seed. In the *aba2* mutant of *N. plumbaginifolia*, ABA deficiency is due to a mutation in the *ABA2* gene, encoding zeaxanthin epoxidase, a key step in ABA biosynthesis (Marin *et al.*, 1996). Antisense- and sense-*ABA2* transformation of *N. plumbaginifolia* resulted in decreased and increased ABA biosynthesis and seed dormancy, respectively (Frey *et al.*, 1999). The onset of dormancy in *Nicotiana tabacum* is correlated with a peak in ABA content at approximately 15–20 days after pollination (DAP); a rapid decline in ABA content follows during further seed maturation; and dormancy has been established when seeds are harvested after 25 DAP (Phillips *et al.*, 1997; Leubner-Metzger and Meins, 2000). Seed dormancy is not established and precocious germination occurs in transgenic tobacco expressing an anti-ABA antibody that causes deficiency in free ABA (Phillips *et al.*, 1997).

Dormancy of *Nicotiana* seeds can be released during after-ripening, i.e. a period of dry storage of freshly harvested, mature seeds (Grappin *et al.*, 2000; Leubner-Metzger and Meins, 2000, 2001). The work of Grappin *et al.* (2000) demonstrated that a further decline in ABA content and decreased sensitivity to ABA are involved in the after-ripening-mediated transition from the dormant to the non-dormant state of *N. plumbaginifolia*. In addition, *de novo* ABA biosynthesis occurs in imbibed fresh (dormant) seeds, but not in after-ripened (non-dormant) seeds. The after-ripening-mediated promotion of *N. tabacum* germination is due to the promotion of both testa and subsequent endosperm rupture (Leubner-Metzger and Meins, 2000, 2001). Addition of

ABA to the medium during imbibition resembles the effects of maternal ABA during seed development and residual ABA in mature seeds. Imbibition of fresh or after-ripened tobacco seeds in medium with 10 μM ABA greatly delays endosperm rupture, but does not affect the kinetics of testa rupture of fresh or after-ripened tobacco seeds.

Involvement of β-1,3-Glucanase in the After-ripening-mediated Promotion of Tobacco Testa and Endosperm Rupture

Class I β-1,3-glucanase (βGlu I) is transcriptionally induced in germinating tobacco seeds just prior to endosperm rupture but after testa rupture (Leubner-Metzger *et al.*, 1995, 1998). βGlu I induction is highly localized in the micropylar endosperm at the site of radicle emergence. Light, GA and ethylene promote βGlu I expression and endosperm rupture. ABA inhibits βGlu I expression and endosperm rupture of wild-type seeds and transgenic TCIB1 seeds, which originate from empty-vector-transformed tobacco lines (TCIB1) and serve as proper controls in sense and antisense experiments (Leubner-Metzger and Meins, 2000, 2001). A chimeric ABA-inducible βGlu I transgene was used for the transformation of tobacco and yielded independent sense-βGlu I lines (TKSG7). Sense-βGlu I transformation caused over-expression of βGlu I in TKSG7 seeds and promoted endosperm rupture of mature seeds and of ABA-treated after-ripened seeds.

Recently, we discovered that the after-ripening-mediated promotion of tobacco seed germination is mainly due to a promotion of testa rupture and a similar promotion of subsequent endosperm rupture (Leubner-Metzger, 2002a). Furthermore, over-expression of βGlu I in sense-βGlu I-transformed TKSG7 seeds replaces the after-ripening effects and promotes testa rupture and endosperm rupture of fresh TKSG7 seeds. Reciprocal crosses between wild-type tobacco and sense-βGlu I transformant lines showed that βGlu I over-expression in the seed covering layers can replace the promoting effect of after-ripening on testa rupture and endosperm rupture, but only if the mother plant is a sense-βGlu I line (Fig. 12.1). This maternal effect supports a model of two sites for βGlu I action: (i) βGlu I contribution to the after-ripening-mediated release of dormancy in the dry seed state, which is manifested in the promotion and ABA-insensitivity of testa rupture during imbibition (Fig. 12.2); and (ii) ABA-sensitive expression of βGlu I in the micropylar endosperm, which contributes to endosperm rupture (Fig. 12.2). In contrast to endosperm rupture, which seems to be caused by enzymatic degradation of the micropylar endosperm tissue during radicle extension, testa rupture appears to be achieved by a different mechanism that is characterized by a spread of increasing cracks in random directions along the ridges of the testa. In agreement with a role of βGlu I during after-ripening, a delay in testa rupture and a similar delay of subsequent endosperm rupture was found in after-ripened antisense-βGlu I seeds (Leubner-Metzger and Meins, 2001). Fresh TCIB1 and antisense-βGlu I seeds do not differ in their kinetics of testa rupture. The delay in testa rupture of after-ripened antisense-βGlu I seeds must be established during after-ripening of the dry,

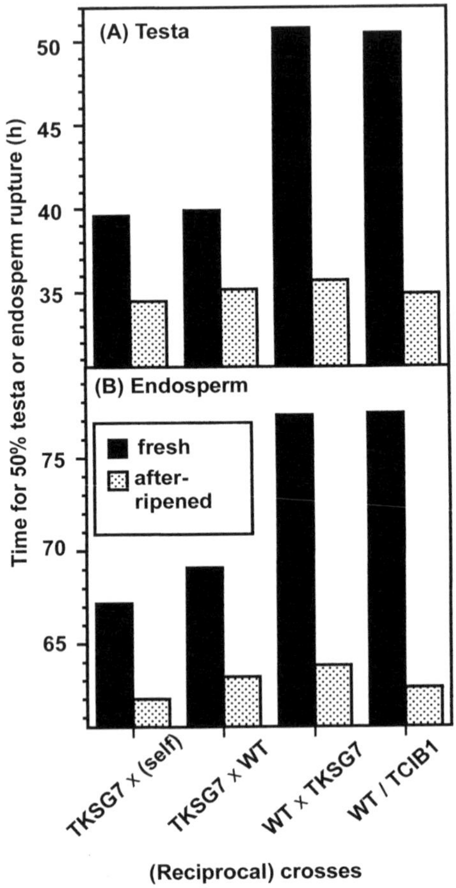

Fig. 12.1. The effect of after-ripening and sense-βGlu I transformation on (A) testa rupture and (B) endosperm rupture during the germination of tobacco seeds. Fresh or after-ripened progeny seeds from reciprocal crosses of a homozygous monogenic sense-βGlu I line (TKSG7) or an empty-vector (TCIB1) line with wild-type (WT) were compared. The female parent in the cross is on the left and the male parent is not shown in self-crosses. The incidence of testa and endosperm rupture expressed as percentage was scored over time from the start of imbibition in continuous light and the time needed for 50% rupture was determined. Mean values presented are based on the results from three independent TKSG7 lines; mean values ± SE for each single TKSG7 line and further details are published in Leubner-Metzger (2002a).

mature seed. These findings support the view that βGlu I is expressed and susceptible to antisense inhibition in antisense-βGlu I seeds. The after-ripening-mediated release of tobacco dormancy is also correlated with a decrease in GA requirement for testa rupture during dark imbibition (Leubner-Metzger, 2002a). Thus, as in *Arabidopsis* and tomato, testa-imposed dormancy of tobacco appears to be regulated by an indirect ABA-GA interaction. The importance of testa characteristics appears to be a common feature during

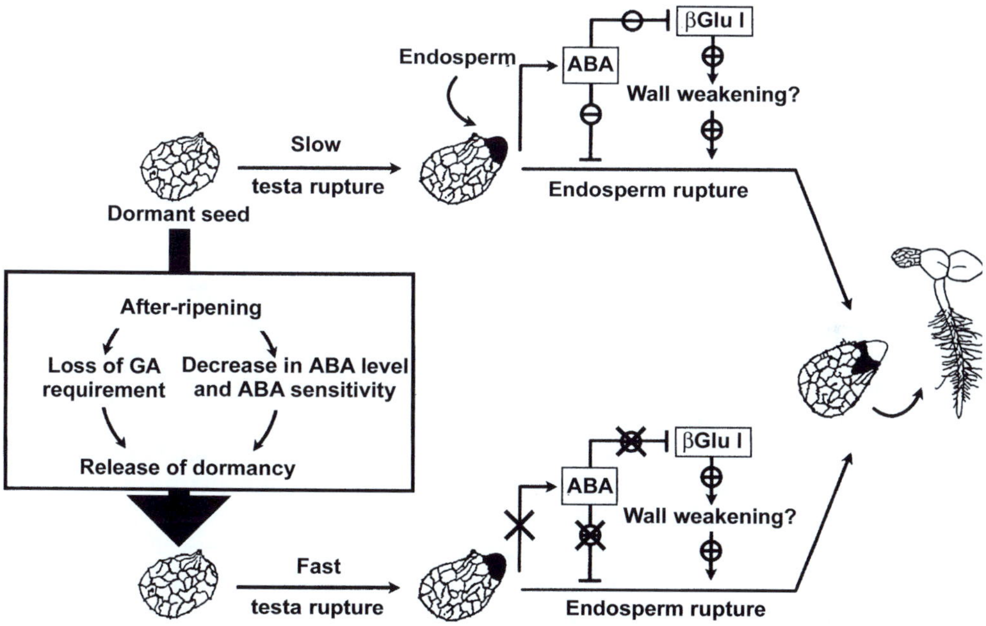

Fig. 12.2. A speculative model relating after-ripening, βGlu I and endosperm rupture of tobacco. According to the model, βGlu I, which is transcriptionally down-regulated by ABA, contributes to endosperm wall weakening and breaking of coat-imposed dormancy. ABA accumulated during seed maturation is sufficient to delay βGlu I induction and endosperm rupture. After-ripening decreases ABA levels and possibly sensitivity to ABA, allowing the induction of βGlu I during imbibition, which helps to modulate coat-imposed dormancy. After-ripening promoted testa rupture of seeds imbibed in the light and is also correlated with loss of GA requirement for dark-germination. The model does not exclude the possibility that other factors contribute to after-ripening and modulation of coat-imposed dormancy, or that ABA has additional functions, e.g. in photodormancy. (Modified from G. Leubner-Metzger and F. Meins, 2000; © Blackwell Publishing, reprinted with permission.)

the after-ripening-mediated release of coat-enhanced dormancy in endospermic and non-endospermic seeds.

Endosperm-imposed Dormancy is Regulated by Antagonistic Interactions of GA, Ethylene and BR with ABA

In addition to the testa, in many endospermic species the usually triploid (two-thirds of its genome originates from the mother plant) micropylar endosperm confers coat-imposed dormancy. Thus, in endospermic seeds the contributions of both the testa and the endosperm layers to the degree of coat-imposed dormancy have to be considered (Hilhorst, 1995; Hilhorst and Downie, 1995; Bewley, 1997a; Nonogaki *et al.*, 2000; Toorop *et al.*, 2000; Wu *et al.*, 2000). Endosperm rupture is the main germination-limiting process in members of the *Asteraceae* (e.g. lettuce) and *Solanaceae* (e.g. tomato and tobacco). In these cases of endosperm-limited germination, weakening of the micropylar endosperm surrounding the radicle tip seems to be required for radicle protrusion and is likely to involve cell wall hydrolysis by the action of GA-induced hydrolytic enzymes. It is possible, but not proven, that tomato endosperm weakening is a biphasic process and only the second phase is inhibited by ABA (Hilhorst and Downie, 1995; Bewley, 1997a; Nonogaki *et al.*, 2000; Toorop *et al.*, 2000; Wu *et al.*, 2000). βGlu I is induced by GA just prior to tomato endosperm rupture and is inhibited by ABA, which also inhibits germination (Wu *et al.*, 2000). The close correlation between βGlu I induction and the onset of endosperm rupture under a variety of physiological conditions support the hypothesis that βGlu I contributes to endosperm rupture of tobacco. ABA inhibits the induction of the βGlu I genes, specifically delays endosperm rupture (Fig. 12.3) and results in the formation of a novel structure, consisting of the enlarging radicle with a sheath of greatly elongated endosperm tissue (Leubner-Metzger *et al.*, 1995). Direct evidence for a causal role of βGlu I during endosperm rupture comes from sense transformation with a chimeric ABA-inducible βGlu I transgene (Leubner-Metzger and Meins, 2000). ABA down-regulates the βGlu I host genes in TCIB1 and wild-type seeds, but due to the ABA-inducible βGlu I-transgene it causes high-level βGlu I expression in TKSG7 seeds. ABA treatment delays endosperm rupture of after-ripened TCIB1 and TKSG7 seeds, but due to the sense-βGlu I transformation this delay is significantly reduced in TKSG7 seeds. βGlu I over-expression reduces the ABA-mediated delay in endosperm rupture of fresh and after-ripened seeds. These results support the view that a threshold βGlu I content is required, but not sufficient, for endosperm rupture. In the presence of ABA βGlu I becomes a limiting factor for endosperm rupture, and removal of this block due to expression of the ABA-inducible βGlu I-transgene in TKSG7 seeds promotes endosperm rupture until other ABA-sensitive processes become limiting. While these results do not show how βGlu I promotes endosperm rupture, they directly show that βGlu I is causally involved and that it substantially contributes to endosperm rupture.

Tobacco germination is accompanied by ethylene evolution and endogenous ethylene is required for the promotion of endosperm rupture and high-level βGlu I expression of light-imbibed seeds (Fig. 12.3) and of non-photodormant dark-imbibed seeds (Leubner-Metzger *et al.*, 1998). Ethylene does not affect the spatial and temporal pattern of βGlu I expression and does not break photodormancy or affect the kinetics of testa rupture. A promoter deletion analysis of a tobacco βGlu I gene in germinating tobacco seeds suggests that the distal region, which contains the positively acting ethylene-responsive element (ERE), is required for high-level, ethylene-sensitive expression; that the proximal region is necessary and sufficient for low-level micropylar-endosperm specific expression; and, that both regions contribute to down-regulation by ABA (Leubner-Metzger *et al.*, 1998). These promoter regions contain several highly conserved *cis*-acting elements for regulation by tissue-specific factors, GA, ABA and ethylene (Leubner-Metzger *et al.*, 1998; Leubner-Metzger, 2001). Enhancer activity and ethylene responsiveness of βGlu I depend on the AGCCGCC box present as two copies in the ERE. They are the binding site of ERE binding proteins (EREBPs), which are transcription factors mediating ethylene responses. Transcripts of the EREBPs showed a novel pattern of expression during tobacco seed germination (Leubner-Metzger *et al.*, 1998). A direct antagonistic interaction of ethylene and GA with ABA in regulating βGlu I accumulation in the micropylar endosperm and endosperm rupture (Fig. 12.3) may therefore be mediated, at least in part, by the EREBP-type transcription factors.

Finally, brassinosteroids (BR) and GA seem to promote tobacco (Fig. 12.3) and *Arabidopsis* seed germination by distinct signal transduction pathways and distinct mechanisms (Leubner-Metzger, 2001; Steber and McCourt, 2001). GA and light act in a common pathway to release tobacco photodormancy, whereas BR does not release photodormancy. βGlu I induction in the micropylar endosperm and release of coat-imposed dormancy seem to be associated with the GA/light pathway but not with BR sig-

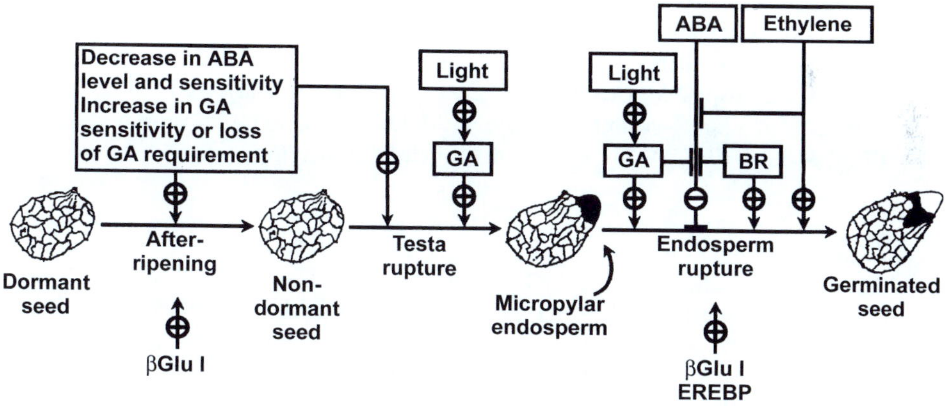

Fig. 12.3. Hormonal interactions during tobacco seed after-ripening, dormancy release and germination and their effects on testa rupture and endosperm rupture. Expression of the abscisic acid (ABA)-inhibited βGlu I genes contribute to the release of coat-imposed dormancy and the promotion of germination by acting at two sites. First, decreases in ABA level and sensitivity eventually permit βGlu I expression in seeds during after-ripening. This βGlu I contributes to the release of coat-imposed dormancy and promotes testa rupture in the light. Second, βGlu I is induced by the light/gibberellin (GA) pathway in the micropylar endosperm and facilitates endosperm rupture. Endosperm-specific βGlu I expression and endosperm rupture are inhibited by ABA and promoted by light, GA and ethylene. The light/GA pathway also counteracts ABA effects by promoting ABA degradation. Ethylene and brassinosteroids (BR) counteract ABA effects and promote endosperm rupture, but do not affect testa rupture. EREBPs (ethylene-responsive element binding proteins) are transcription factors that mediate hormonal regulation of βGlu I expression and endosperm rupture. BR and light/GA promote tobacco endosperm rupture by distinct signal transduction pathways. A 'plus' sign means promotion and a 'minus' sign inhibition of a process. (After Leubner-Metzger, 2002b, reprinted with permission.)

nalling. These findings suggest a model for the endosperm-limited germination of tobacco:

1. Photodormancy is released exclusively by the GA/light-pathway.
2. Promotion of subsequent endosperm rupture by the BR and the GA/light signal transduction pathways is achieved by independent and distinct mechanisms.
3. Ethylene promotes endosperm rupture by enhancing βGlu I expression.
4. ABA inhibits endosperm rupture by interfering with these three hormones.
5. The GA/light pathway and ethylene regulate βGlu I induction in the micropylar endosperm and seem to control endosperm weakening.
6. The BR pathway seems to promote endosperm rupture of non-dormant seeds by directly enhancing the growth potential of the embryo.

Acknowledgements

My research is supported by a grant from the Deutsche Forschungsgemeinschaft (LE 720/3), which is gratefully acknowledged. Permissions for the modified reprints of Fig. 12.2 (© 2000, *The Plant Journal* 23, 215–221, Blackwell Publishing, Oxford, UK) and Fig. 12.3 (© 2002, 'Hormonal interactions during seed dormancy release and germination' by G. Leubner-Metzger, in: *Handbook of Seed Science*, A. Basra (ed.), The Haworth Press, Inc., Binghamton, NY, USA) are also gratefully acknowledged.

References

Arcila, J. and Mohapatra, S.C. (1983) Development of tobacco seedling. 2. Morphogenesis during radicle protrusion. *Tobacco Science* 27, 35–40.

Benech-Arnold, R.L., Giallorenzi, M.C., Frank, J. and Rodriguez, V. (1999) Termination of hull-imposed dormancy in developing barley grains is correlated with changes in embryonic ABA levels and sensitivity. *Seed Science Research* 9, 39–47.

Bewley, J.D. (1997a) Breaking down the walls – a role for endo-β-mannanase in release from seed dormancy? *Trends in Plant Science* 2, 464–469.

Bewley, J.D. (1997b) Seed germination and dormancy. *The Plant Cell* 9, 1055–1066.

Bihlmeier, M. (1927) Der Einfluss der Vorquellung und der Samenschale auf die Keimung lichtgeförderter Samen. *Jahrbücher für wissenschaftliche Botanik* 67, 702–732.

Debeaujon, I. and Koornneef, M. (2000) Gibberellin requirement for *Arabidopsis* seed germination is determined both by testa characteristics and embryonic abscisic acid. *Plant Physiology* 122, 415–424.

Debeaujon, I., Léon-Kloosterziel, K.M. and Koornneef, M. (2000) Influence of the testa on seed dormancy, germination, and longevity in *Arabidopsis*. *Plant Physiology* 122, 403–413.

Downie, B. and Bewley, J.D. (1996) Dormancy in white spruce (*Picea glauca* [Moench.] Voss.) seeds is imposed by tissues surrounding the embryo. *Seed Science Research* 6, 9–15.

Frey, A., Audran, C., Marin, E., Sotta, B. and Marion-Poll, A. (1999) Engineering seed dormancy by the modification of zeaxanthin epoxidase gene expression. *Plant Molecular Biology* 39, 1267–1274.

Grappin, P., Bouinot, D., Sotta, B., Miginiac, E. and Jullien, M. (2000) Control of seed dormancy in *Nicotiana plumbaginifolia*: post-imbibition abscisic acid synthesis imposes dormancy maintenance. *Planta* 210, 279–285.

Groot, S.P.C. and Karssen, C.M. (1987) Gibberellins regulate seed germination in tomato by endosperm weakening: a study with gibberellin-deficient mutants. *Planta* 171, 525–531.

Groot, S.P.C. and Karssen, C.M. (1992) Dormancy and germination of abscisic acid-deficient tomato seeds. *Plant Physiology* 99, 952–958.

Hilhorst, H.W.M. (1995) A critical update on seed dormancy. I. Primary dormancy. *Seed Science Research* 5, 61–73.

Hilhorst, H.W.M. and Downie, B. (1995) Primary dormancy in tomato (*Lycopersicon esculentum* cv. Moneymaker): studies with the *sitiens* mutant. *Journal of Experimental Botany* 47, 89–97.

Karssen, C.M. and Laçka, E. (1986) A revision of the hormone balance theory of seed dormancy: studies on gibberellin and/or abscisic acid-deficient mutants of *Arabidopsis thaliana*. In: Bopp, M. (ed.) *Plant Growth Substances 1985*. Springer-Verlag, Berlin, pp. 315–323.

Kincaid, R.R. (1935) The effects of certain environmental factors on the germination of Florida cigar-wrapper tobacco seeds. *Technical Bulletin University of Florida, Agricultural Experiment Stations* 277, 5–47.

Koornneef, M. and Karssen, C.M. (1994) Seed dormancy and germination. In: Meyerowitz, E.M. and Somerville, C.R. (eds) *Arabidopsis*. Cold Spring Harbor Laboratory Press, New York, pp. 313–334.

Koornneef, M., Bentsink, L. and Hilhorst, H. (2002) Seed dormancy and germination. *Current Opinion in Plant Biology* 5, 33–36.

Leubner-Metzger, G. (2001) Brassinosteroids and gibberellins promote tobacco seed germination by distinct pathways. *Planta* 213, 758–763.

Leubner-Metzger, G. (2002a) Seed after-ripening and over-expression of class I β-1,3-glucanase confer maternal effects on tobacco testa rupture and dormancy release. *Planta* 215, 659–698. (Online first: DOI 10.1007/s00425-002-0837-y).

Leubner-Metzger, G. (2002b) Hormonal interactions during seed dormancy release and germination. In: Basra, A. (ed.) *Handbook of Seed Science*. The Haworth Press, Inc., Binghampton, New York.

Leubner-Metzger, G. and Meins, F. Jr (2000) Sense transformation reveals a novel role for class I β-1,3-glucanase in tobacco seed germination. *The Plant Journal* 23, 215–221.

Leubner-Metzger, G. and Meins, F. Jr (2001) Antisense-transformation reveals novel roles for class I β-1,3-glucanase in tobacco seed afterripening and photodormancy. *Journal of Experimental Botany* 52, 1753–1759.

Leubner-Metzger, G., Fründt, C., Vögeli-Lange, R. and Meins, F. Jr (1995) Class I β-1,3-glucanase in the endosperm of tobacco during germination. *Plant Physiology* 109, 751–759.

Leubner-Metzger, G., Petruzzelli, L., Waldvogel, R., Vögeli-Lange, R. and Meins, F. Jr (1998) Ethylene-responsive element binding protein (EREBP) expression and the transcriptional regulation of class I β-1,3-glucanase during tobacco seed germination. *Plant Molecular Biology* 38, 785–795.

Liptay, A. and Schopfer, P. (1983) Effect of water stress, seed coat restraint, and abscisic acid upon different germination capabilities of two tomato lines at low temperature. *Plant Physiology* 73, 935–938.

Lopez-Molina, L., Mongrand, S. and Chua, N.-H. (2001) A postgermination developmental arrest checkpoint is mediated by abscisic acid and requires ABI5 transcription factor in *Arabidopsis. Proceedings of the National Academy of Sciences USA* 98, 4782–4787.

Marin, E., Nussaume, L., Quesada, A., Gonneau, M., Sotta, B., Hugueney, P., Frey, A. and Marion-Poll, A. (1996) Molecular identification of zeaxanthin epoxidase of *Nicotiana plumbaginifolia*, a gene involved in abscisic acid biosynthesis and corresponding to the ABA locus of *Arabidopsis thaliana. European Molecular Biology Organization Journal* 15, 2331–2342.

Nonogaki, H., Gee, O.H. and Bradford, K.J. (2000) A germination-specific endo-β-mannanase gene is expressed in the micropylar endosperm cap of tomato seeds. *Plant Physiology* 123, 1235–1245.

Petruzzelli, L., Coraggio, I. and Leubner-Metzger, G. (2000) Ethylene promotes ethylene biosynthesis during pea seed germination by positive feedback regulation of 1-aminocyclopropane-1-carboxylic acid oxidase. *Planta* 211, 144–149.

Phillips, J., Artsaenko, O., Fiedler, U., Horstmann, C., Mock, H.P., Müntz, K. and Conrad, U. (1997) Seed-specific immunomodulation of abscisic acid activity induces a developmental switch. *European Molecular Biology Organization Journal* 16, 4489–4496.

Schopfer, P. and Plachy, C. (1984) Control of seed germination by abscisic acid. II. Effect on embryo water uptake in *Brassica napus* L. *Plant Physiology* 76, 155–160.

Steber, C.M. and McCourt, P. (2001) A role for brassinosteroids in germination in Arabidopsis. *Plant Physiology* 125, 763–769.

Toorop, P.E., van Aelst, A.C. and Hilhorst, H.W.M. (2000) The second step of the biphasic endosperm cap weakening that mediates tomato (*Lycopersicon esculentum*) seed germination is under control of ABA. *Journal of Experimental Botany* 51, 1371–1379.

Wu, C.-T., Leubner-Metzger, G., Meins, F. Jr and Bradford, K.J. (2000) Class I β-1,3-glucanase and chitinase are expressed in the micropylar endosperm of tomato seeds prior to radicle emergence. *Plant Physiology* 126, 1299–1313.

13 Involvement of Energy Metabolism and ABA in Primary and Secondary Dormancies in Oat (*Avena sativa* L.) Seeds – a Physiological Approach

FRANÇOISE CORBINEAU AND DANIEL CÔME

Physiologie Végétale Appliquée, Université Pierre et Marie Curie, Tour 53, 1er étage, 4 place Jussieu, 75252 Paris cedex 05, France

Introduction

At harvest time, seeds of cereals originating from temperate climates, such as barley, oat and wheat, show a primary dormancy which corresponds to an inability to germinate at temperatures above 15–20°C (Côme *et al.*, 1984; Corbineau and Côme, 2000). Dry storage in ambient conditions progressively breaks this dormancy, allowing germination at temperatures up to 30–35°C (Côme *et al.*, 1984; Simpson, 1990).

A common occurrence in dormant seeds of many species is secondary dormancy which develops under conditions unfavourable for germination. Although secondary dormancy has been reported in domestic oat (Corbineau *et al.*, 1993), it is not as common a phenomenon as it is in closely related grasses (Simpson, 1990).

Numerous studies, mainly from 1975 to 1985, have been devoted to the metabolic regulation of cereal seed dormancy, but nothing is really clear in this field (Roberts and Smith, 1977; Côme and Corbineau, 1990). Enhancement of germination of dormant seeds by respiratory inhibitors has suggested that the cyanide-insensitive pathway or, more probably, the pentose phosphate pathway might play a role in the regulation of the germination of dormant cereal seeds. Stimulation of glycolysis through an increase in fructose 2,6-bisphosphate (Fru-2,6-P_2) (Larondelle *et al.*, 1987; Cohn and Footitt, 1993) might also be involved.

Abscisic acid (ABA) is generally considered to be involved in the onset of dormancy during seed development (Karssen, 1995) but there exists no clear relationship between the ABA content of mature dry seeds and their dormancy depth in cereals (Walker-Simmons, 1987). In addition, ABA synthesis during imbibition seems to be associated with the inability of dormant seeds to germinate when placed at suboptimal temperatures (Wang *et al.*, 1995; Yoshioka *et al.*, 1998). However, the regulation of germination by ABA

 The Biology of Seeds: Recent Research Advances (eds G. Nicolás, K.J. Bradford, D. Côme and H.W. Pritchard)

must take into account not only its endogenous level but also the sensitivity of the seeds to this growth regulator. In fact, the sensitivity of the embryo to ABA is highly correlated with dormancy intensity in seeds of various species, including cereals such as wheat (Walker-Simmons, 1987; Corbineau *et al.*, 2000), barley (Wang *et al.*, 1995; Benech-Arnold *et al.*, 1999; Corbineau and Côme, 2000) and oat (Corbineau *et al.*, 1993).

The aim of the present chapter was to examine from a physiological point of view some aspects of the regulation of dormancy in cereal seeds as exemplified by oat (*Avena sativa* L.) seeds. In particular, we have investigated whether energy metabolism and ABA synthesis or embryo sensitivity to ABA are involved in the germination of dormant seeds and the induction of secondary dormancy.

Primary and Secondary Dormancies in Oat Seeds

At harvest time, oat seeds are considered to be dormant because they do not germinate, or germinate poorly, in environmental conditions that are apparently favourable. As in other cereals (Corbineau and Côme, 2000), oat seed primary dormancy is a relative phenomenon; it is hardly expressed at low temperatures (5–15°C) but it increases as the temperature rises (Corbineau *et al.*, 1986).

After harvest, oat seeds become progressively able to germinate at high temperatures during dry storage. This breaking of dormancy, often called dry postmaturation or after-ripening, is expressed by a widening of the temperature range compatible with good germination (Corbineau *et al.*, 1986).

Incubation of primary dormant oat seeds for periods up to 1–3 days at 30°C, a temperature at which they do not germinate, results in a loss of their ability to germinate subsequently at 20°C (Fig. 13.1A). This phenomenon can be considered as a reinforcement of primary dormancy, i.e. a secondary dormancy or a thermodormancy. In the example given in Fig. 13.1.A, only about 20% of seeds remained capable of germinating at 20°C after 1 day at 30°C. It is interesting to note that a very short treatment (2–4 h) at 30°C is sufficient to induce this secondary dormancy in 25–30% of the seeds. It must be noted also that this thermodormancy is progressively released during incubation of the seeds at 30°C for periods longer than 2–3 days (Fig. 13.1A). By incubating primary dormant seeds at 30°C on polyethylene glycol (PEG) solutions at various concentrations, it was shown that the inductive effect of a 2-day incubation at this temperature requires a moisture content of the embryo higher than about 40% (dry weight basis), and that this effect increases as the embryo moisture content increases up to 80% (Fig. 13.1B).

Studies performed with primary and secondary dormant oat seeds have shown that dormancy in this species is mainly due to an inhibition of germination by the grain-covering structures and perhaps the endosperm, since isolated embryos germinate easily in a wide range of temperatures (Corbineau *et al.*, 1986, 1993).

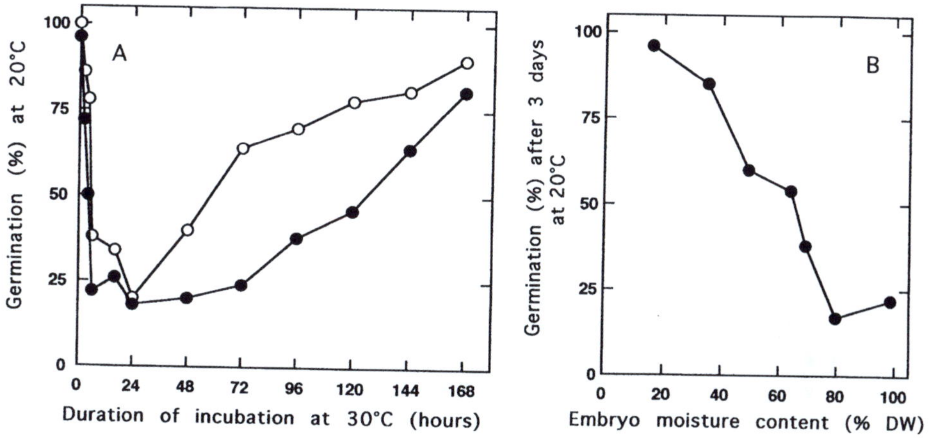

Fig. 13.1. Induction of secondary dormancy in oat seeds. (A) Effects of duration of pre-incubation of primary dormant seeds at 30°C in water on the percentages of germination recorded 3 (○) and 7 (●) days after their transfer to 20°C. (B) Effects of moisture content of the embryo of seeds pretreated for 2 days at 30°C in PEG solutions at various concentrations on germination 3 days after transfer to water at 20°C.

Metabolic Regulation of Primary and Secondary Dormancies in Oat Seeds

Metabolic events characterizing the germination of oat seeds

The inability of primary dormant seeds to germinate at high temperatures is not due to an inability of the embryo to synthesize ATP from ADP and AMP. During incubation at 30°C, embryos from primary dormant and non-dormant seeds exhibit exactly the same increase in ATP/ADP ratio and in energy charge (EC), expressed as (ATP + 0.5 ADP)/(ATP + ADP + AMP), as long as the non-dormant seeds have not yet germinated (Côme *et al.*, 1988).

It has been shown in *A. sativa* seeds that activation of glycolysis through Fru-2,6-P_2 might be involved in the germination process (Larondelle *et al.*, 1987). Fru-2,6-P_2 is a very potent stimulator of PP_i-fructose-6-phosphate 1-phosphotransferase (PP_i-PFK) and an inhibitor of fructose-1,6-bisphosphatase. Its accumulation therefore strongly stimulates glycolysis. In non-dormant seeds, the concentration of Fru-2,6-P_2 sharply increases until 16 h of incubation at 30°C, and then decreases during a period that roughly corresponds to radicle protrusion. In dormant seeds, however, which do not germinate at this high temperature, Fru-2,6-P_2 content increases during the first few hours of imbibition and then remains constant. Germination of dormant seeds at 10°C or at 30°C in the presence of ethanol is also associated with a rise in Fru-2,6-P_2 content (Larondelle *et al.*, 1987). Similar elevation of Fru-2,6-P_2 occurs in red rice placed in the presence of propionaldehyde that breaks dormancy (Cohn and Footitt, 1993).

Effects of respiratory inhibitors

Respiratory inhibitors (KCN, NaN_3) greatly stimulate the germination of dormant cereal seeds (Roberts and Smith, 1977), particularly that of oat seeds (Côme *et al.*, 1988). In the presence of these inhibitors, the EC value remains at about 0.55 (Corbineau, unpublished data).

Breaking of primary dormancy during dry after-ripening is ATP-independent, since the EC value in dry seeds remains around 0.2 (Côme *et al.*, 1988). In contrast, by incubating primary dormant seeds for 2 days at 30°C in polyethylene glycol (PEG) or NaN_3 solutions at various concentrations, it has been found that induction of thermodormancy requires that the EC of the embryo is at least about 0.6 (Fig. 13.2), i.e. that it is an active phenomenon. The dormancy inductive effect of incubation at 30°C increases as the embryo EC increases up to around 0.8, and there exists a linear relationship between the embryo EC value and the degree of seed thermodormancy (Fig. 13.2).

Various authors (Roberts and Smith, 1977; Côme and Corbineau, 1990) have considered that respiratory inhibitors stimulate germination through the cyanide insensitive pathway or more probably an activation of the pentose phosphate pathway. In oat, the cyanide insensitive pathway does not play a real role since salicylhydroxamic acid, an inhibitor of this pathway, does not inhibit germination of non-dormant seeds and does not alter the stimulatory effect of NaN_3 or KCN on germination of primary dormant seeds placed at 30°C (Côme *et al.*, 1988). Although the hypothesis implicating the pentose phosphate pathway is the most likely, it depends only on indirect arguments (Côme and Corbineau, 1990).

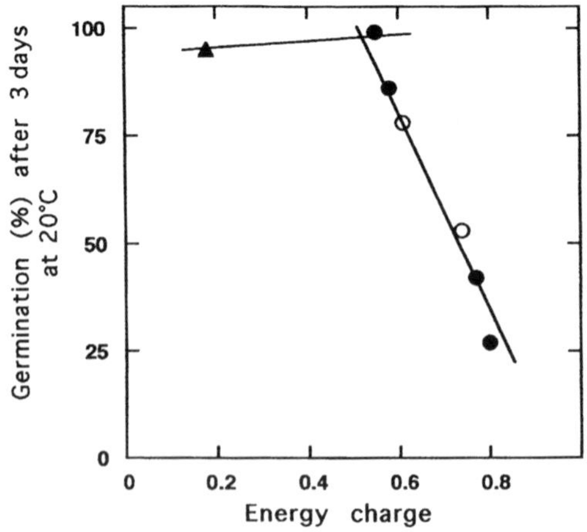

Fig. 13.2. Relationship between the energy charge of the embryo of primary dormant seeds incubated for 2 days at 30°C in PEG (o) or NaN_3 (●) solutions and subsequent seed germination after 3 days at 20°C; ▲, seeds maintained dry for 2 days at 30°C.

ABA and Expression of Primary and Secondary Dormancies in Oat Seeds

Involvement of ABA biosynthesis

Numerous studies using ABA-deficient and/or ABA-insensitive mutants, or performed with fluridone, an inhibitor of ABA biosynthesis, applied during seed development have demonstrated that this plant hormone is involved in the onset of primary dormancy during seed maturation (Karssen, 1995). Incubation of primary dormant oat seeds at 30°C in the presence of fluridone (0.1 mM) results in a lower embryo ABA content (Table 13.1) and a strong stimulation of germination at this high temperature (data not shown). Therefore, as in other cereals (Ried and Walker-Simmons, 1990; Wang *et al.*, 1995), ABA biosynthesis occurring during the first hours of imbibition is probably involved in the inability of dormant oat seeds to germinate at high temperatures. Fluridone also prevents the induction of thermodormancy (Table 13.1), thus indicating that ABA biosynthesis is required in this phenomenon. Interestingly, NaN_3, which allows germination of dormant oat seeds at 30°C and prevents the induction of secondary dormancy, results in a decrease in embryo ABA content (Table 13.1).

Exogenous ABA applied to primary dormant seeds for 2 days at 30°C induces a deeper secondary dormancy, i.e. reduces the germination percentages obtained after transfer of the seeds to water at 20°C (Table 13.1). When

Table 13.1. Effects of ABA, fluridone and NaN_3, applied alone or in combination for 2 days at 30°C, on ABA content of the embryo (expressed as percentage of ABA content in the embryo of dry untreated seeds) and on the subsequent germination percentages obtained after 3 and 7 days at 20°C (means ± SD of four replications of 25 seeds) for primary dormant oat seeds. Control seeds were preincubated on water for 2 days at 30°C.

Conditions of incubation at 30°C	Embryo ABA content (%)	Germination of seeds (%) at 20°C on water	
		After 3 days	After 7 days
Water	74.2	8 ± 2	44 ± 2
ABA (0.1 mM)	–	7 ± 3	23 ± 4
ABA (1 mM)	–	0	11 ± 2
Fluridone (0.1 mM)	11.2	95 ± 2	100
Fluridone (0.1 mM)			
+ ABA (0.1 mM)	–	84 ± 8	98 ± 1
+ ABA (0.5 mM)	–	62 ± 9	100
+ ABA (1 mM)	–	44 ± 6	94 ± 2
NaN_3 (1 mM)	32.8	88 ± 1	95 ± 1
NaN_3 (1 mM)			
+ ABA (0.1 mM)	–	92 + 4	98 ± 1
+ ABA (1 mM)	–	45 ± 9	96 ± 3

applied simultaneously with fluridone or NaN_3, exogenous ABA does not completely reverse the effects of the two compounds (Table 13.1). These results indicate that mainly endogenous ABA is involved in the regulation of germination of dormant oat seeds.

Sensitivity of the embryo to ABA as related to dormancy

As in many seeds, exogenous ABA strongly inhibits the germination of embryos excised from primary dormant oat caryopses, and its inhibitory effect increases with increasing concentration (Fig. 13.3). This inhibitory effect of ABA is temperature-dependent. In oat, it is less pronounced at 10°C than at 30°C (Corbineau *et al.*, 1991). The sensitivity of the embryo to ABA markedly decreases after the breaking of primary dormancy (Fig. 13.3). Involvement of the responsiveness of embryo to ABA in the regulation of seed dormancy has also been suggested for other cereals, such as wheat (Walker-Simmons, 1987; Corbineau *et al.*, 2000) and barley (Wang *et al.*, 1995; Benech-Arnold *et al.*, 1999; Corbineau and Côme, 2000), and other species (Karssen, 1995).

Embryos isolated from thermodormant seeds, i.e. primary dormant seeds incubated for 2–3 days at 30°C in water, are much more sensitive to ABA than those isolated from dry primary dormant seeds (Fig. 13.3). A longer incubation (7 days) at 30°C, which results in a release of thermodormancy, is associated with a decrease in sensitivity of the embryo to ABA (Fig. 13.3) and a decline in embryo ABA content (data not shown).

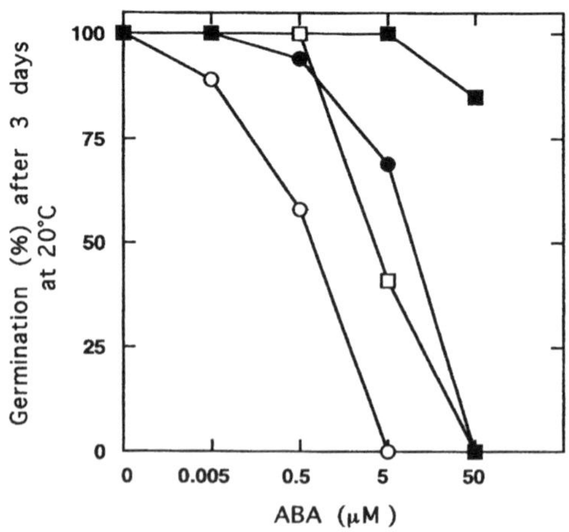

Fig. 13.3. Effects of ABA concentration on germination after 3 days at 20°C for embryos isolated from non-dormant seeds (■), primary dormant seeds (●) and primary dormant seeds preincubated in water at 30°C for 3 days (○) (thermodormant seeds) and 7 days (□) (seed thermodormancy has been released).

Conclusion

As with other cereals, primary dormancy of oat seeds corresponds to the inability of caryopses to germinate at high temperatures, and this inability originates mainly in the covering structures and perhaps also the endosperm. However, the embryo itself is involved in this phenomenon. In particular, embryos isolated from primary dormant seeds are much more sensitive to exogenous ABA and water potential of the germination medium than those isolated from non-dormant seeds. The reinforcement of dormancy by a short (1–3 days) incubation of primary dormant caryopses at 30°C suggests that dormancy is a physiological state maintained by dynamic processes. Interestingly, thermodormant seeds can revert to a physiological state similar to primary dormancy with prolonged incubation at 30°C.

Two main complementary approaches to the study of the regulation of germination and dormancy in cereal seeds have been developed. The first one is an investigation of the role played by some metabolic pathways, such as glycolysis and the pentose phosphate pathway, and the second one corresponds to the analysis of the possible involvement of ABA biosynthesis and/or embryo sensitivity to ABA. The stimulatory effect of fluridone on the germination of dormant seeds suggests that ABA biosynthesis during imbibition is involved in the lack of germination. In oat, ABA synthesis seems also to be required for induction of secondary dormancy. Expression of dormancy is also related to responsiveness of the embryo to ABA. This relation between embryo responsiveness to ABA and seed dormancy depth is probably related to changes in ABA perception and/or in the ABA signal transduction pathway in which protein phosphorylation through ABA-regulated protein kinases might play a key role (Walker-Simmons, 1998).

References

Benech-Arnold, R.L., Giallorenzi, M.C., Frank, J. and Rodriguez, V. (1999) Termination of hull-imposed dormancy in developing barley grains is correlated with changes in embryonic ABA levels and sensitivity. *Seed Science Research* 9, 39–47.

Cohn, M.A. and Footitt, S. (1993) Initial signal transduction steps during the dormancy breaking process. In: Côme, D. and Corbineau, F. (eds) *Fourth International Workshop on Seeds. Basic and Applied Aspects of Seed Biology.* ASFIS, Paris, pp. 599–605.

Côme, D. and Corbineau, F. (1990) Some aspects of metabolic regulation of seed germination and dormancy. In: Taylorson, R.B. (ed.) *Recent Advances in the Development and Germination of Seeds.* Plenum Press, New York, pp. 165–179.

Côme, D., Lenoir, C. and Corbineau, F. (1984) La dormance des céréales et son élimination. *Seed Science and Technology* 12, 629–640.

Côme, D., Corbineau, F. and Lecat, S. (1988) Some aspects of metabolic regulation of cereal seed germination and dormancy. *Seed Science and Technology* 16, 175–186.

Corbineau, F. and Côme, D. (2000) Dormancy of cereal seeds as related to embryo sensitivity to ABA and water potential. In: Viémont, J.D. and Crabbé J. (eds) *Dormancy in Plants: from Whole Plant Behaviour to Cellular Control.* CAB International, Wallingford, UK, pp. 183–194.

Corbineau, F., Lecat, S. and Côme, D. (1986) Dormancy of three cultivars of oat seeds (*Avena sativa* L.). *Seed Science and Technology* 14, 725–735.

Corbineau, F., Poljakoff-Mayber, A. and Côme, D. (1991) Responsiveness to abscisic acid of embryos of dormant oat (*Avena sativa*) seeds. Involvement of ABA-inducible proteins. *Physiologia Plantarum* 83, 1–6.

Corbineau, F., Black, M. and Côme, D. (1993) Induction of thermodormancy in *Avena sativa* seeds. *Seed Science Research* 3, 111–117.

Corbineau, F., Benamar, A. and Côme, D. (2000) Changes in sensitivity to abscisic acid of the developing and maturing embryo of two wheat cultivars with different sprouting susceptibility. *Israel Journal of Plant Sciences* 48, 189–197.

Karssen, C.M. (1995) Hormonal regulation of seed development, dormancy, and germination studied by genetic control. In: Kigel, J. and Galili, G. (eds) *Seed Development and Germination*. Marcel Dekker, New York, pp. 333–350.

Larondelle, Y., Corbineau, F., Dethier, M., Côme, D. and Hers, H.G. (1987) Fructose 2,6-bisphosphate in germinating oat seeds. A biochemical study of seed dormancy. *European Journal of Biochemistry* 166, 605–610.

Ried, J.L. and Walker-Simmons, M.K. (1990) Synthesis of abscisic acid-responsive, heat-stable proteins in embryonic axes of dormant wheat grain. *Plant Physiology* 93, 662–667.

Roberts, E.H. and Smith, R.D. (1977) Dormancy and the pentose phosphate pathway. In: Khan, A.A. (ed.) *The Physiology and Biochemistry of Seed Dormancy and Germination*. Elsevier, North-Holland Biochemical Press, Amsterdam, pp. 385–411.

Simpson, G.M. (1990) *Seed Dormancy in Grasses*. Cambridge University Press, Cambridge, UK, 297 pp.

Walker-Simmons, M. (1987) ABA levels and sensitivity in developing wheat embryos of sprouting resistant and susceptible cultivars. *Plant Physiology* 84, 61–66.

Walker-Simmons, M. (1998) Protein kinases in seeds. *Seed Science Research* 8, 193–200.

Wang, M., Heimovaara-Dijkstra, S. and Van Duijn, B. (1995) Modulation of germination of embryos isolated from dormant and non-dormant barley grains by manipulation of endogenous abscisic acid. *Planta* 195, 586–592.

Yoshioka, T., Endo, T. and Satoh, S. (1998) Restoration of seed germination at suboptimal temperatures by fluridone, an inhibitor of abscisic acid biosynthesis. *Plant Cell Physiology* 39, 307–312.

14 Mobilization of the Galactomannan-containing Cell Walls of Tomato Seeds: Where Does β-Mannosidase Fit into the Picture?

Beixin Mo and J. Derek Bewley

Department of Botany, University of Guelph, Guelph, Ontario N1G 2W1, Canada

Introduction

The major cell wall components of the endosperm that surround the embryo in seeds of tomato (*Lycopersicon esculentum* Mill.) are galactomannans (Groot *et al.*, 1988). Three enzymes are involved in their hydrolysis: endo-β-mannanase (EC 3.2.1.78), an endo-hydrolase that cleaves the mannan backbone; α-galactosidase (EC 3.2.1.22), which simultaneously removes the galactose side chains; and β-mannosidase (also called β-mannoside mannohydrolase or exo-β-mannanase, EC 3.2.1.25) which releases mannose from oligomers produced by prior endo-β-mannanase activity (McCleary, 1982).

Endo-β-mannanase activity has been studied extensively in the endosperms of tomato; it is synthesized in the micropylar endosperm prior to germination, and in the lateral endosperm following the completion of germination (Nomaguchi *et al.*, 1995; Toorop *et al.*, 1996). Two different isozymes of endo-β-mannanase have been identified in the endosperm: one exclusive to the micropylar region, produced during germination; and the other to the lateral region, produced post-germinatively. cDNA clones for each have been obtained and, using these, gene expression has been shown to follow the temporal pattern predicted by the appearance of the enzyme (Bewley *et al.*, 1997; Nonogaki *et al.*, 2000). There has been some debate as to the requirement for endo-β-mannanase to weaken the micropylar region of the endosperm to allow for radicle emergence (Bewley, 1997); the developing consensus is that it may be involved, but is not exclusively responsible

 The Biology of Seeds: Recent Research Advances (eds G. Nicolás, K.J. Bradford, D. Côme and H.W. Pritchard)

for this event (Bradford *et al.*, 2000). Its requirement for the mobilization of the cell walls of the lateral endosperm, perhaps as an early source of carbohydrate for the growing seedling, is less contentious.

α-Galactosidase is present in the dry tomato seed and retains a steady activity and amount of transcript throughout germination and early seedling growth in the embryo and endosperm (Feurtado *et al.*, 2001). It is synthesized during seed development and is sequestered in protein storage vacuoles (Bassel *et al.*, 2001; Feurtado *et al.*, 2001), although *de novo*-synthesized enzyme released into the cell wall, particularly following germination, might play a role in the removal of galactose from the galactomannan polymers.

The role of β-mannosidase in these events has not been studied in intact tomato seeds but it is present in other germinated seeds, e.g. as a soluble enzyme released from the aleurone layer of fenugreek (Reid and Meier, 1973), as a high-salt-soluble enzyme in the micropylar endosperm of *Datura ferox* (Sánchez *et al.*, 2002) and in association with the cell walls of the cotyledons of lettuce (Ouellette and Bewley, 1986). The following is a summary of our recent research on tomato seeds to purify this enzyme and to obtain its cDNA and genomic clones. This has allowed us to follow β-mannosidase activity and synthesis during and following germination, and to achieve a comparison with the activities of the other two major enzymes involved in tomato endosperm cell wall galactomannan mobilization.

β-Mannosidase Purification, and its cDNA and Genomic Clones

Purification of the enzyme was achieved from 72 h germinated tomato seeds (cv. Glamour) using a four-step process (Mo and Bewley, 2002) following high-salt (0.5 M NaCl) buffer extraction: 40–85% ammonium sulphate precipitation, anion-exchange chromatography (Whatman DE52), gel filtration (Bio-Gel 100) and chromatofocusing (Pharmacia LKB PBE 94 with pH 7–5 Polybuffer exchanger). This final step yielded a single protein using SDS-PAGE with a 0.3% recovery of enzyme activity and a 108-fold purification. Chromatofocusing showed the presence of two isoforms with very similar pIs; that of pI 6.8 was amino acid sequenced at the N-terminal end; in addition three internal tryptic peptides of the pI 6.3 form were sequenced (subsequent analyses showed that the two isoforms are identical in the regions sequenced). A Basic Local Alignment Search Tool (BLAST), National Center of Biotechnology Information (NCBI), National Institutes of Health (NIH) search of an Expressed Sequence Tag (EST) library for *Pseudomonas*-susceptible tomato lines (Cornell University) provided sequences identical to two of the peptides for an unknown protein derived from the 5′-end of a partial cDNA sequence. Based on these, sense and antisense primers were developed and a full-length cDNA clone for β-mannosidase (*LeMSide1*, GenBank accession AF403444) was obtained using PCR on RT-cDNA derived from a poly-A^+ RNA extracted from germinated tomato seed. This sequence has little homology with other animal, bacterial or fungal β-mannosidases, with the greatest amino acid similarity being 26% with the enzyme from the fungus *Pyrococcus furiosus*. A search of *Arabidopsis* cDNA sequences showed one with 76% identity (AB020749); this was incorrectly identified as a putative β-glucosidase.

Southern hybridization of genomic DNA cut with four restriction enzymes and probed with *LeMSide1* revealed a single gene for β-mannosidase in tomato. Using PCR and 3′- and 5′-end sequences of the *LeMSide1*

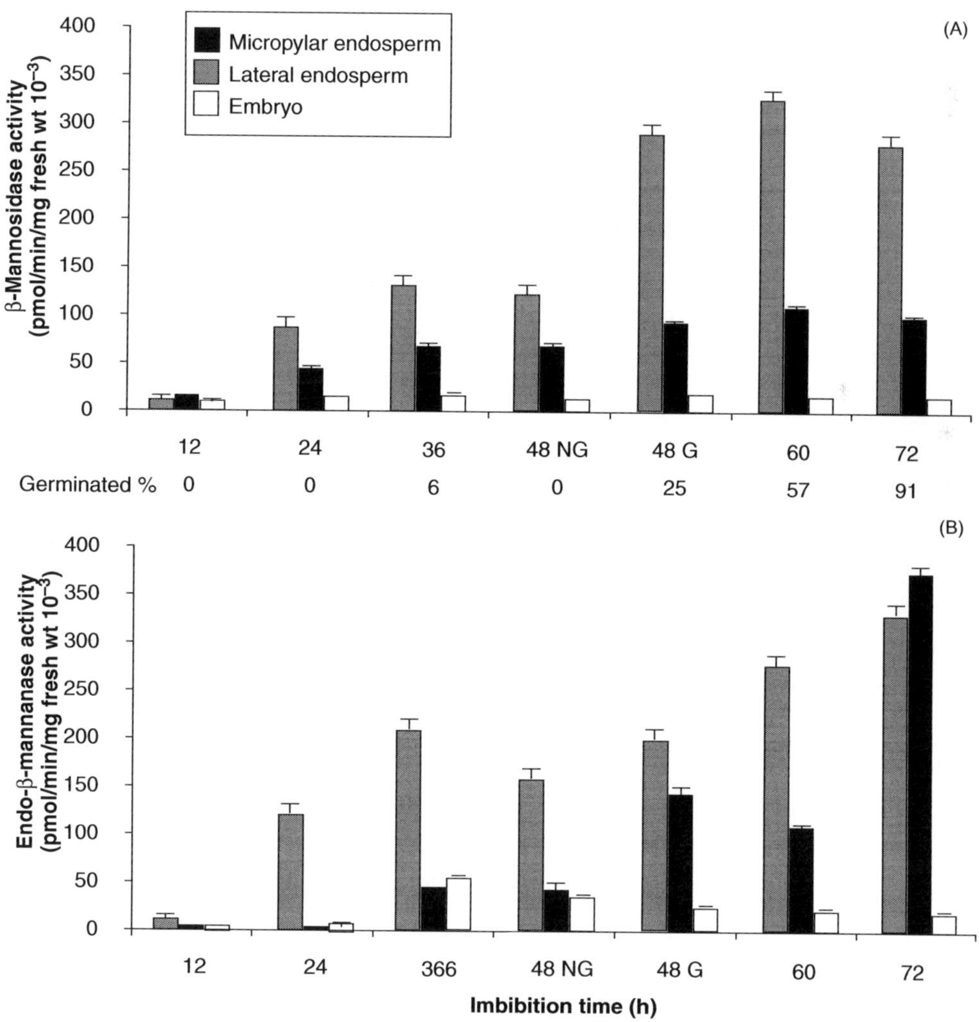

Fig. 14.1. (A) β-Mannosidase and (B) endo-β-mannanase activity in intact tomato seeds in relation to completion of germination (percentage shown below A). Activity of endo-β-mannanase was determined in low-salt-buffer extracts, in which it is totally soluble, and β-mannosidase in high-salt-buffer (0.5 M NaCl) extracts, which is necessary for its solubilization. At 48 h seeds were separated into germinated (G) and non-germinated (NG) and assayed separately. Error bars show variability between duplicate assays of duplicate extracts.

cDNA as primers, and genomic DNA as template, a 2856 bp genomic sequence for β-mannosidase was obtained (*LeMSide 2*, AF413204). The gene includes 1050 bp in ten introns, varying in length from 81 to 208 bp.

Activity and Gene Expression During and Following Germination of Tomato

A low amount of β-mannosidase activity is present in the dry seed but it increases in both the micropylar and lateral endosperm by 36 h, on a tissue fresh weight basis, at the time of radicle emergence in the first seeds to complete germination (Fig. 14.1A). Thereafter, there is a marked increase in the micropylar region as germination is completed and seedling growth ensues. This increase occurs in the micropylar endosperm punctured by the emerged radicle, and so is associated with either the remnant cells of this region, or the immediately adjacent lateral endosperm. A further increase in enzyme activity also occurs in the lateral endosperm after the completion of germination (Fig. 14.1A). Little activity is present in the embryo and this does not change with time. This pattern of β-mannosidase activity is similar to that of endo-β-mannanase (Fig. 14.1B). The substantial and coincidental increases in activity of the two enzymes, particularly within the micropylar endosperm, are indicative of their cooperation in cell wall galactomannan degradation.

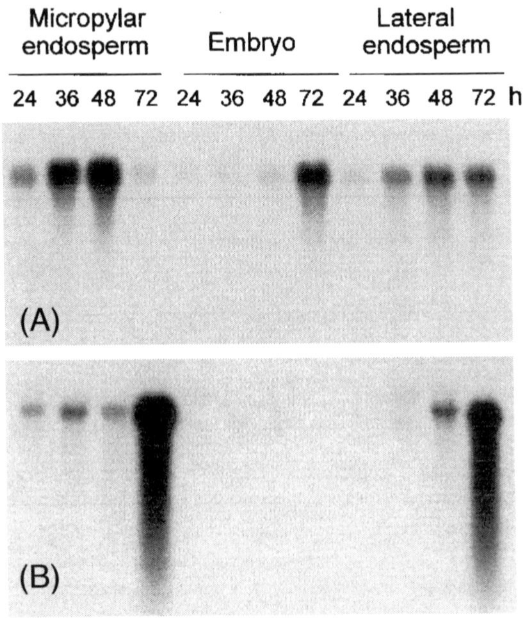

Fig. 14.2. Northern hybridization of total RNA from germinating and germinated tomato seed parts using cDNAs for (A) β-mannosidase (*LeMSide1*) and (B) endo-β-mannanase (*LeMan1*). The same membrane was used for both probes. Equal amounts of RNA were loaded in all lanes.

The genes for β-mannosidase and endo-β-mannanase are expressed in the micropylar endosperm during and after germination (Fig. 14.2), indicative of a coupling between their transcription and translation. There is a continued increase in β-mannosidase transcripts in the micropylar endosperm up until 48 h, and then a sharp decline; a plateau of enzyme activity of this enzyme is reached at this time also (Fig. 14.2A). Transcripts for endo-β-mannanase are abundantly produced in the micropylar endosperm at 72 h, even though this region is now a residual tissue (Fig. 14.2B). It is likely that these transcripts are from the adjacent lateral endosperm tissue which was excised along with the micropylar tissue.

In situ localization of expression of β-mannosidase in the tomato seed at 24 and 72 h was achieved using tissue blots hybridized to a full-length antisense cDNA to *LeMSide1*. At 24 h, prior to the completion of germination, expression of the β-mannosidase gene is low but more or less equally dispersed between the micropylar and lateral endosperm (Fig. 14.3A,B). In con-

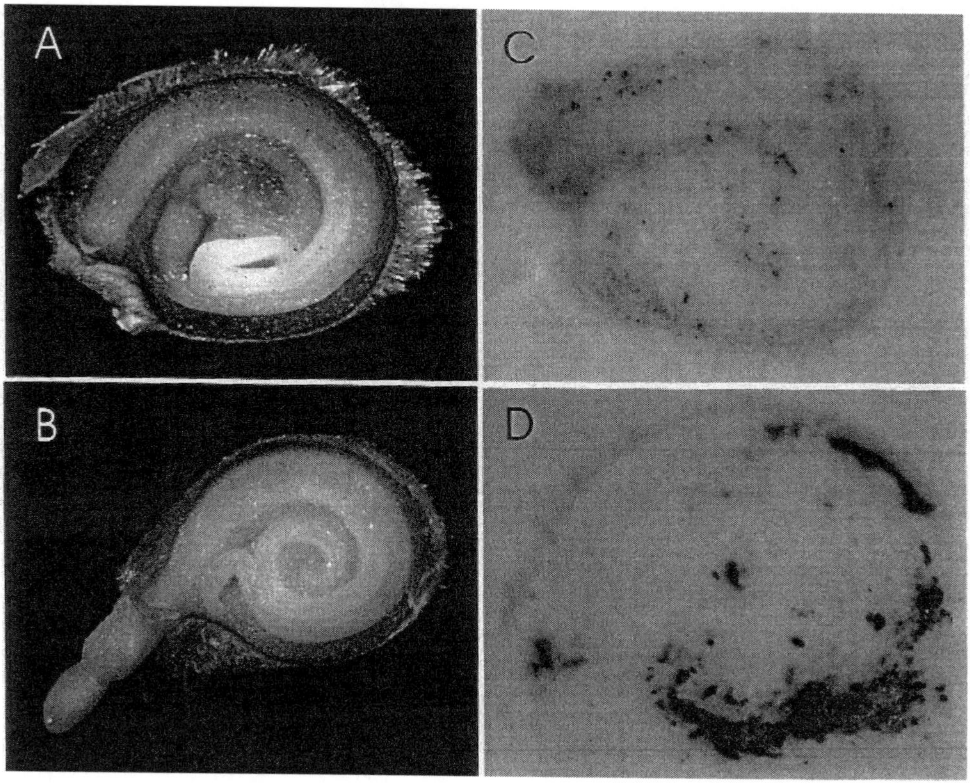

Fig. 14.3. *In situ* localization of β-mannosidase in tomato seeds at (A, B) 24 h and (C, D) 72 h after the start of imbibition. Tomato seeds were bisected and their cut surfaces pressed on to positively charged membrane (Hybond-N$^+$, Amersham Pharmacia Biotech). Linearized plasmid containing tomato seed β-mannosidase cDNA was used as antisense RNA template labelled with digoxygenin (DIG) (Roche). Location of hybridization was achieved colorimetrically using alkaline-phosphatase-conjugated anti-DIG antibody (Roche).

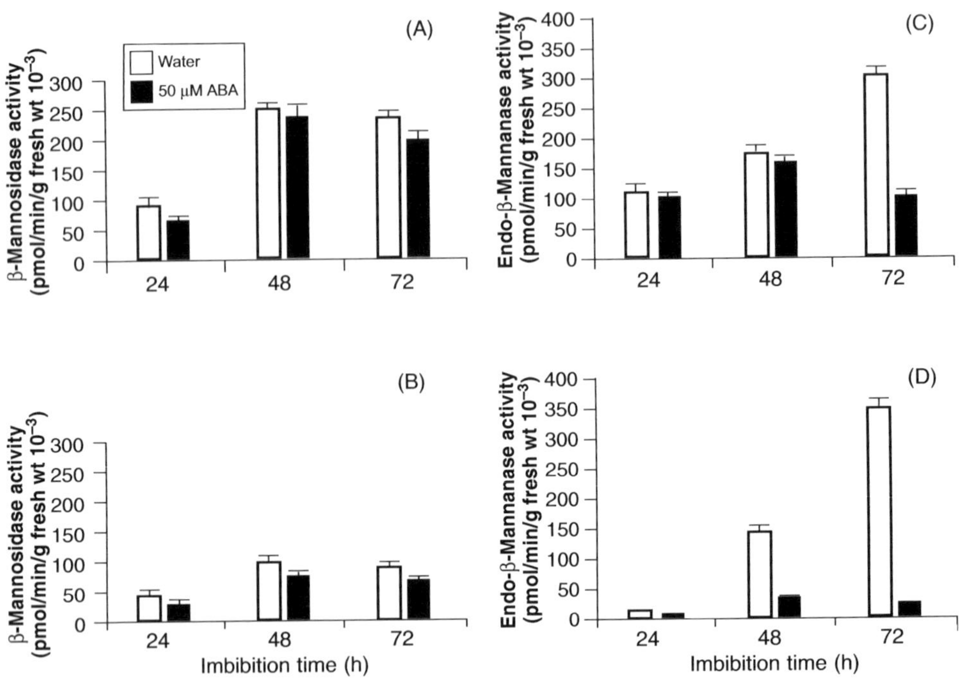

Fig. 14.4. The effect on (A, B) β-mannosidase and (C, D) endo-β-mannanase activities of imbibing tomato seeds for up to 72 h in 50 μM ABA, compared with activities in water-imbibed seeds. (A, C) micropylar endosperm, (B, D) lateral endosperm. Error bars as Fig. 14.1.

trast, in the germinated seed at 72 h, expression is very strong in the lateral endosperm, with some residual activity in the remnants of the micropylar endosperm (Fig. 14.3C,D). Nonogaki *et al.* (2000) have shown by tissue printing that endo-β-mannanase is expressed exclusively in the micropylar region of the endosperm during germination, and in the lateral endosperm after germination. Thus, β-mannosidase expression is different in that it is initially present in both regions of the endosperm, but similar in that after germination it is most abundantly expressed in the lateral endosperm. There is no obvious region of high expression of the β-mannosidase gene in the embryo at 72 h, despite the high degree of hybridization with its cDNA to RNA extracted from this seed part (see Fig. 14.2).

Sensitivity of β-Mannosidase Activity to ABA and Gibberellin (GA)

ABA at 50 μM completely inhibits germination but its effect on β-mannosidase activity in the micropylar endosperm is small, even after 72 h (Fig. 14.4A). The increase in enzyme activity is almost equal to that in seeds that completed germination on water (25% germinated at 48 h and 90% at 72 h). Endo-β-mannanase activity in this region is not sensitive to ABA (Toorop *et al.*, 1996) (Fig. 14.4C), although by 72 h its activity is over 50% lower compared with that in germinated seeds on water. Activity of endo-β-mannanase in the lateral endosperm is strongly inhibited by ABA (Fig. 14.4D), because completion of germination is required for activity to increase in this region (Toorop *et al.*, 1996). While β-mannosidase activity does not increase to the same extent in the lateral endosperm of seeds germinated on water, neither is it as sensitive to ABA, reaching about 50% activity at 72 h in seeds prevented from germinating (Fig. 14.4B) compared with those that have germinated.

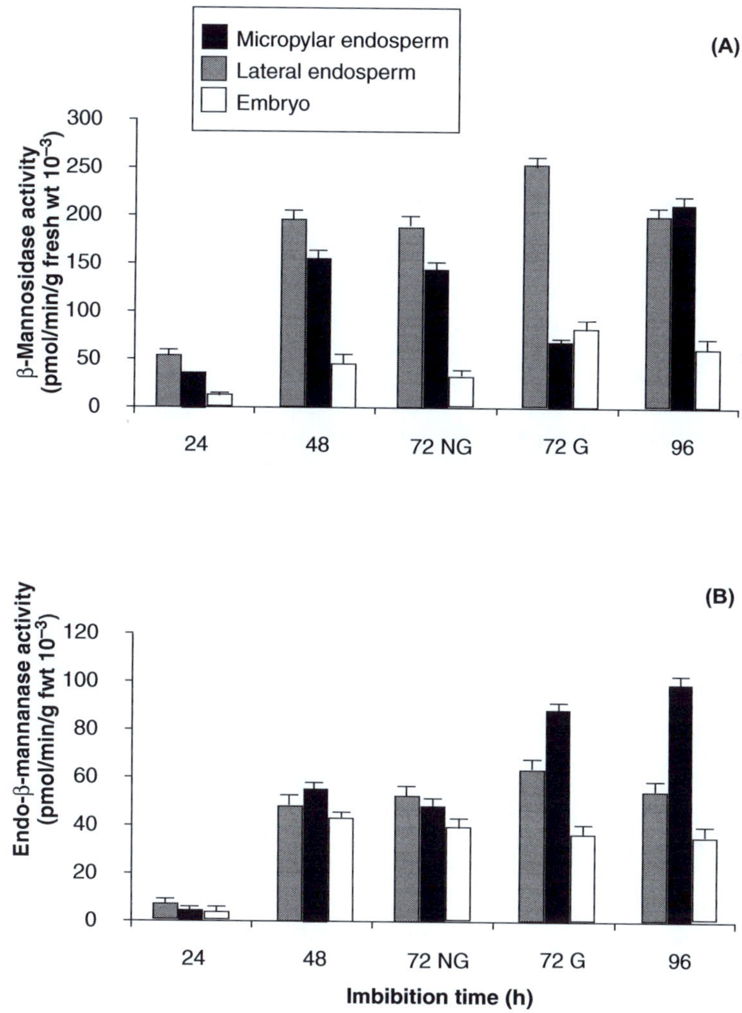

Fig. 14.5. Promotion of (A) β-mannosidase and (B) endo-β-mannanase activities in different tissues of germinating and germinated *gib-1* mutant tomato seeds by 100 μM GA. Seeds were imbibed intact in GA for the times shown before being dissected and assayed for enzyme activity. Enzyme activities in intact *gib-1* seeds imbibed in water for up to 96 h were negligible. At 72 h seeds were separated into germinated (G) and non-germinated (NG) and assayed separately. Error bars as Fig. 14.1.

Gib-1 mutants of tomato seed do not contain sufficient endogenous gibberellin to allow them to complete germination (Groot and Karssen, 1987). These seeds produce little β-mannosidase or endo-β-mannanase but both increase in activity in GA-treated seeds. The GA-stimulated increase in β-mannosidase activity (Fig. 14.5A) is similar quantitatively to that in non-mutant seeds during and after germination (see Fig. 14.1A), whereas the activity of endo-β-mannanase (Fig. 14.5B) is only about one-third even after 96 h (see Fig. 14.1B). Activity of both enzymes in the embryo is generally low and is stimulated little by GA (Fig. 14.5).

Conclusion

The behaviour of β-mannosidase and endo-β-mannanase in germinating and germinated tomato seeds shows considerable similarities from a qualitative standpoint, but differences quantitatively. There are much larger differences compared with α-galactosidase, however. A summary of the relationships between β-mannosidase and endo-β-mannanase is presented in Table 14.1. From these, it is reasonable to conclude that they work in collaboration in the mobilization of the galactomannan-containing cell walls of tomato seeds. Whether they have controlling processes or elements in common remains to be determined.

Table 14.1. Comparison of β-mannosidase and endo-β-mannanase in tomato seeds.[a]

Properties	Region[b]	β-Mannosidase (EC 3.2.1.25)	Endo-β-mannanase (EC 3.2.1.78)
Enzyme, message and gene			
Molecular mass (kDa)		56.7	38.9
pI		6.8	5.2–5.8
Solubility		High-salt buffer	Low-salt buffer, water
Substrate		Mannose oligomers	Mannose polymers
Number of genes		1	4+
Expression and activity			
Time of synthesis	me	During germination[c]	During germination[c]
	le	During, after germination	After germination
Control by ABA	me	Little or none	Little or none
	le	Weak negative	Strong negative
Control by GA in gib-1 mutant	me	Strong positive	Positive
	le	Strong positive	Positive

[a]α-Galactosidase is not included, since it is synthesized during development and is sequestered in the dry seed. New synthesis of this enzyme occurs during and following germination, but its substrates may include cytoplasmic raffinose-series oligosaccharides as well as cell wall galactomannans.

[b]Regions: me, micropylar endosperm; le, lateral endosperm.

[c]The micropylar endosperm is essentially destroyed at the completion of germination when it is penetrated by the radicle. Residual cells in this endosperm region, and intact cells of the immediately adjacent lateral endosperm, from which they cannot be separated, account for any activity observed following germination.

References

Bassel, G.W., Mullen, R.T. and Bewley, J.D. (2001) α-Galactosidase is synthesized in tomato seeds during development and is localized in the protein storage vacuoles. *Canadian Journal of Botany* 79, 1417–1424.

Bewley, J.D. (1997) Breaking down the walls – a role for endo-β-mannanase in release from seed dormancy? *Trends in Plant Science* 2, 464–469.

Bewley, J.D., Burton, R.A., Morohashi, Y. and Fincher, G.B. (1997) Molecular cloning of a cDNA encoding a (1-4)-β-mannan endohydrolase from the seeds of germinated tomato (*Lycopersicon esculentum*). *Planta* 203, 454–459.

Bradford, K.J., Chen, F., Cooley, M.B., Dahal, P., Downie, B., Fukunaga, K.K., Gee, O.H., Gurushinge, S., Mella, R.A., Nonogaki, H., Wu, C.-T., Yang, H. and Yim, K.-O. (2000) Gene expression prior to emergence in imbibed tomato seeds. In: Black, M., Bradford, K.J. and Vazquez-Ramos, J. (eds) *Seed Biology. Advances and Applications.* CAB International, Wallingford, UK, pp. 231–251.

Feurtado, J.A., Banik, M. and Bewley, J.D. (2001) The cloning and characterization of α-galactosidase present during and following germination of tomato (*Lycopersicon esculentum* Mill.) seed. *Journal of Experimental Botany* 52, 1239–1249.

Groot, S.P.C. and Karssen, C.M. (1987) Gibberellins regulate seed germination in tomato by endosperm weakening: a study with gibberellin-deficient mutants. *Planta* 171, 525–532.

Groot, S.P.C, Kieliszewska-Rokicka, B., Vermeer, E. and Karssen, C.M. (1988) Gibberellin-induced hydrolysis of endosperm cell walls in gibberellin-deficient tomato seeds prior to radicle protrusion. *Planta* 174, 500–504.

McCleary, B.V. (1982) Purification and properties of a β-mannosidase mannohydrolase from guar. *Carbohydrate Research* 101, 75–92.

Mo, B. and Bewley, J.D. (2002) β-Mannosidase (EC3.2.1.25) activity during and following germination of tomato (*Lycopersicon esculentum* Mill.) seeds. Purification, cloning and characterization. *Planta* 215, 141–152.

Nomaguchi, M., Nonogaki, H. and Morohashi, Y. (1995) Development of galactomannan-hydrolyzing activity in the micropylar endosperm tip of tomato seed prior to germination. *Physiologia Plantarum* 94, 105–109.

Nonogaki, H., Gee, O.H. and Bradford, K.J. (2000) A germination-specific endo-β-mannanase is expressed in the micropylar endosperm cap of tomato seeds. *Plant Physiology* 123, 1235–1245.

Ouellette, B.F.F. and Bewley, J.D. (1986) β-Mannoside mannohydrolase and the mobilization of the endosperm cell wall of lettuce seeds, cv. Grand Rapids. *Planta* 169, 333–338.

Reid, J.S.G. and Meier, H. (1973) Enzymatic activities and galactomannan mobilization in germinating seeds of fenugreek (*Trigonella foenum-graecum* L. *Leguminosae*). Secretion of α-galactosidase and β-mannosidase by the aleurone layer. *Planta* 112, 301–308.

Sánchez, R.A., de Miguel, L., Lima, C. and de Lederkremer, R.M. (2002) Effect of low water potential on phytochrome-induced germination, endosperm softening and cell-wall mannan degradation in *Datura ferox* seeds. *Seed Science Research* 12, 155–163.

Toorop, P.E., Bewley, J.D. and Hilhorst, H.W.M. (1996) Endo-β-mannanase isoforms are present in the endosperm and embryo of tomato seeds, but are not essentially linked to the completion of germination. *Planta* 200, 153–158.

15 Molecular Cloning and Expression of a Gene Encoding a Germin-like Protein in Common Bean (*Phaseolus vulgaris* L.)

A.L. Chateigner-Boutin, C. Aubry, M.C. Morère-Le Paven and R. Jalouzot

Laboratoire de Génétique, UMR Physiologie Moléculaire des Semences, UFR Sciences, 2 Bvd Lavoisier, 49045 Angers, France

Introduction

Germination begins with imbibition and concludes with radicle emergence. With imbibition, metabolism quickly recommences. Much is known about germination-specific genes associated with reserve mobilization during seedling growth but less information is available about genes functionally related to the initial processes resulting in the completion of germination.

Wheat germin was first described in a search for germination-specific proteins (Thompson and Lane, 1980). Germins have been identified in other *Gramineae* but they have never been detected in dicots. They are present as 25–30 kDa apoplastic glycoproteins that form either homopentamers (Lane *et al.*, 1993) or homohexamers (Woo *et al.*, 1998). Wheat and barley germins are hydrogen peroxide-generating oxalate oxidases (Dumas *et al.*, 1993; Lane *et al.*, 1993). The barley germin was also shown to have a manganese superoxide dismutase (SOD) activity (Woo *et al.*, 2000). In these plants, germin could be involved in the regulation of embryo growth by controlling the cross-linking of plant cell wall polymers (Showalter, 1993) and in defence against abiotic stresses and pathogen attack (Dumas *et al.*, 1995; Hurkman and Tanaka, 1996; Berna and Bernier, 1999). Current evidence indicates that germin is probably involved in post-germinative events.

Proteins with sequence identity to germin, called germin-like proteins (GLPs), have been found in all angiosperm families, including *Gramineae*, and also in gymnosperms and mosses. About 100 different sequences related to germin are now known (Bernier and Berna, 2001). While wheat and barley germin/oxalate oxidase share more than 95% identity at the amino acid level, GLPs share between 30 and 70% identity with the wheat germin

(eds G. Nicolás, K.J. Bradford, D. Côme and H.W. Pritchard)

encoded by the reference gene *gf-2.8*. Phylogenetic analyses defined four GLP subgroups distinct from the true germin subfamily (Carter *et al.*, 1999; Carter and Thornburg, 2000). Despite this heterogeneity, all germins and GLPs bear in their sequences three conserved oligopeptides, named A, B and C. Systematic genome sequence scanning programs have found complex gene families encoding proteins of all GLP subgroups (Bernier and Berna, 2001).

Germins and GLPs would be members of a superfamily that also includes seed storage globulins and sucrose-binding proteins designated 'cupins' because of their similar β-barrel structure (Dunwell, 1998; Dunwell *et al.*, 2000). Most of the plant GLPs are not germination specific. Some of them have been detected in plant leaves, roots or floral tissues under various physiological conditions (Membré *et al.*, 1997) or in response to stress and to pathogen attack (Michalowski and Bohnert, 1992; Wojtaszek *et al.*, 1997). This hypothesis is supported by the finding of SOD activity in a moss GLP (Yamahara *et al.*, 1999) and in the tobacco Nectarin I (Carter and Thornburg, 2000). A GLP from barley (HvGLP1; Vallelian-Bindschedler *et al.*, 1998) was recently shown to have ADP-glucose pyrophosphatase/phosphodiesterase activity (Rodríguez-López *et al.*, 2001). Nevertheless, no enzyme activity has been associated with most known GLPs, not even oxalate oxidase activity, except for a GLP identified in *Silene vulgaris* (Bringezu *et al.*, 1999). Some of them act as cell receptors, such as a root receptor for rhicadhesin in pea (Swart *et al.*, 1994) and two auxin-binding proteins in peach (Ohmiya *et al.*, 1998), whereas others could be structural cell wall proteins. Thus, a wide range of functions have been associated with GLPs. Here, we report the molecular cloning of the first GLP cDNA from common bean (*Phaseolus vulgaris* L.), the tissue-specific expression of the corresponding gene and the characterization of its 5′ flanking region.

Molecular Cloning of a cDNA Encoding a Germin-like Protein in Common Bean

Common bean seeds (*P. vulgaris* L. cv. 'Fin de Bagnols') were grown in a dark growth chamber at 22°C. The percentage of germinated seeds was determined from 0 to 36 hours after imbibition (HAI) (Fig. 15.1) and total RNA was isolated from axes of germinating seeds at different times. In order to search for markers of germination, RT-PCR was conducted with degenerate primers previously designed to isolate a GLP cDNA from *Pinus caribaea* (Neutelings *et al.*, 1998) and corresponding to the A and C conserved boxes of GLPs. A partial cDNA sequence of 387 bp was highly amplified from 17 HAI samples. This PCR product was cloned and sequenced and this sequence was compared to the GENBANK database using the FASTA program (Pearson and Lipman, 1998). The best scores were found with *GLP* cDNAs from *Prunus persica* (*PpABP19*) and *Arabidopsis thaliana* (*AtGLP1*) with 65.4% and 63.4% identity, respectively. A rapid amplification of cDNA ends (RACE) (Frohman *et al.*, 1988) was performed and all PCR products of interest were cloned and sequenced. The cDNA named *PvGLP1* is 838 bp

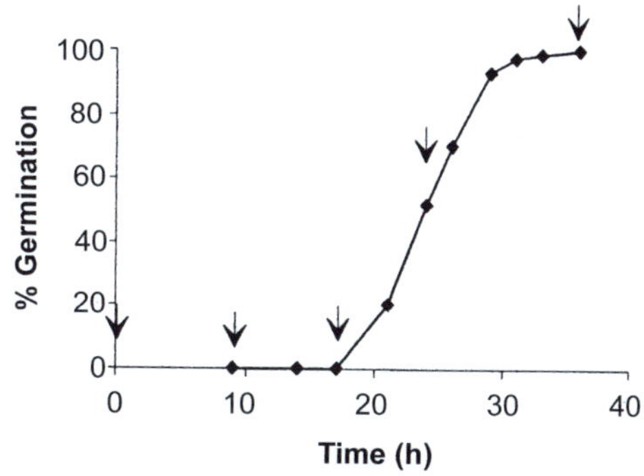

Fig. 15.1. Seed germination time-course of *Phaseolus vulgaris* L. The percentage of germination was determined from 30 seeds of cv. 'Fin de Bagnols' grown on sterile moist sand (8% w/v distilled water) in a dark growth chamber at 22°C. Arrows indicate the stages where total RNA was purified from axes for further *PvGLP1* expression studies.

long with a 5′-untranslated region (UTR) of 44 bp, an open reading frame (ORF) of 620 bp and a 3′-UTR of 173 bp terminated by a poly-A region (accession number PVU276491, EMBL data bank). The amino acid sequence deduced from the ORF revealed that *PvGLP1* encodes a protein of 206 amino acids with a calculated molecular weight of 21,833 Da and a theoretical isoelectric point of 7.77. The three highly conserved oligopeptides A, B and C are present in the PvGLP1 sequence. A phylogenetic tree deduced from the PvGLP1 sequence alignment of 70 GLPs showed that it was a new member of the GLP subfamily 3 (Carter and Thornburg, 2000). A single potential *N*-glycosylation site was identified in the PvGLP1 sequence at a highly conserved location. Like most GLPs, a putative signal peptide of 17 amino acids was found on the PvGLP1 sequence (Nielsen *et al.*, 1997). The mature protein of 189 amino acids has a calculated molecular weight of 19,887 Da and a theoretical isoelectric point of 7.99. Hence, PvGLP1 may be an apoplastic glycoprotein.

The identity between the entire PvGLP1 sequence and two auxin binding proteins, ABP19 and ABP20 from peach, reached 60.4% and 59.9% respectively. In particular, inside a region (96–113 bp) coinciding with a putative binding site for auxin, five of the six highly conserved amino acids found in different auxin-binding proteins (Ohmiya *et al.*, 1998) were present (data not shown). Thus, PvGLP1 might be an auxin-binding protein. We have investigated this hypothesis using two approaches in parallel: one was to follow the expression of the *PvGLP1* gene during germination at the mRNA and the protein levels and to study the effect of auxin on its expression; the other was to identify the promoter sequence of the gene in order to search for elements potentially implicated in auxin responses.

Expression of *PvGLP1* Gene During Germination and in Different Tissues

Regulation at the transcriptional level

Expression of *PvGLP1* was determined in embryo axes throughout germination, ranging from dry seed to 36 HAI (Fig. 15.2). Low amounts of *PvGLP1* mRNA were detected up to 9 HAI. A strong hybridization signal appeared in the axes of seeds from 17 HAI onwards, showing an increase in *PvGLP1* expression a few hours before radicle emergence.

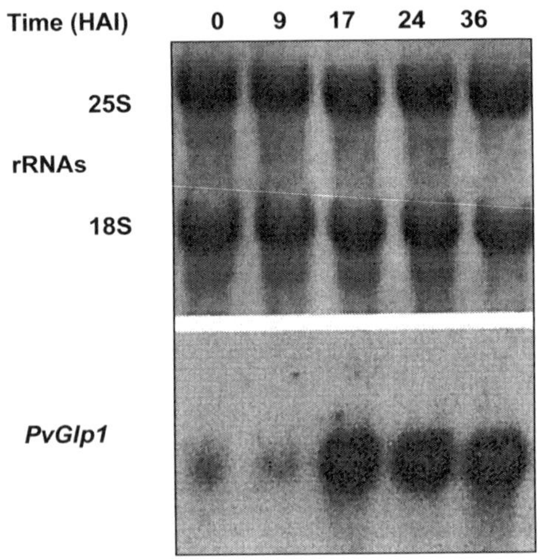

Fig. 15.2. Northern blot analysis of the expression of the *PvGLP1* gene during and following germination. Total RNA was extracted from axes of dry and imbibed seeds (from 0 to 36 HAI). RNA samples (10 µg per lane) were separated on 1.2% formaldehyde gel, blotted and hybridized with ^{32}P-labelled *PvGLP1* cDNA. Equal loading was checked by direct visualization of 25S and 18S rRNAs on the membrane under UV exposure.

The same experiment performed with RNA from mature vegetative tissues failed to detect any hybridization signal. So, RT-PCR with *PvGLP1* sequence-specific primers was performed on RNA from roots, stems, leaves and pods (Fig. 15.3). The expression of *PvGLP1* was very low in pods and not detected in the other tissues. Hence, *PvGLP1* gene expression was specific to early stages of germinating axes growth.

In order to test the effect of auxin on *PvGLP1* expression, seeds were imbibed in the presence of 100 µM of indole acetic acid (IAA). The time when 50% of seeds have germinated (T_{50}) was delayed by 2 h in IAA-treated seeds (27 HAI) in comparison with untreated seeds (25 HAI, Fig. 15.1). RNAs purified from axes at 4, 9 and 14 HAI were assayed by RT-PCR with the *PvGLP1*-specific primers in parallel with RNA from axes of untreated seeds (Fig. 15.4). The results showed a marked decrease of the mRNA level

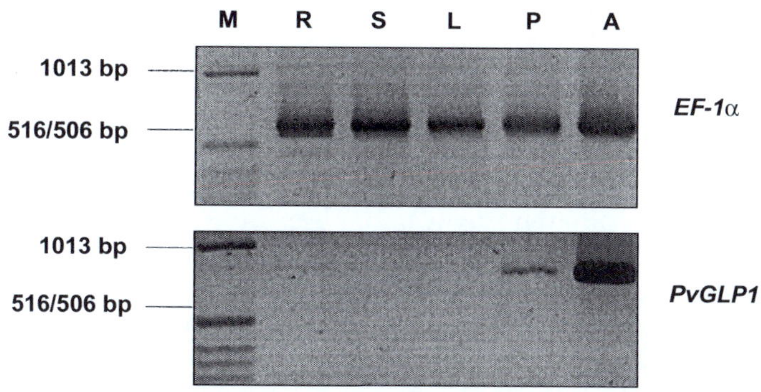

Fig. 15.3. *PvGLP1* expression in mature vegetative tissues and embryonic axes detected by RT-PCR. One μg of total RNA purified from roots (R), stems (S), leaves (L), pods (P) and seed axes at 24 HAI (A) was used for reverse transcription and cDNAs were submitted to 35 cycles of PCR with *PvGLP1* sequence-specific primers. In parallel, RT-PCR was performed with primers specific for the translation elongation factor *EF-1α* gene as a control. PCR products were analysed on a 1.4% agarose gel. M is 1 kb DNA ladder (Life Technologies). Some molecular-length markers are given in base pairs.

of *PvGLP1* in axes of IAA-treated seeds, particularly at the stage 14 HAI, compared with axes of untreated seeds. In the same conditions, the expression of the gene encoding the translation elongation factor *EF-1α* was not affected by IAA. Unlike *gf.2.8* encoding wheat germin, expression of which was induced by auxin (Berna and Bernier, 1997), our results suggest that *PvGLP1* expression would be negatively regulated by auxin at the transcriptional level. It seems, then, that germins and GLPs, which are already known to differ in several aspects (e.g. germins exhibit an enzyme activity

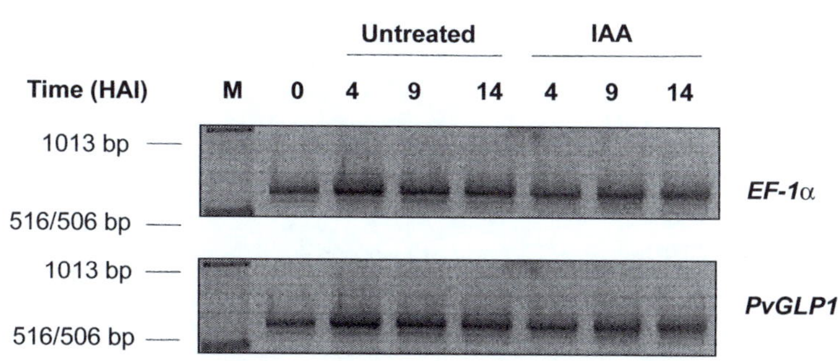

Fig. 15.4. Effect of IAA on *PvGLP1* expression. One μg of total RNA purified from axes of untreated or 100 μM IAA-treated seeds at 0, 4, 9 and 14 HAI was used for RT-PCR with *PvGLP1* sequence-specific primers. In parallel, the same experiment was performed with primers specific for the translation elongation factor *EF-1α* gene. PCR products were analysed on a 1.4% agarose gel. M is 1 kb DNA ladder (Life Technologies). Some molecular-length markers are given in base pairs.

while most GLPs do not), are also different in terms of regulation of gene expression.

Regulation at the post-transcriptional level

In order to search for the presence of the PvGLP1 protein during germination and in different plant tissues, an immunoblot analysis was performed using polyclonal antibodies (kindly provided by F. Bernier, IBMP, Strasbourg, France) raised against an oligopeptide specific to AtGER3, a GLP of *A. thaliana* (Membré *et al.*, 2000). Whereas mRNAs from *PvGLP1* increased dramatically before radicle emergence, a protein cross-reacted with the antiserum at all the stages of germination at a steady state level (0–24 HAI) (Fig. 15.5). Its molecular weight, estimated to be approximately 23 kDa compared with the molecular weight marker and the AtGER3 monomer detected in *A. thaliana* (Membré *et al.*, 1997), is similar to the molecular weight calculated from the PvGLP1 peptide sequence (19,887 Da), considering that GLPs are glycosylated proteins. The same pattern was previously reported concerning the expression of a GLP gene from mustard (Heintzen *et al.*, 1994): whereas the mRNA level underwent drastic fluctuations during dark/light cycles, the amounts of protein were maintained at a fairly constant level. The auxin effect at the protein level remains to be studied.

Concerning leaves, stems and roots, no protein was detected by the antiserum (data not shown). Hence, the specific expression of *PvGLP1* in the axis during the early stages of germination suggests that the protein could play an exclusive role in the germination process.

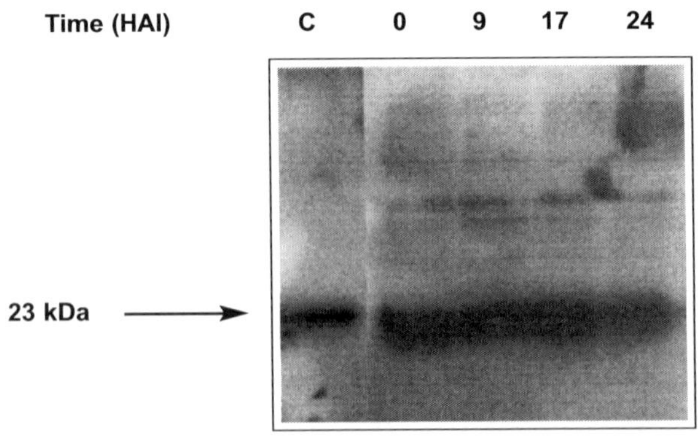

Fig. 15.5. Detection of PvGLP1 by Western blot analysis. Twenty μg of proteins from bean seed axes (0, 9, 17, 24 HAI) previously heat-denatured were separated on a 12% SDS-PAGE gel, blotted and reacted with antibodies raised against the oligopeptide DPKGPQSPSGYS (amino acids 29–40) specific of AtGER3 (Membré *et al.*, 2000). In parallel, proteins extracted from leaves of *Arabidopsis thaliana* and heat-denatured were used as control (lane C).

Molecular Cloning of the *PvGLP1* Promoter

To understand the regulatory mechanisms of *PvGLP1* gene expression, the 5′-flanking region of the coding sequence was isolated using a DNA genomic walking PCR method described by Siebert *et al.* (1995). A special adaptor was ligated to the blunt ends of digested bean genomic DNA. Nested PCR reactions were carried out, using for each of them an adaptor primer and a 5′ *PvGLP1*-specific primer. The amplified fragments were cloned and sequenced, leading to identification of 1718 bp of genomic DNA upstream from the ATG start codon. Analysis of the sequence by PlantCare (Lescot *et al.*, 2002) pointed out several putative TATA boxes located from 67 to 140 nucleotides (nt) upstream from the ATG. However, the nearest from the ATG shared the best sequence homology with the consensus sequence for TATA box and flanking regions in plants (Joshi, 1987) with ten conserved nucleotides (ACGCTATAAATAG) out of 13. Moreover, a motif located 24 nt downstream from this putative TATA box and 44 nt upstream from the ATG matched by 5 nt out of 7 with the plant gene consensus transcription start site described by Joshi (1987) and contained an adenine which could be the transcription initiation site, as for most plant genes. Like most of the leader sequences in plant genes, the 44 nt sequence is A+T rich (67.4%).

The PlantCare analysis (Lescot *et al.*, 2002) of the *PvGLP1* 5′ flanking sequence pointed out many sequence homologies with modules involved in light responsiveness. The role of these motifs in the *PvGLP1* expression regulation remains for the moment unknown.

This promoter was also examined for putative auxin-reponsive elements. There were two sequences similar to the consensus TGTCTC (one mismatch per sequence) found in D1 and D4 elements known to be involved in auxin inducibility (Liu *et al.*, 1994). However, additional nucleotides upstream from this sequence are required for auxin inducibility (Ulmasov *et al.*, 1995), unless TGTCTC is multimerized with appropriate spacing and orientation (Guilfoyle *et al.*, 1998). This is not the case in the promoter of *PvGLP1*. Another sequence matched with six of the eight nucleotides of the highly conserved TGTCCCAT motif found in many genes that respond to auxin. Three nearly perfect repeats of this element were also identified in the promoter of the wheat germin *gf-2.8* gene that is auxin regulated (Berna and Bernier, 1997). Finally, two direct repeats matched perfectly with several sequences localized in the promoter of *gf-2.8* previously defined as 'legumin-boxes' (Chamberland *et al.*, 1992). These types of sequence are also part of a greater group identified as a potential regulatory element of auxin-inducible genes (Berna and Bernier, 1997).

Thus, several sequences of the promoter of *PvGLP1* share homologies with different putative elements leading to auxin inducibility. However, our preliminary experimental results tended to show a decrease in *PvGLP* expression during germination in the presence of exogenous auxin. The role of these elements in the promoter of *PvGLP1*, if there is one, remains to be established.

Conclusion

PvGLP1 was the first gene encoding a germin-like protein described in *P. vulgaris* and its expression was almost exclusively associated with the germination process. Like the wheat germin, the protein could play a role in promoting cell expansion (Berna and Bernier, 1997). Moreover, PvGLP1 shared sequence homologies with two auxin-binding proteins identified in peach and the characterization of the 5′ flanking region pointed out sequences homologous to different auxin-response elements. However, *PvGLP1* expression would be, in our conditions, down-regulated at the transcriptional level by exogenous IAA. The use of transcriptional fusions of the 5′ flanking region to a reporter gene in transformation experiments will be helpful in elucidating the regulation of *PvGLP1* gene expression.

References

Berna, A. and Bernier, F. (1997) Regulated expression of a wheat germin gene in tobacco: oxalate oxidase activity and apoplastic localization of the heterologous protein. *Plant Molecular Biology* 33, 417–429.

Berna, A. and Bernier, F. (1999) Regulation by biotic and abiotic stress of a wheat germin gene encoding oxalate oxidase, a H_2O_2-producing enzyme. *Plant Molecular Biology* 39, 539–549.

Bernier, F. and Berna, A. (2001) Germins and germin-like proteins: plant do-all proteins. But what do they do exactly? *Plant Physiology and Biochemistry* 39, 545–554.

Bringezu, K., Lichtenberger, O., Leopold, I. and Neumann, D. (1999) Heavy metal tolerance of *Silene vulgaris*. *Journal of Plant Physiology* 154, 536–546.

Carter, C. and Thornburg, R.W. (2000) Tobacco Nectarin I. Purification and characterization as a germin-like, manganese superoxide dismutase implicated in the defense of floral reproductive tissues. *Journal of Biological Chemistry* 275, 36726–36733.

Carter, C., Graham, R.A. and Thornburg, R.W. (1999) Nectarin I is a novel, soluble germin-like protein expressed in the nectar of *Nicotiana* sp. *Plant Molecular Biology* 41, 207–216.

Chamberland, S., Daigle, N. and Bernier, F. (1992) The legumin boxes and the 3′ part of soybean β-conglycinin promoter are involved in seed gene expression in transgenic tobacco plants. *Plant Molecular Biology* 19, 937–949.

Dumas, B., Sailland, A., Cheviet, J.P., Freyssinet, G. and Pallett, K. (1993) Identification of barley oxalate oxidase as a germin-like protein. *Compte-Rendus de l'Académie des Sciences Paris Life Sciences* 316, 793–798.

Dumas, B., Freyssinet, G. and Pallett, K. (1995) Tissue-specific expression of germin-like OxO during development and fungal infection of barley seedlings. *Plant Physiology* 107, 1091–1096.

Dunwell, J.M. (1998) Cupins: a new superfamily of functionally diverse proteins that include germins and plant storage proteins. *Biotechnology and Genetic Engineering Reviews* 15, 1–32.

Dunwell, J.M., Khuri, S. and Gane, P.J. (2000) Microbial relatives of the seed storage proteins of higher plants: conservation of function during evolution of the cupin superfamily. *Microbiology and Molecular Biology Reviews* 64, 153–179.

Frohman, M.A., Dush, M.K. and Martin, G.R. (1988) Rapid production of full-length cDNAs from rare transcripts: amplification using a single gene-specific oligonucleotide primer. *Proceedings of the National Academy of Sciences USA* 85, 8998–9002.

Guilfoyle, T., Hagen, G., Ulmasov, T. and Murfett, J. (1998) How does auxin turn on genes? *Plant Physiology* 118, 341–347.

Heintzen, C., Fischer, R., Melzer, S., Kappeler, S., Apel, K. and Staiger, D. (1994) Circadian oscillations of a transcript encoding a germin-like protein that is associated with cell walls in young leaves of the long-day plant *Sinapis alba* L. *Plant Physiology* 106, 905–915.

Hurkman, W.J. and Tanaka, C.K. (1996) Germin gene expression is induced in wheat leaves by powdery mildew infection. *Plant Physiology* 111, 735–739.

Joshi, C.P. (1987) An inspection of the domain between putative TATA box and translation start site in 79 plant genes. *Nucleic Acids Research* 15, 6643–6653.

Lane, B.G., Dunwell, J.M., Ray, J.A., Schmitt, M.R. and Cuming, A.C. (1993) Germin, a protein marker of early plant development, is an oxalate oxidase. *Journal of Biological Chemistry* 268, 12239–12242.

Lescot, M., Dehais, P., Thijs, G., Marchal, K., Moreau, Y., Van de Peer, Y., Rouze, P. and Rombauts, S. (2002) PlantCARE, a database of plant *cis*-acting regulatory elements and a portal to tools for *in silico* analysis of promoter sequences. *Nucleic Acids Research* 30, 325–327.

Liu, Z.B., Ulmasov, T., Shi, X., Hagen, G. and Guilfoyle, T.J. (1994) The soybean GH3 promoter contains multiple auxin-inducible elements. *The Plant Cell* 6, 645–657.

Membré, N., Berna, A., Neutelings, G., David, A., David, H., Staiger, D., Saez Vasquez, J., Raynal, M., Delseny, M. and Bernier, F. (1997) cDNA sequence, genomic organization and differential expression of three *Arabidopsis* genes for germin/oxalate oxidase-like proteins. *Plant Molecular Biology* 35, 459–469.

Membré, N., Bernier, F., Staiger, D. and Berna, A. (2000) *Arabidopsis thaliana* germin-like proteins: common and specific features point to a variety of functions. *Planta* 211, 345–354.

Michalowski, C.B. and Bohnert, H.J. (1992) Nucleotide sequence of a root-specific transcript encoding a germin-like protein from the halophyte *Mesembryanthemum crystallinum*. *Plant Physiology* 100, 537–538.

Neutelings, G., Domon, J.M., Membré, N., Bernier, F., Meyer, Y., David, A. and David, H. (1998) Characterization of a germin-like protein expressed in somatic and zygotic embryos of pine (*Pinus caribaea* Morelet). *Plant Molecular Biology* 38, 1179–1190.

Nielsen, H., Engelbrecht, J., Brunak, S. and Von Heijne, G. (1997) Identification of prokaryotic and eucaryotic signal peptides and prediction of their cleavage sites. *Protein Engeneering* 10, 1–6.

Ohmiya, A., Tanaka, Y., Kadowaki, K.I. and Hayashi, T. (1998) Cloning of genes encoding auxin-binding proteins (ABP19/20) from peach; significant peptide sequence similarity with germin-like proteins. *Plant Cell Physiology* 39, 492–499.

Pearson, W.R. and Lipman, D.J. (1988) Improved tools for biological sequence comparison. *Proceedings of the National Academy of Sciences USA* 85, 2444–2448.

Rodríguez-López, M., Baroja-Fernández, E., Zandueta-Criado, A., Moreno-Bruna, B., Muñoz, F.J., Akazawa, T. and Pozueta-Romero, J. (2001) Two isoforms of a nucleotide-sugar pyrophosphatase/phosphodiesterase from barley leaves (*Hordeum vulgare* L.) are distinct oligomers of HvGLP1, a germin-like protein. *Federation of European Biochemical Societies Letters* 490, 44–48.

Showalter, A.M. (1993) Structure and function of plant cell wall proteins. *The Plant Cell* 5, 9–23.

Siebert, P.D., Chenchik, A., Kellogg, D.E., Lukyanov, K.A. and Lukyanov, S.A. (1995)

An improved PCR method for walking in uncloned genomic DNA. *Nucleic Acids Research* 23, 1087–1088.

Swart, S., Logman, T.J.J., Smit, G., Lugtenberg, B.J.J. and Kijne, J.W. (1994) Purification and partial characterization of a glycoprotein from pea (*Pisum sativum*) with receptor activity for rhicadhesin, an attachment protein of Rhizobiaceae. *Plant Molecular Biology* 24, 171–183.

Thompson, E.W. and Lane, B.G. (1980) Relation of protein synthesis in imbibing wheat embryos to the cell-free translational capacities of bulk mRNA from dry and imbibing embryos. *Journal of Biological Chemistry* 255, 5965–5970.

Ulmasov, T., Liu, Z.B., Hagen, G. and Guilfoyle, T.J. (1995) Composite structure of auxin response elements. *The Plant Cell* 7, 1611–1623.

Vallelian-Bindschedler, L., Mösinger, E., Métraux, J.P. and Schweizer, P. (1998) Structure, expression and localization of a germin-like protein in barley (*Hordeum vulgare* L.) that is insolubilized in stressed leaves. *Plant Molecular Biology* 37, 297–308.

Wojtaszek, P., Stobiecki, M. and Bolwell, G.P. (1997) Changes in the composition of exocellular proteins of suspension-cultured *Lupinus albus* cells in response to fungal elicitors or $CuCl_2$. *Journal of Experimental Botany* 48, 2015–2021.

Woo, E.J., Dunwell, J.M., Goodenough, P.W. and Pickersgill, R.W. (1998) Barley oxalate oxidase is a hexameric protein related to seed storage proteins: evidence from X-ray crystallography. *Federation of European Biochemical Societies Letters* 437, 87–90.

Woo, E.J., Dunwell, J.M., Goodenough, P.W., Marvier, A.C. and Pickersgill, R.W. (2000) Germin is a manganese containing homohexamer with oxalate oxidase and superoxide dismutase activities. *Nature Structural Biology* 7, 1036–1040.

Yamahara, T., Shiono, T., Suzuki, T., Tanaka, K., Takio, S., Sato, K., Yamazaki, S. and Satoh, T. (1999) Isolation of a germin-like protein with manganese superoxide dismutase activity from cells of a moss, *Barbula unguiculata*. *Journal of Biological Chemistry* 274, 33274–33278.

16 Isolation and Characterization of Genes Related to the Breaking of Beechnut Dormancy and Putatively Involved in Ethylene Signal Perception and Transduction

ANGEL P. CALVO, JESÚS A. JIMÉNEZ, CARLOS NICOLÁS, GREGORIO NICOLÁS AND DOLORES RODRÍGUEZ

Departamento de Fisiología Vegetal, Facultad de Biología, Universidad de Salamanca, Plaza de los Doctores de la Reina s/n, 37007 Salamanca, Spain

Introduction

Ethylene is known to play a key role in several aspects of growth and development of many plants, from seed germination to abscission of plant organs, leaf and flower senescence and fruit ripening (Kepczynski and Kepczynska, 1997). In recent years, experiments on *Arabidopsis thaliana* have led to the discovery of genes that encode ethylene receptors, proteins that bind ethylene when expressed in yeast (Schaller and Bleecker, 1995), and possible transcription factors involved in ethylene signalling that specifically interact with the ethylene-responsive element (ERE), known as the GCC box (Ohme-Takagi and Shinshi, 1995).

There is increasing evidence that the ethylene signal transduction pathway is highly conserved across many plant species. Genes homologous to the *Arabidopsis* ethylene receptor ETR1 have been cloned and divided into two subfamilies based on DNA and protein sequence similarity. Both subfamilies of receptors have been shown to be important in the perception of ethylene, since mutations in each class result in ethylene-insensitive plants (Hua *et al.*, 1998). Other ethylene receptors homologous to ETR1 have been reported in a number of plants, especially in tomato (Chang and Shockey, 1999).

The signal is transduced from the membrane to the nucleus, where ethylene-responsive transcription factors (ERF, previously known as EREBP)

 The Biology of Seeds: Recent Research Advances (eds G. Nicolás, K.J. Bradford, D. Côme and H.W. Pritchard)

lead to the expression of effector genes involved in a variety of responses to the hormone (Fujimoto *et al.*, 2000).

The molecular mechanism of gene regulation by ethylene has been studied in vegetative tissues, senescence of flowers and fruit ripening (Johnson and Ecker, 1998), but there are few reports about ethylene-regulated genes during seed dormancy or germination.

Ethylene has been shown to be involved in seed germination of many plant species, being detected during imbibition and before visible germination, and its production stimulated by factors that break dormancy (Kepczynski and Kepczynska, 1997). Gibberellins are also clearly related to the onset of germination in many seeds.

Our research is focused on the study of the mechanisms of ethylene action and its relation to gibberellins during the breaking of dormancy in *Fagus sylvatica* seeds in particular, by analysing the expression of specific genes regulated by these two hormones and involved in this process. Beechnut represents a suitable model to study seed dormancy of woody plants where little is known about the mechanisms involved in the transition from dormancy to germination. We have shown that both ethylene and GA_3 produce a fast release from dormancy and increase the germination percentages from the first week of treatment. In contrast, ABA and paclobutrazol (PCB, a well-known gibberellin biosynthesis inhibitor) maintain the seeds in a dormant state.

In this work we report the isolation and characterization of two cDNA clones coding for an ethylene receptor and an ethylene transcription factor, respectively. They are up-regulated by ethephon (ethylene producing compound) and GA_3 (in the case of FsERS1) and seem to be correlated with the breaking of dormancy and the onset of germination in beechnuts. A possible relationship between ethylene and gibberellins in the regulation of this process is discussed.

Materials and Methods

Plant material and germination conditions

Seeds of *F. sylvatica* were obtained from the Danish State Forestry Improvement Station, dried to a moisture content of 21% and stored in sealed jars. Their viability was approximately 97% (checked by standard procedures recommended by the suppliers). Seed germination was carried out as described in Nicolás *et al.* (1997). The imbibition media used were: 100 μM ABA, 100 μM GA_3, 700 μM ethephon (ETP), 10 μM paclobutrazol (PCB), 100 μM aminooxyacetic acid (AOA), 100 μM AOA + 100 μM GA_3, 700 μM ETP + 10 μM PCB and sterile water used as a control. For all the experiments the pericarp was manually removed before imbibition and seeds were maintained at 4°C in the dark for 1–6 weeks in the different media.

Isolation of cDNA for ethylene receptor homologues

To isolate ethylene receptor fragments, two degenerate oligonucleotide primers [ETRS: 5´ATGGA(A,G)G(C,T)(A,C,T)TGC(A,G)A(C,T)TG-(C,T)(A,T)T-3´ and ETRAS: 5´-GA(A,G,C)AGATC(C,T)A(A,G)(A,T,C,G)AC-(A,G)TC(A,G)TT-3´] were synthesized, based on conserved amino acid regions of ETR1. For RT-PCR, cDNA was synthesized from 1 μg of poly (A^+) RNA prepared from ethephon-treated seeds using ETRAS as primer. After PCR, the amplified fragment was cloned into the pGEM-T Vector (Promega Inc.) and sequenced. The full-length cDNA clone was isolated by screening a cDNA library constructed from poly (A^+) RNA of *F. sylvatica* seeds (Nicolás *et al.*, 1997) using the FsERS1 PCR fragment as probe. The predicted gene product encoded by this clone revealed homology to an ethylene receptor and was named FsERS1.

Isolation of ethylene-responsive transcription factor

By differential screening of a cDNA library from *F. sylvatica* and using poly (A^+) RNA from seeds imbibed in either 100 μM ABA for 2 weeks or 0.7 mM ethephon for 2 weeks, one full-length clone showing homology with ethylene-responsive transcription factors and called FsEREBP1, was isolated.

DNA sequencing

Plasmid DNA templates were isolated by the Wizard *Plus* Minipreps DNA Purification System (Promega). Determination of the nucleotide sequence of the cDNA clones was performed on a ABI 377 sequencer (Perkin-Elmer Applied Biosystems) using the *Taq* DyeDeoxy Terminator Cycle Sequencing kit. Both the DNA and deduced protein sequences were compared with other sequences in the EMBL GenBank and SwissProt databases, respectively, using the FASTA algorithm.

RNA extraction and Northern blot

Total RNA was extracted using Qiagen pack-500 cartridge (Qiagen Inc.), following the manufacturer's protocol. Poly (A^+) RNA was purified from total RNA by affinity chromatography in oligo(dT)-cellulose columns using the mRNA Purification Kit (Pharmacia Biotech).

Northern analysis was performed as previously described by Nicolás *et al.* (1997). Blotted membranes were hybridized with either FsERS1 or FsERBP1 probes and exposed to X-Omat films (Kodak), using rRNAs stained with ethidium bromide as an RNA loading control.

Results

Isolation and characterization of a cDNA clone encoding a putative ethylene receptor in *F. sylvatica* L. seeds

By means of RT-PCR, using degenerated oligonucleotides corresponding to two conserved amino acid sequences conserved among the ethylene receptors, we obtained a cDNA fragment encoding an ethylene receptor from *F. sylvatica*. The full-length cDNA clone was isolated from a cDNA library constructed using mRNA from *F. sylvatica* seeds (Nicolás *et al.*, 1997). This clone contained 2553 bp, with an open reading frame of 1905 nucleotides encoding a 635 amino acid protein of 71.07 kDa, a long leader of 429 bp and a 3′-untranslated region of 219 bp with a putative polyadenylation signal (AAATA) (EMBL accession number AJ420194).

This clone shows high identity with members of the ethylene receptor family, particularly with those of *Carica papaya* (82.2%), *Pisum sativum* (81.7%) and *Phaseolus aureus* (81.3%). The deduced amino acid sequence has three specific motifs described in all ethylene receptors (ETR/ERS): a sensor region containing three hydrophobic domains and two conserved cysteine residues, the GAF motif (cGMP binding domain) and a histidine kinase with their conserved regions, although it lacks the carboxyl-terminal response-regulator-like (RR) domain present in the ETR1 receptor (Hua *et al.*, 1998), indicating that this protein is a homologue of ERS1 receptors and, therefore, the corresponding gene was designated FsERS1 (Fig. 16.1).

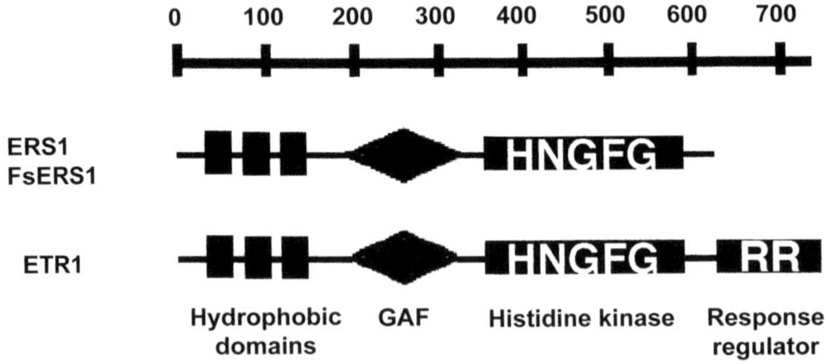

Fig. 16.1. Schematic diagram of ERS1 and ETR1 from *Arabidopsis* and FsERS1 from *Fagus sylvatica*, showing the conserved domains within the ethylene receptors.

The analysis of FsERS1 in *F. sylvatica* by genomic DNA Southern blot showed that this gene was present as a single copy in the genome of *F. sylvatica* (data not shown).

Expression of FsERS1

The effect of the different treatments described above in 'Materials and Methods' was analysed by Northern blot. Transcript levels were low or

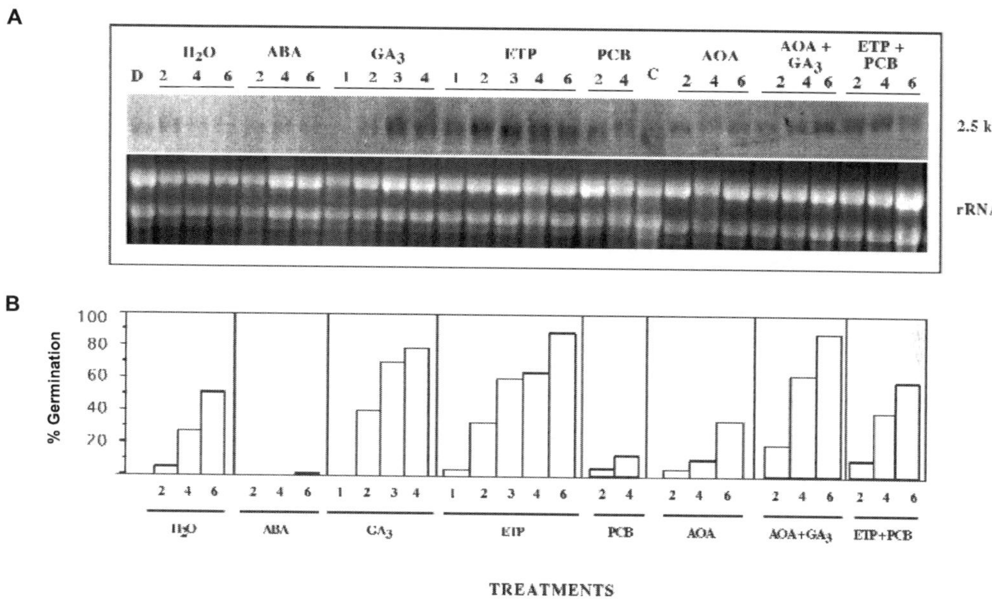

Fig. 16.2. (A) Northern blot analysis of total RNA isolated from *F. sylvatica* dormant seeds (D) and dormant seeds imbibed from 1 to 6 weeks at 4°C in H_2O as a control and in the presence of abscisic acid (ABA), gibberellic acid (GA_3), ethephon (ETP), paclobutrazol (PCB), aminooxy-acetic acid (AOA), AOA + GA_3 and ETP + PCB. Ten μg RNA were used per lane and hybridized with FsERS1 cDNA probe. Bottom panel: ethidium bromide-stained gel showing rRNAs. (B) Germination percentages of *F. sylvatica* seeds obtained under the indicated treatments.

undetectable in dry dormant seeds and after 6 weeks of stratification in water, ABA, PCB or AOA (Fig. 16.2A) where dormancy was maintained or slowly released. However, in treatments with GA_3 or ethephon, which produced a fast release from dormancy, there was a clear increase in the level of expression as the stratification period proceeded and germination percentages increased up to 70–80% (Fig. 16.2B). The addition of either PCB with ethephon or AOA with GA_3, which delay the breaking of dormancy, reduced the levels of FsERS1 transcript compared with the effect of both hormones by themselves (Fig. 16.2A).

Isolation and characterization of a cDNA coding for a putative ethylene-responsive transcription factor in *F. sylvatica* seeds

By means of differential screening of a cDNA library constructed using mRNA from *F. sylvatica* seeds, we isolated a cDNA encoding a putative ethylene-responsive transcription factor. This clone, named FsEREBP1 (Accession number, AJ420195), contains 1650 bp with an open reading frame of 1125 nucleotides encoding a polypeptide of 375 amino acids and an estimated molecular mass of 41.81 kDa.

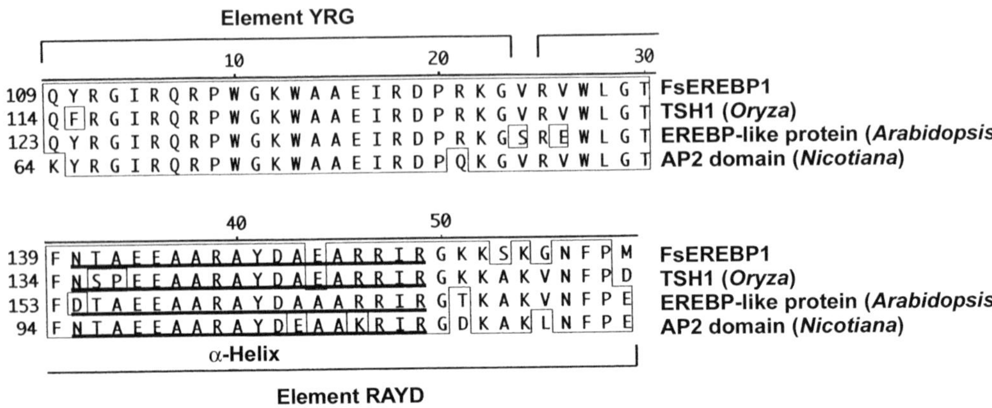

Fig. 16.3. Alignment of the derived amino acid sequences from FsEREBP1 AP2/ERF domain, with others AP2/ERF domains.

Comparison of the amino acid sequence deduced for the cDNA sequence with the databases revealed high homology in a conserved region of 59 amino acids present in other DNA-binding proteins and designated as the ERF domain. The DNA-binding domain of FsEREBP1 has nearly 90% amino acid identity with ERF domains of other plant species (Fig. 16.3) and includes two subdomains: YRG, involved in DNA binding, and the RAYD element, which is probably involved in DNA binding through the interaction of its hydrophobic face with the major groove of DNA (Okamuro *et al.*, 1997).

Southern blot analysis suggests the presence of a single copy for FsEREBP1 gene in the *F. sylvatica* genome, although some homologous genes in the *Fagus* genome cannot be discarded (data not shown).

Expression of FsEREBP1

Using FsEREBP1 clone as a probe, we analysed the expression of this gene in the seeds under the different treatments used in this work by Northern blotting (Fig. 16.4). The addition of ethephon clearly increased the transcript level compared with the control in water and this effect was not modified by the joint addition of PCB (inhibitor of gibberellin biosynthesis). Imbibition in the presence of ABA or PCB, treatments that maintain the dormant state in beechnuts, had no effect in the expression of this gene; even under treatment with GA_3, which releases dormancy in these seeds (Fig. 16.4B), FsEREBP1 transcripts were almost undetectable. These results suggest that this gene is specifically regulated by ethylene during the breaking of beechnut dormancy.

Discussion

Ethylene has been involved in the regulation of dormancy and germination in several seeds (Matilla, 2000) but little is known about the molecular mech-

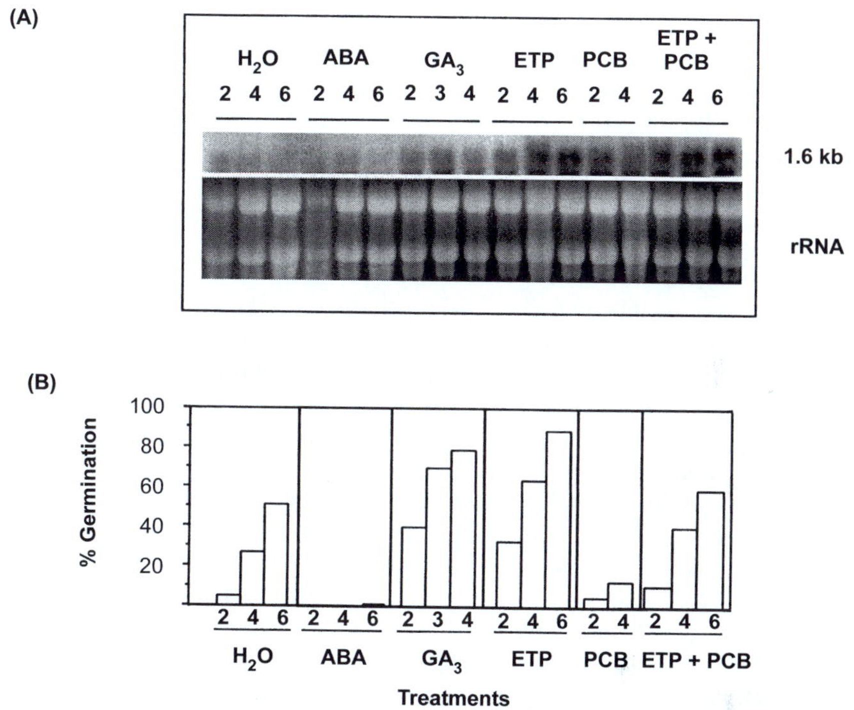

Fig. 16.4. (A) Northern blot analysis of total RNA isolated from *F. sylvatica* dormant seeds imbibed from 1 to 6 weeks at 4°C in H_2O as a control and in the presence of abscisic acid (ABA), gibberellic acid (GA_3), ethephon (ETP), paclobutrazol (PCB) and ETP + PCB. Ten μg RNA were used per lane and hybridized with FsEREBP1 cDNA probe. Bottom panel: ethidium bromide-stained gel showing rRNAs. (B) Germination percentages of *F. sylvatica* seeds obtained under the indicated treatments.

anisms of these effects. To try to get a deeper insight into the mechanism of ethylene action during the transition from dormancy to germination in beechnuts, we studied some genes and the corresponding proteins involved in the perception and signalling of this hormone.

In this work we have isolated and characterized two cDNA clones, FsERS1 and FsEREBP1, coding for an ethylene receptor and a transcriptional factor implicated in ethylene responses, respectively (Figs 16.1 and 16.3). To understand better the regulation of these genes, we have studied their expression under different treatments that maintain or break dormancy in *F. sylvatica* seeds, and also have attempted to find a relationship between ethylene and gibberellins in the breaking of dormancy of these seeds.

The amino acid sequences of ethylene receptors in plants are clearly related to those of proteins involved in the two component signal transduction system in bacteria and contain all the conserved residues required for histidine kinase activity (Wilkinson *et al.*, 1995). FsERS1 exhibits all the features of ethylene receptors (ETR/ERS) (Fig. 16.1). The protein sequence of *Arabidopsis* ETR1 contains a response regulator domain at its C-terminus, while the protein sequence of ERS lacks this domain (Hua *et al.*, 1995). The putative ethylene receptor, FsERS1, isolated from *F. sylvatica*, also lacks the response regulator; thus it may be included in the ERS group.

Expression of FsERS1 is significantly increased by ethephon and GA_3 (Fig. 16.2A), and correlates with the onset of germination (Fig. 16.2B) and the ethylene production in the seeds (data not shown). A correlation between the expression of ACC oxidase genes and the expression of the ethylene receptor has been shown in melon fruit (Yamamoto *et al.*, 1995) and mango (Gutiérrez-Martínez *et al.*, 2001). However, AOA (an ethylene biosynthesis inhibitor) counteracts the promoting effect of GA_3, and paclobutrazol (a GA biosynthesis inhibitor) also reduces the effect of ethephon. These results suggest a relationship between ethylene and gibberellins in the regulation of this gene and the breaking of dormancy of beech seeds. It appears that GAs could activate the synthesis of endogenous ethylene, as previously suggested by Kaneta *et al.* (1997), and then trigger the ethylene signal transduction cascade involved in the release of seed dormancy.

The isolation and characterization of FsEREBP1 revealed that the corresponding protein contains all the domains described for EREBPs of *Arabidopsis*, tobacco and other species (Ohme-Takagi and Shinshi, 1995; Hao *et al.*, 1998). These transcriptional factors are induced by various types of biotic and abiotic stress, and by the addition of ethephon (Ohme-Takagi and Shinshi, 1995) or ethylene (Fujimoto *et al.*, 2000).

EREBPs are novel DNA-binding proteins that are unrelated to the bZIP and zinc finger families. Also they exhibit no sequence homology with other known transcription factors or DNA-binding proteins. The DNA-binding domain of EREBPs (Fig. 16.3) is conserved among various species, suggesting that it is evolutionarily conserved in plants and that these proteins represent a new class of DNA-binding proteins essential for the regulation of transcription (Ohme-Takagi and Shinshi, 1995).

Expression of FsEREBP1 is induced by ethephon (Fig. 16.4), but it is not affected by GA_3 nor PCB, suggesting that this gene is regulated by ethylene and that this effect is independent of gibberellin biosynthesis.

Our results indicate that ethylene is very important in the transition from dormancy to germination in *F. sylvatica* seeds by regulating the expression of the genes and proteins involved in its own signalling cascade. Moreover, GAs are also necessary for breaking seed dormancy. The evidence suggests that both hormones interact and influence the dormancy-breaking and germination processes.

Acknowledgements

This work was supported by grants BFI2000-1361 from the Ministerio de Ciencia y Tecnología (Spain) and SA010/02 from Junta de Castilla y León.

References

Chang, C. and Shockey, J.A. (1999) The ethylene-response pathway: signal perception to gene regulation. *Current Opinion in Plant Biology* 2, 352–358.

Fujimoto, S.Y., Ohta, M., Usui, A., Shinshi, H. and Ohme-Takagi, M. (2000) *Arabidopsis* ethylene-responsive element binding factors act as transcriptional activators or repressors of GCC box-mediated gene expression. *Plant Cell* 12, 393–404.

Gutiérrez-Martínez, P., López-Gómez, R. and Gómez-Lim, M.A. (2001) Identification of an ETR1-homologue from mango fruit expressing during fruit ripening and wounding. *Journal of Plant Physiology* 158, 101–108.

Hao, D., Ohme-Takagi, M. and Sarai, A. (1998) Unique mode of GCC box recognition by the DNA-binding domain of ethylene-responsive element-binding factor (ERF domain) in plant. *Journal of Biological Chemistry* 273, 26857–26861.

Hua, J., Chang, C., Sun, Q. and Meyerowitz, E.M. (1995) Ethylene insensitivity conferred by *Arabidopsis* ERS gene. *Science* 269, 1712–1714.

Hua, J., Sakai, H., Nourizadeh, S., Chen, Q.G., Bleecker, A.B., Ecker, J.R. and Meyerowitz, E.M. (1998) EIN4 and ERS2 are members of the putative ethylene receptor gene family in *Arabidopsis*. *Plant Cell* 10, 1321–1332.

Johnson, R. and Ecker, J.R. (1998) Ethylene gas signal transduction pathway: Molecular perspective. *Annual Review of Genetics* 32, 227–254.

Kaneta, T., Kakimoto, T. and Shibaoka, H. (1997) Gibberellin A_3 causes a decrease in the accumulation of mRNA for ACC oxidase and in the activity of the enzyme in azuki bean (*Vigna angularis*) epicotyls. *Plant and Cell Physiology* 38, 1135–1141.

Kepczynski, J. and Kepczynska, E. (1997) Ethylene in seed dormancy and germination. *Physiologia Plantarum* 101, 720–726.

Matilla, A.J. (2000) Ethylene in seed formation and germination. *Seed Science Research* 10, 111–126.

Nicolás, C., Rodríguez, D., Poulsen, F., Eriksen, E.N. and Nicolás, G. (1997) The expression of an abscisic acid responsive glycine-rich protein coincides with the level of seed dormancy in *Fagus sylvatica*. *Plant and Cell Physiology* 38, 1303–1310.

Ohme-Takagi, M. and Shinshi, H. (1995) Ethylene inducible DNA binding proteins that interact with an ethylene-responsive element. *Plant Cell* 7, 173–182.

Okamuro, J.K., Caster, B., Villarroel, R., Van Montagu, M. and Jofuku, K.D. (1997) The AP2 domain of APETALA2 defines a large new family of DNA binding proteins in *Arabidopsis*. *Proceedings of the National Academy of Sciences USA* 94, 7076–7081.

Schaller, G.E. and Bleecker, A.B. (1995) Ethylene-binding sites generated in yeast expressing the *Arabidopsis* ETR1 gene. *Science* 270, 1809–1811.

Wilkinson, J.A., Lanahan, M.B., Yen, H.C., Giovannoni, J.J. and Klee, H.J. (1995) An ethylene-inducible component of signal transduction encoded by *never-ripe*. *Science* 270, 1807–1809.

Yamamoto, M., Miki, T., Ishiki, Y., Fujinami, K., Yanagisawa, Y., Nakagawa, H., Ogura, N., Hirabayashi, T. and Sato, T. (1995) The synthesis of ethylene in melon fruit during the early stage of ripening. *Plant and Cell Physiology* 36, 591–596.

17 Developing Weedy Rice for Map-based Cloning of Seed Dormancy Genes

Michael E. Foley[1], Xingyou Gu[2] and Durga Nadella[2]

[1]USDA-Agricultural Research Service, Biosciences Research Laboratory, Fargo, ND 58105-5674, USA; [2]Department of Plant Sciences, North Dakota State University, Fargo, ND 58105–5674, USA

Introduction

Dormancy is the failure of a viable seed to germinate, after a specific length of time, in a particular set of environmental conditions that allow germination after the restrictive state has been terminated by either natural or artificial conditions (Simpson, 1990). Despite much research, knowledge is limited about mechanisms that govern dormancy. Dormancy and germinability (which is the propensity for immediate, intermediate or much delayed germination) are controlled by genetic and environmental factors. Characterizing genes that directly regulate germinability will allow the development of new fundamental knowledge about dormancy, after-ripening and germination.

The goal of our research, using rice (*Oryza sativa* L.) as a model system, is to map-based clone and characterize genes that directly regulate germinability in grasses. Many grass species, including the important crops barley, rice, sorghum and wheat, and many serious weedy grasses, such as weedy rice and wild oat, display seed dormancy (Simpson, 1990). However, map-based cloning is not practical with all species, particularly those that have a relatively large complement of nuclear DNA, such as wild oat (11,000 Mbp). Rice is the model experimental system for grasses because: many genetic resources are available; it is a diploid with a small complement of nuclear DNA (430 Mbp); it is used as the base genome for comparative genetics in grasses; and its genome has been sequenced (Gale *et al.*, 1996; Goff *et al.*, 2002). Moreover, several groups have successfully utilized a map-based approach to clone genes from rice.

 The Biology of Seeds: Recent Research Advances (eds G. Nicolás, K.J. Bradford, D. Côme and H.W. Pritchard)

Weedy rice (*O. sativa*) accompanies cultivated rice production worldwide (Oka, 1988). Dormancy in weedy rice is much stronger than in cultivated rice. A strong dormancy phenotype is critical for this research and it is likely that weedy rice has novel alleles that influence germinability that are not present in domesticated cultivars. Thus, we shall focus on weedy rice.

Little research has been done beyond the origin and differentiation of weedy rice (Oka, 1988; Suh *et al.*, 1997) but rice geneticists have investigated seed dormancy in cultivated varieties to improve their resistance to preharvest sprouting (Chang and Tagumpay, 1973; Tripathi and Rao, 1982; Tomar, 1984; Seshu and Sorrells, 1986; Shenoy, 1993). Heritability for seed dormancy in cultivars, estimated under field conditions, ranges from 0.12 to 0.42 (Chang and Yen, 1969). We characterized the types and levels of seed dormancy in weedy and cultivated rice accessions obtained from throughout the world, evaluated inheritance for hull- and pericarp/testa-imposed dormancy, and began to develop some populations for further research.

Materials and Methods

Germinability was evaluated for 45 accessions, including 17 varieties of *O. sativa* indica, seven of japonica, four of *O. glaberrima* and 17 weedy rice strains. Accessions were self-pollinated for one or more generations before yielding seeds for assays. Plants were maintained in a greenhouse with 29/21°C day/night temperatures, 59 ± 12.4% relative humidity (RH) and a 14 h photoperiod, except when a shorter photoperiod was used to synchronize flowering. Panicles were bagged 10 days after flowering (DAF). Seeds were harvested at 40 DAF and dried in the greenhouse for 3 days. Dried seeds were stored at –20°C or after-ripened at room temperature (approximately 25°C) for different numbers of days after harvest (DAH).

Germinability was evaluated using intact seeds, caryopses, caryopses with the pericarp/testa scraped, and excised embryos. Twenty-four seeds or caryopses were placed in a 24-well tissue culture plate lined with filter paper, wetted and incubated at 30°C and 100% RH in the dark. Embryos were excised from caryopses that had been imbibed for 2 h with the aid of a surgical knife under a stereomicroscope. Excised embryos were cultured on tissue culture media without sugar and considered germinated when the radicle or coleoptile expanded by > 3 mm. For dormant accessions, the duration of dormancy was also measured by germinating partially after-ripened seeds. Forty seeds were placed in a Petri dish lined with a filter paper, wetted, and incubated for 7 days as described above.

Weedy strains 'SS18' and 'LD' were selected as dormancy parents to cross with non-dormant varieties ('CO39', 'EM93-1' and 'WYJ') for genetic analysis of seed covering-imposed dormancy. Forty to 50 seeds at different DAH or 30 non-after-ripened caryopses were germinated as described above. The sample sizes for 'SS18'- and 'LD'-derived F_2 populations were about 200 and 140 plants, respectively. Two parameters – average degree of dominance (ADD) and broad-sense heritability (h^2_b) – were used to compare the genetic difference in degree of dormancy between crosses and between germination assays at different DAH. The parameters were calculated based on principles described by Mather and Jinks (1972).

The F_2 of the 'CO39'/'SS18' cross from the above experiment was selected to map the genes or quantitative trait loci (QTLs) for hull-imposed seed dormancy. Phenotypic evaluation was repeated using the seeds from the ratoon plants of the population. The ratoon population was obtained by maintaining the F_2 plants in greenhouse after harvest. The young tillers from each ratoon plant were split and transplanted into new pots. Microsatellite (Temnykh *et al.*, 2000) and amplified fragment length polymorphism (AFLP) markers (Vos *et al.*, 1995) were used to determine the polymorphism between the parents.

Backcross (BC) and phenotypic selection techniques are being used to develop near-isogenic lines (NILs) for major dormancy genes. F_2 plants with strongly dormant seeds and without photoperiod sensitivity and seed shattering were selected from the mapping population as donors of dormancy genes. The non-dormant line 'EM93-1' was chosen as the recurrent parent, because it is insensitive to photoperiod and has an extremely short growth duration (90 days). The most dormant plants in a BC population were selected to generate the next generation of BC population.

Results and Discussion

Levels and types of seed dormancy

Most of the weedy strains had seed dormancy as judged by $< 5\%$ germination (Table 17.1 and data not shown). Most of the weedy strains or cultivars that displayed seed dormancy had non-dormant (data not shown), or weakly to moderately dormant caryopses. However, weedy strains 'C9589', 'C9541' and 'LD' showed relatively strong caryopsis dormancy (Table 17.1). Endosperm- and embryo-imposed dormancy were not detected (Table 17.1). Based on the accessions tested, dormancy was imposed by the maternal tissue(s) of the hull, pericarp/testa, or both.

The duration of dormancy with intact seeds was much longer in dormant weedy strains than in dormant cultivars as judged by the days of after-ripening required to achieve 50% seed germination (Table 17.1). There was no correlation between the germination percentage of non-after-ripened seeds or caryopses and the duration of dormancy (days to 50% seed germination). Based on the above results, we conclude that weedy rice harbours ideal donors of genes that control a high degree of hull- and pericarp/testa-imposed dormancy.

Inheritance of seed covering-imposed dormancy

The ADD for seed germination in the 'CO39'/'SS18' and 'LD'/'WYJ' crosses changed from > 0.8 (highly positive) to 0 (no dominance), then to < 0 (negative partial dominance) during after-ripening (Table 17.2). The period of after-ripening required to change from positive to negative dominance was

Table 17.1. Germination of non-after-ripened seeds, caryopses (I), caryopses (II) with pericarp removed, and excised embryos, and period of after-ripening required to achieve 50% germination.

Accession	Germination (1)				Days to 50% seed
	Seeds[a]	Caryopses (I)[b]	Caryopses (II)[a]	Embryos[c]	germination[a]
O. sativa weedy strains					
'C9589'	0	6	90	100	> 90
'C9541'	3	15	97	100	50
'LD'	0	22	98	100	60
'S434'	3	35	96	100	50
'C9588'	0	40	92	100	> 90
'C9520'	3	41	98	95	80
'Hapcheon3'	6	60	100	100	50
'TKN 5-3'	0	65	90	100	80
'BT8Ab'	0	71	100	100	50
'W1670-14'	0	75	96	95	> 90
'TKN12'	0	77	90	100	60
'SS18'	3	84	97	100	80
O. sativa cultivars					
'N22'	3	44	100	100	40
'Peta'	2	55	95	100	20
'CO39'[ND]	81	95			
'EM93-1'[ND]	88	96			
'WYJ'[ND]	85	94			
O. glaberrima cultivars					
'NSGC5943'	2	72	100	100	30
'Kabre 80'	20	65	90	100	20

[a,b,c]Germination was rated at (a) 7, (b) 21 or (c) 4 days after imbibition.
[ND]Non-dormant lines.

longer in the 'SS18'- than in the 'LD'-derived crosses in seeds from F_1 plants.

The ADD for germination of non-after-ripened caryopses was highly positive in the 'LD'/'WYJ' and highly negative in the 'CO39'/'SS18' crosses (Table 17.2). Seed dormancy has been considered to be dominant in some cultivated varieties (Seshu and Sorrells, 1986; Shenoy, 1993). The change of ADD, with the process of after-ripening or by the genotypes used for the crosses, indicates that a highly positive dominance of dormancy occurs only in non-after-ripened seeds or caryopses from the crosses with a strongly dormant parent. Homozygous and heterozygous genotypes for dormancy can be distinguished from each other by germinating partially after-ripened seeds even in the cross where the dormancy is completely dominant.

Heritability (h^2_b) for seed germination varied with the duration of after-ripening (Table 17.3). Estimates of h^2_b were lower (0.64–0.76) at 0 DAH and highest (> 0.9) at 20 DAH in the F_2 populations. For seed germination of the

Table 17.2. Average degree of dominance (ADD) for seed and caryopsis germination.

Cross	Seeds at days after harvest (DAH) 0	20	40	60	80	Caryopses at 0 DAH
'CO39'/'SS18'	0.81	0.71	0.39	0.00	–0.16	–0.75
'LD'/'WYJ'	0.88	0.79	0.12	–0.50		0.90

ADD = $[M_{F1} - (M_{PD} + M_{PND})/2]/(M_{PD} - M_{PND})/2$, where M_{PD}, M_{PND} and M_{F1} are means of germination percentages in parental (dormant), parental (non-dormant) and F_1 generations, respectively.

F_2 populations from the 'LD'/'WYJ' cross at 60 DAH, the genetic variation was almost identical to environmental variation ($h^2_b = 0$). The F_2 populations derived from 'SS18', the parent with the longer duration of seed dormancy, maintained a higher h^2_b (> 0.8) at 40 and 60 DAH. Heritability was much higher for caryopsis germination in the F_2 population from the 'LD'/'WYJ' cross than in the 'SS18'-derived F_2 population (Table 17.3). These data from the F_1 and F_2 generations are consistent with a large genetic difference within the dormant lines for pericarp/testa-imposed dormancy.

The F_2 distribution was bimodal for the germination of seeds at 20 DAH in the 'CO39'/'SS18' cross (Fig. 17.1A) and for the germination of non-after-ripened caryopses in the 'LD'/'WYJ' cross (Fig. 17.1B). A similar distribution to that in Fig. 17.1A was also observed at 40 and 60 DAH for the 'CO39'/'SS18' cross (data not shown). The bimodal distributions coincided with the high h^2_b estimates (Table 17.3), supporting the hypothesis that there are major genes regulating the hull- and pericarp/testa-imposed dormancy in the crosses, respectively. The ratios of plants in the lower and higher germination groups were 120:71 and 82:49 in the 'SS18'- and 'LD'-derived F_2 populations, respectively, and this does not fit any monogenic or digenic model reported in cultivated varieties (Tripathi and Rao, 1982; Tomar, 1984; Seshu and Sorrells, 1986; Shenoy, 1993). It is possible that there are more than

Table 17.3. Broad-sense heritability (h^2_b) for seed and caryopsis germination.

Cross	Seeds at days after harvest (DAH) 0	20	40	60	Caryopses at 0 DAH
'CO39'/'SS18'	0.64[a]	0.93	0.82	0.84	0.34[a]
'LD'/'WYJ'	0.76[a]	0.95	0.55	0.00	0.82[a]

[a]Values are calculated with the germination data transformed by $\sin^{-1}(x)^{-0.5}$.
$h^2_b = [V_{F2} - (V_{PD} + V_{PND} + V_{F1})/3]/V_{F2}$, where V_{PD}, V_{PND}, V_{F1} and V_{F2} are the variances of germination percentages in parental (dormant), parental (non-dormant), F_1 and F_2 generations, respectively.

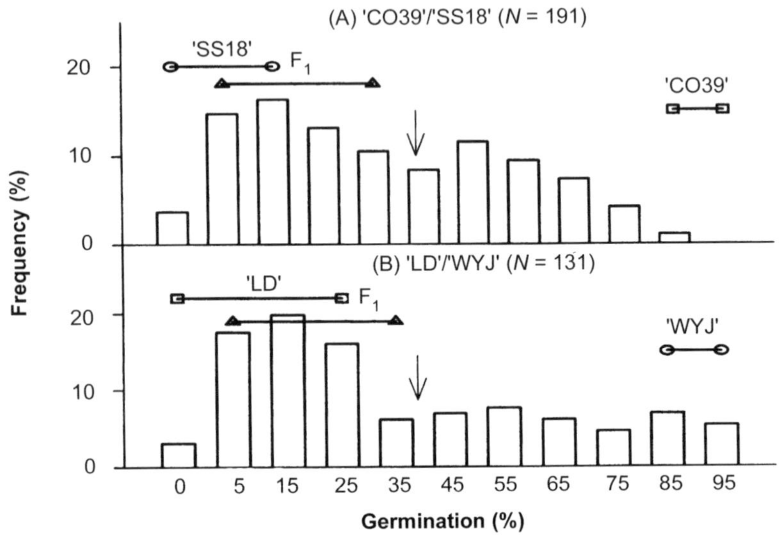

Fig. 17.1. Distribution of F_2 populations and range of parental and F_1 generations. (A) Germination of seeds at 20 days after harvest (DAH) in the 'CO39'/'SS18-2' cross. (B) Germination of non-after-ripened caryopses from the 'LD'/'WYJ' cross. The arrow (↓) is an assumed boundary segregating the groups with lower and higher germination. *N* is the sample size.

two major genes involved in the genetic control of the hull- or pericarp/testa-imposed dormancy in weedy rice.

Development of populations for the mapping of genes/QTLs and the selection of NILs

Phenotypic evaluation for seed dormancy in the mapping population was done by germinating the seeds from the original and ratoon F_2 plants harvested in 2000 and 2001, respectively. The germination of seeds harvested in these two years was highly correlated. For example, the linear correlation coefficient (*r*) was 0.6436** at 60 DAH. The *r* estimate indicates that there is still a large variation in seed germination between the years. It is necessary to repeat phenotypic assays for seed dormancy to identify genes/QTLs less affected by environmental factors.

The most dormant F_2-derived backcross has been advanced to the BC_2F_1 generation. There were several traits that displayed a wide range of variation in the weedy rice-derived F_2 populations. These included seed shattering, flowering date and partial sterility that may affect seed germination or significantly increase phenotypic variation for seed dormancy. After two generations of backcross and phenotypic selection, seed shattering and sterility had been eliminated, and many morphological traits such as flowering time and plant height were synchronized to a large extent (data not shown). Distribution of BC_2F_1 plants for seed germination was bimodal

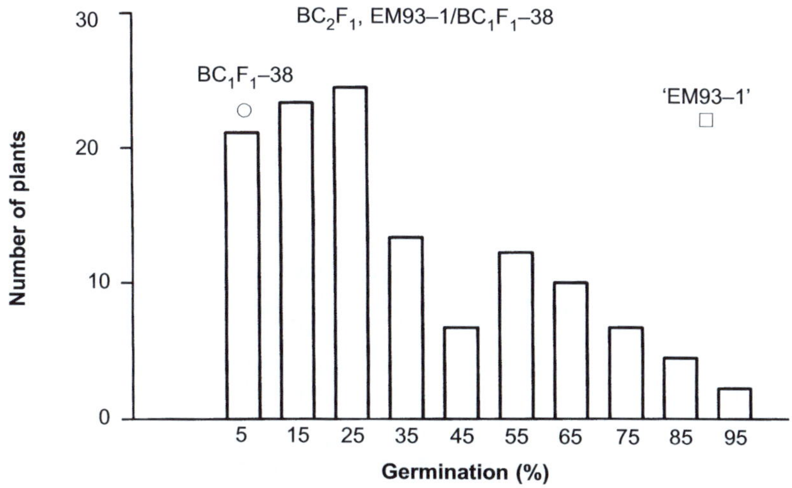

Fig. 17.2. Distribution of plants in BC_2F_1 ('EM93-1'///'EM93-1'/F_2-14) population for germination of seeds at 12 days after harvest. The F_2-14 was a dormant F_2 plant selected from the 'CO39'/'SS18' cross. BC_1F_1-38 was a dormant plant selected from the last generation.

(Fig. 17.2), which was similar to that in the BC_1F_1 population (data not shown). The distinct bimodal distribution indicates that the major genes for hull-imposed dormancy have been successfully retained in this population.

Polymorphic rates for AFLP and microsatellite markers between parental lines are shown in Table 17.4. The polymorphism for AFLP markers was 4–5% higher between 'SS18' and non-dormant lines 'CO39' and 'EM93-1' than between two non-dormant lines. This difference suggests a larger genetic distance between the weedy strain and cultivated varieties. High polymorphism at the DNA level will allow us to scan the genome for genes/QTLs for hull-imposed dormancy from the mapping population and

Table 17.4. Polymorphic rates of AFLP or microsatellite markers between parental lines.

Parent pairs	Number of primer combinations	Number of polymorphic alleles	Polymorphic rate (%)
'CO39' vs. 'SS18'	35[a]	25	71
'CO39' vs. 'SS18'	30[b]	190	17[c]
'EM93-1' vs. 'SS18'	7[b]	51	16[c]
'CO39' vs. 'EM93-1'	7[b]	39	12[c]

[a]Microsatellite primers selected from 12 rice chromosomes.
[b]AFLP primer combinations of *Pst*I(3)/*Mse*I(3); the figures in parentheses are the numbers of selective nucleotides at the 3′ ends of the primers.
[c]The percentage is the number of polymorphic alleles to the total number of amplified alleles.

to do a fine mapping of major genes/QTLs in the advanced backcross populations.

Conclusion

Seed dormancy in selected *O. sativa* weedy and cultivated strains and *O. glaberrima* cultivars was imposed by the seed coverings, i.e. hull and pericarp/testa. Genotypes with a higher level or a much longer duration of seed covering-imposed dormancy were present in weedy rice strains. Genetic analysis was conducted using weedy rice strains 'SS18' and 'LD' as parents with strong hull- and pericarp/testa-imposed dormancy, respectively. A highly positive dominance and a high level of heritability detected in the F_1 and F_2 generations confirmed that these two strains are appropriate gene donors for map-based cloning of the QTLs for seed covering-imposed dormancy. Seed dormancy, as judged by seed germination, was affected by environmental factors and the genetic background. We used a ratooning technique to repeat the phenotypic evaluation for a primary mapping population to reduce the environmental variation. We have used backcrossing and phenotypic selection techniques to begin development of NILs for major seed dormancy genes. In the near future, we shall also use marker-assisted selection techniques to develop NILs.

Acknowledgements

Weedy strain 'SS18' was kindly provided by Dr H.S. Suh. This work was supported by grants from USDA-National Research Initiative (9804919 and 0200668).

References

Chang, T.T. and Tagumpay, O. (1973) Inheritance of grain dormancy in relation to growth duration in 10 rice crosses. *SABRAO Newsletter* 5, 87–94.

Chang, T.T. and Yen, S.T. (1969) Inheritance of grain dormancy in four rice crosses. *Botanical Bulletin of Academia Sinica* 10, 1–9.

Gale, M.D., Devos, K.M. and Moore, G. (1996) Rice as the pivotal genome in the new era of grass comparative genetics. In: Khush, G.S. (ed.) *Rice Genetics III*. IRRI, Manila, Philippines, pp. 77–84.

Goff, S.A., Ricke, D., Lan, T.H., Presting, G., Wang, R.L., Dunn, M., Glazebrook, J., Sessions, A., Oeller, P., Varma, H., Hadley, D., Hutchinson, D., Martin, C., Katagiri, F., Lange, B.M., Moughamer, T., Xia, Y., Budworth, P., Zhong, J.P., Miguel, T., Paszkowski, U., Zhang, S.P., Colbert, M., Sun, W.L., Chen, L.L., Cooper, B., Park, S., Wood, T.C., Mao, L., Quail, P., Wing, R., Dean, R., Yu, Y.S., Zharkikh, A., Shen, R., Sahasrabudhe, S., Thomas, A., Cannings, R., Gutin, A., Pruss, D., Reid, J., Tavtigian, S., Mitchell, J., Eldredge, G., Scholl, T., Miller, R.M., Bhatnagar, S., Adey, N., Rubano, T., Tusneem, N., Robinson, R., Feldhaus, J., Macalma, T., Oliphant, A. and Briggs, S. (2002) A draft sequence of the rice genome (*Oryza sativa* L. ssp. *japonica*). *Science* 296, 92–100.

Mather, K. and Jinks, J.L. (1972) *Biometrical Genetics: the Study of Continuous Variation*. Cornell University Press, Ithaca, New York.

Oka, H.I. (1988) *Origin of Cultivated Rice.* Japan Scientific Societies Press, Tokyo.

Seshu, D.V. and Sorrells, M.E. (1986) Genetic studies on seed dormancy in rice. In: *Rice Genetics.* IRRI, Manila, Philippines, pp. 369–382.

Shenoy, V.V. (1993) Genetics of hull-imposed dormancy in rice seeds. *Rice Genetics Newsletter* 10, 108–109.

Simpson, G.M. (1990) *Seed Dormancy in Grasses.* Cambridge University Press, Cambridge, UK, pp. 3–113.

Suh, H.S., Sato, Y.I. and Morishima, H. (1997) Genetic characterization of weedy rice *(Oryza sativa* L.) based on morph-physiology, isozymes and RAPD markers. *Theoretical and Applied Genetics* 94, 316–321.

Temnykh, S., Park, W.D., Ayres, N., Cartinhour, S., Hauck, H., Lipovich, L., Cho, Y.G., Ishii, T. and McCouch, S.R. (2000) Mapping and genome organization of microsatellite sequences in rice (*Oryza sativa* L.). *Theoretical and Applied Genetics* 100, 697–712.

Tomar, J.B. (1984) Genetics of grain dormancy in rice (*Oryza sativa* L.). *Genetica Agraria* 38, 443–446.

Tripathi, R.S. and Rao, M.J.B. (1982) Inheritance of grain dormancy in rice. *Oryza* 19, 17–19.

Vos, P., Hogers, R., Bleeker, M., Reijans, M., Lee, T. van de, Hornes, M., Frijters, A., Pot, J., Peleman, J., Kuiper, M. and Zabeau, M. (1995) AFLP: a new technique for DNA fingerprinting. *Nucleic Acids Research* 23, 4407–4414.

18 PCNA-associated Proteins During Maize Germination

Jorge M. Vázquez-Ramos and María de la Paz Sánchez

Departamento de Bioquímica, Facultad de Química, Universidad Nacional Autónoma de México, Avenida Universidad y Copilco, 04510 México DF, México

Introduction

Dry seed embryos contain cell populations that are mostly in the G1 phase of the cell cycle (Deltour and Jacqmard, 1975; Conger and Carabia, 1976; Bewley and Black, 1994), possibly because the morphogenetical programme for embryo formation imposes a halt in proliferation at the end of mitosis or in early G1. There might be a developmental starvation of mitogenic signals that would lead cells to a G0-like stage. Cells blocked at any other stage of the cell cycle (with the possible exception of cells in G2) may not be stable and may not be able to survive the period of seed maturation and drying.

Upon imbibition, embryonic cells will be devoted to repairing cell structures and macromolecules damaged or disassembled during drying, with the purpose of re-establishing metabolism. It is very likely that the perception of healthy structures and metabolism is the initiator of the germination process *sensu stricto*, and cells will trigger the biochemical pathways towards proliferation and establish the G1 phase. The S-phase delay suffered by germinating aged or deteriorated seeds exemplifies the importance of a proper G1 phase taking place during early germination (Sen and Osborne, 1974; Elder *et al.*, 1987; Gutiérrez *et al.*, 1993). Inhibition of seed germination in the presence of abscisic acid, with cells blocked in the G1 phase (Bewley and Black, 1994), would also point to its relevance.

Previous work in our laboratory with maize (*Zea mays* L.) seed osmopriming indicated that meristematic cells were blocked at a stage previous to DNA replication. Release of the germination block caused a rapid entry into the S phase, corroborating the need for establishment of the G1 phase for successful germination (Cruz-García *et al.*, 1995).

Searches for cell cycle markers in cells of dry maize seed embryo axes have indicated that several G1, S and G2/M markers are present, as if cells were ready to initiate proliferation (García *et al.*, 1997; Cruz-García *et al.*,

1998; Herrera-Teigeiro *et al.*, 1999). Interestingly, markers like a cyclin D-type protein and proliferating cell nuclear antigen (PCNA) are soon degraded after imbibition and new proteins are synthesized (Cruz-García *et al.*, 1998; Herrera *et al.*, 2000). This evidence suggests that cell cycle proteins stored in dry embryos may not be functional and that newly made proteins would be necessary, explaining the reported inhibition of germination by inhibitors of transcription/translation (Vázquez-Ramos and Reyes-Jiménez, 1990; Zúñiga-Aguilar *et al.*, 1995).

Cyclin D, together with a cyclin-dependent kinase (Cdk), controls the time of entry into the G1 phase (Morgan, 1997). Mammalian cyclin D, complexed with Cdk4 (or Cdk6), receives information proceeding from external stimuli and, activated as a kinase, phosphorylates and thus stimulates the pathway to initiate an irreversible cell cycle (Xiong *et al.*, 1992). In plants, several D-type cyclins have been described and their expression may respond not only to proliferation conditions but also to differentiation and developmental conditions (Meijer and Murray, 2001).

In mammal cells the cyclin D protein, in addition to its association with a Cdk protein, has also been found associated with PCNA (Xiong *et al.*, 1992). The latter is the processivity factor for replicative DNA polymerases, but in recent years PCNA has been found associated with different cyclin/kinase complexes, suggesting that PCNA might participate in the recognition of target proteins by cyclin/Cdks (Kelman, 1997).

PCNA Protein and Cyclin D During Maize Germination

PCNA is a 29 kDa protein that associates as a homotrimeric ring encircling the DNA in the transition double strand–single strand that represents the growing replication fork. In this way, PCNA serves as a sliding clamp for replicative δ or ε DNA polymerases (Kelman, 1997). Maize PCNA can be found in dry seeds as a monomer and in high-molecular-size complexes of 150 and more than 200 kDa, as determined by Western blots of native polyacrylamide gels. As germination advances, the homotrimeric form of PCNA starts to appear, showing a maximum by the time the S phase is established (15 h after imbibition); by this time, the amount of the high-molecular-size complexes diminishes (Sánchez *et al.*, 2002). Similar results have been produced following a different methodology: gel filtration through Superdex 200. Therefore, PCNA in dry seeds and during early germination is associated with proteins forming complexes, the nature of which was a goal pursued here.

As indicated above, mammalian cyclin D associates with PCNA. Maize PCNA was also found associated with the D-type maize cyclin in different ways. Firstly, the putative cyclin D was found to co-elute, through Superdex 200, with PCNA in the same fractions. Secondly, the PCNA-containing high-molecular-size protein complexes in native gels, described above, when denatured and separated by polyacrylamide electrophoresis contained a 52 kDa protein, previously identified as a maize D-type cyclin (Cruz-García *et al.*, 1998), as detected with a heterologous antibody against human cyclin D.

Thirdly, these same anti-cyclin D antibodies co-precipitated maize PCNA from maize embryo protein extracts.

Association between PCNA and cyclin D was dependent on germination progression. The proteins co-immunoprecipitated only during the first 15 h of germination. This correlates well with the previous finding that cyclin D is only stable during the first 15 h of germination and is then degraded (Herrera *et al.*, 2000). Addition of phytohormones to germinating seeds changed the timing of association. Both cytokinins (benzyladenine, BA) and abscisic acid (ABA) caused the complex to dissociate after 6 h of germination – a puzzling result, since cytokinins accelerate germination and ABA blocks germination. Under the influence of BA, cyclin D disappears after only 6 h of germination and thus it would only be natural not to find cyclin D in immunoprecipitates; however, cyclin D is stabilized by treating embryo axes with ABA and can be found even after 24 h of germination (Herrera *et al.*, 2000). Therefore, the physiological process that the complex of PCNA and cyclin D performs during the G1 phase of the cell cycle must be disrupted by ABA .

PCNA, Cyclin D and Cdk-A During Maize Germination

Kinase activity in cyclin/Cdk complexes is due to the Cdk subunit. Neither the cyclin nor the Cdk moieties alone can phosphorylate substrates. It was therefore necessary to show that the association of PCNA and cyclin D in maize cells also included a Cdk-type protein. In mammal cells cyclin D associates with the Cdk4 kinase, implying that there are other Cdks. In fact, Cdk1 was the first Cdk discovered and its original name was (and still is) $p34^{Cdc2}$ (Simanis and Nurse, 1986). In yeast cells, this kinase participates both in G1/S and in G2/M transitions, associating with different cyclins (Nurse, 1994). In higher eukaryotes, the equivalent protein to $p34^{Cdc2}$ is Cdk1, a G2/M kinase. Cdk1 is distinguished from other Cdks in that it contains in the cyclin-binding box a conserved sequence, PSTAIRE, that is present in Cdk2 and Cdk3 but not in Cdk4 or Cdk6 (Morgan, 1997). Searches for a G1, Cdk4(6)-type of kinase in plant cells have been unsuccessful; instead, two general types of Cdks have been found; Cdk-A and Cdk-B. The A-type Cdk is a G1 and G2/M kinase whereas Cdk-B is a G2/M kinase. Furthermore, Cdk-A is a PSTAIRE kinase, whereas Cdk-B has a variation of this sequence (Mironov *et al.*, 1999).

Antibodies developed against the PSTAIRE motif helped us to identify a Cdk-A type kinase in maize embryo axis cells. In fact, two PSTAIRE-containing proteins of 36 and 32 kDa were found. Interestingly, the lower-molecular-size protein was present only in the period of 3–15 h of germination, whereas the 36 kDa protein was present all the time during germination, showing little variation (Sánchez *et al.*, 2002). The existence of two genes coding for $p34^{Cdc2}$-like proteins in maize had been reported before (Colasanti *et al.*, 1991).

Using an antibody developed against the PSTAIRE sequence in Cdk1, it was demonstrated that the two Cdk-A-type proteins co-immunoprecipitate

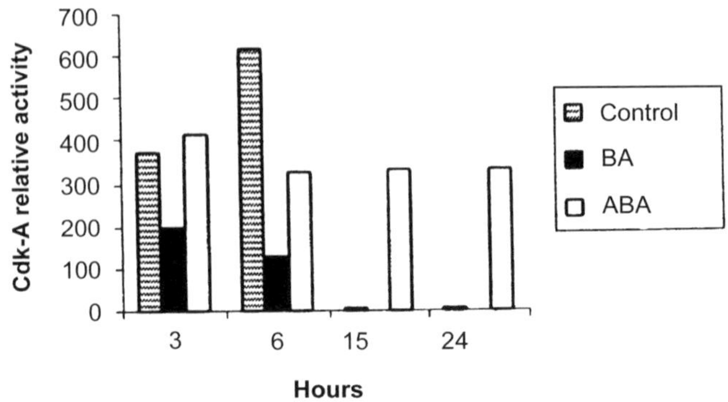

Fig. 18.1. PCNA-associated Cdk-A kinase activity during maize germination. Effect of phytoregulators: BA, benzyladenine; ABA, abscisic acid.

with maize PCNA. The 32 kDa protein was associated only during the first 6 h of germination, whereas the 36 kDa protein remained associated during the whole period measured. More relevant was the finding that kinase activity in PCNA immunoprecipitates was high only in the period of 0–6 h of germination (Fig. 18.1), corresponding with the time of association to the lower molecular form of the kinase, even though the 36 kDa kinase was still associated with PCNA (Sánchez *et al.*, 2002). The substrate used in these kinase assays was the maize retinoblastoma-related protein (ZmRBR), homologous to the natural target of G1 cyclin complexes in mammal cells (Brehm and Kouzarides, 1999). For comparison, the A-type Cdks were assayed under a different protocol. In yeast cells, the complex formed by $p34^{Cdc2}$ and cyclin B often contains another protein known as $p13^{Suc1}$, which strongly binds the Cdk moiety (Brizuela *et al.*, 1987). This protein (recently demonstrated in mammal and plant cells) has been used in affinity columns to bind Cdk proteins. Thus, $p13^{Suc1}$ was used to pull down Cdk-A from protein extracts derived from germinating maize seed axes and then associated Cdks were identified and kinase activity was measured. It was surprising to find that $p13^{Suc1}$ only binds the 36 kDa protein and that kinase activity could only be measured after 10–15 h of germination – the exact opposite behaviour of kinase activity in PCNA immunoprecipitates (Sánchez *et al.*, 2002).

These results are indicative of the existence of at least two protein complexes to which PCNA is bound during germination: one containing the 32 kDa protein and the other containing the 36 kDa protein. The behaviour of the 32 kDa kinase would strongly suggest that PCNA-associated kinase activity during early germination is the responsibility of this kinase, although other possibilities cannot be eliminated; the 36 kDa associated kinase may have other protein targets and be active in the S/G2 phases. Another possibility is that PCNA complexes may contain different associated cyclins: besides the D-type cyclin, PCNA appears to bind a cyclin A-type protein during germination, which would strengthen the idea that different PCNA complexes may phosphorylate different substrates.

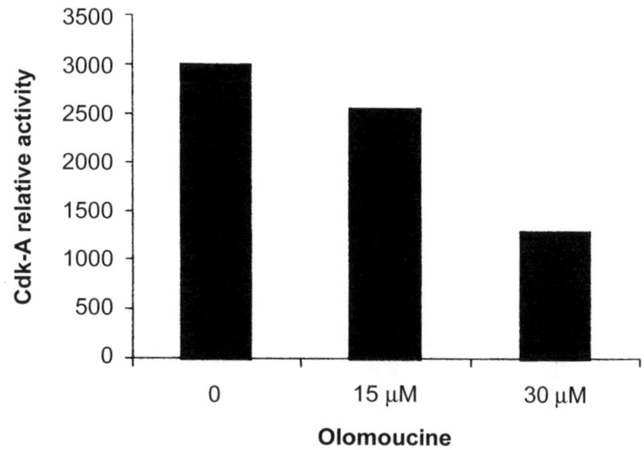

Fig. 18.2. Inhibition of PCNA-associated Cdk-A kinase activity by olomoucine, a Cdk-specific inhibitor.

Further corroboration that the kinase activity displayed by PCNA immunoprecipitates is of Cdk-A type has been obtained using the specific Cdk inhibitors roscovitin (data not shown) and olomoucine, which reduced kinase activity down to 40% of control activity (Fig. 18.2).

Phytoregulators and PCNA Complexes

As mentioned above, plant hormones such as cytokinins and abscisic acid influence seed germination, the former by accelerating the time of entry into S phase, presumably by shortening the G1 phase (Reyes-Jiménez *et al.*, 1991). ABA blocks germination in what appears to be a pre-S state, although the mechanism has not been elucidated. PCNA complexes seem to be mainly G1 complexes and thus these may be the target of hormonal action.

PCNA-associated kinase activity was measured in immunoprecipitates from embryo axes germinated for different times in the presence/absence of phytoregulators during a 24 h experimental period. Similar to the control, BA caused an early reduction in kinase activity by 6 h of germination (Fig. 18.1); however, this happened only if the substrate was ZmRBR. In this case, none the less, both the 32 and the 36 kDa PSTAIRE proteins were PCNA-bound all the time and there was phosphorylation of an endogenous protein of around 60 kDa that co-precipitates with PCNA and which is the target of the Cdk-A kinase up to 24 h of germination.

The effect of ABA was even more revealing. In its presence, the PCNA-associated kinase from all germination times, 3–24 hours, phosphorylated ZmRBR equally well (Fig. 18.1). The endogenous 60 kDa protein was also phosphorylated. These results indicate that the complex formed by PCNA is a true G1 complex and that the activity of the associated kinase must somehow be reduced or eliminated for cells to go into the S phase, as demonstrated by its faster disappearance in BA-treated seed axes and the continuous kinase activity in ABA-treated seed axes. In addition, the PCNA-associated kinase has a developmentally regulated capacity to recognize substrates. The easiest way to explain this is by considering a change in the associated cyclin protein, i.e. the cyclin D-like protein present in the early hours of germination may be substituted by another D-type cyclin or else by an A-type cyclin, conferring this new kinase complex with a different substrate specificity. As was mentioned above, PCNA can form complexes with a cyclin A-like protein during germination.

PCNA, Cytokinins and the S Phase

The S phase starts 12–15 h after maize seed imbibition and the activity of a replicative polymerase increases gradually as germination advances; this increased activity correlates with a concomitant gradual phosphorylation of some of its subunits (Baíza *et al.*, 1989; Coello and Vázquez-Ramos, 1995). The S phase start time is dramatically shortened if seeds are imbibed in the presence of cytokinins; under these conditions the S phase starts by 4–6 h after imbibition and at the same time the activity of the α-type DNA polymerase increases several-fold (Reyes-Jiménez *et al.*, 1991; Vázquez-Ramos, 2000).

To find out if DNA polymerase α is the target of the PCNA-associated kinase(s), and if phosphorylation increases polymerase activity, polymerase extracted from dry seed embryo axes (in a non-phosphorylated state; Coello and Vázquez-Ramos, 1995) was mixed with PCNA immunoprecipitates from axes imbibed for different times in the presence/absence of BA. The result showed that phosphorylation of the polymerase catalytic subunit (103 kDa) was more evident in immunoprecipitates of 13 h of germination under control conditions whereas under BA conditions, the peak of phosphorylation was obtained in immunoprecipitates of 3 hours of germination (Gómez and Vázquez-Ramos, unpublished observations). This result indicates that the PCNA-associated kinase(s) is either activated, or its subunit composition modified, by the action of BA. However, phosphorylation did not modify enzyme activity, suggesting that the BA-promoted phosphorylation and stimulation of DNA polymerase has more to do with the way this enzyme establishes contacts with the template and with other protein factors at the replication fork.

Conclusions

In mammal cells and in yeast, the G1 cyclin/Cdk kinase complex is essential for cell cycle triggering. This kinase complex is the receptor and the interpreter of mitogenic signals elicited by external stimuli. The binding of the nuclear-localized cyclin/Cdk complex to PCNA (also a nuclear protein) could be understood in terms of the need of the kinase to locate its target substrates near the chromosome. PCNA could act as a dynamic platform.

The existence in plants of the corresponding orthologues indicates that these proteins are evolutionarily conserved. It is most relevant that a similar

PCNA/cyclin/Cdk complex can be isolated in seeds, and that this complex is functional during early germination, indicating that the basic mechanism is also evolutionarily conserved. The results presented, including phosphorylation of the native substrate, the RBR protein, protein kinase inhibition by specific inhibitors, and association to PCNA, all validate the definition of the kinase moiety as a true G1 cyclin/Cdk.

Several features point to the importance of this complex in seed germination: both cyclin D and PCNA, although present in dry seed cells, are produced *de novo* during early germination. The kinase in PCNA immunoprecipitates is only active, with RBR as substrate, during the G1 phase period of germination. Modulation of kinase presence and activity by phytohormones is consistent with its role in G1: with cytokinins, activity on RBR disappears soon due to cell cycle activation, while with ABA, the G1-phase kinase is stabilized for at least 24 h, cells cannot go into S phase and germination is inhibited. Figure 18.3 shows a model illustrating these events during germination.

All these results strengthen the idea that initiation of the cell cycle is an important event for germination establishment and completion. Much more molecular and physiological work should be done to substantiate and verify its validity.

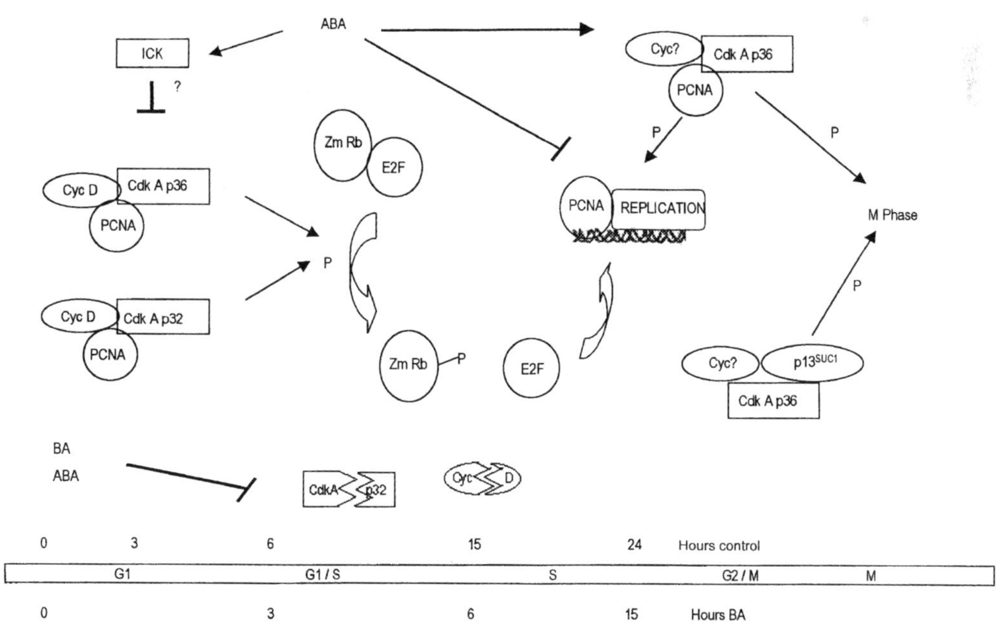

Fig. 18.3. Schematic view of cell cycle events taking place during maize germination. During the G1 stage in germination, the two PCNA-associated Cdks can phosphorylate ZmRBR protein and, very likely, release an E2F-like protein, allowing onset of the S phase. Cdk activity and the S phase would be regulated by phytohormones. PCNA also forms protein complexes with Cdk-A in S and G2 phases, probably including different cyclins, and these may influence the G2/M phases. Cdk-A also seems to form complexes without PCNA that would participate in G2/M. P, inorganic phosphate.

Acknowledgements

We thank PAPIIT for grant IN 206799.

References

Baíza, A.M., Vázquez-Ramos, J.M. and Sánchez de Jiménez, E. (1989) DNA synthesis and cell division in embryonic maize tissues during germination. *Journal of Plant Physiology* 135, 416–421.

Bewley, J.D. and Black, M. (eds) (1994) *Seeds. Physiology of Development and Germination*, 2nd edn. Plenum Press, New York.

Brehm, A. and Kouzarides, T. (1999) Retinoblastoma protein meets chromatin. *Trends in Biochemical Sciences* 24,142–145.

Brizuela, L., Draetta, G. and Beach, D. (1987) $p13^{suc1}$ acts in the fission yeast cell division cycle as a component of the $p34^{cdc2}$ protein kinase. *EMBO Journal* 6, 3507–3514.

Coello, P. and Vázquez-Ramos, J.M. (1995) Maize DNA polymerase 2 is a phosphoprotein with increasing activity during germination. *European Journal of Biochemistry* 231, 99–103.

Colasanti, J., Tyers, M. and Sundaresan, V. (1991) Isolation and characterization of cDNA clones encoding a functional $p34^{cdc2}$ homologue from *Zea mays*. *Proceedings of the National Academy of Sciences USA* 88, 3377–3381.

Conger, V.V. and Carabia, J.V. (1976) Microspectrophotometric determination of the 2C and 4C nuclear complement in the root and shoot of the dormant maize embryo. *Environmental and Experimental Botany* 16, 171–175

Cruz-García, F., Jiménez, L.F. and Vázquez-Ramos, J.M. (1995) Biochemical and cytological studies in osmoprimed maize seeds. *Seed Science Research* 5, 15–23.

Cruz-García, F., Zúñiga-Aguilar, J.J. and Vázquez-Ramos, J.M. (1998) Effect of stimulating maize germination on cell cycle proteins. *Physiologia Plantarum* 102, 573–581.

Deltour, R. and Jacqmard, M. (1975) Relation between water stress and DNA synthesis during germination of *Zea mays* L. *Annals of Botany* 38, 329–334.

Elder, R.H., Dell'Aquila, A., Mezzina, M., Sarasin, A. and Osborne, D.J. (1987) DNA ligase in repair and replication in the embryos of rye, *Secale cereale*. *Mutation Research* 181, 61–71.

García, E., Orjuela, D., Camacho, Y., Zúñiga, J.J., Plasencia, J. and Vázquez-Ramos, J.M. (1997) Comparison among DNA polymerases 1, 2 and 3 from maize embryo axes. A DNA primase activity copurifies with DNA polymerase 2. *Plant Molecular Biology* 33, 445–455.

Gutiérrez, G., Cruz, F., Moreno, J., González-Hernández, V. and Vázquez-Ramos, J.M. (1993) Natural and artificial seed ageing in maize: germination and DNA metabolism. *Seed Science Research* 3, 279–285.

Herrera, I., Sanchez, M.P., Molina, J., Plasencia, J. and Vázquez-Ramos, J.M. (2000) Proliferating cell nuclear antigen expression in maize seed development and germination: regulation by phytohormones and its association with putative cell cycle proteins. *Physiologia Plantarum* 110, 127–134.

Herrera-Teigeiro, I., Jiménez-García, L.F. and Vázquez-Ramos, J.M. (1999) Benzyladenine promotes early activation of $p34^{cdc2}$-like kinase(s) during maize germination. *Seed Science Research* 9, 55–62.

Kelman, Z. (1997) PCNA: structure, functions and interactions. *Oncogene* 14, 629–640.

Meijer, M. and Murray, J.A.H. (2001) Cell cycle controls and development of plant

form. *Current Opinion in Plant Biology* 4, 44–49.
Mironov, V., De Veylder, L., Van Montagu, M. and Inzé, D. (1999) Cyclin-dependent kinases and cell division in plants: the nexus. *The Plant Cell* 11, 509–521.
Morgan, O.D. (1997) Cyclin-dependent kinases: engines, clocks and microprocessors. *Annual Review of Cell Developmental Biology* 13, 261–291.
Nurse, P. (1994) Ordering S phase and M phase in the cell cycle. *Cell* 79, 547–550.
Reyes-Jiménez, J., Jiménez-García, L.F., González, M.A. and Vázquez-Ramos, J.M. (1991) Benzyladenine stimulation of nuclear DNA synthesis and cell division in germinating seeds. *Seed Science Research* 1, 113–117.
Sánchez, M.P., Torres, A., Boniotti, M.B., Gutierrez, C. and Vázquez-Ramos, J.M. (2002) PCNA protein associates to Cdk-A type protein kinases in germinating maize. *Plant Molecular Biology* 50, 167–175.
Sen, S. and Osborne, D.J. (1974) Germination of rye embryos following hydration-dehydration treatments: enhancement of protein and RNA synthesis and earlier induction of DNA replication. *Journal of Experimental Botany* 25, 1010-1019.
Simanis, V. and Nurse, P. (1986) The cell cycle control gene cdc2+ of fission yeast encodes a protein kinase potentially regulated by phosphorylation. *Cell* 45, 261–268.
Vázquez-Ramos, J.M. (2000) Cell cycle control during maize germination. In: Black, M., Bradford, K.J. and Vázquez-Ramos, J. (eds) *Seed Biology. Advances and Applications*. CAB International, Wallingford, UK, pp. 261–269.
Vázquez-Ramos, J.M. and Reyes-Jiménez, J. (1990) Stimulation of DNA synthesis and DNA polymerase activity by benzyladenine during early germination of maize axes. *Canadian Journal of Botany* 68, 2590–2594.
Xiong, Y., Zhang, H. and Beach, D. (1992) D-Type cyclins associate with multiple protein kinases and the DNA replication and repair factor PCNA. *Cell* 71, 505–514.
Zúñiga-Aguilar, J.J., López, I., Gómez, A. and Vázquez-Ramos, J.M. (1995) Does benzyladenine stimulate DNA metabolism by modifying gene expression during maize germination? *Seed Science Research* 5, 219–226.

19 Tissue Printing for Localization of mRNA Expression in Seeds

HIROYUKI NONOGAKI[1] AND KENT J. BRADFORD[2]

[1]Department of Horticulture, Oregon State University, Corvallis, OR 97331-7304; USA; [2]Department of Vegetable Crops, One Shields Avenue, University of California, Davis, CA 95616-8631, USA

Introduction

Recent advances in plant genome projects and mutant research have identified a number of interesting genes that are potentially involved in seed development, germination and post-germinative seedling growth (Debeaujon *et al.*, 2000, 2001; Luo *et al.*, 2000; Yadegari *et al.*, 2000; Penfield *et al.*, 2001; Stone *et al.*, 2001; Borisjuk *et al.*, 2002; Lee *et al.*, 2002; Raz and Koornneef, 2002; Tzafrir *et al.*, 2002; Xiong *et al.*, 2002). To understand the physiological roles of the gene products and the regulation of gene expression, it is necessary to characterize the timing and localization of gene expression in seeds. Information on the temporal and spatial patterns of gene expression will shed light on the functions of the seed-specific genes.

RNA gel blotting can provide information on gene expression but requires careful extraction and manipulation and is relatively time-consuming. To determine the localization of gene expression in seeds, different seed parts, such as radicle, hypocotyls and cotyledons of the embryo and the micropylar and the lateral parts of the endosperm, must be dissected and RNA extracted from the small seed parts separately. To analyse the manner of regulation of gene expression, such as developmental and hormonal regulation, repeated dissection of seed tissues is required for different developmental stages or different chemical treatments to prepare a complete set of RNA samples for a single RNA blot.

In situ hybridization is used to detect specific mRNAs to determine the tissue localization of gene expression. However, in many cases it is difficult to obtain high-quality thin tissue sections from seeds. In the case of developing seeds in which the tissues are still fresh and soft, sectioning of the tissues is relatively less problematic but, even when mature seeds are fully imbibed, it is often difficult to obtain thin sections from them consistently. The difference in hardness between the outer layers (e.g. pericarp and testa) and the inner layers (e.g. endosperm, embryo) of mature seed makes it difficult to produce uniform and reproducible tissue sections from seeds. Seed tissues tend to dislodge from paraffin or plastic during sectioning because of separation between the embedding material and seed tissue sections. The fixatives such as glutaraldehyde and paraformaldehyde are not sufficiently infiltrated into seed tissues because of the high density of seed storage reserves such as proteins and starch in the cells. Therefore, it is difficult to fix all of the cells in seed tissues. This is especially the case for the endosperm tissues of the seeds of dicotyledonous species that have thick cell walls and many lipid and protein bodies in the cells. Although dissecting a seed into thin slices before fixation enhances the efficiency of fixation, the embryos can be lost from the seed slices before the tissues are embedded.

As an alternative method, we applied tissue printing (Ye and Varner, 1991) for determining the localization of gene expression in seeds (Nonogaki *et al.*, 2000). Seed tissue printing produces clear results with high reproducibility. Since this method uses a colorimetric reaction for detecting gene expression, the localization of gene expression can be visualized directly on the printed images of seeds and considerable anatomical detail is retained. In this chapter, we describe the detailed protocol of our tissue printing method with examples of application of this technique in our research on tomato seed germination.

Printing of Half-seeds

Bisected or half-seeds are usually used for tissue printing. Imbibed seeds are bisected with a sharp razor blade. A double-edged razor blade is more desirable because it is thinner than a single-edged razor blade and produces a cleanly cut surface for printing. It is very important to keep the cut surface of the half-seed intact for obtaining a clear image of the tissues and RNA signal. As shown in Fig. 19.1, half-seeds are placed with the cut surface down on a positively charged membrane such as Hybond N^+ (Amersham Pharmacia Biotech, Piscataway, USA) supported on multiple sheets of paper towels. Using soft layers of paper towels under the membrane is important to obtain a deep print of the seed. If the membrane is supported on a hard stage, the printed image will not be deep enough to visualize seed morphology. To avoid contamination by RNases, seed halves are covered with plastic film and pressed by a gloved thumb or finger for 15 s (Fig. 19.1). Based on our experience with several different gene probes, 15 s are enough to transfer a detectable amount of mRNA molecules (Chen and Bradford, 2000; Nonogaki *et al.*, 2000; Wu *et al.*, 2001; Chen *et al.*, 2002). When printing is done for too long, the prints typically result in fuzzy images because half-seeds are slightly moved during a prolonged period of printing. Seed must also be imbibed sufficiently or transfer of RNA and proteins to the membrane is poor.

Since we developed our tissue printing method to detect gene expression in the micropylar region of the endosperm of tomato seeds, we used longitudinally bisected half-seeds. It is also possible to obtain the prints of

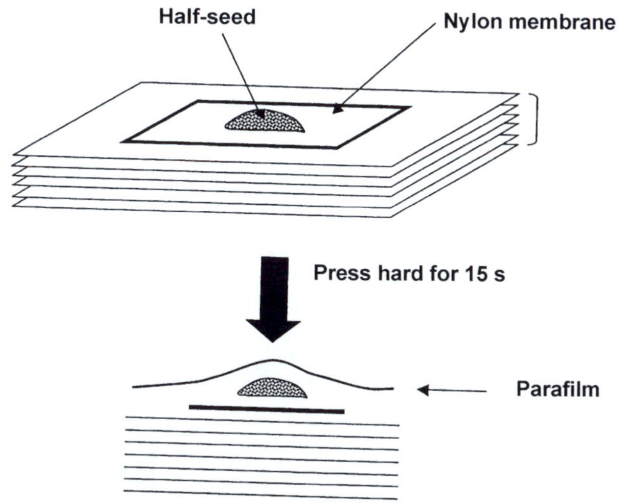

Fig. 19.1. A scheme for printing half-seeds. Seeds are bisected into halves and pressed with the cut surface down on to a positively charged nylon membrane supported by several layers of paper towels. Printing is done for 15 s and then the tissue is removed from the membrane. To avoid RNase contamination, the half-seed is covered by plastic film when printed.

transversely dissected seed tissues. The prints of transversely bisected tomato seeds are clear enough to distinguish the embryonic and endospermic tissues (Fig. 19.2A). In tissue prints of carrot seeds, the individual cells in the endosperm are visible (Fig. 19.2B). Tissue printing of transversely dissected seeds will be useful, for example, for determining the distribution of gene expression in the endosperm in response to hormones secreted by the embryo, such as gibberellic acid. The printing method described above can be applied for many species of seeds.

Probe Synthesis and Hybridization

We have used digoxigenin (DIG)-labelled RNA probes for hybridization. For synthesis of RNA probes, specific cDNAs were subcloned into pBSIIKS vector (Stratagene, La Jolla, California) and transcribed using T3 or T7 promoters in the vector to make antisense RNA probes. The DIG-labelled UTPs (Amersham Pharmacia Biotech, Piscataway, New Jersey) are incorporated into the RNA molecules when the antisense probes are synthesized. Hybridization is done at 60°C with a hybridization buffer containing 50% (v/v) deionized formamide, 4% (w/v) blocking reagent (Roche, Indianapolis, Indiana), 0.2% (w/v) SDS, 5× SSC, and approximately 100 ng/ml RNA probe after 1 h pre-hybridization at the same temperature. After hybridization, the membranes are washed once for 20 min with 2× SSC, 0.1% (w/v) SDS at 70°C and twice for 20 min with 0.2× SSC, 0.1% (w/v) SDS at 70°C. The combination of RNA probes and high stringency

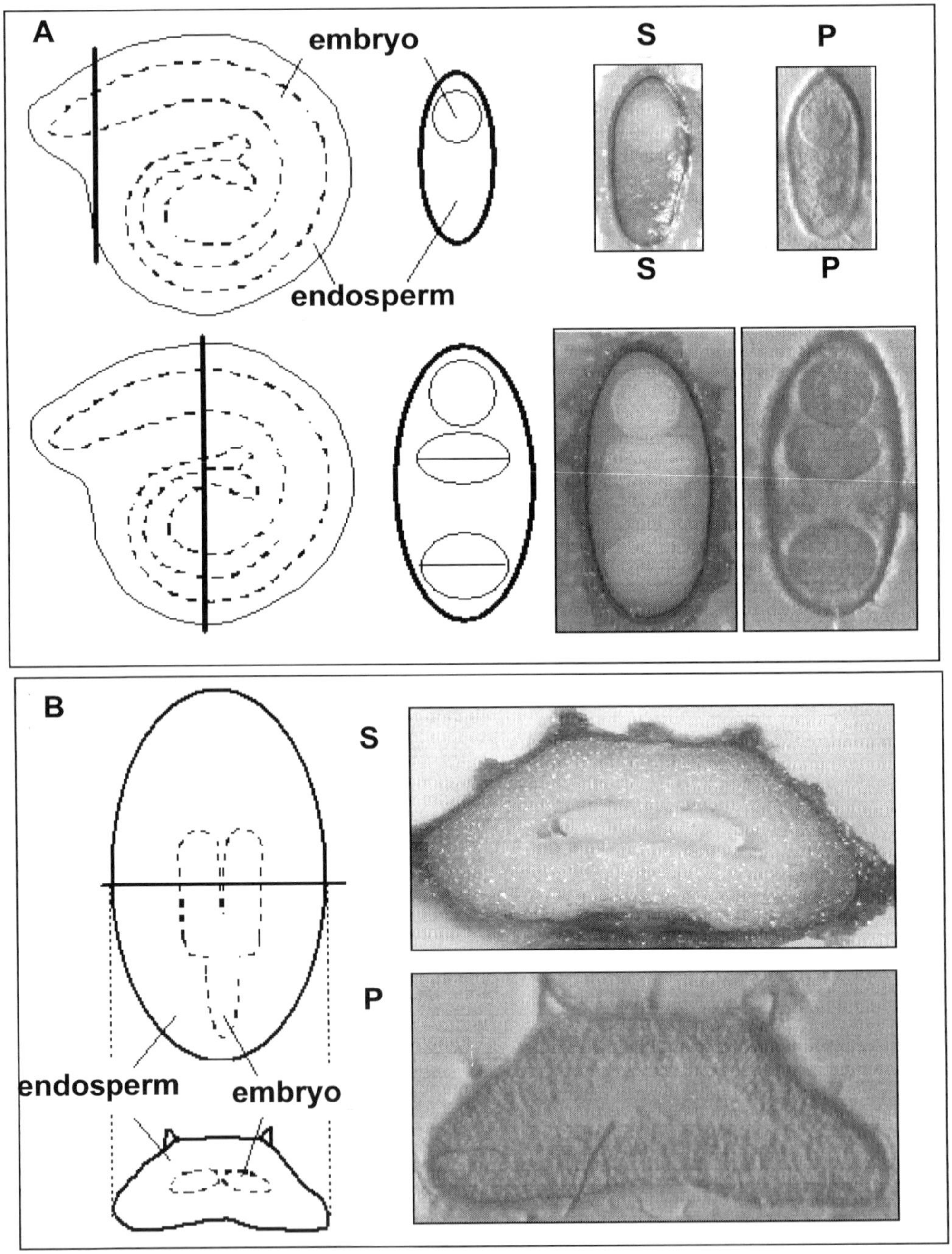

Fig. 19.2. Tissue prints of (A) tomato and (B) carrot seeds. Seeds were transversely dissected and printed. S, photographs of seed sections; P, tissue prints. Schematic presentations of tomato and carrot seeds are shown next to the photographs.

of washing gives highly specific signals with very low background in tissue prints.

Detection of Signal

The DIG molecules incorporated into the antisense RNA probes are detected using alkaline phosphatase-conjugated anti-DIG antibody (Fig. 19.3). After being washed at 60°C as described above, the membrane is briefly rinsed at room temperature with 0.1 M maleic acid buffer, pH 7.5, containing 0.15 M NaCl and 0.3% (v/v) Tween 20 (buffer A). The membrane is blocked with 5% (w/v) non-fat milk in buffer A. The antibody is added to the blocking solution at a dilution of 1:15,000 and the membrane is incubated at 25°C for 1 h. The membrane is washed for 20 min three times to remove the excess antibody.

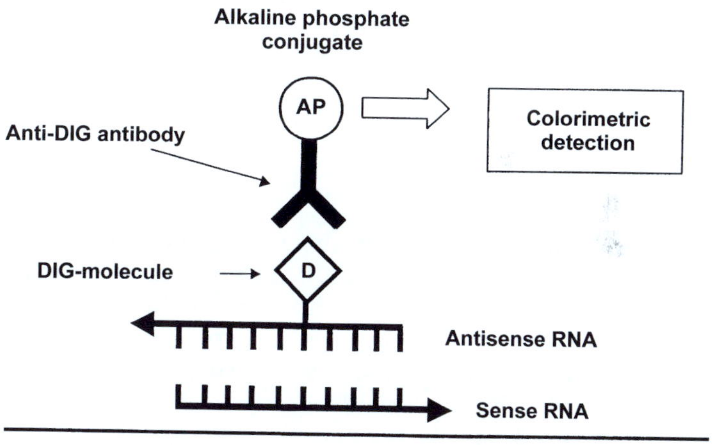

Fig. 19.3. Diagram for colorimetric detection of DIG-labelled RNA probe in tissue printing. The target sense-RNA printed on a membrane is hybridized by the antisense-RNA probe that incorporates DIG molecules. The anti-DIG antibody conjugated with alkaline phosphatase binds to the DIG molecules. The signal is detected colorimetrically by using a substrate for alkaline phosphatase (BCIP-NBT).

A chemiluminescent substrate such as Lumi-Phos 530 (Lumigen, Southfield, Michigan) can be used for detection of signals in RNA gel blots but, for tissue printing, seed morphology cannot be visualized clearly on X-ray film. Thus, we use a colorimetric substrate for alkaline phosphatase activity to detect hybridization of the RNA probes. The membrane is incubated in 0.18 M Tris-HCl buffer, pH 8.8, containing 0.025 mg/ml 5-bromo-4-chloro-3-indolyl-phosphate (BCIP), 0.1 mg ml nitroblue tetrazolium (NBT), and 2 mM $MgCl_2$. As most species of seeds have uncoloured embryo and endosperm tissues, the prints of seeds usually have a white background that

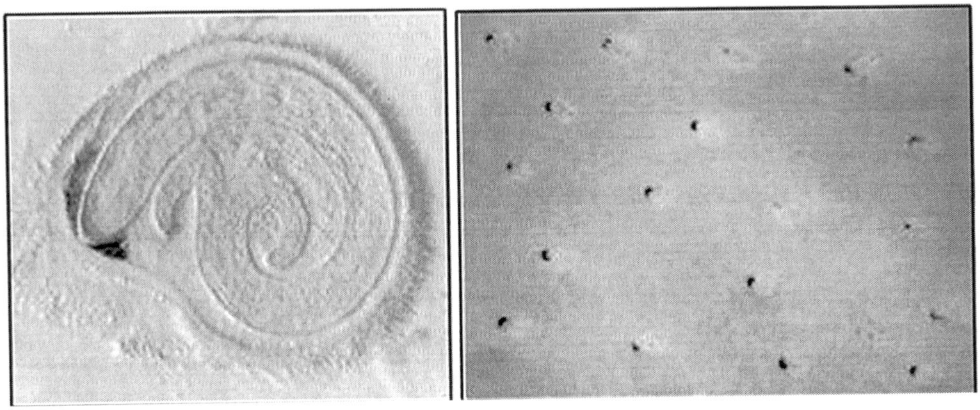

Fig. 19.4. Typical examples of detection of gene expression in seeds by tissue printing. (A) The expression of a xyloglucan endotransglycosylase gene (*LeXET4*) in imbibed tomato seeds. The colour reaction is detected in the micropylar region of the endosperm (endosperm cap). (From Chen et al., 2002.) (B) The expression of a mannanase gene (*LeMAN2*) in the endosperm caps of tomato seeds. Multiple seeds are printed on a single membrane.

does not interfere with the colorimetric (blue to purple) reaction of BCIP-NBT. The signals are detected as a dark area in seed prints (Fig. 19.4A). The final enzyme reaction for detecting signals generally takes a few hours. However, it varies depending on the amount of transcripts transferred to the membranes and the concentration of the specific probes used for hybridization. The enzyme reaction can be kept overnight to enhance the signals, because we rarely observe unspecific signals in tissue printing. After reaction, the membranes are washed with water to remove the substrates and dried at room temperature. The dried membranes can be kept for more than 2 years without losing the original resolution of the images.

Data Analysis and Application

The resolution of RNA signals in tissue prints is high enough to identify the localization of gene expression in seeds (Fig. 19.4A). Best resolution can be obtained by keeping membranes wet to see the dark areas of the RNA signals very clearly. However, if the membrane is submerged in a solution, it is difficult to see the seed prints. A 'semi-dry' condition of the membranes is best for taking pictures. The membranes are washed thoroughly with water, blotted between two sheets of filter papers and observed and photographed under a dissecting microscope. To visualize the seed prints, it is helpful to give light from both sides of the membrane rather than lighting it from directly above. A dissection microscope with flexible light arms is desirable for taking pictures. We have used a dissecting microscope with a CCD camera that is connected to a computer with an imaging program (*PIXERA*, Pixera Corp., Los Gatos, California).

As shown in Fig. 19.4B, multiple seed samples can be printed on a membrane and the signals from individual seeds for a specific probe can be compared side by side. This approach will be useful for single-seed assays for seed population studies (Still and Bradford, 1997). In addition, each cut surface from a single seed can be printed on a separate membrane, creating mirror image prints of the same seed (Fig. 19.5). By hybridizing one print with an antisense probe and the other with a sense probe, the specificity of the hybridization signals from tissue printing can be confirmed. Two different probes can also be hybridized to opposing halves from a single seed (Fig. 19.5). To quantify expression of a certain gene on a single-seed basis, a gene-specific probe (e.g. Probe 1) can be used for one half and another probe of a constitutively expressed gene or ribosomal RNA (e.g. Probe 2) for the other half. The intensity of the specific signal in individual seeds can be normalized based on the intensity of the signal from the constitutive probe for the same seed. This approach can also be applied for comparing mRNA and protein expression. One seed print can be exposed to a specific antibody and the mirror image print to a probe for the mRNA coding for that protein (A. Mella and K.J. Bradford, unpublished results). Alternatively, relative protein amounts can be determined by exposing one half-seed print to an antibody

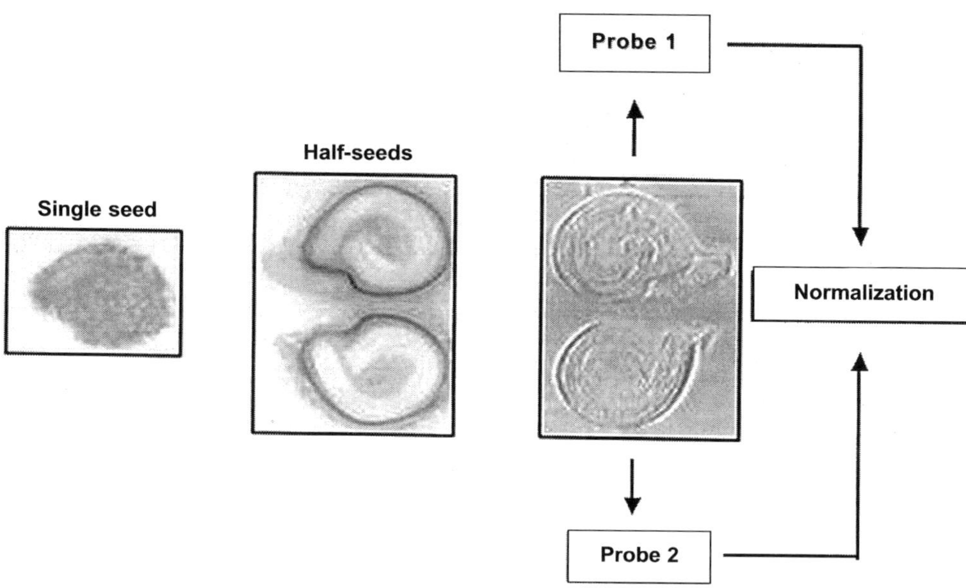

Fig. 19.5. A scheme for application of tissue printing for single-seed assays. A single seed is bisected into opposing halves, which are printed on membranes as mirror images. Two different probes can be used for the same seed to confirm the specificity of the signal or to quantify the signal (see text for details).

and the other to Coomassie blue or amido black for total protein staining. It is also possible to process many seeds at a time via tissue printing (Fig. 19.4B), making it feasible to screen for seed mutants as well.

Acknowledgements

We thank Dr Feng Chen for providing the original picture of tissue printing of a xyloglucan endotransglycosylase gene (*LeXET4*).

References

Borisjuk, L., Wang, T.L., Rolletschek, H., Wobus, U. and Weber, H. (2002) A pea seed mutant affected in the differentiation of the embryonic epidermis is impaired in embryo growth and seed germination. *Development* 129, 1595–1607.

Chen, F. and Bradford, K.J. (2000) Expression of an expansin is associated with endosperm weakening during tomato seed germination. *Plant Physiology* 124, 1265–1274.

Chen, F., Nonogaki, H. and Bradford, K.J. (2002) A gibberellin-dependent xyloglucan endotransglycosylase gene is expressed in the endosperm cap during tomato seed germination. *Journal of Experimental Botany* 53, 1–9.

Debeaujon, I., Leon-Kloosterziel, K.M. and Koornneef, M. (2000) Influence of the testa on seed dormancy, germination, and longevity in *Arabidopsis*. *Plant Physiology* 122, 403–413.

Debeaujon, I., Peeters, J.M., Leon-Kloosterziel, K.M. and Koornneef, M. (2001) The *TRANSPARENT TESTA12* gene of *Arabidopsis* encodes a multidrug secondary transporter-like protein required for flavonoid sequestration in vacuoles of the seed coat endothelium. *Plant Cell* 13, 853–871.

Lee, S., Cheng, H., King, K.E., Wang, W., He, Y., Hussain, A., Lo, J., Harberd, N.P. and Peng, J. (2002) Gibberellin regulates *Arabidopsis* seed germination via *RGL2*, a *GAI/RGA*-like gene whose expression is up-regulated following imbibition. *Genes and Development* 16, 646–658.

Luo, M., Bilodeau, P., Dennis, E.S., Peacock, W.J. and Chaundhury, A. (2000) Expression and parent-of-origin effects for *FIS2*, *MEA*, and *FIE* in the endosperm and the embryo of developing *Arabidopsis* seeds. *Proceedings of the National Academy of Sciences USA* 97, 10637–10642.

Nonogaki, H., Gee, O.H. and Bradford, K.J. (2000) A germination-specific endo-β-mannanase gene is expressed in the micropylar endosperm cap of tomato seeds. *Plant Physiology* 123, 1235–1245.

Penfield, S., Meissner, R.C., Shoue, D.A., Carpita, N.C. and Bevan, M.W. (2001) *MYB61* is required for mucilage deposition and extrusion in the *Arabidopsis* seed coat. *Plant Cell* 13, 2777–2791.

Raz, V., Bergervoet, J.H.W. and Koornneef, M. (2002) Sequential steps for developmental arrest in *Arabidopsis* seeds. *Development* 128, 243–252.

Still, D.W. and Bradford, K.J. (1997) Endo-β-mannanase activity from individual tomato endosperm caps and radicle tips in relation to germination rates. *Plant Physiology* 113, 21–29.

Stone, S.L., Kwong, L.W., Matusdaira-Yee, K., Pelletier, J., Lepiniec, L., Fisher, R.L., Goldberg, R.B. and Harada, J.J. (2001) *LEAFY COTYLEDON2* encodes a B3 domain transcription factor that induces embryo development. *Proceedings of the National Academy of Sciences USA* 98, 11806–11811.

Tzafrir, I., McElver, J.A., Liu, C.-M., Yang, L.J., Wu, J.Q., Martinez, A., Patton, D.A. and Meinke, D.W. (2002) Diversity of *TITAN* functions in *Arabidopsis* seed development. *Plant Physiology* 128, 38–51.

Wu, C.-T., Leubner-Metzger, G., Meins, F. Jr and Bradford, K.J. (2001) Class I β-1,3-glucanase and chitinase are expressed specifically in the micropylar endosperm of tomato seeds prior to radicle emergence. *Plant Physiology* 126, 1299–1313.

Xiong, L., Lee, B.-H., Ishitani, M., Lee, H., Zhang, C. and Zhu, J.-K. (2002) *FIERY1* encoding an inositol polyphosphate 1-phosphatase is a negative regulator of abscisic acid and stress signaling in *Arabidopsis*. *Genes and Development* 15, 1971–1984.

Yadegari, R., Kinoshita, T., Lotan, O., Cohen, G., Katz, A., Choi, Y., Katz, A., Nakashima, K., Harada, J.J., Goldberg, R.B., Fisher, R.L. and Ohad, N. (2000) Mutations in the *FIE* and *MEA* genes that encode interacting polycomb proteins cause parent-of-origin effects on seed development by distinct mechanisms. *Plant Cell* 12, 2367–2381.

Ye, Z.H. and Varner, J.E. (1991) Tissue-specific expression of cell wall proteins in developing soybean tissues. *Plant Cell* 3, 23–37.

20 Translational Control by Differential CAP-dependency in Selected Subpopulations of Maize-stored mRNAs

TZVETANKA D. DINKOVA, RAÚL AGUILAR AND ESTELA SÁNCHEZ DE JIMÉNEZ

Departamento de Bioquímica, Facultad de Química, Universidad Nacional Autónoma de México, 04510 México DF, México

Introduction

Germination of cereal seeds could be considered as the period of time elapsing between initial seed imbibition and radicle protrusion. One of the first molecular events taking place during this process is protein synthesis reinitiation. Since formation of active polysomes is observed a few minutes after imbibition, it was considered that, during the first stages of germination, translation relies on the pool of stored mRNAs in the seed (Bewley and Black, 1994). At the time of resuming transcription (3–6 h after imbibition), newly synthesized transcripts will also account for translation; however, at each time, only a fraction from the total mRNA pool is translated. The remaining transcripts are probably stored inactive for translation until needed, or are degraded. Therefore, expression regulation of specific proteins is under translational control, which is required to allow the germination process to proceed.

Several regulatory pathways are involved in the precise selection of transcripts for translation (Raught *et al.*, 2000). As a general rule, the first point in selecting an eukaryotic mRNA for translation is the binding of its 5′ end catabolite activator protein (CAP) structure (^{7}mGpppN, where N is any nucleotide) by the eukaryotic initiation factor (eIF) 4F complex. In plants, this complex is composed of the CAP-binding protein subunit, either eIF4E or eIFiso4E, and of the adaptor protein, either eIF4G or eIFiso4G (Browning, 1996). Other initiation factors are recruited throughout interactions with eIF4G or eIFiso4G to allow the ribosome scanning on the 5′ mRNA untranslated region (UTR) until the AUG initiation codon is found.

 The Biology of Seeds: Recent Research Advances (eds G. Nicolás, K.J. Bradford, D. Côme and H.W. Pritchard)

An alternative CAP-independent mechanism has been described for some viral mRNAs which contain a signal sequence for internal ribosome entry: IRES (Jang *et al.*, 1988). A growing number of cellular eukaryotic mRNAs translated by this means are now being reported (Holcik *et al.*, 2000). The CAP-independent mechanism is based on initial binding of the IRES in the 5′ UTR of these mRNAs, instead of the CAP structure at their 5′ end. Other regulatory mechanisms for mRNA translation involve specific sequence elements in their 5′ UTRs, such as 5′ TOP (tract of oligopyrimidines), 3′ UTR, and mRNA stability, among others (Raught *et al.*, 2000).

In maize axes, stored mRNAs are selectively translated during germination, as revealed by *in vivo* specific patterns of protein expression in the absence of transcription (Sánchez de Jiménez and Aguilar, 1984). Messenger RNA recognition by translation initiation factors might have a predominant role on selective mRNA translation in this system, since the eIF4E and eIFiso4E factors have been shown to have specific expression patterns within this period (Dinkova and Sánchez de Jiménez, 1999). Based on the antecedents mentioned, this research proposes that a mechanism underlying selective translation in maize during germination will be the recognition of transcript-specific CAP structure or the absence of CAP in the mRNA. Current efforts are centred on identification of the proteins synthesized under CAP-dependent and CAP-independent conditions, in order that the basic proteomics for successful germination can be further elucidated.

Results and Discussion

Expression of eIF4E isoforms during maize germination

Plants have been shown to contain two main different isoforms of the CAP-binding protein, named eIF4E and eIFiso4E (Browning, 1996). Each of these isoforms is capable of binding a corresponding isoform of the adaptor subunit eIF4G (eIF4G and eIFiso4G), to form the eIF4F and eIFiso4F complexes. Differential expression patterns at protein level have been found for these factors at specific developmental stages (Gallie *et al.*, 1998; Dinkova and Sánchez de Jiménez, 1999). The eIFiso4E protein is present at higher levels with respect to eIF4E in quiescent tissues, for example in wheat germ (Browning *et al.*, 1990) and maize embryonic axes (Dinkova and Sánchez de Jiménez, 1999). Following seed imbibition, eIFiso4E maintains constant and high levels until completion of germination, whereas eIF4E has relatively very low levels at the initial germination stages, and increases to a level similar to eIFiso4E at 24 h of seed germination (Fig. 20.1). Early during maize germination, significant *de novo* synthesis of the eIFiso4E, but not of the eIF4E protein has been demonstrated, indicating differential regulatory mechanisms for the expression of these isoforms at this developmental stage (Dinkova *et al.*, 2000).

Considering the patterns of eIF4E isoform expression, it has been proposed that regulation by the relative abundance of the two eIF4E isoforms would account for the *in vivo* differential protein synthesis patterns observed during seed germination (Sánchez de Jiménez and Aguilar, 1984). The pres-

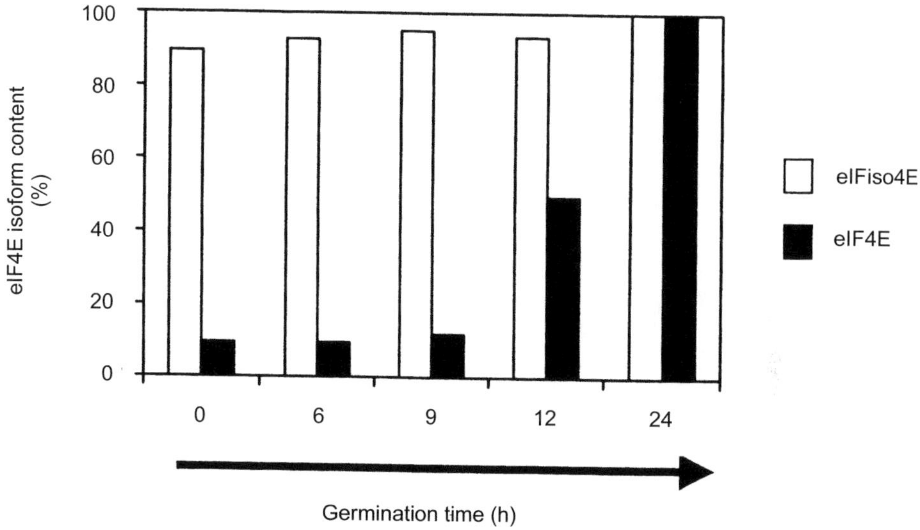

Fig. 20.1. Content of translation initiation factors during maize germination. Maize seeds (*Zea mays* L. cv. Chalqueno) were germinated for the indicated times; eIF4E and eIFiso4E proteins were purified as described previously (Dinkova and Sánchez de Jiménez, 1999) and were revealed by Western blot. The graph represents densitometric analysis for each isoform at the selected germination time. Protein content of each eIF4E isoform at 24 h of germination was taken as 100%.

ent research provides further information about this specific mRNA recognition system.

CAP dependency for translation of stored and 24 h germinated mRNA pools in maize

We have theoretically classified the maize axes mRNA into three main subpopulation groups, according to their translation efficiency dependent on the eIF4E isoform (Fig. 20.2).

Group I is composed of capped mRNAs preferentially recognized by the eIFiso4E isoform within the eIFiso4F CAP-binding complex; Group II is composed of capped mRNAs preferentially recognized by eIF4E within the eIF4F complex; and Group III is composed of mRNAs that are capable of undergoing translation by CAP-independent conditions and which probably possess IRES elements in their 5′ UTR.

To investigate the CAP dependency of specific mRNA subpopulations present in the axes during maize germination, total mRNA was tested on *in vitro* translation experiments using wheat germ extract (WGE) or rabbit reticulocyte lysate (RRL). To set the CAP-dependent system (CD), depletion of endogenous eIF4E isoforms from the WGE was achieved by treatment with ^{7}mGTP-sepharose. The absence of these isoforms in the extract was verified by Western blot analysis. Such extract was reconstituted with either of the maize eIF4E isoforms and labelled as CD. CAP-independent condition (CI) was achieved by RRL treatment with a viral Lb protease, an enzyme that

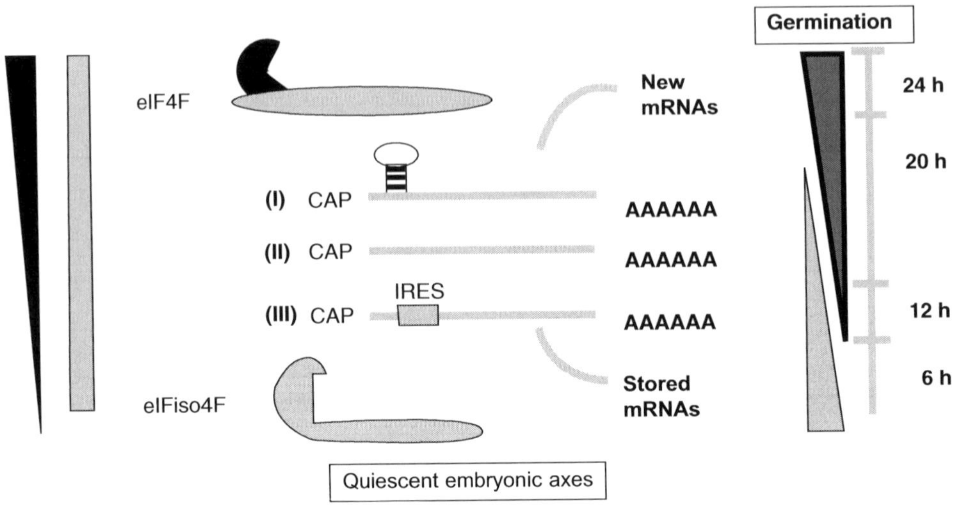

Fig. 20.2. Proposed model for mRNA translation during germination according to CAP dependency and eIF4E isoform availability. In quiescent embryonic axes eIFiso4E is several times more abundant than eIF4E and remains at a constant level throughout germination, whereas the eIF4E isoform significantly increases toward the end of germination. This model proposes that maize transcripts are selected for translation, from the stored ■ or newly synthesized ■ mRNA pools, according to internal availability of active CAP-binding proteins. (I) mRNAs recognized mostly by eIF4E; (II) mRNAs preferentially recruited for translation by eIFiso4E; (III) mRNAs translated under CAP-independent conditions.

cleaves the adaptor eIF4G protein into two peptides (Lamphear *et al.*, 1995). The resulting system is inactive for CAP-dependent translation due to disruption of the eIF4F initiation complex, but still promotes CAP-independent translation of specific mRNAs that bind internally to the ribosome (CI). The presence of CAP-independently translated mRNAs in maize was suggested by the results shown in Table 20.1.

Previous *in vitro* translation experiments performed with mRNAs from different sources, under full or eIF4E isoform-depleted translation system, showed distinctive translation efficiencies. Considering that translation of tobacco mosaic virus (TMV) mRNA is totally CAP dependent (translation under CI conditions near 0%), the pool of stored mRNAs in maize displayed a marked CAP independence for translation (33%), much larger than mRNAs from 24 h germinated embryonic axes (15%), or even mRNAs from other sources, such as yeast (17%) (Sánchez de Jiménez *et al.*, 1997).

Distinctive translation patterns of mRNA subsets from quiescent and 24 h germinated embryonic axes

Even though general translation inhibition is observed under CAP-independent conditions (Table 20.1), analysis of the remaining *in vitro* translated products revealed specific changes in the protein synthesis patterns according to the CAP dependency of the mRNA subpopulations. The *in vitro* trans-

Table 20.1. In vitro translation of mRNAs from different sources in a CAP-independent system (CI condition).

Source of mRNA	Native WGE[a] (pmol [35S]-Met ml^{-1})	WGE-eIF4E depleted[b] (pmol [35S]-Met ml^{-1})	Remaining translation (%)
Tobacco mosaic virus (TMV)	52 (± 5.4)	2.1 (± 0.3)	4 (± 0.2)
Yeast	31 (± 2.6)	5.4 (± 0.2)	17 (± 0.4)
24 h germinated maize axes	35 (± 4.5)	5.5 (± 0.4)	15 (± 0.1)
Quiescent maize axes	27 (± 4.1)	9 (± 0.3)	33 (± 3)

[a]Translation reactions were set in wheat germ extract (Promega Corp.), supplemented with [^{35}S]-methionine and the selected mRNAs. After 60 min of incubation at 30°C, the labelled proteins were quantified by trichloroacetic acid precipitation and liquid scintillation counting.
[b]Previous to the in vitro translation, the wheat germ extract was treated for 45 min with 2 volumes of m^{7}GTP-sepharose to deplete endogenous eIF4E isoforms (CI condition). Native WGE values are taken as 100%.

lated mRNAs from quiescent (0 h) and germinated (24 h) maize embryonic axes showed quite different protein patterns under standard conditions (Fig. 20.3, CD lanes), in accordance with previous work that demonstrated that axes gene expression experiences important changes during the germination period (Sánchez de Jiménez and Aguilar, 1984).

Depletion of eIFiso4E from the *in vitro* translation system drastically reduced translation of most of the mRNAs from both subpopulations, though translation of some specific proteins remained (Fig. 20.3, 4E lanes). On the other hand, many of the proteins translated by the CD system were still synthesized in the presence of eIF4E alone, though at different rates in each case (Fig. 20.3, i4E vs. 4E lanes). This reveals that competition for translation of their corresponding mRNAs would probably be relevant *in vivo*, where the eIFiso4E/eIF4E ratio changes during the germination period.

The *in vitro* CAP-independent translation system (CI) was achieved by cleavage of the adaptor protein (eIF4G) with the consequent disruption of the contact between the CAP (eIF4E interaction) and the rest of the translation initiation complex (Lamphear *et al.*, 1995). In this system, translation of most of the mRNAs was inhibited in both quiescent and 24 h germinated axes (Fig. 20.3, CI lanes). However, some specific transcripts were translated, particularly from the stored mRNAs (quiescent axes). According to their molecular masses, the main proteins synthesized under CAP-independent conditions are apparently different for 0 h and 24 h germinated mRNA subpopulations. The CAP-independently translated mRNAs from quiescent axes mainly rendered protein products of approximately 14, 16, 17, 19 and 20 kDa, whereas those of 24 h germinated axes rendered proteins of approximately 16, 17, 24, 70 and 80 kDa.

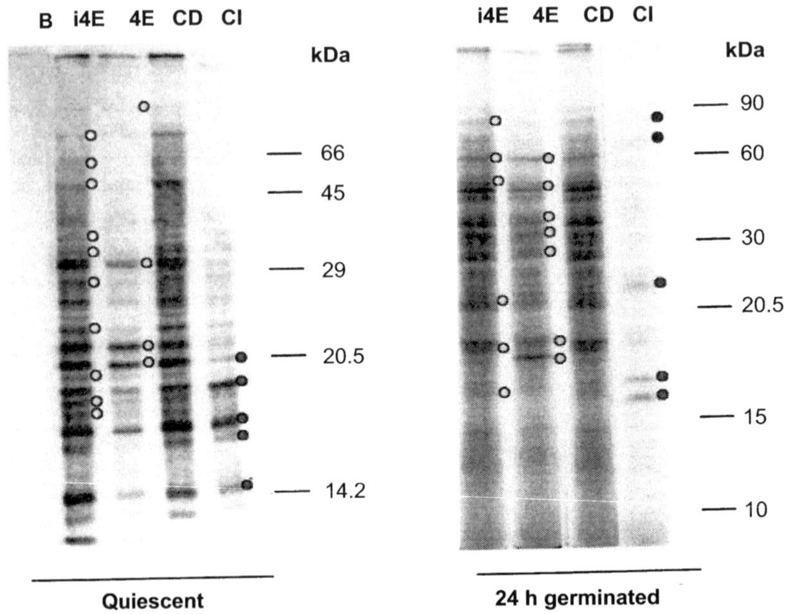

Fig. 20.3. SDS-PAGE patterns of *in vitro* synthesized proteins under different CAP-dependent conditions using as the mRNA source quiescent and 24 h germinated embryonic axes. B, control translation system without mRNA source; i4E, translation in the presence of maize eIFiso4E; 4E, translation in the presence of maize eIF4E; CD, CAP-dependent translation; CI, CAP-independent translation; (●), protein synthesis selectively enhanced under specific translation conditions.

Characterization of specific maize proteins synthesized under CAP-independent conditions

Several of the stored transcripts present in maize axes have been previously identified and the translational regulatory mechanisms involved in their expression elucidated (Sánchez de Jiménez, 2000) but, to date, no IRES elements have been described for plant mRNAs. Recently, however, a study has demonstrated that reporter genes with IRES-driven translation can be expressed *in vivo* in plants (Urwin *et al.*, 2000). In non-plant eukaryotes, an increasing number of cellular transcripts with IRES elements in their 5′ UTR, which are efficiently translated under CAP-independent conditions, has been described. Some of the proteins encoded by these mRNAs are involved in processes such as growth and differentiation, apoptosis, stress tolerance and other important events (Holcik *et al.*, 2000).

CAP-independently translated mammal mRNAs have been reported, which include eIF4G, ornithine descarboxylase, the c-Myc protein and other proliferation-related proteins (Pyronnet *et al.*, 2000), as well as several heat shock proteins (Hsps) in *Drosophila* (Song *et al.*, 1995). Taking this information into account, we tested the presence of some potential IRES mRNAs among the above-mentioned transcripts in the 0 h and 24 h germinated mRNA subpopulations of maize, by immunoprecipitation of the correspondent products from the *in vitro* translation reaction under CD and CI conditions (Table 20.2). As control for CD and CI translation, a bicistronic construct of two reporter genes separated by a known IRES (Martínez-Salas, 1999) was assayed in the *in vitro* translation systems.

Results from these experiments indicated that maize eIFiso4G mRNA could be equally well translated under CD or CI conditions. This transcript is more abundant in 24 h germinated than in quiescent embryonic axes. The other isoform, eIF4G, seems to be CAP-dependently translated in both the 0 h and 24 h germinated mRNA subpopulations. Other transcripts that have been previously found not to be IRES dependent, such as eIFiso4E and the ribosomal protein rpS6 (Dinkova *et al.*, 2000), showed CAP-dependent translation in our system. On the other hand, two candidates for CAP-independent translation were revealed by these experiments: the mRNA encoding the c-Myc protein, which is required for cell division after radicle protrusion, and Hsp101, a heat shock protein whose mRNA is present in quiescent axes and rapidly declines during germination (Dinkova *et al.*, unpublished results). Current experiments on maize Hsp101 in our laboratory are devoted to testing the presence of an IRES sequence within the 5′ UTR of its mRNA.

Conclusions

From the data presented here, it is evident that there are mechanisms of translational control that specifically determine gene expression through mRNA CAP dependency during seed germination. Several mRNAs from two different subpopulations during germination (0 h and 24 h) were observed to produce distinctive translation patterns according to the CAP

Table 20.2. Identification of some maize transcripts and their CAP-dependency requirements.

Identified protein	Proposed translation regulation		Translation system	
	Mammals	Maize	CD	CI
c-Myc	IRES	IRES	+	+
eIF4G	IRES	Unknown	+	–
eIFiso4G	No	IRES	+	+
eIFiso4E	No	5′ TOP	+	–
rpS6	5′ TOP[a]	5′ TOP	+	–
Hsp101	No	IRES	–	+

[a]mRNAs from quiescent or germinated maize embryonic axes were in vitro translated under CAP-dependent or CAP-independent conditions. The proteins of interest were immunoprecipitated with specific antibodies and the labelled bands were detected by electrophoresis and fluorography.

IRES, internal ribosome entry site; No, not reported for mammals; 5′ TOP, tract of oligopyrimidines; CD, CAP-dependent; CI, CAP-independent.

dependency of the *in vitro* translation system. The data suggest that early in maize germination sets of mRNAs, probably corresponding to constitutive proteins, are recruited for translation in a CAP-dependent fashion by the eIFiso4E factor, whereas other transcripts, specifically recognized by eIF4E, would be most likely translated later, when the eIF4E isoform level increases. In addition, enhanced synthesis of specific proteins is observed under CAP-independent conditions. Selected transcripts (i.e. eIFiso4G, Hsp101 and c-Myc) appear to be translated via a CAP-independent mechanism most probably in order to fulfil cell requirements under specific physiological conditions (cell division, heat stress, etc.).

Acknowledgements

This work has been supported by Dirección General de Asuntos del Personal Académico (DGAPA), Universidad Nacional Autónoma de México (UNAM), Grant IN202900. We also wish to thank Dr Encarnación Martínez Salas from Universidad Autónoma de Madrid for the bicistronic construct used as control for CD and CI translation, and Dr Jorge Nieto Sotelo from Instituto de Biotecnología, UNAM, for the Hsp101 plasmid construct.

References

Bewley, J.D. and Black, M. (1994) Cellular events during germination and seedling growth. In: Bewley, J.D. (ed.) *Seeds: Physiology of Development and Germination*, 2nd edn. Plenum Press, New York, pp. 147–196.

Browning, K.S. (1996) The plant translational apparatus. *Plant Molecular Biology* 32, 107–144.

Browning, K.S., Humphreys, J., Hobbs, W., Smith, G.B. and Ravel, J.M. (1990) Determination of the amounts of the protein synthesis initiation and elongation factors in wheat germ. *Journal of Biological Chemistry* 265, 17967–17973.

Dinkova, T.D. and Sánchez de Jiménez, E. (1999) Differential expression and regulation of translation initiation factors –4E and –iso4E during maize germination. *Physiologia Plantarum* 107, 419–425.

Dinkova, T.D., Aguilar, R. and Sánchez de Jiménez, E. (2000) Expression of maize initiation factor eIFiso4E is regulated at translational level. *Biochemical Journal* 351, 825–831.

Gallie, D.R., Le, H., Tanguay, R.L. and Browning, K.S. (1998) Translation initiation factors are differentially regulated in cereals during development and following heat shock. *The Plant Journal* 14, 715–722.

Holcik, M., Sonenberg, N. and Korneliuk, R.G. (2000) Internal ribosome initiation of translation and the control of cell death. *Trends in Genetics* 16, 469-473.

Jang, S.K., Krausslich, H.-G., Nicklin, M.J.H., Duke, G.M., Palmemberg, A.C. and Wimmer, E. (1988) A segment of the 5´ non translated region of encephalomyocarditis virus RNA directs internal entry of ribosomes during *in vitro* translation. *Journal of Virology* 63, 2636–2643.

Lamphear, B.J., Kirchweger, R., Skern, T. and Rhoads, R.E. (1995) Mapping of functional domains in eukaryotic protein synthesis initiation factor 4G (eIF4G) with picornaviral proteases. *Journal of Biological Chemistry* 270, 21975–21983.

Martínez-Salas, E. (1999) Internal ribosome entry site biology and its use in expression vectors. *Current Opinion on Biotechnology* 10, 458–464.

Pyronnet, S., Paradayrol, L. and Sonenberg, N. (2000) A cell cycle-dependent internal ribosome entry site. *Molecular Cell* 5, 607–616.

Raught, B., Gingras, A.C. and Sonenberg, N. (2000) Regulation of ribosomal recruitment in eukaryotes. In: Sonenberg, N., Hershey, J.W.B. and Mathews, M.B. (eds) *Translational Control of Gene Expression.* CSHL Press, New York, pp. 245–294.

Sánchez de Jiménez, E. (2000) Protein synthesis in seed germination: connecting external cell signals with the ribosome. In: Black, M., Bradford, K.J. and Vázquez-Ramos, J. (eds) *Seed Biology: Advances and Applications.* CAB International, Wallingford, UK, pp. 11–25.

Sánchez de Jiménez, E. and Aguilar, R. (1984) Protein synthesis patterns: relevance of old and new messenger RNA in germinating maize embryos. *Plant Physiology* 75, 231–234.

Sánchez de Jiménez, E., Aguilar, R. and Dinkova, T.D. (1997) S6 ribosomal protein phosphorylation and translation of stored mRNA in maize. *Biochimie* 79, 187–194.

Song, H.-J., Gallie, D.R. and Duncan, R.F. (1995) m7GpppG CAP dependence for efficient translation of *Drosophila* 70-kDa heat-shock-protein (Hsp70) mRNA. *European Journal of Biochemistry* 232, 778–788.

Urwin, P., Yi, L., Martin, H., Atkinson, H. and Gilmartin, P.M. (2000) Functional characterization of the EMCV IRES in plants. *The Plant Journal* 24, 583–589.

21 Expression of a Ribosomal Protein Gene During Germination of cabbage (*Brassica oleracea* f. *oleracea*) Seeds

PETER E. TOOROP[1*], STEVEN P.C. GROOT[2] AND HENK W.M. HILHORST[1]

[1] Laboratory of Plant Physiology, Wageningen University, Arboretumlaan 4, 6703 BD Wageningen, The Netherlands; [2] Plant Research International, PO Box 16, 6700 AA Wageningen, The Netherlands

Introduction

Ribosomal proteins form an essential piece of the translation machinery in living cells. Together with the ribosomal RNA they are assembled into ribosomes, responsible for translation of mRNAs into proteins. Upon completion of the *Arabidopsis* genome sequencing project, 80 ribosomal proteins have been identified encoded by 249 genes (Barakat *et al.*, 2001). Despite the occurrence of multiple copies for all genes, often interpreted as pseudo-genes, mutants for some genes coding for ribosomal proteins show a phenotype (Van Lijsebettens *et al.*, 1994; Tsugeki *et al.*, 1996; Weijers *et al.*, 2001). These results, and differential expression patterns in *Arabidopsis* tissues for homologous genes coding for ribosomal protein L16 (Williams and Sussex, 1995), suggest that each gene has a specialized function that cannot be complemented by a homologous gene.

Expression in the embryo was confirmed by embryo lethality found in mutants for ribosomal proteins S5 and S16 (Tsugeki *et al.*, 1996; Weijers *et al.*, 2001). Induction of transcription was observed for ribosomal protein genes in germinating tobacco seeds (Gao *et al.*, 1994). In *Arabidopsis* ribosomal protein genes are up-regulated during germination and down-regulated during dormancy (Toorop *et al.*, unpublished results). One might hypothesize that expression of ribosomal protein genes is correlated with germination performance. This hypothesis is tested here with cabbage seeds, using temperature and osmotic stress to vary germination.

*See list of contributors for new address.

 The Biology of Seeds: Recent Research Advances (eds G. Nicolás, K.J. Bradford, D. Côme and H.W. Pritchard)

Materials and Methods

Seed germination

Seeds of cabbage (*Brassica oleracea* f. *oleracea* L.) of variety Bartolo were used. Seedlots were selected for this research project. Three replicates of approximately 100 seeds were germinated in 9 cm Petri dishes on two layers of Whatman no. 1 filter paper and 5 ml distilled water. Seeds were incubated at 20 or 35°C (±1°C) in darkness or in the presence of red light (652 nm, 26 W/m^2) obtained from a red LED (Kingbright Electronic Europe GmbH, Geldern, Germany). Seeds were either imbibed in water or in polyethylene glycol (PEG).

RNA isolation

An RNA isolation protocol was followed that was modified after Wan and Wilkins (1994). Embryos from 20 seeds were frozen in liquid nitrogen, ground in a Micro-Dismembrator U (B. Braun Biotech International, Melsungen, Germany) at 2000 rpm for 2 min, and used for RNA isolation.

Northern analysis

RNA was separated on a 1.5% (w/v) MOPS-buffered agarose gel. The gel was stained with ethidium bromide and a digital image of the total RNA was produced with a Kodak EDAS120 camera and a UV transilluminator. The gel was blotted on to positively charged nylon membrane (GeneScreen Plus; NEN Life Science Products) with 10× SSC, baked in a vacuum oven at 80°C for 30 min, and hybridized with ^{32}P-labelled fragments. DNA probes were labelled with α-^{32}P-dATP, using the Random Primed DNA Labeling Kit (Roche Diagnostics) and purified over a Sephadex G50 column (Boehringer Mannheim). The Northern blots were pre-hybridized for 2 h at 42°C in a solution of 50% formamide, 6× SSC, 5× Denhardt's reagent and 0.5% SDS and hybridized overnight in the same solution after addition of labelled probe. The Northern blots were washed for 5 min at room temperature in 2.0× SSC + 0.1% SDS, 20 min at 60°C in 1.0× SSC + 0.1% SDS, 20 min at 60°C in 0.2× SSC + 0.1% SDS and rinsed briefly in 0.2× SSC + 0.1% SDS before autoradiography. Digital images of autoradiographs were produced with a Kodak EDAS120 camera and a white light transilluminator. Total RNA was shown as a control for equal loading.

RT-PCR and PCR analysis

cDNA was prepared of RNA samples, independent of the ones used for differential display analysis, using the Gibco BRL kit 'Superscript Preamplification System for First Strand cDNA Synthesis' with the oligo(dT) primer provided. For quantitative PCR assays, reactions were performed with 1:5, 1:25, and 1:125 dilutions of cDNA with 21 cycles typically, to ensure that amplifications were within the linear range. Only the results of the 1:25 dilutions were presented, since linearity was confirmed consistently.

For cloning and amplification of the ribosomal protein gene Bo*RPL27* the following PCR primers were used: forward primer 5′-GCCG-GCAAAAAGGCAGTGA-3′ and reversed primer 5′-GAGCGTGTAACGTG-TAGGCATCAG-3′. Primers were designed with the PrimerSelect program (DNAStar Inc.). Samples were separated on a 1.5% (w/v) TAE-buffered agarose gel, excised from the gel and cloned into the pGEM-T Easy vector (Promega). Samples were sequenced and similarity searches were performed with the NCBI BLAST server (www.ncbi.nlm.nih.gov/BLAST). For the two PCR primers the amplified product length is 207 bp. PCR-products were separated on a 1.5% (w/v) TAE-buffered agarose gel, blotted on to positively charged nylon membrane with 10× SSC and baked and hybridized as described above for Northern blots. Digital images of autoradiographs were produced with a Kodak EDAS120 camera and a white light transilluminator, and bands were quantified with the Kodak 1D Image Analysis software.

Results

Two germination-associated cDNAs were cloned in *Arabidopsis* seeds, coding for ribosomal proteins *At*RPL27B and *At*RPL36B (results not shown). Expression of these two genes was studied in cabbage seeds in relation to germination performance, using the *Arabidopsis* genes as heterologous probes in Northern analysis. Both genes displayed marginal expression in the dry seed, and high expression was induced during germination in water (Fig. 21.1). When exposed to an increasing osmotic stress expression was proportionally lower. Prolonged incubation at a high osmotic stress (9 days at –1.0 MPa) did not allow completion of germination, whereas expression was slightly increased compared with 18 h incubation at –1.0 MPa (Fig. 21.1). Expression of both genes after 18 h in either water or PEG correlated with maximum germination and correlated inversely with the period to reach 50% germination (T_{50}). These results show that expression of genes homologous to At*RPL27B* and At*RPL36B* in cabbage seeds correlate with germination performance, and indicate that the mRNA abundance of homologues of the two ribosomal proteins could function as a potential marker for germination.

A cabbage homologue of At*RPL27B* was cloned using consensus primers. For this purpose, homologous sequences coding for RPL27 were aligned from *Arabidopsis, Pisum sativum, Panax ginseng, Solanum tuberosum* and *Pyrobotrys stellata* that were retrieved from the NCBI site. Other homologues were retrieved as well but were not considered for designing the consensus primers, because they were too distant in phylogeny (Fig. 21.2). The cabbage clone Bo*RPL27* showed high nucleotide similarity to (putative) *Arabidopsis* RPL27 homologues located at chromosome 3 (90%), chromosome 4 (86%) and chromosome 2 (79%; Fig. 21.3). Expression was induced during germination and reached a maximum shortly before radicle protrusion at

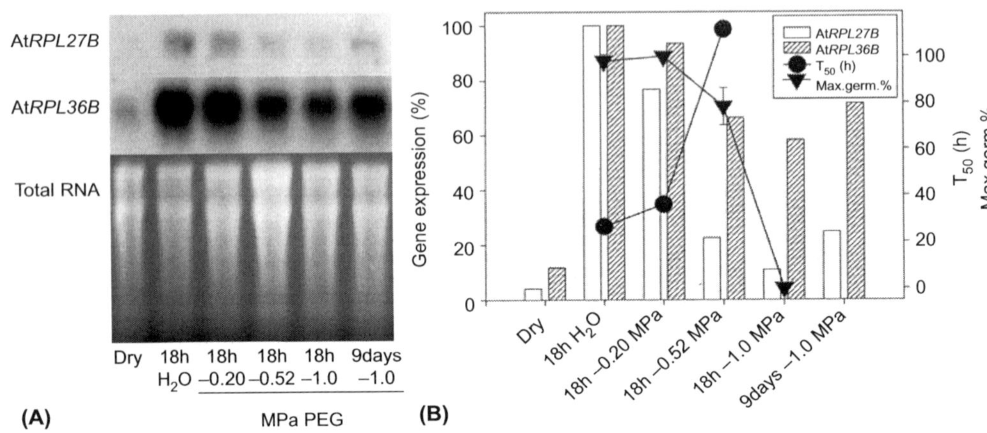

Fig. 21.1. (A) Northern blot with RNA samples of cabbage seeds that were either dry or imbibed in water, –0.20, –0.52 or –1.0 MPa PEG for 18 h, or in –1.0 MPa PEG for 9 days. The blot was hybridized with the heterologous probe At*RPL27B*, washed and hybridized with the heterologous probe At*RPL36B*. The lower panel shows the ethidium bromide-stained gel with equal loading of total RNA. (B) Expression (% of maximum, with 18 h H_2O as 100%) of At*RPL27B* (open bars) and At*RPL36B* (hatched bars) in cabbage seeds versus T_{50} (circles) and maximum germination (triangles).

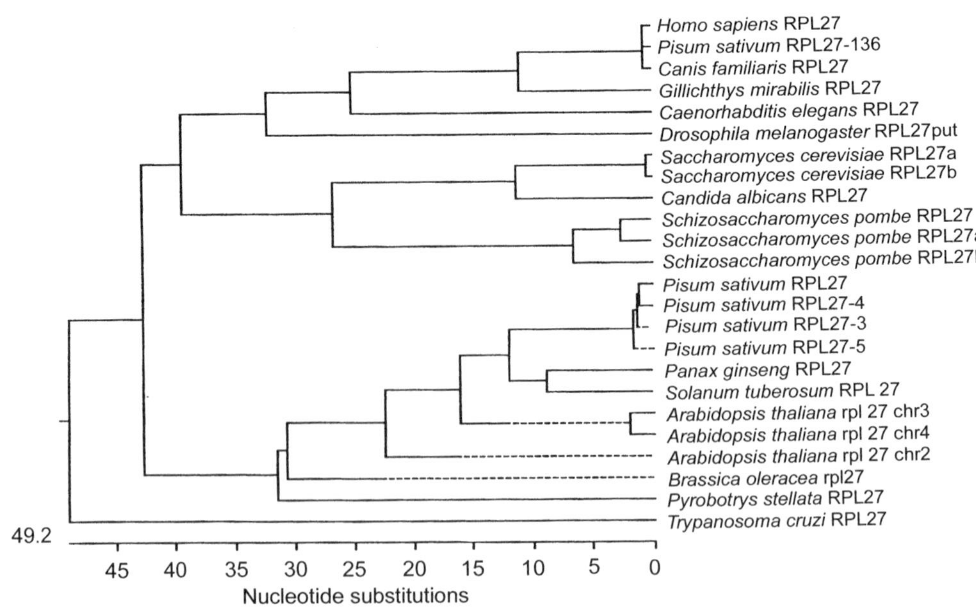

Fig. 21.2. Phylogenetic tree generated with RPL27 sequences. The alignment was made using the Clustal V algorithm in the MEGALIGN software of the DNAStar package.

Bo*RPL27*	GCCGGCAAAAAGGCAGTGATCATCCGCTCCTTCGACGACGGAAACCGTGAGCG
At*RPL27B* chr3	GCCGGCAAAAAGGCAGTGATCATCAAATCCTTCGATGACGGAACCAGTGACCG
At*RPL27C* chr4	GCCGGAAAGAAAGCCGTCATCATCAAATCCTTCGACGACGGTAACCGTGATCG
At*RPL27A* chr2	ACCGGTAAAAAGGCGGTGATCGTGAAATCCTTCGATGACGGTACCGTTGAGAA
Bo*RPL27*	TCCTTACGGACACTGCCTCGTCGCGGGACTGAAGAAGTATCCGAGCAAGGTCA
At*RPL27B* chr3	TCGTTACGGACACTGCCTCGTCGCCGGACTGAAGAAGTACCCGAGCAAAGTCA
At*RPL27C* chr4	TCCTTACGGACACTGCCTCGTCGCCGGACTCAAGAAGTACCCGAGCAAAGTCA
At*RPL27A* chr2	GAAGTACGGTCACTGTCTCGTCGCCGGATTGAAGAAGTACCCAAGCAAAGTCA
Bo*RPL27*	TCCGCAAGGACTCAGCCAAGAAGACGGCGAAGAAATCTCGAGTCAAGTGCTTC
At*RPL27B* chr3	TCCGCAAAGACTCAGCGAAGAAGACGGCGAAGAAATCAAGAGTGAAATGTTTC
At*RPL27C* chr4	TCCGCAAAGACTCAGCTAAGAAGACAGCTAAGAAATCTAGGGTTAAGTGTTTC
At*RPL27A* chr2	TTCGCAAGGATTCGGCAAAGAAGACGGCTAAGAAATCGCGCGTGAAATGCTTC
Bo*RPL27*	ATAAAGGTCGTGAACTACCAGCATCTGATGCCTACACGTTACACGCTC
At*RPL27B* chr3	ATCAAGCTCGTGAATTACCAGCATCTGATGCCTACACGTTACACGCTC
At*RPL27C* chr4	ATCAAGCTTGTTAATTACCAGCATCTGATGCCTACTCGTTACACACTC
At*RPL27A* chr2	TTCAAGGTCATCAATTACCAGCACGTGATGCCTACGCGCTACACTCTT

Fig. 21.3. Multiple alignment of partial RPL27 sequences of cabbage (Bo*RPL27*) and three *Arabidopsis* homologues (At*RPL27*) located on chromosomes 2, 3 and 4. Sequences were aligned using the MegAlign program (DNAStar Inc.) using the ClustalW algorithm. Identical nucleotides are shaded light, similarity between Bo*RPL27* and only one or two of the other sequences are shaded dark. Sequences to which PCR primers bind are underlined.

25 h (Fig. 21.4A). Upon radicle protrusion, expression decreased slightly (Fig. 21.4A). This decline in expression was also observed in *Arabidopsis* seeds (data not shown) and is probably caused by a burst in production of other mRNAs, causing a decline of the ribosomal protein mRNA relative to total RNA levels. Good linear correlation was found between expression of Bo*RPL27* and imbibition time (r^2 = 0.97) (Fig. 21.4A). Variation in germination speed and maximum germination were obtained by applying osmotic stress (–0.2 MPa), temperature stress (35°C) and temperature stress in the presence of red light (35°C R; Fig. 21.4B). A good negative log-linear correlation between Bo*RPL27* expression and T_{50} (r^2 = 0.98) was found. This result confirmed that expression of ribosomal protein gene *L27* is induced during germination in cabbage seeds and that the level of expression is indicative of germination speed.

Discussion

Ribosomal protein genes seem to be expressed in tissues with high meristematic activity or cell division. Expression of ribosomal proteins is also higher in growing axillary buds than in dormant buds (Stafstrom and Sussex, 1992). Growth of axillary buds and germination of seeds are considered analogous systems, and indeed comparable expression patterns of ribosomal protein genes were observed. Using two *Arabidopsis* genes coding for ribosomal protein L27B and L36B as heterologous probes, expression in cab-

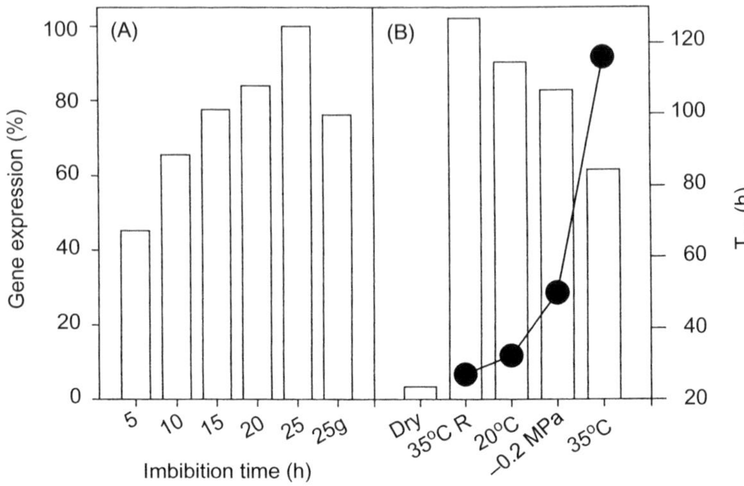

Fig. 21.4. Expression of Bo*RPL27* of cabbage seeds (A) during and upon (25g; radicle protrusion occurred in this sample) germination in water and (B) in dry seeds and during germination under conditions varying in temperature and osmotic stresses.

bage seeds was found to correlate with germination performance. A cabbage cDNA clone, highly homologous to At*RPL27B*, showed a similar result and confirmed this finding. Transcripts of ribosomal protein genes *L27B* and *L36B* are hardly present in the dry cabbage seed, although seeds are known to store mRNAs during development (Bewley and Black, 1994). Both presence (S4 and S6) and absence (L3 and L16) of stored transcripts of ribosomal protein genes have been reported in maize embryonic axes (Beltran-Pena *et al.*, 1995). Induction of transcription of ribosomal protein genes is likely to be involved in growth, since a variety of growing tissues display increased expression of ribosomal protein genes (Stafstrom and Sussex, 1992). Induction of transcription of At*RPL27B* and At*RPL36B* in germinating cabbage seeds is therefore not surprising, and can be explained by anticipating growth upon completion of germination.

Changes in ribosomal protein gene transcription have been found to be irrelevant to protein abundance in *Brassica napus* plants, suggesting post-transcriptional control (Sáez-Vásquez *et al.*, 2000). Our findings that transcription of gene Bo*RPL27* correlates with germination performance may not have any effect on the protein levels of RPL27. Nevertheless, since *de novo* protein synthesis occurs in germinating seeds, translation of the newly synthesized transcripts seems most likely (Bewley and Black, 1994). A correlation of expression of ribosomal protein genes and germination has been described before. Induction of transcription was observed for ribosomal protein genes *L25* and *L34* in germinating seeds of tobacco (Gao *et al.*, 1994). Moreover, our results for *L27B* and *L36B* indicate a correlation of ribosomal protein gene transcript abundance with germination performance. Applying temperature or osmotic stress slowed down germination and decreased maximum germination, and also decreased expression of At*RPL27B* and At*RPL36B*. These results imply that expression of ribosomal protein genes is a marker for seed germination.

Acknowledgements

This research was supported by the European Community (Fisheries, Agriculture and Agro-Industrial project grant no. CT97-3711, 'Genetic and molecular markers for seed quality'. Bejo Zaden BV (Warmenhuizen, The Netherlands) is acknowledged for generously providing cabbage seeds.

References

Barakat, A., Szick-Miranda, K., Chang, I.F., Guyot, R., Blanc, G., Cooke, R., Delseny, M. and Bailey-Serres, J. (2001) The organization of cytoplasmic ribosomal protein genes in the Arabidopsis genome. *Plant Physiology* 127, 398–415.

Beltran-Pena, E., Ortiz-Lopez, A. and Sanchez de Jimenez, E. (1995) Synthesis of ribosomal proteins from stored mRNAs early in seed germination. *Plant Molecular Biology* 28, 327–336.

Bewley, J.D. and Black, M. (1994) *Seeds. Physiology of Development and Germination*, 2nd edn. Plenum Press, New York, 445 pp.

Gao, J., Kim, S.R., Chung, Y.Y., Lee, J.M. and An, G. (1994) Developmental and environmental regulation of two ribosomal protein genes in tobacco. *Plant Molecular Biology* 25, 761–770.

Sáez-Vásquez, J., Gallois, P. and Delseny, M. (2000) Accumulation and nuclear targeting of BnC24, a *Brassica napus* ribosomal protein corresponding to a mRNA accumulating in response to cold treatment. *Plant Science (Limerick)* 156, 35–46.

Stafstrom, J.P. and Sussex, I.M. (1992) Expression of a ribosomal protein gene in axillary buds of pea seedlings. *Plant Physiology* 100, 1494–1502.

Tsugeki, R., Kochieva, E.Z. and Fedoroff, N.V. (1996) A transposon insertion in the Arabidopsis SSR16 gene causes an embryo-defective lethal mutation. *Plant Journal* 10, 479–489.

Van Lijsebettens, M., Vanderhaeghen, R., De Block, M., Bauw, G., Villarroel, R. and Van Montagu, M. (1994) An S18 ribosomal protein gene copy at the Arabidopsis PFL locus affects plant development by its specific expression in meristems. *EMBO Journal* 13, 3378–3388.

Wan, C.-Y. and Wilkins, T.A. (1994) A modified hot borate method significantly enhances the yield of high-quality RNA from cotton (*Gossypium hirsutum* L.). *Analytical Biochemistry* 223, 7–12.

Weijers, D., Franke-van Dijk, M., Vencken, R.J., Quint, A., Hooykaas, P. and Offringa, R. (2001) An Arabidopsis Minute-like phenotype caused by a semi-dominant mutation in a RIBOSOMAL PROTEIN S5 gene. *Development* 128, 4289–4299.

Williams, M.E. and Sussex, I.M. (1995) Developmental regulation of ribosomal protein L16 genes in *Arabidopsis thaliana*. *Plant Journal* 8, 65–76.

22 Proteomics of *Arabidopsis* Seed Germination and Priming

Karine Gallardo[1], Claudette Job[1], Steven P.C. Groot[2], Magda Puype[3], Hans Demol[3], Joel Vandekerckhove[3] and Dominique Job[1]

[1]Laboratoire mixte CNRS-INRA-Bayer CropScience, 14/20 rue Pierre Baizet, 69263 Lyon, France; [2]Plant Research International, PO Box 16, 6700 AA Wageningen, The Netherlands; [3]Flanders Interuniversity Institute for Biotechnology and Department of Biochemistry, Gent University, Gent, Belgium

Introduction

Upon hydration under suitable conditions the seed, if not dormant, reactivates its metabolism and commences germination, giving rise to a new plant (Bewley and Black, 1994; Bewley, 1997). An important step in this complex process is referred to as germination *sensu stricto*, which incorporates those events that start with the uptake of water by the quiescent dry seed and terminate with the protrusion of the radicle and the elongation of the embryonic axis. Despite intensive work, the actual cellular and molecular mechanisms that lead to the initiation of germination are still not known.

From a biochemical and molecular point of view, studying germination is difficult because a population of seeds is biochemically and developmentally heterogeneous (Still and Bradford, 1997) and does not complete the process synchronously. Priming treatments are used to synchronize the germination of individual seeds (Bourgne *et al.*, 2000). They initiate germination-related processes but prevent emergence of the radicle and are followed by drying for storage and marketing of the treated seeds. Seed priming generally causes faster germination and faster field emergence, which have practical agronomic implications, notably under adverse germination conditions (Heydecker *et al.*, 1973; Bradford, 1986; McDonald, 2000; Seed Savers' Network, 2001). Actually, optimization of such treatments rests on carrying out subsequent germination assays, which only provide retrospective indications of the effectiveness of the priming conditions.

Proteomics and cDNA microarray technology (Girke *et al.*, 2000; Bove *et al.*, 2002; Koornneef *et al.*, 2002; van der Geest, 2002) may prove valuable by

(eds G. Nicolás, K.J. Bradford, D. Côme and H.W. Pritchard)

providing simultaneous information over a multitude of processes, as occurs in complex developmental processes such as germination. *Arabidopsis thaliana* contains a small genome (120 Mb) of which sequencing is now completed (*Arabidopsis* Genome Initiative, 2000) and is, therefore, the model of choice for studying plant genetics. Recently, we developed a proteome analysis of seed germination and priming using this model plant (Gallardo, 2001; Gallardo *et al.*, 2001, 2002a,b). Our long-term objective is to provide reference maps of seed proteins to focus on the effects of genetic and environmental factors during seed maturation, desiccation, dormancy release and germination. Below, we highlight some of the specific features of these proteomics approaches.

Proteome Analyses During Germination

Under optimal conditions (25°C), dry mature *Arabidopsis* seeds (ecotype Landsberg *erecta*) started to germinate at 1.6 days of imbibition and it took almost 2.2 days for 50% of the seeds to germinate (T_{50}) on water. Total soluble proteins were extracted from various seed samples (dry mature seeds, 1-day and 2-days imbibed seeds, corresponding to germination *sensu stricto* and radicle protrusion, respectively), and analysed by two-dimensional (2D) gel electrophoresis as described (Gallardo *et al.*, 2002b). A comparison of 2D gels for the various protein extracts allowed classifying seed proteins from their specific accumulation patterns (Fig. 22.1).

The type-0 proteins (about 1250 proteins detected in 2D gels; 110 identified) correspond to proteins present in dry mature seeds and whose abundance remains constant throughout the germination process. Type-1 and -2 proteins (33 of the type-1 and 13 type-2 proteins have been identified by MALDI-TOF analysis) correspond to proteins whose abundance varies (up- and down-regulation, respectively) during germination *sensu stricto*. Type-3 and -4 proteins (11 type-3 and 13 type-4 proteins have been identified) show a specific increased or decreased accumulation, respectively, during radicle protrusion. Identified proteins can be globally classified according to the groups defined in Fig. 22.2. Amongst the type-0 proteins, the group of storage proteins and enzymes involved in their mobilization dominate in the dry mature seeds. However, their representation is strongly reduced amongst the proteins showing varying accumulation levels during germination (protein types 1–4). Here, there is a significant increase in the representation of proteins playing a role in metabolism and cellular division. This is consistent with increased metabolic activity during germination and also with experimental evidence, compelling the view that resumption of cell cycle activity is a specific feature of early germination (Liu *et al.*, 1994; Özbingöl *et al.*, 1999; de Castro *et al.*, 2000; see below). Also, accumulation level within the group of stress-response proteins increases strongly during germination (Fig. 22.2). Presumably, this occurs at least in part in response to oxidative stress and in order to prevent the emerging plant from pathogen/herbivore attack (e.g. as exemplified by the induction of enzymes and proteins involved in the catabolism of glucosinolates) (Gallardo *et al.*, 2001).

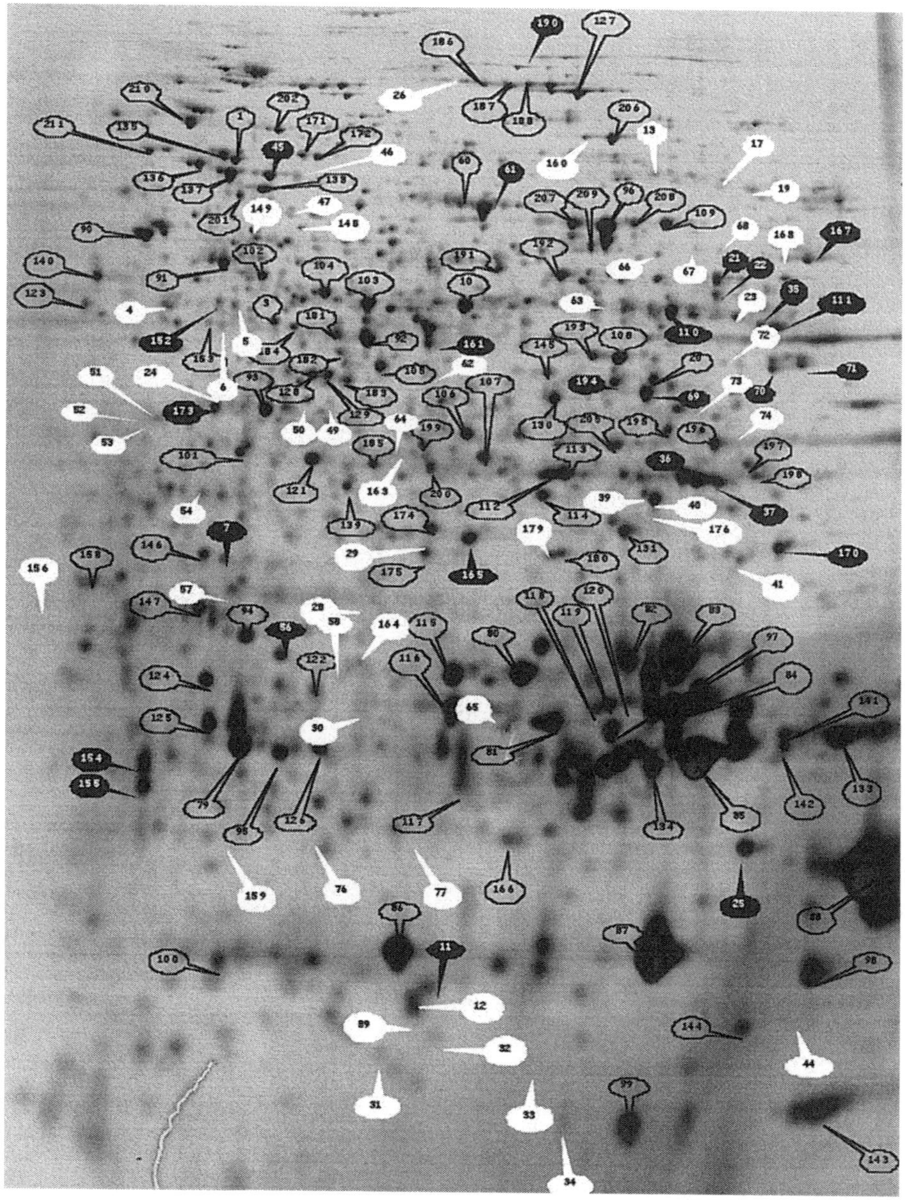

Fig. 22.1. Characterization of *Arabidopsis* seed proteins. The figure shows a 2D gel of total soluble proteins from dry mature seeds. Colour code for labelled proteins identified by MALDI-TOF (matrix assisted laser desorption ionization time of flight) analysis (Gevaert *et al.*, 1998): transparent, constant level during germination (type-0 proteins); white, varying level during germination *sensu stricto* (type-1 and -2 proteins); black, varying level during radicle emergence (type-3 and -4 proteins). Protein numbers refer to proteins listed in Gallardo *et al.* (2001, 2002a).

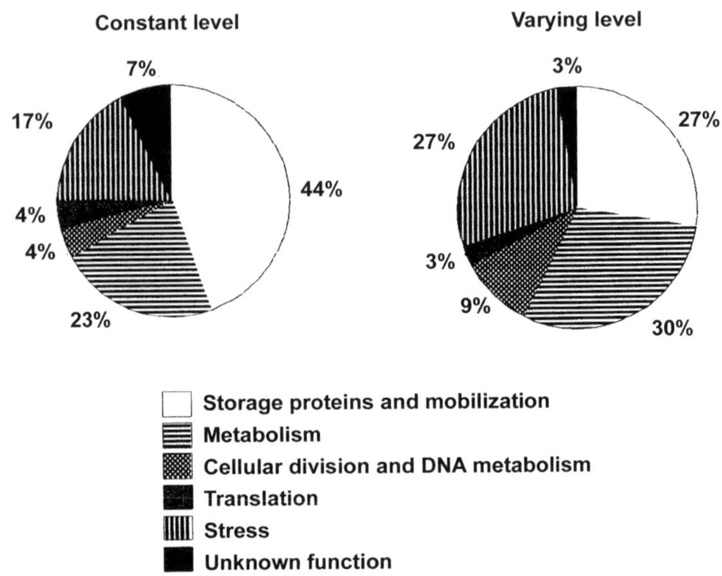

Fig. 22.2. Classification of *Arabidopsis* seed proteins. Following 2D gel electrophoresis, proteins were identified by MALDI-TOF analysis. Left: proteins of type 0 whose level remains constant during germination; right: proteins whose level varies during germination (types 1, 2, 3 and 4) (see Fig. 22.1).

Progressive Build-up of Metabolic Pathways

Mobilization during germination of stored triacylglycerols involves many enzymes in several subcellular compartments. The conversion of fatty acids to succinate takes place within the glyoxysomes, which contain enzymes for fatty acid β-oxidation and the glyoxylate cycle (Beevers, 1980, 1982). We observed that several enzymes in this pathway, such as phosphoglycerate kinase and citrate synthase, could already be detected from the dry mature stage. Others accumulate during later stages of seed germination, before radicle emergence (aconitase, malate synthase, catalase and phosphoenolpyruvate carboxykinase) or following this event (isocitrate lyase, phosphoglyceromutase and mitochondrial malate dehydrogenase) (Gallardo *et al.*, 2001). Such a progressive build-up of the pathway is presumably required for proper timing of storage lipid mobilization. A recent characterization of two allelic *Arabidopsis* mutants, *icl-1* and *icl-2*, which lack the glyoxylate cycle because of the absence of the key enzyme isocitrate lyase, demonstrated that the glyoxylate cycle is not essential for germination but is important for seedling establishment and survival (Eastmond *et al.*, 2000).

Another example of progressive build-up of a complex metabolic pathway was observed for the biosynthesis of methionine and *S*-adenosylmethionine (AdoMet) (Gallardo *et al.*, 2002b). Met synthase, which catalyses the last step in the plant Met biosynthetic pathway (Ravanel *et al.*, 1998), is present at low level in dry mature seeds and its level increases strongly at

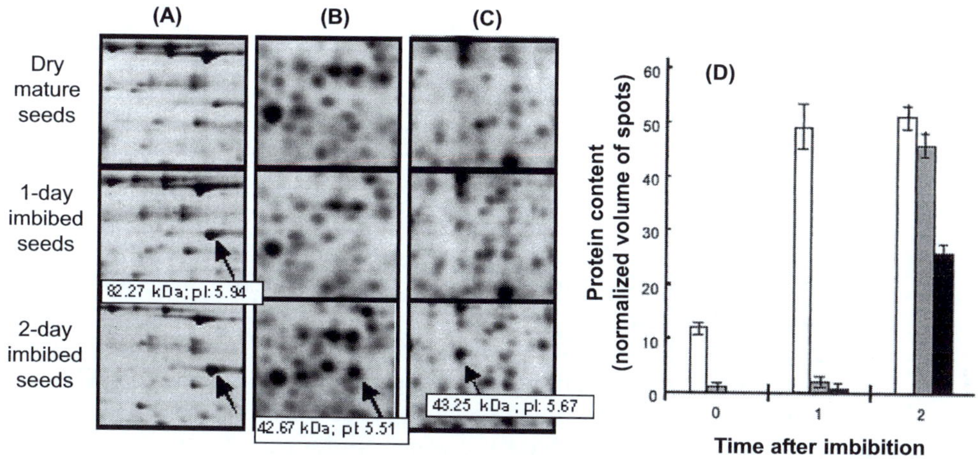

Fig. 22.3. Accumulation levels of methionine synthase and AdoMet synthetase (from Gallardo *et al*., 2002b). (A, B, C) Enlarged windows of 2D gel profiles as shown in Fig. 22.1 at various stages of *Arabidopsis* germination. From top to bottom: dry mature seeds, 1-day imbibed seeds corresponding to germination *sensu stricto* (prior to radicle emergence), 2-day imbibed seeds corresponding to the radicle emergence step. The arrows denote the positions of methionine synthase (A) and AdoMet synthetase (B, C). Experimental M_r and pI are: (A) methionine synthase (M_r = 82.27 kDa, pI = 5.94); (B) AdoMet synthetase (M_r = 42.67 kDa, pI = 5.51; accession number 9229983); (C) AdoMet synthetase (M_r = 43.25 kDa, pI = 5.67; accession number 127045). (D) Quantitation of the accumulation level of methionine synthase and AdoMet synthetase during germination. The data are from panels A, B and C. Seed samples were analysed at time 0 (dry mature seeds) and after 1 day and 2 days of imbibition in water. The protein levels are expressed as normalized volumes of the spots ± SD. White bars, methionine synthase; black and grey bars, AdoMet synthetase accession numbers 9229983 and 127045, respectively.

1 day of imbibition, prior to radicle protrusion. Its level is not increased further at 2 days of imbibition. In contrast, the two isoforms of AdoMet synthetase, which catalyse the synthesis of the ethylene precursor AdoMet from methionine and ATP, are absent in dry mature seeds but specifically accumulate at 2 days of imbibition, coincident with the moment of radicle protrusion (Fig. 22.3).

Thus, Met synthase and AdoMet synthetase are fundamental components controlling methionine metabolism in the transition from a quiescent to a highly active state during germination. Moreover, the observed temporal patterns of accumulation of these proteins are consistent with a role of endogenous ethylene in *Arabidopsis* only after radicle protrusion and not during germination *sensu stricto* (Gallardo *et al*., 2002b). It is striking that AdoMet synthetase is required for the initiation of germination of *Saccharomyces cerevisiae* ascospores (Choih *et al*., 1977). Furthermore, the supplementation of AdoMet to germination medium stimulated germination and outgrowth of ascospores of *S. cerevisiae* (Brawley and Ferro, 1980). Thus, the same basic mechanisms are involved in initiation of germination of yeast spores and plant seeds, which highlights the importance of metabolic control in the transition from a quiescent to a highly active state during germination.

Cell Cycle Activity

Our proteomics approach revealed an accumulation of five proteins associated with cell cycle events during germination *sensu stricto* of *Arabidopsis* seeds (Gallardo *et al.*, 2001, 2002a). They were identified as actin 7, α-2,4 tubulin, α-3,5 tubulin, β-tubulin and a WD-40 repeat protein. An accumulation of β-tubulin during early germination has repeatedly been observed in many species (de Castro *et al.*, 2000).

Liu *et al.* (1994) demonstrated a gibberellic acid (GA) requirement for resumption of cell cycle activity during germination of tomato seeds. We assessed the role of GAs during *Arabidopsis* seed germination by proteomics using the seeds from the GA-deficient *ga1* mutant (Gallardo *et al.*, 2002a). This analysis indicated that, in *Arabidopsis*, GAs do not participate into many processes involved in germination *sensu stricto* (prior to radicle protrusion), as for example the initial mobilization of seed protein and lipid reserves. Out of 46 protein changes detected during germination *sensu stricto* (1 day of incubation on water; type-1 and -2 proteins), only one, corresponding to the presumed cytoskeleton component α-2,4 tubulin, appears to depend on the action of GAs. This protein strongly accumulates in the wildtype seeds during 1 day of imbibition in water but not in the *ga1* mutant seeds (Fig. 22.4).

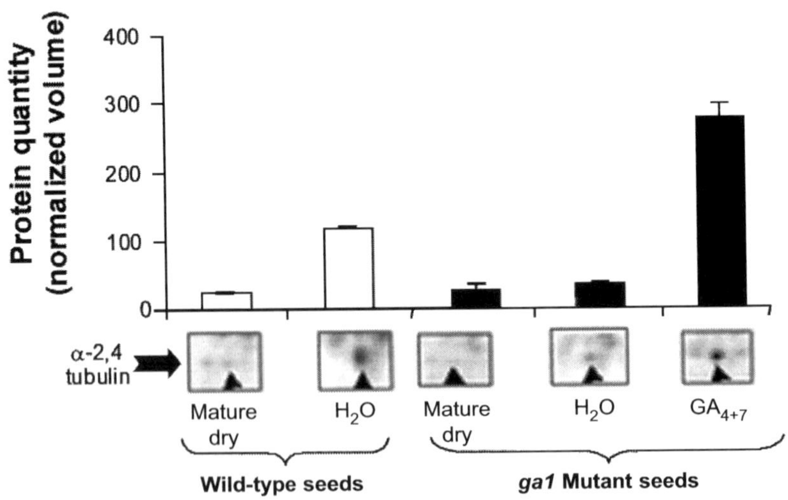

Fig. 22.4. Quantitation of the accumulation level of α-2,4 tubulin during 1 day of imbibition of *Arabidopsis* seeds (from Gallardo *et al.*, 2002a). The results are expressed as normalized volumes of the α-2,4 tubulin spot (± SD) after 1 day of imbibition in water or in the presence or 100 μM GA_{4+7}. A portion of an area of silver-stained 2D gels is shown under the graph. The arrows point to the position of the α-2,4 tubulin spot.

Thus, in *Arabidopsis* the accumulation of only part of the cytoskeleton components is under GA control.

Imbibition and Desiccation

Desiccation tolerance of seeds is a complex multifactorial trait (Kermode, 1995). To understand better the mechanisms involved in adaptation/tolerance of seeds to water stress, we analysed changes in 2D protein patterns that occur in seeds incubated for 1 day in water and then re-dried. Two types of proteins were identified (Gallardo *et al.*, 2001). One corresponds to imbibition-specific polypeptides, whose levels increase during the 1 day of imbibition and then decrease during subsequent drying. The other corresponds to desiccation-specific polypeptides, whose levels decrease during the 1 day of imbibition and then re-increase during subsequent drying. Thus, dehydration stress induces specific and reversible protein changes in seeds. Figure 22.5 shows that isoforms of cytosolic glyceraldehyde-3-phosphate dehydrogenase (GAPDHc) behave as desiccation-specific proteins in *Arabidopsis* seeds. Interestingly, dehydration strongly increases the GAPDHc protein level in leaves of the resurrection plant, *Craterostigma plantagineum*, which can withstand very severe desiccation (Velasco *et al.*, 1994). Thus, the induction of GAPDHc during desiccation is a conserved feature among different tissues and organs in plants.

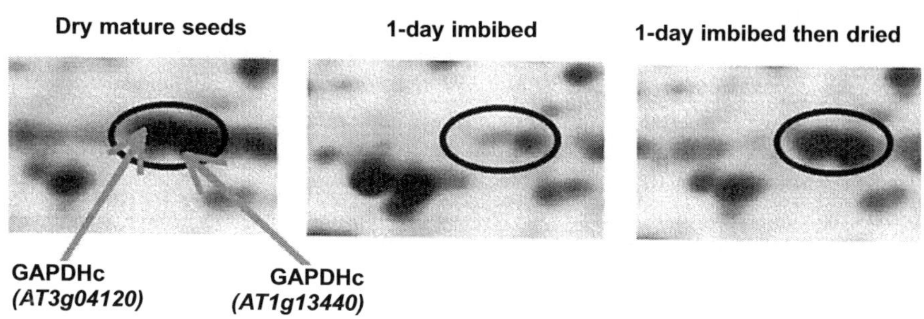

Fig. 22.5. Isoforms of cytosolic GAPDH behave as desiccation-specific proteins in *Arabidopsis* seeds. A portion of an area of silver-stained 2D gels is shown for the dry mature seeds, the seeds incubated for 1 day in water, and the seeds incubated for 1 day in water then re-dried. The arrows point to the position of the cytosolic GADPH spots.

In contrast, actin7, which is the sole source of actin in seed tissues (McDowell *et al.*, 1996), proved to be an imbibition-specific protein in *Arabidopsis* seeds (Gallardo *et al.*, 2001). Our data suggest that all actin-based processes in seed germination are highly sensitive to the hydration level in seed tissues.

As shown in Fig. 22.3, the accumulation level of methionine synthase was maximum after 1 day of imbibition, prior to radicle emergence. This

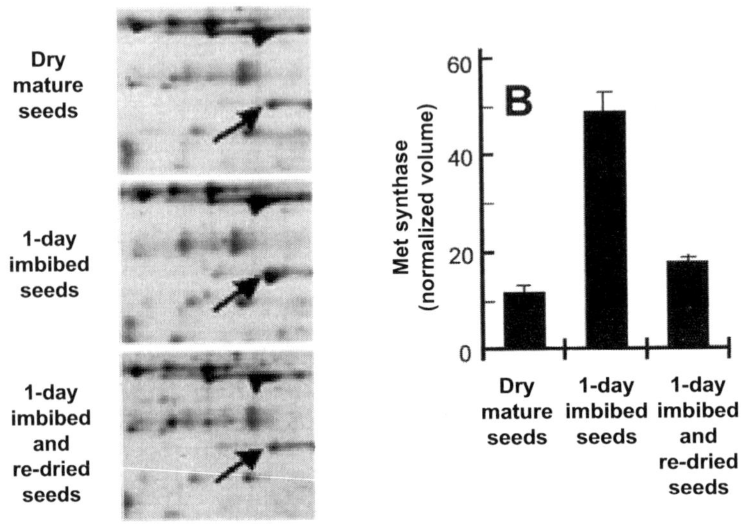

Fig. 22.6. Characterization of methionine synthase during imbibition and drying of *Arabidopsis* seeds (from Gallardo *et al.*, 2002b). (A) 2D gel profile of total proteins from dry mature seeds, 1-day imbibed seeds (germination *sensu stricto*) and 1-day imbibed then re-dried seeds. The figure shows portions of 2D gels as shown in Fig. 22.1. The arrows point to the position of methionine synthase. (B) Quantitation of the accumulation level of methionine synthase.

increased level proved to be reversible, following subsequent drying of the 1 day imbibed seeds back to the original water content of the dry mature seeds (Fig. 22.6). Thus, in accordance with metabolic control of seed germination and seedling establishment, the accumulation of methionine synthase correlates with the metabolic status of the seeds (Gallardo *et al.*, 2002b).

Priming

We designed two priming treatments for *Arabidopsis* seeds, a hydro- and an osmopriming treatment (Gallardo *et al.*, 2001). They both improved *Arabidopsis* seed performance (T_{50}=1.58 days for both treatments compared with 2.2 days for the untreated seeds). Priming-associated polypeptides, whose abundance increases during both treatments, were detected. Some of them were identified as degradation products of 12S-cruciferin B-subunits. The same behaviour has been observed during priming of sugar beet seeds (Job *et al.*, 1997; Capron *et al.*, 2000) and tomato seeds (Job, C., Job, D. and Groot, S.P.C., in preparation). This highlights the similarity, concerning storage protein mobilization during priming, between seeds from different plant families.

Tubulin subunits were also found to accumulate during priming. This is the case for both α- and β-chains. Such an accumulation of β-tubulin during

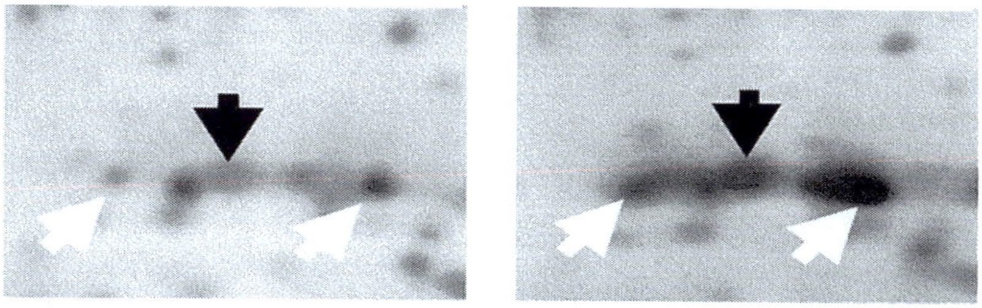

Fig. 22.7. Induction of stress proteins during osmopriming of *Arabidopsis* seeds. A portion of an area of silver-stained 2D gels is shown for the dry mature seeds (untreated seeds), and for the seeds osmoprimed in –0.75 MPa PEG (polyethylene glycol 6000) solution then re-dried (osmoprimed seeds). The white and black arrows point to the positions of LMW HSPs and dehydrin (RAB18) spots, respectively.

priming has repeatedly been observed in many species, in relation with reactivation of cell cycle activity (de Castro *et al.*, 2000).

The abundance of low-molecular-weight heat-shock proteins (LMW HSPs) was found to increase specifically in osmoprimed seeds (Fig. 22.7). Since LMW HSPs have molecular chaperone activity (Lee *et al.*, 1995), these data suggest that they act by maintaining the proper folding of other proteins during the incomplete hydration resulting from osmopriming.

The detection of such proteins also confirms that water stress generated by high osmotic potential induces specific changes in protein synthesis (Jin *et al.*, 2000). The accumulation of other stress proteins such as dehydrin (Fig. 22.7), MLO (*Mycoplasma*-like organism) protein (Devoto *et al.*, 1999), HSP 60 and catalase is also induced in primed seeds (Gallardo, 2001; Gallardo *et al.*, 2001), supporting the view that stress responses are fully activated during priming. This observation might account for the fact that priming not only causes faster germination but also improves seed vigour (see Bruggink *et al.*, 1999, and references therein).

Conclusion and Future Research

Our results provide clear reference maps of *Arabidopsis* seed proteins at various stages of germination. These maps will be useful for analysing mutants that are affected in seed germination, as we have shown for the *ga1* GA-deficient mutant of *Arabidopsis*. Our work illustrates that proteomics can provide global information over a multitude of processes occurring during seed germination. These data can be analysed further in combination with cDNA microarray technology, which will indicate whether gene regulation is controlled at the level of transcription, or translation and protein accumulation. Protein function can be further studied by a combination of forward and reverse genetics and proteomics, as has already been demonstrated in yeast and *Escherichia coli*. These global expression-profiling approaches may prove useful for providing new information regarding genes involved in seed quality (Groot *et al.*, 2000).

Acknowledgements

This work was supported by grants from the European Community (FAIR project grant no. CT97-3711, 'Genetic and molecular markers for seed quality'), the Région Rhône-Alpes (Biotechnology program 'The seed') and by the Fund for Scientific Research of Flanders.

References

Arabidopsis Genome Initiative (2000) Analysis of the genome sequence of the flowering plant *Arabidopsis thaliana*. *Nature* 408, 796–815.

Beevers, H. (1980) The role of the glyoxylate cycle. In: Stumpf, P.K. and Conn, E.E. (eds) *The Biochemistry of Plants*, Vol. 4. Academic Press, New York, pp. 117–130.

Beevers, H. (1982) Glyoxysomes in higher plants. *Annals of the New York Academy of Sciences* 386, 243–251.

Bewley, J.D. (1997) Seed germination and dormancy. *Plant Cell* 9, 1055–1066.

Bewley, J.D. and Black, M. (1994) *Seeds. Physiology of Development and Germination*. Plenum Press, New York.

Bourgne, S., Job, C. and Job, D. (2000) Sugar beet seed priming: solubilization of the basic subunit of 11-S globulin in individual seeds. *Seed Science Research* 10, 153–161.

Bove, J., Jullien, M. and Grappin, P. (2002) Functional genomics in the study of seed germination. *Genome Biology* 3, Reviews 1002.

Bradford, K.J. (1986) Manipulation of seed water relations via osmotic priming to improve germination under stress conditions. *HortScience* 21, 1105–1112.

Brawley, J.V. and Ferro, A.J. (1980) Stimulation of yeast ascospore germination and outgrowth by *S*-adenosylmethionine. *Journal of Bacteriology* 142, 608–614.

Bruggink, G.T., Ooms, J.J.J. and van der Toorn, P. (1999) Induction of longevity in primed seeds. *Seed Science Research* 9, 49–53.

Capron, I., Corbineau, F., Dacher, F., Job, C., Côme, D. and Job, D. (2000) Sugar beet seed priming: effects of priming conditions on germination, solubilization of 11-S globulin and accumulation of LEA proteins. *Seed Science Research* 10, 243–254.

Choih, S.J., Ferro, A.J. and Shapiro, S.K. (1977) Function of *S*-adenosylmethionine in germination of yeast ascospores. *Journal of Bacteriology* 131, 63–68.

de Castro, R.D., van Lammeren, A.A.M., Groot, S.P.C., Bino, R.J. and Hilhorst, H.W.M. (2000) Cell division and subsequent radicle protrusion in tomato seeds are inhibited by osmotic stress but DNA synthesis and formation of microtubular cytoskeleton are not. *Plant Physiology* 122, 327–335.

Devoto, A., Piffanelli, P., Nilsson, I.M., Wallin, E., Panstruga, R., von Heijne, G. and Schulze-Lefert, P. (1999) Topology, subcellular localization, and sequence diversity of the MLO family in plants. *Journal of Biological Chemistry* 274, 34993–35004.

Eastmond, P.J., Germain, V., Lange, P.R., Bryce, J.H., Smith, S.M. and Graham, I.A. (2000) Postgerminative growth and lipid catabolism in oilseeds lacking the glyoxylate cycle. *Proceedings of the National Academy of Sciences USA* 97, 5669–5674.

Gallardo, K. (2001) De l'analyse protéomique des graines d'*Arabidopsis thaliana* vers une compréhension du processus de germination. PhD thesis, University of Aix-Marseilles III, France.

Gallardo, K., Job, C., Groot, S.P.C., Puype, M., Demol, H., Vandekerckhove, J. and Job, D. (2001) Proteomic analysis of *Arabidopsis* seed germination and priming. *Plant Physiology* 126, 835–848.

Gallardo, K., Job, C., Groot, S.P.C., Puype, M., Demol, H., Vandekerckhove, J. and Job, D. (2002a) Proteomics of *Arabidopsis* seed germination. A comparative study of wild-type and GA-deficient seeds. *Plant Physiology* 129, 823–837.

Gallardo, K., Job, C., Groot, S.P.C., Puype, M., Demol, H., Vandekerckhove, J. and Job, D. (2002b) Importance of methionine biosynthesis for *Arabidopsis* seed germination and seedling growth. *Physiologia Plantarum* 116, 238–247.

Gevaert, K., Demol, H., Sklyarova, T., Vandekerckhove, J. and Houthaeve, T. (1998) A peptide concentration and purification method for protein characterization in the subpicomole range using matrix assisted laser desorption/ionisation-post-source decay (MALDI-MS) sequencing. *Electrophoresis* 19, 909–917.

Girke, T., Todd, J., Ruuska, S., White, J., Benning, C. and Ohlrogge, J. (2000) Microarray analysis of developing *Arabidopsis* seeds. *Plant Physiology* 124, 1570–1581.

Groot, S.P.C., van der Geest, A.H.M., Tesnier, K., Alonso-Blanco, C., Bentsink, L., Donkers, H., Koornneef, M., Vreugdenhil, D. and Bino, R.J. (2000) Molecular analysis of *Arabidopsis* seed quality. In: Black, M., Bradford, K.J. and Vázquez-Ramos, J. (eds) *Seed Biology: Advances and Applications*. CAB International, Wallingford, UK, pp. 123–132.

Heydecker, W., Higgins, J. and Gulliver, R.L. (1973) Accelerated germination by osmotic seed treatment. *Nature* 246, 42–44.

Jin, S., Chen, C.C.S. and Plant, A.L. (2000) Regulation by ABA of osmotic-stress-induced changes in protein synthesis in tomato roots. *Plant Cell and Environment* 23, 51–60.

Job, C., Kersulec, A., Ravasio, L., Chareyre, S., Pépin, R. and Job, D. (1997) The solubilization of the basic subunit of sugarbeet seed 11-S globulin during priming and early germination. *Seed Science Research* 7, 225–243.

Kermode, A.R. (1995) Regulatory mechanisms in the transition from seed development to germination: interactions between the embryo and the seed environment. In: Kigel, J. and Galili, G. (eds) *Seed Development and Germination*. Marcel Dekker, New York, pp. 273–332.

Koornneef, M., Bentsink, L. and Hilhorst, H. (2002) Seed dormancy and germination. *Current Opinion in Plant Biology* 5, 33–36.

Lee, G.J., Pokala, N. and Vierling, E. (1995) Structure and *in vitro* molecular chaperone activity of cytosolic small heat shock proteins from pea. *Journal of Biological Chemistry* 270, 10432–10438.

Liu, Y., Bergervoet, J.H.W., De Vos, C.H.R., Hilhorst, H.W.M., Kraak, H.L., Karssen, C.M. and Bino, R.J. (1994) Nuclear replication activities during imbibition of abscisic acid- and gibberellin-deficient tomato (*Lycopersicon esculentum* Mill.) seeds. *Planta* 194, 368–373.

McDonald, M.B. (2000) Seed priming. In: Black, M. and Bewley, J.D. (eds) *Seed Technology and its Biological Basis*. Sheffield Academic Press, Sheffield, UK, pp. 287–325.

McDowell, J.M., An, Y.-Q., Huang, S., McKinney, E.C. and Meagher, R.B. (1996) The *Arabidopsis ACT7* actin gene is expressed in rapidly developing tissues and responds to several external stimuli. *Plant Physiology* 111, 699–711.

Özbingöl, N., Corbineau, F., Groot, S.P.C., Bino, R.J. and Côme, D. (1999) Activation of the cell cycle in tomato (*Lycopersicon esculentum* Mill.) seeds during osmoconditioning as related to temperature and oxygen. *Annals of Botany* 84, 245–251.

Ravanel, S., Gakière, B., Job, D. and Douce, R. (1998) The specific features of methionine biosynthesis and metabolism in plants. *Proceedings of the National Academy of Sciences USA* 95, 7805–7812.

Seed Savers' Network (2001) Priming seed to improve establishment. Available at: www.seedsavers.net

Still, D.W. and Bradford, K.J. (1997) Endo-β-mannanase activity from individual tomato endosperm caps and radicle tips in relation to germination rates. *Plant Physiology* 113, 21–29.

Van der Geest, A.H.M. (2002) Seed genomics: germinating opportunities. *Seed Science Research* 12, 143–153.

Velasco, R., Salamini, F. and Bartels, D. (1994) Dehydration and ABA increase mRNA levels and enzyme activity of cytosolic GAPDH in the resurrection plant *Craterostigma plantagineum*. *Plant Molecular Biology* 26, 541–546.

23 Involvement of Oxidative Stress and ABA in CN-mediated Elimination of Embryonic Dormancy in Apple

Renata Bogatek[1], Helena Gawrońska[2] and Krystyna Oracz[1]

[1]*Department of Plant Physiology, Warsaw Agricultural University, Rakowiecka 26–30, 02-528 Warsaw, Poland;* [2]*Department of Pomology and Basic Sciences in Horticulture, Warsaw Agricultural University, Nowoursynowska 166, 02-787 Warsaw, Poland*

Introduction

Dormant apple embryos germinate slowly and the seedlings show several growth and developmental abnormalities (Bogatek *et al.*, 1991). Dormancy and its removal are under temperature and light control and several plant growth regulators (gibberellins, cytokinins, ethylene and abscisic acid) are involved in the regulation of these processes (Lewak, 1981). Our particular attention was addressed to the free hydrogen cyanide, which is produced by the embryo during dormancy removal (Dziewanowska *et al.*, 1979). It has been demonstrated that a short pretreatment of dormant apple embryos with gaseous HCN leads to their increased germinability and disappearance of growth abnormalities of young seedlings (Bogatek *et al.*, 1991). Data collected so far indicate that HCN may act on dormancy through regulation of sugar catabolism (Bogatek, 1995; Bogatek *et al.*, 1999).

The aim of the present work was to answer whether there are other, parallel modes of action of HCN on apple embryo dormancy and in particular whether the HCN pretreatment: (i) provokes oxidative stress and/or activates an antioxidant system, (ii) induces changes in abscisic acid (ABA) level, and (iii) whether these changes are related to CN-mediated elimination of embryonic dormancy. Therefore, studies were undertaken to determine changes in: (i) the level of reactive oxygen species (ROS), (ii) activities of enzymes involved in ROS scavenging, and (iii) the level of ABA.

Material and Methods

Seeds of apple (*Malus domestica* Borb.) cv. Antonówka, 2000 harvest, were used in the studies. Isolated embryos were treated for 6 h with 1 mM HCN as described by Bogatek and Lewak (1988). Lots of 30 embryos were cultured for 7 days at 25°C/20°C (day/night) with 12 h photoperiod and under 150 μmol/m^2/s of photosynthetic active radiation (PAR). Germinating embryos were counted daily and at selected days they were collected for analysis. ABA was assayed by the ELISA method (Weiler, 1982). Level of H_2O_2 and activities of superoxide dismutase (SOD), catalase (CAT) and glutathione reductase (GR) were determined as described by Bailly *et al.* (1996). All experiments were repeated at least twice with three to four replications in each. Results are expressed as means ± SE.

Results and Discussion

Germinability of control dormant embryos achieved 15% after 7 days of culture and the developing seedlings exhibited abnormalities as described earlier. Pretreatment of embryos with HCN resulted in increased germination, reaching 85% after 7 days (Fig. 23.1), with normal seedling development. These results confirm earlier data (Bogatek *et al.*, 1999).

The level of ABA, which was high in dormant embryos, decreased during the culture under control conditions by more than ten times. However, during the first 3 days its drop was much sharper than during the last days of the experiment (Fig. 23.1). It is evident that there is no correla-

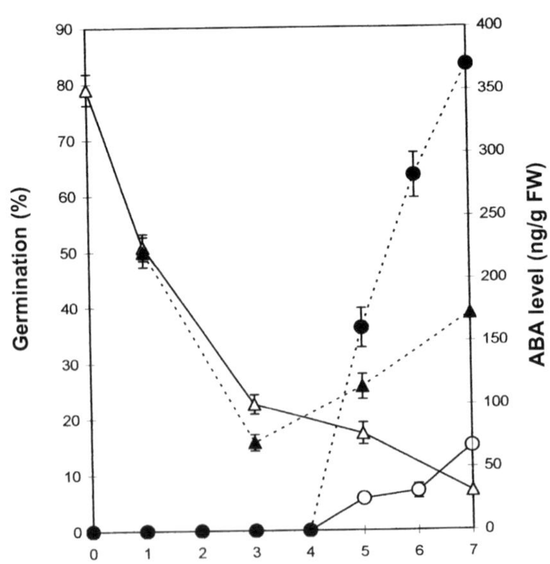

Fig.23.1. Effect of a 6 h pretreatment with 1 mM HCN on germination (○,●) and on the level of ABA (△,▲) during culture of apple embryos. Control (open symbols), HCN pretreated (filled symbols and broken lines). Vertical bars indicate ±SE.

tion between the decrease in ABA level and still very low embryo germinability, indicating that ABA is not the sole factor responsible for dormancy in apple embryos. The lack of a good correlation between ABA level in the seeds and their germinability has been noted for some other species (Walton, 1980–1981).

Pretreatment of embryos with HCN enhanced to a small degree the sharp decrease in ABA content during the first 3 days of culture (Fig. 23.1), but later during the culture a marked increase in this hormone level was noted.

The finding that a higher level of ABA occurs in well-germinating (CN-stimulated) embryos than in untreated dormant ones agrees with the opinion of Corbineau *et al.* (2002) that ABA level is not decisive in the breaking of dormancy. It might also indicate that in apple embryos the alleviation of dormancy coincides with a decrease in sensitivity to ABA. A similar suggestion has been expressed by Gazzarini and McCourt (2001) that in *Arabidopsis* mutants the sensitivity to ABA decreases along with germination and seedling development. Lack of a good correlation between lowering of ABA level and germinability and differences in sensitivity to ABA has been observed in dormant seeds of several other species (Walker-Simmons, 1987; Wang *et al.*, 1995; Corbineau *et al.*, 2000).

ROS generation in apple embryos was also affected by HCN. The level of H_2O_2 peaked on the first day of culture, in both control and HCN pretreated embryos but was threefold greater in the pretreated ones (Fig. 23.2), thus indicating an oxidative stress induction by HCN.

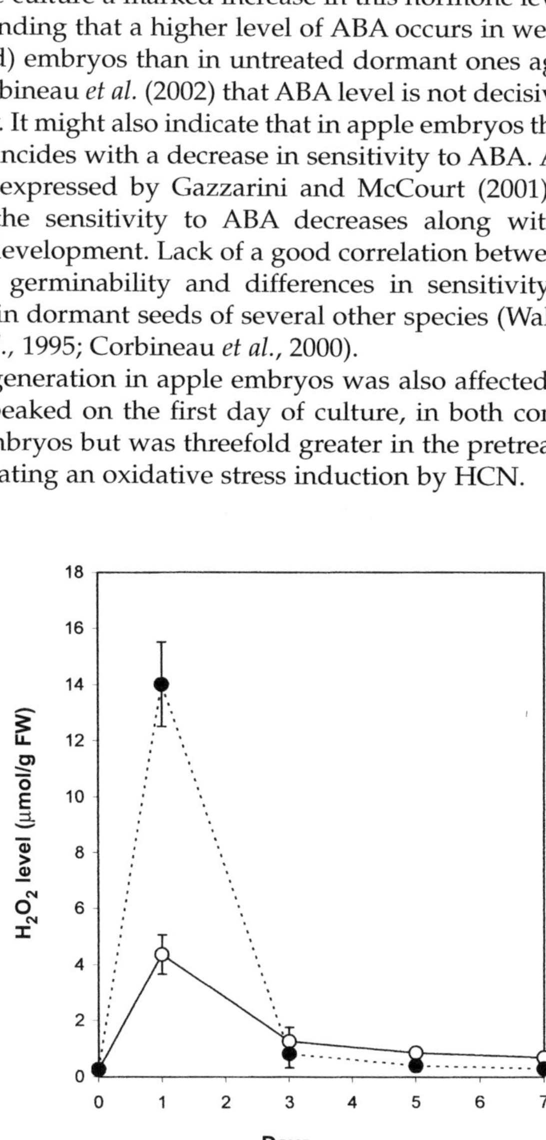

Fig. 23.2. Effect of HCN pretreatment (1 mM) on the level of H_2O_2 in germinating embryos and developing seedlings of apple. Control (○), HCN pretreated (●). Vertical bars indicate ±SE.

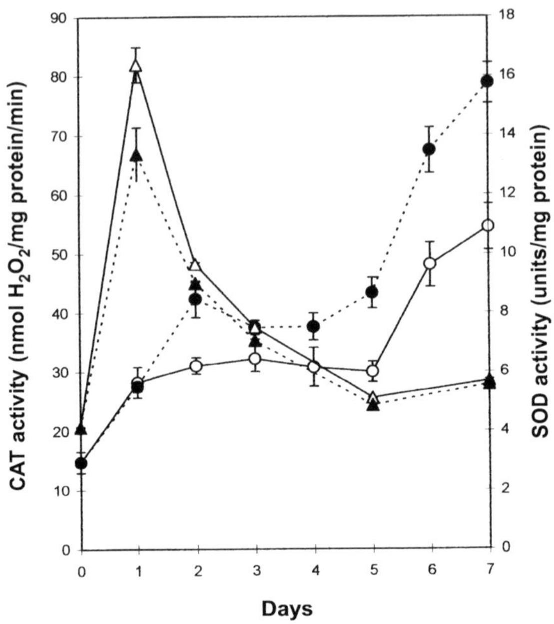

Fig. 23.3. Effect of HCN pretreatment (1 mM) on activity of SOD (△,▲) and CAT (○,●) in germinating embryos and developing seedlings of apple. Control (open symbols), HCN pretreated (filled symbols and broken lines). Vertical bars indicate ±SE.

Wang *et al.* (1995) demonstrated that the treatment of dormant barley seeds with H_2O_2 leads to a decrease in ABA level. It is not excluded that a similar relation between changes in H_2O_2 and ABA occurs in apple embryos; we have also observed simultaneous increase of H_2O_2 and a decrease of ABA levels. Therefore, the observed sharp rise in H_2O_2 during the early period of apple embryo culture could be responsible for the decrease of ABA level recorded in this study.

Induction of oxidative stress by HCN pretreatment resulted in an increase in the activity of all the antioxidative enzymes studied. Accumulation of H_2O_2 corresponded well with maximum SOD activity noted both in control and in HCN pretreated embryos on the first day of culture (Fig. 23.3). In contrast, CAT activity was relatively low during the first few days of culture, then it increased sharply at the end of the experiment (Fig. 23.3), showing a good correlation with the kinetics of germination. CAT activity increase could probably be related to mobilization of lipid reserves, which is also CN-stimulated and is involved in processes of dormancy removal, as has been previously documented in apple (Bogatek and Lewak, 1991).

The most important effect of cyanide, among all studied enzymes, was recorded for GR activity. Throughout the whole experiment, GR activity in HCN pre-treated embryos was almost twofold higher than that of controls

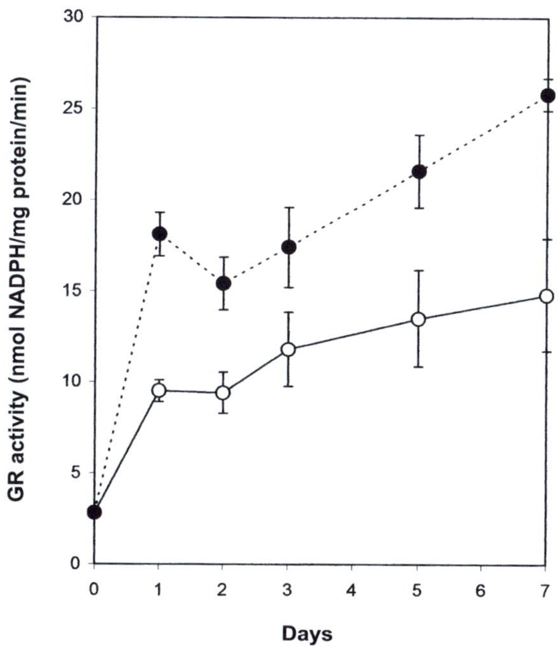

Fig. 23.4. Effect of HCN pretreatment (1 mM) on activity of GR in germinating embryos and developing seedlings of apple. Control (○), HCN pretreated (●). Vertical bars indicate ±SE.

(Fig. 23.4), which indicates that a short treatment with HCN stimulates glutathione-dependent hydrogen peroxide detoxification.

The accumulation of H_2O_2, in response to oxidative stress provoked by cyanide, might induce the pentose phosphate pathway (PPP) through oxidation–reduction of glutathione and of NADP. The crucial role of PPP in dormancy breaking and seed germination has been postulated many times (Côme and Corbineau, 1989; Fontaine *et al.*, 1994) for seeds of different species, including apple (Bogatek and Lewak, 1988).

Results of this work together with data from our previous studies allow us to propose that HCN plays an important signalling role in the events leading to breaking of seed dormancy in apple.

Acknowledgements

This work was supported by the grant of the President of Warsaw Agricultural University No. 50401020011. The authors wish to express their thanks to Professor E.W. Weiler from Ruhr University, Bochum, Germany, for generously supplying antibodies and tracer used for ABA assay and to Mrs M. Dzieciol for her skillful assistance.

References

Bailly, C., Benamar, A., Corbineau, F. and Côme, D. (1996) Changes in malondialdehyde content and in superoxide dismutase, catalase and glutathione reductase activities in sunflower seeds as related to deterioration during accelerated ageing. *Physiologia Plantarum* 97, 104–110.

Bogatek, R. (1995) The possible role of fructose 2,6-bisphosphate in the cyanide-mediated removal of embryonic dormancy in apple. *Physiologia Plantarum* 94, 460–464.

Bogatek, R. and Lewak, S. (1988) Effects of cyanide and cold treatment on sugar catabolism in apple seeds during dormancy removal. *Physiologia Plantarum* 73, 406–411.

Bogatek, R. and Lewak, S. (1991) Cyanide controls enzymes involved in lipid and sugar catabolism in dormant apple embryos during culture. *Physiologia Plantarum* 83, 422–426.

Bogatek, R., Dziewanowska, K. and Lewak, S. (1991) Hydrogen cyanide and embryonal dormancy in apple seeds. *Physiologia Plantarum* 83, 417–421.

Bogatek, R., Côme, D., Corbineau, F., Picard, M.-A., Żarska-Maciejewska, B. and Lewak, S. (1999) Sugar metabolism as related to the cyanide-mediated elimination of dormancy in apple embryos. *Plant Physiology and Biochemistry* 37, 577–585.

Côme, D. and Corbineau, F. (1989) Some aspects of metabolic regulation of seed germination and dormancy. In: Taylorson, R.B. (ed.) *Recent Advances in the Development and Germination of Seeds.* Plenum Press, New York, pp. 165–179.

Corbineau, F., Benamar, A. and Côme, D. (2000) Changes in sensitivity to ABA of the developing and maturing embryos of two wheat cultivars with different sprouting susceptibility. *Israel Journal of Plant Sciences* 48, 189–197.

Corbineau, F., Bianco, J., Garello, G. and Côme, D. (2002) Breakage of *Pseudotsuga menziesii* seed dormancy by cold treatment as related to changes in seed ABA sensitivity and ABA level. *Physiologia Plantarum* 114, 313–319.

Dziewanowska, K., Niedźwiedź, I. and Lewak, S. (1979) Hydrogen cyanide and cyanogenic compounds in seeds. II. Changes in free HCN level in apple seeds during stratification. *Physiologie Vegetale* 17, 681–686.

Fontaine, O., Huault, C., Pavis, N. and Billard, J.-P. (1994) Dormancy breakage of *Hordeum vulgare* seeds: effects of hydrogen peroxide and scarification on glutathione level and glutathione reductase activity. *Plant Physiology and Biochemistry* 32, 677–683.

Gazzarrini, S. and McCourt, P. (2001) Genetic interactions between ABA, ethylene and sugar signalling pathways. *Current Opinion in Plant Biology* 4, 387–391.

Lewak, S. (1981) Regulatory pathways in removal of apple seed dormancy. *Acta Horticulturae* 120, 303–306.

Walker-Simmons, M. (1987) ABA level and sensitivity in developing wheat embryos of sprouting resistant and susceptible cultivars. *Plant Physiology* 84, 61–66.

Walton, D.C. (1980–1981) Does ABA play a role in seed germination? *Israel Journal of Botany* 29, 168–180.

Wang, M., Heimovaars-Dijkstra, S. and Van Duijn, B. (1995) Modulation of germination of embryo isolated from dormant and non-dormant barley grains by manipulation of endogenous abscisic acid. *Planta* 195, 586–592.

Weiler, E. (1982) An enzyme-immunoassay for *cis*-(+)-abscisic acid. *Physiologia Plantarum* 54, 510–514.

24 The Regulation of the Thermoinhibition of Seed Germination in Winter Annual Plants by Abscisic Acid

TOSHIHITO YOSHIOKA*, TAKERU GONAI, SHUSUKE KAWAHARA, SHIGERU SATOH AND TERUYOSHI HASHIBA

Graduate School of Agricultural Science, Tohoku University, 1-1 Amamiya, Aoba-ku, Sendai, Miyagi 981-8555, Japan

Introduction

Obligate winter annuals possess a pattern of life cycle: seeds germinate in autumn, the emerged plants grow vegetatively in winter and set flowers and seeds in the following spring. That is, they escape their growth in summer as soil-buried seeds. Many ecological studies using buried seeds have revealed that germination of winter annuals is prevented during summer because the maximum temperatures at which seeds can germinate are below those occurring in the habitat (for reviews see Baskin and Baskin, 1998; Probert, 2000). However, physiological mechanisms, needed by the seeds to sense that the temperatures are inadequate for their germination, are unclear.

When quiescent imbibed seeds are exposed to high temperatures, their germination is often inhibited. The seeds can germinate again by lowering the temperature. This suppression of germination at supraoptimal temperatures is called thermoinhibition (Vidaver and Hsiao, 1975). Lettuce (*Lactuca sativa* L.) seeds show thermoinhibition at temperatures above 28–30°C (Berrie, 1966). Reynolds and Thompson (1971) noted that an application of abscisic acid (ABA) to the incubation medium reduces the threshold temperatures for lettuce seed germination, i.e. enhances thermoinhibition. In early works, however, no difference in endogenous ABA levels was observed between lettuce seeds incubated at low and high temperatures (Berrie and Robertson, 1976). Even when the ABA levels in the lettuce seeds were significantly changed depending on the temperature, the changes were small and not always correlated with germination behaviour (Braun and Khan, 1975).

*See list of contributors for new address.

(eds G. Nicolás, K.J. Bradford, D. Côme and H.W. Pritchard)

Fluridone is a herbicide that blocks phytoene desaturase in the carotenoid pathway (Bartels and Watson, 1978). Because the carotenoids are precursors of ABA, it is also an inhibitor of ABA biosynthesis. By treating the lettuce seeds with this chemical, we found that continuous ABA biosynthesis is indispensable for both the maintenance of high ABA contents and inhibition of germination at a high temperature (Yoshioka *et al.*, 1998).

Therefore, to resolve the above discrepancies and to elucidate how seeds sense the cues for appropriate timing of germination, we examined seed responsiveness to temperature through ABA using a lettuce cultivar, Grand Rapids, an obligate winter annual, greater chickweed (*Stellaria neglecta* L.) and fluridone.

Materials and Methods

Lettuce seeds were purchased from Sakata Seed Company Ltd (Yokohama, Japan). Seeds of greater chickweed were collected in the suburbs of Sendai, Japan. For germination tests, 50 seeds were sown in a 5 cm Petri dish on two layers of filter paper wetted with 1.8 ml of water or 30 μM fluridone solution, and germinated seeds were scored at given times after the start of incubation.

For measurements of ABA and gibberellin (GA), seeds (1 g for ABA and 20 g for GA) were homogenized in methanol after adding $[^2H_6]$ABA and $[^2H_2]GA_1$, respectively, as internal standards. The extracts were purified and fractioned according to Yoshioka *et al.* (1998) and Endo *et al.* (1989). The fractions containing S-ABA or GA_1 were collected and methylated with diazomethane. The GA_1 fractions were then converted into methyl ester trimethylsilyl ether form. We analysed them by a GC-MS with selected ion monitoring (JMS-700, JEOL). Contents of S-ABA and GA_1 in the seeds were determined from peak area ratios of ions at 190 and 194 and at 506 and 508, respectively.

Physiological Mechanism of Thermoinhibition in Lettuce Seeds

The germination of Grand Rapids lettuce seeds was inhibited at a supraoptimal temperature of 30°C in darkness (Fig. 24.1B, thermoinhibition). The application of the ABA inhibitor, fluridone, to the incubation medium enabled the thermoinhibited seeds to germinate (Fig. 24.1C). At the supraoptimal temperature, ABA content in the seeds was consistently high, but decreased in the presence of fluridone (Fig. 24.1A). This means that the metabolism and synthesis of ABA are balanced in the thermoinhibited seeds and that continuous synthesis is necessary for the continued inhibition of germination at the supraoptimal temperature. However, the induction of seed germination by fluridone application appeared only in a restricted temperature range (Fig. 24.2); that is, the seeds could germinate at 28°C in the presence of a 30 μM fluridone solution, the lower supraoptimal temperature, but they still could not germinate at 33°C, the higher supraoptimal temperature. It is suggested, therefore, that the thermoinhibition at the higher supraoptimal temperature is not caused by enhanced ABA biosynthesis alone.

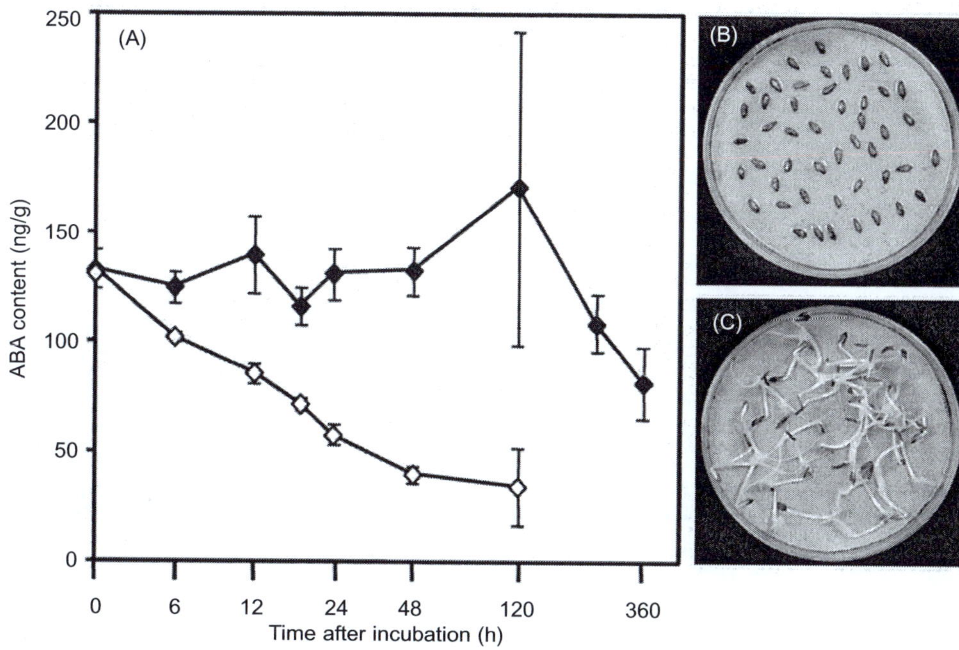

Fig. 24.1. Effects of fluridone on ABA contents and germination of lettuce seeds at a supraoptimal temperature. (A) Seeds (1 g dry weight) were incubated in the presence (◇) or absence (◆) of 30 μM fluridone at 30°C, and their ABA contents were determined at designated times. Note that the abscissa is shown on a log scale. (B) Thermoinhibition of lettuce seed germination. Seeds were incubated without fluridone at 30°C for 10 days. (C) Rescue from thermoinhibition with fluridone. Seeds were incubated in 30 μM fluridone solution at 30°C for 10 days.

As shown in Table 24.1, the application of a GA biosynthesis inhibitor, chlormequat, prevented germination at 28°C even in the presence of fluridone. In addition, a combined treatment with fluridone and GA_3 was needed to induce the germination at 33°C, but neither fluridone nor GA_3 alone was effective. Induction of germination by the application of fluridone in combination with GA_3 was shown also in dormant yellow-cedar seeds (Schmitz *et al.*, 2001). From these findings, we assumed at first that the germination could be inhibited at such a high temperature as 33°C by a low endogenous GA content accompanied by a high ABA content. Hence, we determined the endogenous contents of GA_1, which has been proved by Toyomasu *et al.* (1993) to be the main endogenous bioactive GA in lettuce seeds. As a result, the GA_1 content did not differ between the optimal and supraoptimal temperatures (data not shown). Thus, a high temperature would prevent germination of lettuce seeds by increasing the endogenous ABA contents without affecting the endogenous GA_1 content.

We found that the ABA concentration causing 50% inhibition of lettuce seed germination was 0.67 μM at 33°C, whereas it was 67.8 μM at 18°C. This

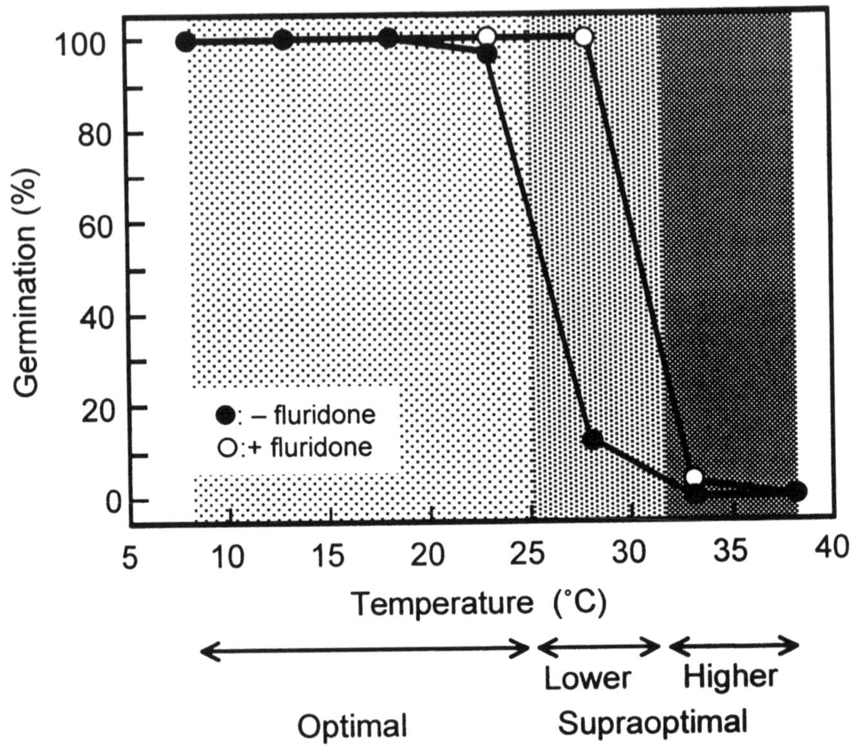

Fig. 24.2. Seed responsiveness to temperature in the presence of fluridone. Lettuce seeds were incubated at various temperatures in the presence (○) or absence (●) of 30 μM fluridone, and germination was scored after 5 days of incubation.

is in line with the previous report by Walker-Simmons (1988) with isolated wheat embryos, in which ABA was 100 times more effective at 30°C than at 15°C in reducing embryonic germination. These results show that the sensitivity of lettuce seeds to ABA increases as the incubation temperature rises in the range of 18–33°C. Thus, the thermoinhibition of lettuce seed germination at the higher supraoptimal temperature would be caused by both high ABA content and high ABA sensitivity.

In another experiment, we observed that GA might promote ABA degradation in lettuce seeds (data not shown). A similar suggestion was proposed by Grappin *et al.* (2000) working with *Nicotiana plumbaginifolia* seeds. In our experiment, GA_3 application or exposure of seeds to red light restored germination at 33°C in the presence of fluridone (Table 24.1). Toyomasu *et al.* (1993, 1998) revealed that red light promotes GA_1 synthesis in lettuce seeds by inducing expression of Ls3h1, a GA-3β-hydroxylase. Red light may enhance ABA degradation by promoting GA biosynthesis. Therefore, it is tempting to suggest that, in association with the inhibition of ABA biosynthesis by fluridone, enhanced ABA degradation by GA reduces the ABA con-

Table 24.1. Effects of fluridone (an inhibitor of ABA biosynthesis), chlormequat (an inhibitor of GA biosynthesis), GA_3 and red light on germination of lettuce seeds at supraoptimal temperatures. Fifty seeds were incubated in water or 30μM fluridone solution in combination with 1 mM chlormequat solution or 2 mM GA_3 solution in continuous darkness, or darkness after exposure to 5.0 μmol/m²/s red light for 180 s. The germination was scored after a 5-day incubation. Data are means of three replicates with standard errors.

Temperature (°C)	Medium	Germination (%)	
		– Fluridone	+ Fluridone
28	Water	7.7 ± 5.2	86.0 ± 5.2
	Chlormequat	0	0
33	Water	0	0
	GA_3	0	95.3 ± 2.4
	Red light	0	100 ± 0.0

tent to a sufficiently low level to allow seed germination at the higher supraoptimal temperature. The GA action on ABA degradation needs to be proved directly by determining levels of ABA metabolites.

Restoration of Germination by Fluridone of Soil-buried Seeds in Greater Chickweed

Soil-buried seeds of greater chickweed were conditionally dormant in June–August; maximum temperatures for germination gradually increased from 17 to 22°C while daily minimum temperatures in the soil were just above the germinable temperatures (Fig. 24.3A). In September, the temperature ranges for germination overlapped with those in the soil, and a flush of seedling emergence occurred. When exhumed seeds were tested for germination in the fluridone solution, upper temperature limits for germination increased by 5–8°C, and were within the temperature ranges in the soil. Actually, even in summer, if the exhumed seeds were immersed in the fluridone solution and sown again in the field, 20–60% of them germinated, whereas water-treated seeds could not germinate (Fig. 24.3B). Therefore, it is likely that the seeds germinated with the fluridone application were previously in a state of thermoinhibition at the lower supraoptimal temperature. The remaining ungerminated seeds might be in a state of thermoinhibition at the higher supraoptimal temperature.

The restoration of germination at supraoptimal temperatures by the fluridone treatment was observed in 17 out of 19 species of winter annuals (Yoshioka *et al.*, 1998). Thus, we conclude that seeds of obligate winter annual plants regulate their ABA contents by sensing supraoptimal temperatures, and inhibit their germination until the appropriate season for germination. This temperature-sensory mechanism mediated by ABA will play a decisive role in the control of life cycles in winter annual plants.

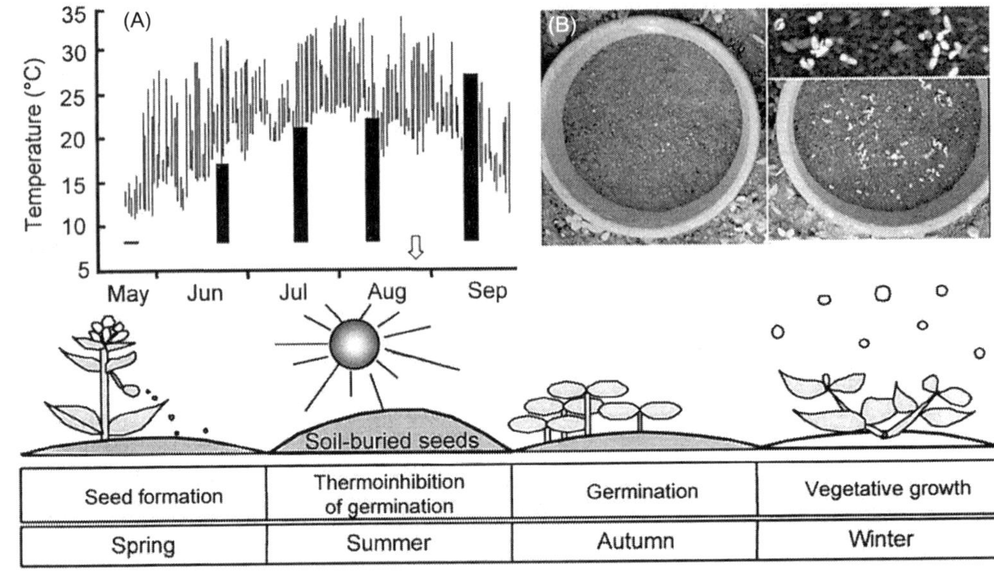

Fig. 24.3. Seasonal changes in soil temperature and temperature range allowing germination of greater chickweed seeds. (A) Freshly harvested seeds were buried in the soil and exhumed monthly from May to September. The exhumed seeds were incubated in water at 8, 13, 18, 23, 28 and 33°C for 14 days in darkness. The columns show the temperature ranges at which more than 5% seeds germinated. The vertical lines indicate the daily soil temperatures at a depth of 1 cm. A flush of seedling emergence is indicated by the arrow. (B) In July, the exhumed seeds were immersed in water (left) or 30 μM fluridone solution (right) for 30 min and buried again in the field, and then their germination was checked 14 days after burial.

Acknowledgements

We are grateful to Mr Kazuma Yamauchi of Dow Chemical Japan Co. Ltd, Dr Yasuo Kamuro of BAL Planning Co. Ltd and Professor Takeshi Sassa of Yamagata University for providing us with fluridone, (*S*)-(+)-ABA and GA_1, respectively. We would like to thank Dr Carol Baskin of Kentucky University and Dr Rei-ichi Miura of Kyoto University for helpful suggestions.

References

Bartels, P.G. and Watson, C.W. (1978) Inhibition of carotenoid synthesis by fluridone and norflurazon. *Weed Science* 26, 198–203.

Baskin, C.C. and Baskin, J.M. (1998) Germination ecology of seeds with nondeep physiological dormancy. In: Baskin, C.C. and Baskin, J.M. (eds) *Seeds: Ecology, Biography, and Evolution of Dormancy and Germination.* Academic Press, San Diego, California, pp. 57–64.

Berrie, A.M.M. (1966) The effect of temperature and light on the germination of lettuce seeds. *Physiolgia Plantarum* 19, 429–436.

Berrie, A.M.M. and Robertson, J. (1976) Abscisic acid as an endogenous component in lettuce fruits, *Lactuca sativa* L. cv. Grand Rapids. Does it control thermodormancy? *Planta* 131, 211–215.

Braun, J.W. and Khan, A.A. (1975) Endogenous abscisic acid levels in germinating and nongerminating lettuce seed. *Plant Physiology* 56, 731–733.

Endo, K., Yamane, H., Nakayama, M., Yamaguchi, I., Murofushi, N., Takahashi, N. and Katsumi, M. (1989) Endogenous gibberellins in the vegetative shoots of tall and dwarf cultivars of *Phaseolus vulgaris* L. *Plant and Cell Physiology* 30, 137–142.

Grappin, P., Bouinot, D., Sotta, B., Miginiac, E. and Jullien, M. (2000) Control of seed dormancy in *Nicotiana plumbaginifolia*: post-imbibition abscisic acid synthesis imposes dormancy maintenance. *Planta* 210, 279–285.

Probert, R.J. (2000) The role of temperature in the regulation of seed dormancy and germination. In: Fenner, M. (ed.) *Seeds: the Ecology of Regeneration in Plant Communities*, 2nd edn. CAB International, Wallingford, UK, pp. 261–292.

Reynolds, T. and Thompson, P.A. (1971) Characteristics of the high temperature inhibition of germination of lettuce (*Lactuca sativa*). *Physiologia Plantarum* 24, 544–547.

Schmitz, N., Xia, J. and Kermode, A.R. (2001) Dormancy of yellow-cedar seeds is terminated by gibberellic acid in combination with fluridone or with osmotic priming and moist chilling. *Seed Science and Technology* 29, 331–346.

Toyomasu, T., Tsuji, H., Yamane, H., Nakayama, M., Yamaguchi, I., Murofushi, N., Takahashi, N. and Inoue, Y. (1993) Light effects on endogenous levels of gibberellins in photoblastic lettuce seeds. *Journal of Plant Growth Regulation* 12, 85–90.

Toyomasu, T., Kawaide, H., Mitsuhashi, W., Inoue, Y. and Kamiya, Y. (1998) Phytochrome regulates gibberellin biosynthesis during germination of photoblastic lettuce seeds. *Plant Physiology* 118, 1517–1523.

Vidaver, W. and Hsiao, A.I. (1975) Actions of gibberellic acid and phytochrome on the germination of Grand Rapids lettuce seeds. *Plant Physiology* 53, 266–268.

Walker-Simmons, M. (1988) Enhancement of ABA responsiveness in wheat embryos by high temperature. *Plant Cell and Environment* 11, 769–775.

Yoshioka, T., Endo, T. and Satoh, S. (1998) Restoration of seed germination at supraoptimal temperatures by fluridone, an inhibitor of abscisic acid biosynthesis. *Plant and Cell Physiology* 39, 307–312.

25 SIMPLE (SIMulation of PLant Emergence): a Model for Predicting Crop Emergence

CAROLYNE DÜRR[1], JEAN-NOËL AUBERTOT[1], PASCAL DUBRULLE[1], GUY RICHARD[1], FRUCK DORSAINVIL[1], ERIC JUSTES[1], DOMINIQUE JOB[2] AND JEAN BOIFFIN[1]

[1]Unité d'Agronomie Laon-Reims-Mons, rue Fernand Christ, 02007 Laon cedex, France; [2]Unité Mixte CNRS/INRA/Aventis CropScience, BP 9163, 69263 Lyon cedex 09, France

Introduction

SIMPLE (SIMulation of PLant Emergence) is a model designed to predict crop emergence. We aimed at developing a simulation tool to aid decision making for soil and tillage operations, and choices between seed characteristics, seed treatments and cultivars. The results of these choices are not easy to predict. They depend on a large number of factors, all of which interact. Moreover, the results very much depend on climatic conditions. A model that can predict variations in crop establishment under a wide range of climatic conditions would help to reduce the number of experiments carried out to test the effects of technical practices.

The Basic Principles of SIMPLE

A complete description of the SIMPLE model is given in Dürr *et al.* (2001). SIMPLE predicts germination and emergence percentages and times according to variations in seed characteristics, soil and climatic conditions, and conditions resulting from soil tillage and sowing conditions (Fig. 25.1). It runs seed by seed with a daily interval of prediction. This model can be adapted to several crops and is currently parameterized for sugarbeet, wheat and catch crops.

A seedbed generator produces a digital 3D representation of clods and seeds in the seedbed. A volume is created whose dimensions are row width, seedbed depth and a given length of the row. The input variables are the size distribution of the aggregates in the seedbed, and their shape and spatial organization (Aubertot *et al.*, 1999). The seedbed generator then uses the

 The Biology of Seeds: Recent Research Advances (eds G. Nicolás, K.J. Bradford, D. Côme and H.W. Pritchard)

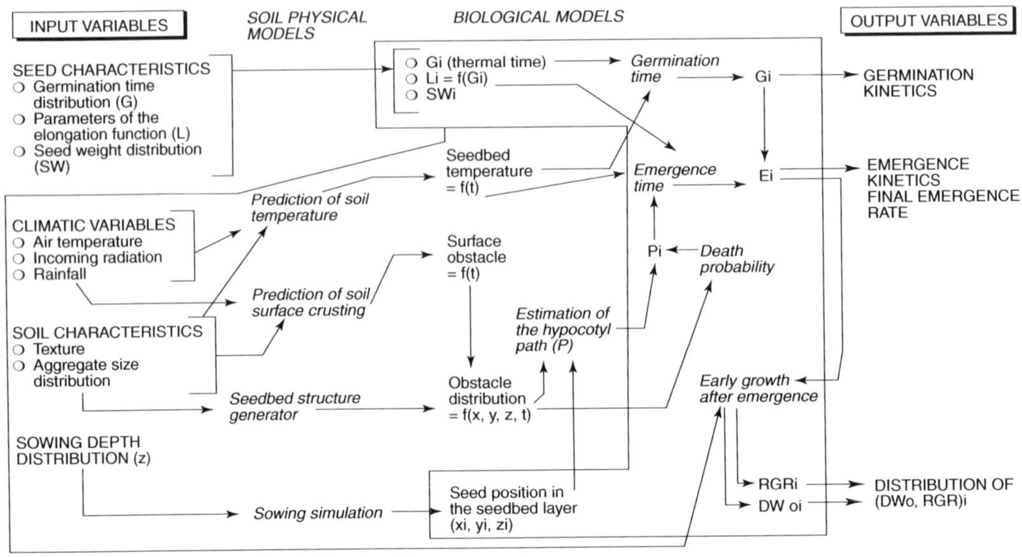

Fig. 25.1. SIMPLE general structure (from Dürr *et al.*, 2001).

sowing depth distribution and places the seeds by choosing their sowing depth at random from the distribution table. Soil temperature and water potential during emergence are predicted from a more or less simplified energy and mass balance (Wittono and Bruckler, 1989; Brisson *et al.*, 1998). This balance uses air temperature, soil evaporation and the proportion of absorbed solar radiation. Biological submodels are run after the soil conditions are defined. They first predict germination on the basis of laboratory tests carried out at different temperatures. This gives a distribution table of germination times, where time is calculated as a cumulative thermal time over a base temperature. Thermal time is cumulated, provided that the soil water potential is greater than the base water potential determined for the seeds (Gummerson, 1986; Dahal and Bradford, 1994; Finch-Savage *et al.*, 1998). The procedure is to choose at random the germination time of seed *i* and use the soil temperature and water potential submodel prediction since sowing. Seed *i* is declared to be germinated when the cumulative thermal time from sowing equals the germination time.

After germination time, the simulator calculates the trajectory and length of the shoot until it reaches the soil's surface, according to the position of each seed and the seedbed structure. The shoot takes a vertical course due to its negative geotropism, as long as it does not meet an aggregate > 5 mm in diameter. This value was determined from experimental results. If it runs into an aggregate > 5 mm in diameter, then the fate of the seedling must be decided. Experiments have demonstrated that seedlings can remain trapped in the small holes at the clod surface. Trapping and death probability depend on clod diameter and on its position in the seedbed (Dürr and Aubertot, 2000). The fate of seedling *i* is chosen at random according to this probability. If it does not die, the shoot path is calculated along the clod. The shoot elongates along the clod, and then it returns to a vertical course until it meets another aggregate. This permits calculation of the total length of the path before reaching the soil surface. The shoot elongation function is then used to calculate the thermal time at which the seedling will reach the soil surface. The parameters of this elongation curve can vary. They are chosen at random from several sets of parameters describing the variability of elongation between the seeds. This allows calculation of the thermal time needed to reach the soil surface. Soil temperature is used to determine the time in days corresponding to each seedling.

Finally, the seedling must in some cases cross a crust when arriving at the soil surface. This is determined for each seedling, one by one, using a semi-empirical submodel (Duval and Boiffin, 1994; Aubertot *et al.*, 2002). The rainfall data are used to determine whether there is a crust. Rainfall data are also used to test whether this crust is wet or dry. If the crust is wet, it is not an obstacle; if it is dry, it is assumed to block a given proportion of seedlings. The fate of seedling *i* is chosen at random within this probability. If the result is that it remains under the crust for the day concerned, the same procedure is reiterated considering what happens the next day. If the cumulative thermal time during this procedure overlaps a given value before emergence, the seedling dies before it can emerge.

SIMPLE Test and Evaluation

Predictions were compared with observations carried out under field conditions to validate the SIMPLE model. Germination and emergence predictions can be evaluated separately. They were tested under several field sowing conditions in the cold, wet spring of the north of France with sugarbeet (Dürr *et al.*, 2001). They were also tested in the case of a catch crop sown in late summer under hot and more or less dry conditions. The predictions were reasonably good when compared with the observations under all the conditions tested, with or without water stress during germination. This indicates that the simplified threshold model with base water potential and temperature values (Finch-Savage *et al.*, 1998) gave good predictions when temperature and soil water potentials varied over a wide range, and from day to day. Emergence changes through time and the final emergence percentages were also well predicted. However, the predictions were in some cases slightly earlier than the values observed (Fig. 25.2). Analysis of these discrepancies showed that certain factors not taken into account in the model might have decreased the elongation rate: high or very low temperatures, water stress and soil impedance. This remains to be improved.

Simulations with SIMPLE

The model was used for simulation studies in the case of changes in seedbed structure (Dürr *et al.*, 2001; Louvigny *et al.*, 2002), and also when changing certain seed characteristics.

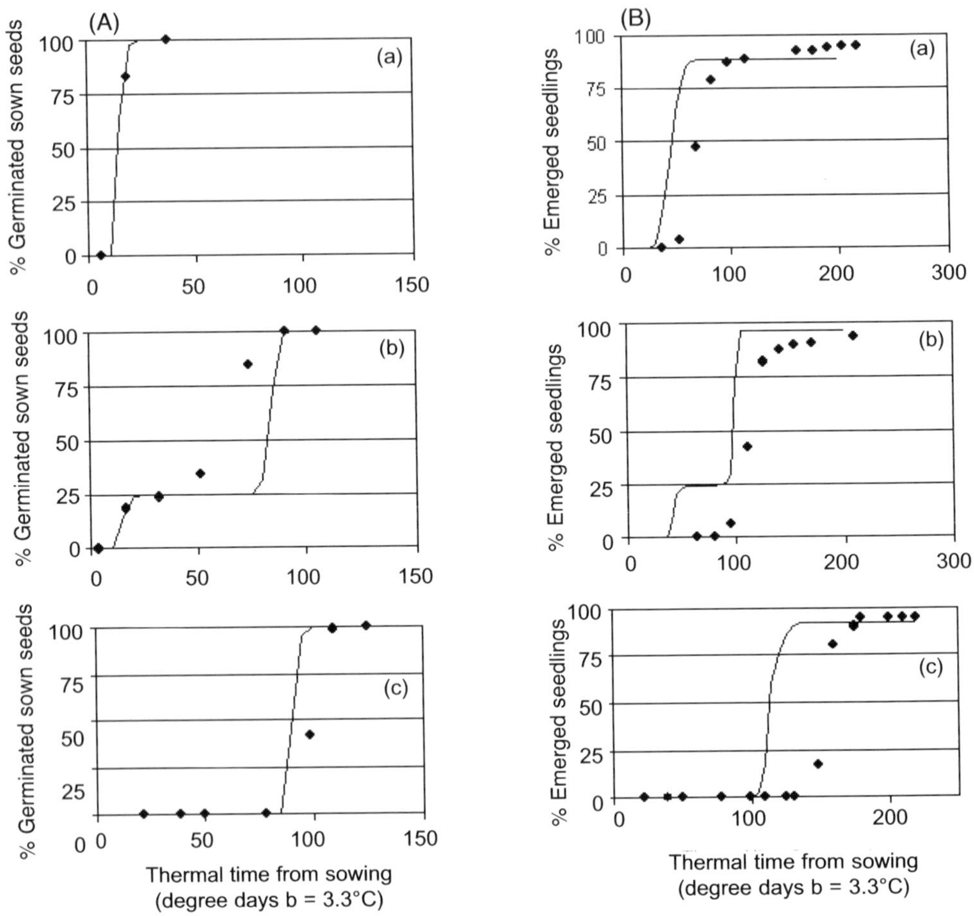

Fig. 25.2. Predictions of the SIMPLE model compared with field observations in the case of a catch crop (white mustard, *Sinapis alba* L.): (A) germination; (B) emergence. Lines = simulations; symbols = observations. Field experiments with (a) no water stress, (b) a deep sowing in dry conditions and rain occurring 4 days after sowing and (c) a shallow sowing in dry conditions and rain occurring 5 days after sowing

We simulated several sowing conditions for sugarbeet. Two seedbed structures were simulated: a fine seedbed with few clods and a coarser seedbed due, for example, to reduced tillage. Several types of drills exist for sugarbeet; they organize the clods differently at the soil surface (Aubertot *et al.*, 1999). These different spatial organizations were simulated. The final emergence rate is high when the seedbed has few clods, with no difference between drills (Fig. 25.3A). However, the simulations indicate large differences in emergence rates when the seedbed is coarse, in relation to the organization of clods at the soil surface by the seed drills. This example illustrates that SIMPLE allows simulation of sowing practices that are still rare and it can evaluate their impact on emergence.

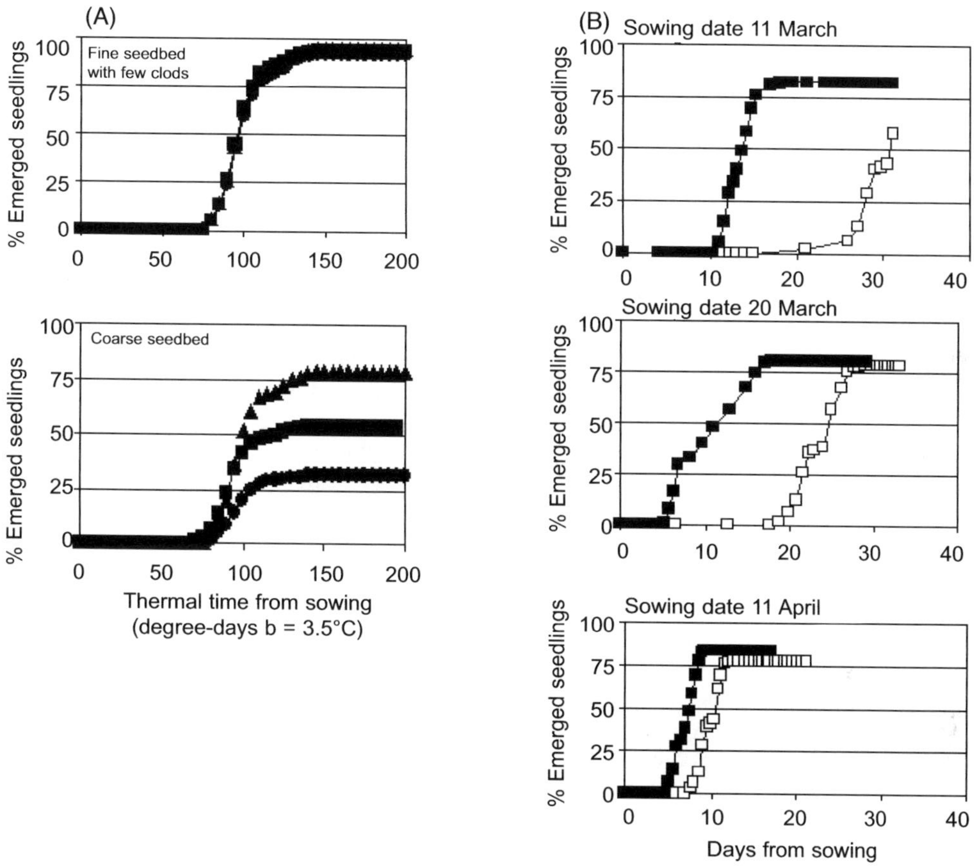

Fig. 25.3. Simulation in the case of sugarbeet sowings. (A) Results for several seedbed preparations and drills. Organization of clods at the soil surface by the drill: ▲, out of the central part of the row; ■, uniform; ●, gathered in the central part. (B) Results with control and primed seeds for several sowing dates in spring (■, primed seeds; □, control).

The second example concerns the evaluation of the effects of priming in the case of sugarbeet seeds (2 days, 20°C in PEG at –1.5 MPa; Capron *et al.*, 2000). The change caused by priming that we simulated for the seeds was a narrower germination time distribution, with lower cumulative values measured in laboratory tests at several temperatures. We also decreased the base temperature from 3.5 to 0°C for seeds, as the slope of the relationship between the germination rate and temperatures was altered. We simulated three sowing dates, spread over 1 month during spring (Fig. 25.3B). The simulations indicated that the effect of priming was much greater in the case of the earlier sowing date, though less when sowing later at warmer temperatures. This effect was observed in field experiments. The interest of the simulations is that they can give a more systematic analysis than that provided by field experiments, under a wider range of climatic conditions and sowing dates.

Conclusions

Modelling is a means of integrating and testing our knowledge regarding the description and prediction of soil conditions, together with the processes describing germination and emergence. The fact that SIMPLE is a stochastic model allows the variability of characteristics among the seeds to be taken into account. Another important feature of SIMPLE is its three-dimensional representation of the seedbed, providing a good representation of soil tillage effects. Modelling complements laboratory analyses and field experiments for better characterization, understanding and prediction of the effects of the seed's characteristics. It makes it necessary to define the variables underlying the emergence process, thus giving the basis for comparing species, cultivars and seedlots. SIMPLE still requires improvement in order to cover a wider range of sowing conditions and crop species but simulation studies can already be developed to give statistical evaluations for technical choices, over several years and for several climates and several soils.

References

Aubertot, J.N., Dürr, C., Kiêu, K. and Richard, G. (1999) Characterization of sugar beet seedbed structure. *Soil Science Society of America Journal* 63, 1377–1384.

Aubertot, J.N., Dürr, C., Richard, G., Souty, N. and Duval, Y. (2002) Are penetrometer measurements useful in predicting emergence of sugar beet (*Beta vulgaris* L.) seedlings through a crust? *Plant and Soil* 242 (2), 177–186.

Brisson, N., Mary, B., Ripoche, D., Jeuffroy, M.H., Ruget, F., Nicoullaud, B., Gate, P., Devienne-Barret, F., Antonioletti, R., Dürr, C., Richard, G., Beaudoin, N., Recous, S., Tayot, X., Plenet, D., Cellier, P., Machet, J.M., Meynard, J.M. and Delecolle, R. (1998) STICS: a generic model for the simulation of crops and their water and nitrogen balance. I. Theory and parametrization applied to wheat and corn. *Agronomie* 18, 311–346.

Capron, I., Corbineau, F., Dacher, F., Job, C., Côme, D. and Job, D. (2000) Sugar beet seed priming: effects of priming conditions on germination, solubilization of 11-S globulin and accumulation of LEA proteins. *Seed Science Research* 10, 243–254.

Dahal, P. and Bradford, K.J. (1994) Hydrothermal time analysis of tomato seed germination at suboptimal temperature and reduced water potential. *Seed Science Research* 4, 71–80.

Dürr, C. and Aubertot, J.N. (2000) Emergence of seedlings of sugarbeet (*Beta vulgaris* L.) as affected by the size, roughness and position of aggregates in the seedbed. *Plant and Soil* 219, 211–220.

Dürr, C., Aubertot, J.N., Richard, G., Dubrulle, P., Duval, Y. and Boiffin, J. (2001) SIMPLE: a model for SIMulation of PLant Emergence predicting the effects of soil tillage and sowing operations. *Soil Science Society of America Journal* 65, 414–423.

Duval, Y. and Boiffin, J. (1994) A daily emergence disturbance index for sugarbeet based on soil crusting. In: *Proceedings 13th ISTRO Conference*. Danish Veterinary Institute, Aalborg, Denmark, pp. 633–638.

Finch-Savage, W.E., Steckel, J.R.A. and Phelps, K. (1998) Germination and post-germination growth to carrot seedling emergence: predictive threshold models and sources of variation between sowing occasions. *New Phytologist* 139, 505–516.

Gummerson, R. (1986) The effect of constant temperatures and osmotic potentials on the germination of sugar beet. *Journal of Experimental Botany* 37, 729–741.

Louvigny, N. de, Dürr, C., Fiès, J.C. and Bruckler, L. (2002) Emergence of sugar beet (*Beta vulgaris* L.) from seedbeds after glass waste deposition on soil. *Soil and Tillage* 66, 35–46.

Wittono, H. and Bruckler, L. (1989) Use of remotely sensed soil moisture content as boundary conditions in soil-atmosphere water transport modelling. I. Field validation of a water flow model. *Water Resources Research* 25, 2423—2435.

26 Metabolic Activity in Germinating Tomato (*Lycopersicon esculentum* Mill.) Seeds Not Related to Distribution of Free Water Within the Seed

HENK W.M. HILHORST[1] AND PATRICK SPOELSTRA[2]

[1]*Laboratory of Plant Physiology, Wageningen University and Wageningen Seed Centre, Arboretumlaan 4, 6703 BD Wageningen, The Netherlands;*
[2]*Syngenta Seeds BV, Westeinde 62, 1601 BK Enkhuizen, The Netherlands*

Introduction

When dry seeds take up water, metabolic activity, as expressed by respiratory activity or ATP turnover, commences almost instantaneously. It is generally accepted that energy metabolism can resume so quickly because all the required components of the pathways are available and only require water to become activated (Bewley and Black, 1994). Besides energy metabolism, protein and nucleic acid synthesis also occur relatively early during imbibition.

Measuring metabolic activity in germinating seeds is usually destructive in that extracts have to be made to determine the desired components. Moreover, destructive techniques lead to loss of information about the location of activity within the seed. Also, to meet limitations in sensitivity of the measuring methods used, it is common to use seed batches (populations). Evidently, such methods overlook the considerable seed-to-seed variations that may occur. Only a few studies have been carried out on germination of single seeds and one of these studies has demonstrated, for example, a seed-to-seed variation with respect to cell-wall-degrading enzyme activity of endo-β-mannanase of four to five orders of magnitude (Still and Bradford, 1997).

In order to monitor single seeds during the progress of germination in a non-destructive manner, we transfected tomato plants with a cauliflower mosaic virus promoter (CaMV 35S) fused to a luciferase reporter gene. The

CaMV 35S promoter is often used for expression studies of foreign genes in plants (Benfey *et al.*, 1989). The firefly luciferase gene encodes a protein that catalyses the oxidative decarboxylation of firefly luciferin, using oxygen, ATP and Mg^{2+}, thereby releasing a single photon (562 nm; de Luca and McElroy, 1974). Luciferin is an amphipathic molecule that can easily penetrate plant tissues. When germinating seeds or plants expressing a luciferase reporter gene are incubated or sprayed with a luciferin solution, photons will be emitted by these seeds or plants (Ow *et al.*, 1986). These emitted photons can be spatially and temporally resolved with a deeply cooled or intensified CCD camera.

Walenta *et al.* (1990) have developed a method based on the luciferase–luciferin system to resolve ATP distributions spatially in cryosections of tumour spheroids. In the present study we used the luciferase–luciferin system to resolve ATP distributions spatially in cryosections of germinating tomato seeds. The distribution of ATP was visualized by monitoring the luminescence of frozen sections of seeds on top of a gel containing all the components of the luciferase reaction but excluding ATP. ATP was imaged in germinating tomato seeds at intervals of 3, 6, 17, 24 and 48 h.

Nuclear magnetic resonance (^1H NMR) imaging has been used to study water distributions and physical properties such as the viscosity of water, membrane water permeabiliy and compartment sizes in plant tissues (McFall and van As, 1996; Chudek and Hunter, 1997; Ishida *et al.*, 2000). Imaging of water in seeds has been demonstrated by several authors (Jenner *et al.*, 1988; Ishida *et al.*, 1995; Fountain *et al.*, 1998). NMR imaging is a non-destructive technique that can reveal possible relations between water uptake and germination of a single seed, which cannot be obtained by destructive techniques. We have studied the distribution and uptake of water by the different tissues of tomato seeds during germination with the use of 3D turbo-spin-echo (TSE) NMR imaging (Scheenen *et al.*, 2000; van der Toorn *et al.*, 2000).

Here we address the question of whether metabolic activity, as measured by the above methods, is related to the distribution of water in germinating tomato seeds.

Materials and Methods

Plant growth conditions and seed harvesting

Tomato plants were soil grown in a greenhouse or in growth chambers on a regime of 16 h light (35 W/m^2), 25°C to 8 h dark at 19°C and a relative humidity (RH) of 70%. Tomato fruits were picked at the red stage and the seeds and locular tissue were removed from the ripe tomatoes by cutting. The locular tissue was digested by adding an equal volume of 2% (v/v) hypochloric acid and stirring for 2 h. Seeds were rinsed thoroughly under running tap water then transferred on to two layers of filter paper and dried for 3 days at 35% RH and 21°C. Dry seeds were stored at either room temperature or 4°C.

Reporter gene construct

The reporter gene construct was composed of the CaMV 35S promoter (–348 to +8 sequence), as described by Gardner *et al.* (1981) and Benfey *et al.* (1989), fused to the original luciferase coding sequence cloned by de Wet *et al.* (1985). In front of the luciferase coding sequence an N-terminal SV40 nuclear localization signal was inserted (van der Krol and Chua, 1991). This construct was ligated into the binary vector pMON721 and transformed to *Agrobacterium tumefaciens* (Strain ABI).

Plant transformation

Transgenic tomato cv. Money Maker (MM) plants were obtained via *A. tumefaciens*-mediated transformation of cotyledon explants, as described by Spoelstra (2002).

Measurement of *in situ* luciferase expression

Transgenic seeds were imbibed in 2.5 ml of a 0.1 mM solution of luciferin (Molecular Probes) in 5 cm diameter plastic containers on three layers of filter paper. To minimize reflection of emitted photons, the top filter paper was black (Whatman No. 9). For *in situ* measurement of luciferase expression, seeds were germinated in a transparent incubator at 25°C (S160 Total Visibility Incubator, Stuart Scientific Co. Ltd, UK). Photons emitted by seeds were spatially resolved with a liquid-nitrogen-cooled back-thinned CCD camera (Versarray 512B, Princeton Instruments) operated via Metamorph 4.1 (Universal Imaging Inc.) software. Sixteen-bit monochrome images were generated by on-chip integration of photons emitted by single seeds for 15 or 30 min. Images of photon emission from seeds were analysed for average pixel value using Metamorph 4.1. The pixel values within those images are a direct measurement of the activity of the luciferase protein present in the seeds. Luciferase activity in all figures is expressed as arbitrary units (AU). Time-lapse measurements were performed by generating one or two images every hour during a 2–3 day period, depending on the nature of the experiment.

Imaging of ATP distribution in cryosections of tomato seeds

Visualization of ATP is based upon the reaction of firefly luciferase and luciferin with ATP, O_2 and Mg^{2+}. The reaction generates oxyluciferin and photons with a wavelength of 562 nm at a pH of 7–8. The number of emitted photons is directly related to the number of ATP molecules converted to AMP and PPi (stoichiometry of 1:1). ATP was visualized in cryosections of imbibing seeds according to Spoelstra *et al.* (2002) based on the method described by Walenta *et al.* (1990).

NMR imaging

NMR imaging experiments were performed with a spectrometer consisting of an SMIS console (SMIS Ltd, Guildford, Surrey, UK) operating at 30.7 MHz, equipped with an electromagnet (Bruker, Karlsruhe, Germany), with a 0.7 tesla (T) field over an air gap of 14 cm, which was stabilized by an external ^{19}F lock (SMIS). A custom-engineered radio frequency (rf) coil/gradient system (Doty Scientific, Columbia, South Carolina, USA) containing a small inserted transmitter–receiver coil with a diameter of 9 mm was used. A glass vial with a diameter of 9 mm was fitted into the cylindrical probe, which was placed into the magnet. Small filter papers were cut with diameters corresponding to the diameter of the glass vial and placed on the bottom of the vial. A glass capillary, containing water, was also inserted into the vial, together with a single seed. The filter paper was wetted and kept moist during measurement by water drained from the glass capillary. The magnet bore was cooled with an airflow at room temperature, which resulted in a bore and sample temperature of 27°C during the NMR measurements.

3D images were obtained by a Fast or Turbo Spin Echo sequence with the following settings: 128 × 128 × 14 matrix (covering a volume of 10 × 10 × 4.26 mm^3 and each voxel representing a volume of 78 × 78 × 304 mm^3), 32 echoes in the train, inter-echo time (*TE*) 5.6 ms (spectral width of 50 kHz), repetition time (*TR*) 1.5 s and 10 averages. Since the first echo is used to sample k = 0 the image intensity mainly represents the signal amplitude or water density (Scheenen *et al.*, 2000).

Data processing

Raw data obtained from the turbo spin-echo experiments were Fourier transformed and the absolute images were used for further analysis. Processing was performed with the use of Interactive data language software (IDL, Research Systems Inc., Boulder, Colorado, USA). The processing resulted in images of the 16 slices from which five to six slices contained actual data of proton density information of that single seed. Proton density images were exported from IDL as 128 × 128 pixels, 8-bit greyscale Tiff images, and read into Metamorph 4.5 software (Universal Imaging Corp.). Regions of interest (ROIs) were fitted around the different tissues of the tomato seed: the endosperm, the endosperm cap, the embryo radicle and the remaining part of the embryo containing the axis and cotyledons. The average pixel intensities for the different ROIs were directly logged into Microsoft Excel 2000.

Results

Luciferase activity during germination

The expression of the 35S::luciferase reporter gene was monitored during the course of germination. Luciferase activity was not detected directly upon imbibition of transgenic 35S::luciferase seeds in 0.1 mM luciferin. The length

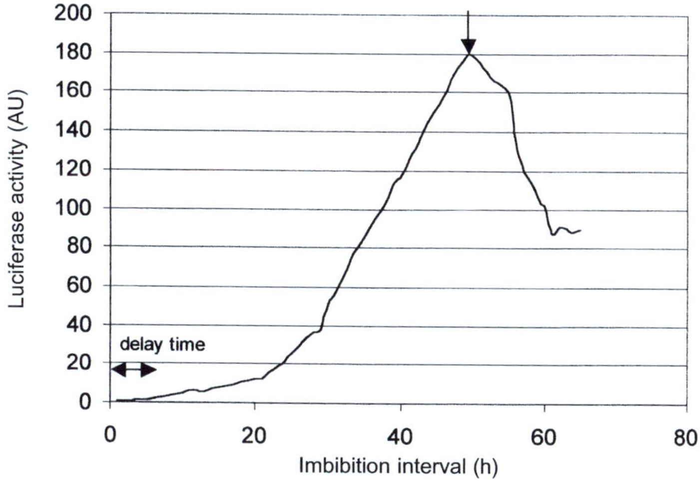

Fig. 26.1. Typical luciferase activity curve during germination of a single tomato seed. Indicated are the delay time (time between start of imbibition and increase of luciferase activity above background levels) and the moment of radicle protrusion (arrow).

of this delay time varied among single seeds of the same seed batch but also the average delay time varied between seed batches (results not shown). The onset of luciferase activity marked the end of the delay time. With the progress of germination of a single seed, luciferase activity showed a continuous increase and reached a maximum at the point at which the radicle tip penetrated the endosperm cap (Fig. 26.1). The shape of the luciferase activity curve during germination was similar among all germinating seeds.

Germination of tomato seeds is generally assumed to be controlled by the radicle tip and endosperm cap opposing it. In this respect radicle protrusion is the net result of the 'thrusting' force of the radicle and the restraint of the endosperm cap. The endosperm cap is anatomically different from the rest of the endosperm and shows tissue-specific gene expression (Bradford *et al.*, 2000). This was also true with respect to luciferase activity, which was strongest in the endosperm cap and of considerably lower intensity in the lateral endosperm and embryo (see Fig. 26.4A).

ATP distribution during germination

ATP concentrations in single seeds were calculated by averaging the ATP concentrations as measured in the three to four middle sections of each seed. The average ATP concentrations found in these single seeds varied over a 2.5- to fivefold range between seeds of the same genotype and imbibition intervals (data not shown). Both the embryo and endosperm contained ATP during germination but the ATP concentration in the embryo was higher compared with that in the endosperm (average of 1.5 times higher at 24 h of

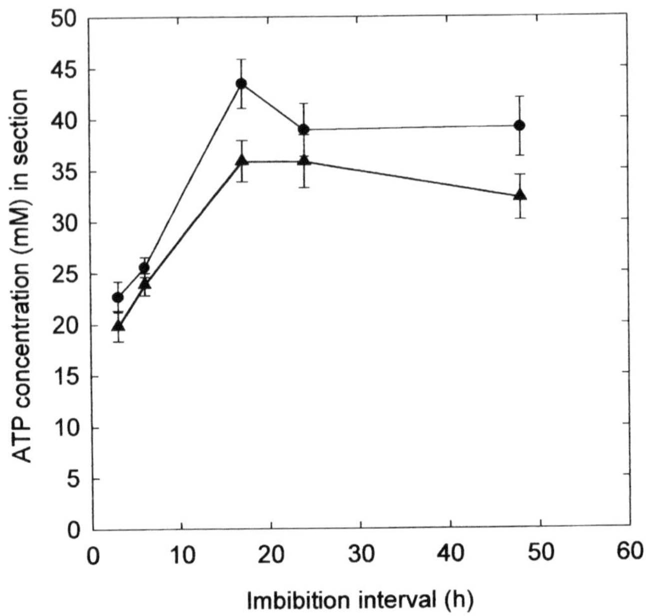

Fig. 26.2. Levels of ATP in whole sections (●) and in cotyledons (▲) during germination of tomato seeds as determined by ATP imaging. ATP concentrations were measured in three to four sections of 10 single seeds and averaged. Bars indicate standard error of mean.

imbibition). Within the embryo the highest concentration of ATP was located in the radicle at all stages of imbibition (see Fig. 26.4B). The ATP concentration in the radicle was compared with the average ATP concentration in the cotyledons. The ATP concentration in the radicle at the different imbibition intervals was 1.4–2.0 ($P < 0.05$) times higher than the ATP concentration in the cotyledons (Fig. 26.2).

Proton density during germination

Water uptake by seeds during germination has often been studied by weighing the imbibed seed and expressing the water content on a dry- or fresh-weight basis (Haigh and Barlow, 1987; Liu, 1996). NMR imaging can also be used to study water uptake or water content in plant material (Scheenen *et al.*, 2000). To verify the accuracy of data on water uptake as assessed by our imaging experiments, we compared data acquired with both methods. Fresh weights of imbibing seeds were determined prior to radicle protrusion at 5, 17, 30 and 42 h of imbibition. Proton densities of six slices containing all data of a single seed were measured and averaged for an integral 'water content' during imbibition. Both methods revealed a similar pattern of water uptake during germination, with a rapid water uptake during phase I, which lasted approximately 12 h, and a stationary or slow increase in water content during phase II (Fig. 26.3). These observations were consistent with the data

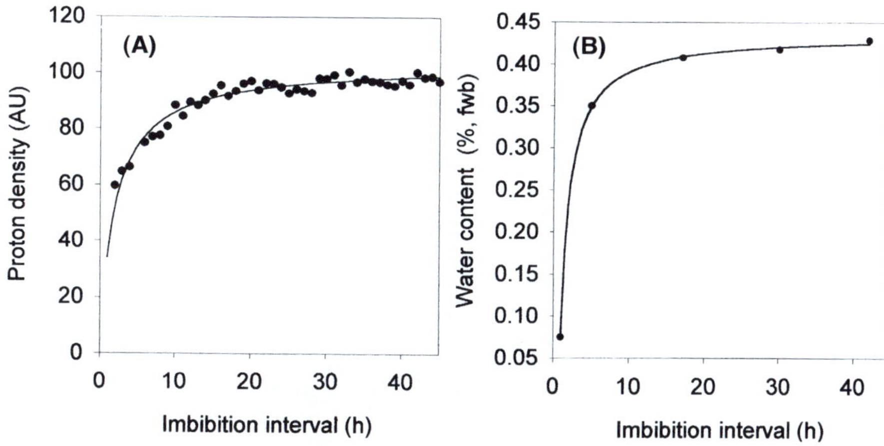

Fig. 26.3. (A) Average proton density (arbitrary units) in a total of six slices of a single tomato seed during germination, determined by 3D TSE NMR imaging. (B) Average water content (percentage of fresh weight) of 15 tomato seeds during germination, determined by weighing.

presented by Haigh and Barlow (1987) and Liu (1996). Phase II was observed in the embryo, radicle and endosperm and showed only a slight increase in water content, if any. The endosperm consistently displayed the highest amount of water compared with the whole embryo, whereas the radicle tip contained the lowest amount of water (see Fig. 26.4C).

Discussion

Luciferase activity was mainly observed in the endosperm cap region throughout the germination process. We do not know the cause of this specific expression. During the transformation procedure and selection of seeds from the progeny, it appeared that luciferase activity was only measurable if two copies of the construct were inserted in the genome. In those lines, luciferase activity was only present in the endosperm cap. From a large number of studies it is known that the enzymatic degradation of the endosperm cap plays a significant role in the germination mechanism of endosperm-retaining seeds, including tomato (Groot and Karssen, 1987; Nonogaki and Morohashi, 1996; Toorop *et al.*, 2000). Thus, a high metabolic activity in this region of the endosperm is plausible, as opposed to the lateral part of the endosperm, in which the degradation does not start until the beginning of radicle protrusion. The highest proton density (resembling free water) is visible in the lateral endosperm (Fig. 26.4). Only upon breakthrough of the radicle did proton density in the endosperm cap increase.

ATP was observed in embryo and endosperm throughout the germination process but the ATP concentration was consistently higher in the embryo. Within the embryo, the total ATP concentration varied by a factor of 3–5 among single seeds of the same batch (results not shown). The ATP

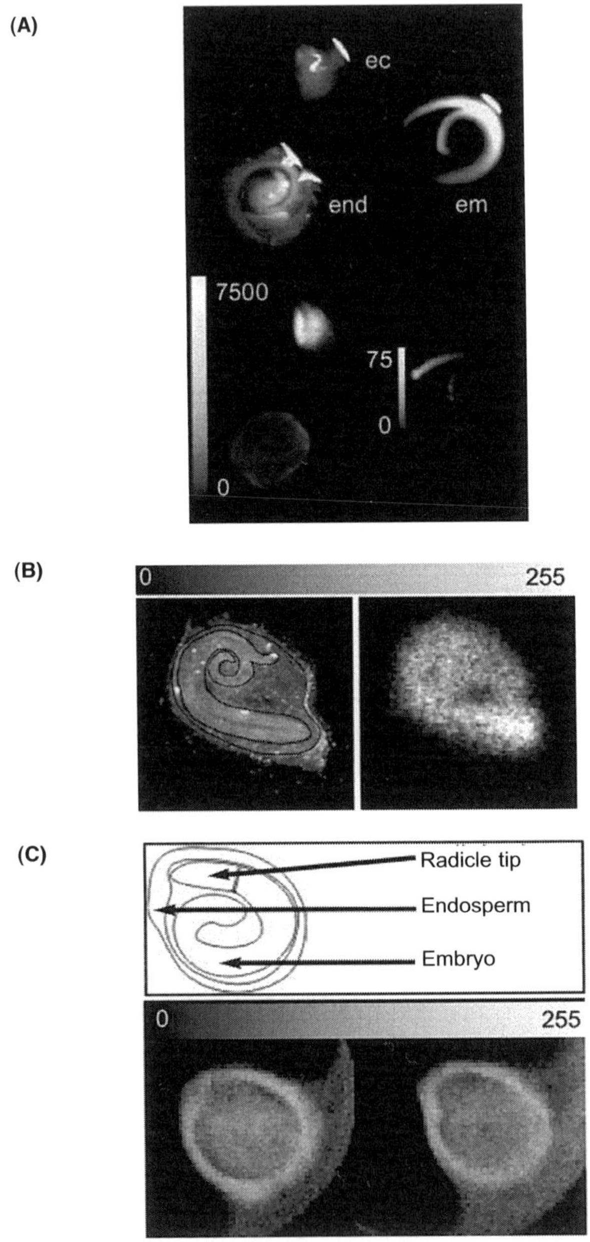

Fig. 26.4. (A) Upper panel: bright field image of dissected tomato seed at 48 h of imbibition, to indicate the orientation of the seed parts of the image in the lower panel (ec, endosperm cap; end, rest of endosperm; em, embryo). Lower panel: luciferase activity in the endosperm cap of a homozygous 35S::luciferase seed imbibed in 0.1 mM luciferin after 48 h of imbibition. (B) Image of typical ATP distribution in a cryosection of a tomato seed after 24 h of imbibition. Bar on top indicates ATP concentration (arbitrary units). (C) Proton density image of two 'slices' of a tomato seed after 45 h of imbibition, acquired by 3D TSE NMR imaging. Top panel shows orientation of the seed.

concentration increased during imbibition and reached a plateau phase after approximately 24 h. We could not find a correlation between ATP concentration and radicle protrusion. Rather, when the distribution of ATP within the embryo was favouring ATP accumulation in the radicle tip, germination could be predicted. From the NMR imaging it appeared that the proton density in the embryo was relatively low, compared with that in the (lateral) endosperm. Apparently, the water content of the embryo was high enough to drive ATP-generating metabolism.

In conclusion, metabolic activity, as measured by luciferase activity and ATP accumulation, is independent of the distribution of proton density, predominantly resembling free water. Metabolic activity must therefore already occur at moisture contents below that of the presence of free water by the method of detection used.

References

Benfey, P.N., Ren, L. and Chua, N.H. (1989) The CaMV 35S enhancer contains at least two domains which can confer different developmental and tissue-specific expression patterns. *EMBO Journal* 8, 2195–2202.

Bewley, J.D. and Black, M. (1994) *Seeds. Physiology of Development and Germination*, 2nd edn. Plenum Press, New York, 445 pp.

Bradford, K.J., Chen, F., Cooley, M.B., Dahal, P., Downie, A.B., Fukunaga, K.K., Gee, O.H., Gurusinghe, S., Mella, R.A., Nonogaki, H., Wu, C.-T., Yang, H. and Yim, K.O. (2000) Gene expression prior to radicle emergence in imbibed tomato seeds. In: Black, M., Bradford, K.J. and Vazquez-Ramos, J. (eds) *Seed Biology: Advances and Applications.* CAB International, Wallingford, UK, pp. 231–251.

Chudek, J.A. and Hunter, G. (1997) Magnetic resonance imaging of plants. *Progress in Nuclear Magnetic Resonance Spectroscopy* 31, 43–62.

de Luca, M. and McElroy, W.D. (1974) Kinetics of the firefly luciferase catalyzed reactions. *Biochemistry* 13, 921–925.

de Wet, J.R., Wood, K.V., Helinski, D.R. and de Luca, M. (1985) Cloning of firefly luciferase cDNA and the expression of active luciferase in *Escherichia coli*. *Proceedings of the National Academy of Sciences USA* 82, 7870–7873.

Fountain, D.W., Forde, L.C., Smith, E.E., Owens, K.R., Bailey, D.G. and Callaghan, P.T. (1998) Seed development in *Phaseolus vulgaris* L. cv. Seminole. 3. NMR imaging of embryos during ethylene-induced precocious germination. *Seed Science Research* 8, 357–365.

Gardner, R.C., Howarth, A.J., Hahn, P., Brown-Luedi, M., Sheperd, R.J. and Messing, J. (1981) The complete nucleotide sequence of an infectious clone of cauliflower mosaic virus by M13mp7 shotgun sequencing. *Nucleic Acids Research* 9, 1871–1880.

Groot, S.P.C. and Karssen, C.M. (1987) Gibberellins regulate seed germination in tomato by endosperm cap weakening: a study with gibberellin-deficient mutants. *Planta* 171, 525–531.

Haigh, A.M. and Barlow, E.W.R. (1987) Water relations of tomato seed germination. *Australian Journal of Plant Physiology* 14, 485–492.

Ishida, N., Ogawa, H. and Kano, H. (1995) Diffusion of cell-associated water in ripening barley seeds. *Magnetic Resonance Imaging* 13, 745–751.

Ishida, N., Koizumi, M. and Kano, H. (2000) The NMR microscope: a unique and promising tool for plant science. *Annals of Botany* 86, 259–278.

Jenner, C.F., Xia, Y., Eccles, C.D. and Callaghan, P.T. (1988) Circulation of water within wheat grain revealed by nuclear magnetic resonance micro-imaging. *Nature* 336, 399–402.

Liu, Y. (1996) Hormones and tomato seed germination. PhD thesis, Wageningen University, Wageningen, The Netherlands.

McFall, J.S. and van As, H. (1996) Magnetic resonance imaging of plants. *Current Topics in Plant Physiology* 16, 33–76.

Nonogaki, H. and Morohashi, Y. (1996) An endo-β-mannanase develops exclusively in the micropylar endosperm of tomato seeds prior to radicle emergence. *Plant Physiology* 110, 555–559.

Ow, D.W., Wood, K.V., de Luca, M., de Wet, J.R., Helinski, D.R. and Howell, S.H. (1986) Transient and stable expression of the firefly luciferase gene in plant cells and transgenic plants. *Science* 234, 856–859.

Scheenen, T.W.J., van Dusschoten, D., de Jager, P.A. and van As, H. (2000) Microscopic displacement imaging with pulsed field gradient turbo spin-echo NMR. *Journal of Magnetic Resonance* 142, 207–215.

Spoelstra, P. (2002) Germination and dormancy of single tomato seeds. A study using non-invasive molecular and biophysical techniques. PhD thesis, Wageningen University, Wageningen, The Netherlands.

Spoelstra, P., Joosen, R.V.L. and Hilhorst, H.W.M. (2002) The distribution of ATP within tomato (*Lycopersicon esculentum* Mill.) embryos correlates with germination whereas total ATP concentration does not. *Seed Science Research* 12, 231–238.

Still, D.W. and Bradford, K.J. (1997) Endo-beta-mannanase activity from individual tomato endosperm caps and radicle tips in relation to germination rates. *Plant Physiology* 113, 21–29.

Toorop, P.E., van Aelst, A.C. and Hilhorst, H.W.M. (2000) The second step of the biphasic endosperm cap weakening that mediates tomato (*Lycopersicon esculentum*) seed germination is under control of ABA. *Journal of Experimental Botany* 51, 1371–1379.

van der Krol, A.R. and Chua, N.H. (1991) The basic domain of plant B-ZIP proteins facilitates import of a reporter protein into plant nuclei. *The Plant Journal* 3, 667–675.

van der Toorn, A., Zemah, H., van As, H., Bendel, P. and Kamenetsky, R. (2000) Developmental changes and water status in tulip bulbs during storage: visualization by NMR. *Journal of Experimental Botany* 51, 1277–1287.

Walenta, S., Doetsch, J. and Mueller Klieser, W. (1990) ATP concentrations in multicellular tumor spheroids assessed by single photon imaging and quantitative bioluminescence. *European Journal of Cell Biology* 52, 389–393.

27 Some Observations on Seed Quality and Mitochondrial Performance

ABDELILAH BENAMAR[1], JOHANN GRELET[1], CHRISTELLE TALLON[1], EMELINE TEYSSIER[1], MICHELLE RENARD[1], PASCALE SATOUR[1], FRÉDÉRIC DUVAL[1], FRANÇOISE MONTRICHARD[1], PASCAL RICHOMME[2] AND DAVID MACHEREL[1]

[1]UMR 1191 Physiologie Moléculaire des Semences (Université d'Angers/INH/INRA), LRPV and [2]SCAS 16 bd Lavoisier, 49045 Angers cedex 01, France

Introduction

Germination is a complex phenomenon that encompasses drastic changes in the metabolic activities and genetic programme of seeds. Such a transition has to be accompanied by efficient transfer of the free energy that was stored in carbohydrates, lipids and protein reserves during seed development and maturation. Mitochondria are semi-autonomous organelles found in all eukaryotes and their function is to provide cellular ATP by the process of oxidative phosphorylation. It is thus likely that mitochondria are major providers of cellular ATP during germination of healthy seeds. This is illustrated by the almost universal requirement of O_2 for germination, although there are variations in the O_2 partial pressure required among species (Al-Ani *et al.*, 1985). In general, poorly differentiated mitochondria observed in quiescent tissue evolve upon imbibition into healthy-looking mitochondria with increased oxidative capacities. Several biogenesis studies that took advantage of translational inhibitors led the pioneering group of Morohashi (1986 and references therein) to propose that mitochondrial biogenesis proceeded from structural development of pre-existing mitochondria in starch-storing seeds and from *de novo* synthesis of mitochondria in lipid-storing seeds. In a detailed review of respiratory metabolism in seeds, Botha *et al.* (1992) pointed out that knowledge concerning the biogenesis of mitochondria and the regulation of mitochondrial metabolism in seeds was still limited. Indeed, very few studies have focused on gene expression for mitochondrial proteins during germination (Ehrenshaft and Brambl, 1990;

 The Biology of Seeds: Recent Research Advances (eds G. Nicolás, K.J. Bradford, D. Côme and H.W. Pritchard)

Logan *et al.*, 2001). In addition, previous studies have been performed with a great variety of seeds and stages of development (in fact many of the studies dealt with post-germinative events) and in most cases have used crude organelle fractions, which render the integration of all the work difficult. We have thus undertaken a research programme designed to gain further insight into mitochondrial functions during the germination of legume seeds, using pea (*Pisum sativum* L., var. Baccara) as the initial material for our investigations. Seeds were grown locally by an agronomic institute (FNAMS, Brain sur l'Authion, France) and, except when stated, all experiments were performed with a seed lot harvested in 1999 and stored for 11 months in appropriate conditions (5°C, 75% relative humidity).

Seed Quality and Resumption of Respiration

It is generally recognized that the resumption of respiration is an early and major event of germination (Bewley, 1997). Although inert in the quiescent state, seeds are alive and thus it appears crucial to take into account their intrinsic physiological quality, which is largely determined by seed storage (provided that the seeds were properly harvested). As a matter of fact, seed quality was not often considered in the numerous studies that dealt with seed mitochondria, although the organelles were pointed out as primary targets in the deterioration of stored seeds (reviewed by Smith and Berjak, 1995). Indeed, lower seed vigour has been correlated with decreased respiratory rates (Leopold and Musgrave, 1980; Amable and Obendorf, 1986; Bettey and Finch-Savage, 1996), which could be attributed to a lower performance of mitochondria isolated from soybean axes (Woodstock *et al.*, 1984). With the aim of exploring the link between mitochondrial performance and seed quality, we settled for priming and accelerated ageing treatments of pea seeds that clearly modulated the kinetics of germination (Fig. 27.1). To estimate respiratory rate in the early period of imbibition, dry seed fragments (approximately 0.3 mm) were directly imbibed inside the oxygen electrode chamber so that their initial oxygen consumption could be monitored (Fig. 27.2). The kinetics were markedly different, with an inverse correlation between seed germinative quality and oxygen consumption rate. When seed fragments were allowed to imbibe on a wet filter paper for 1 h, the oxygen consumption, which reached a steady state, was similar for control and primed seeds (respectively, 96 and 88 nmol O_2/min/g dry weight) while aged seeds still consumed oxygen at a much higher rate (146 nmol O_2/min/g dry weight). The measured oxygen consumption was probably due to mitochondrial respiration, since it was strongly inhibited (70%) by cyanide or antimycin, the residual consumption being affected by the alternative oxidase inhibitor, propylgallate (results not shown). The kinetics of control seed fragments appeared very similar to those obtained by Parrish and Leopold (1977) on fragmented soybean seeds, or those of maize embryo axes (Ehrenshaft and Brambl, 1990). The slower and more linear respiratory rate of primed seed fragments suggests a tighter regulation of respiratory metabolism upon imbibition, while the higher rate observed with aged seed

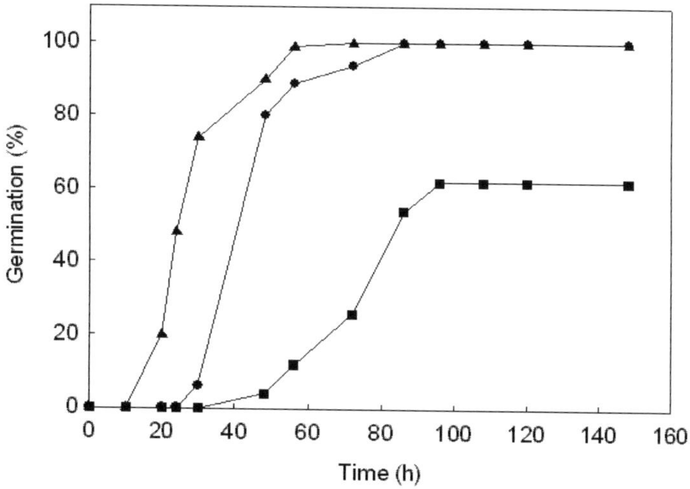

Fig. 27.1. Germination of control, primed and aged pea seeds. Seeds of the 1999 harvest (●, control; ▲, primed; ■, accelerated ageing for 10 days) were germinated at 20°C in pleated paper. The priming treatment consisted of soaking the seeds in aerated deionized water at 20°C for 16 h. Seeds were then rapidly drained and dried on the bench for 1 h. Subsequent fast drying was achieved by keeping the seeds at 20°C in a ventilated desiccation cabinet containing regenerated silica gel. Accelerated ageing was performed by exposing the seeds for 10 days to 45°C in tightly closed boxes containing a saturated NaCl solution that maintained the relative humidity at 76%. Seeds were then rapidly dried in the desiccation cabinet to their initial water content (15–16%).

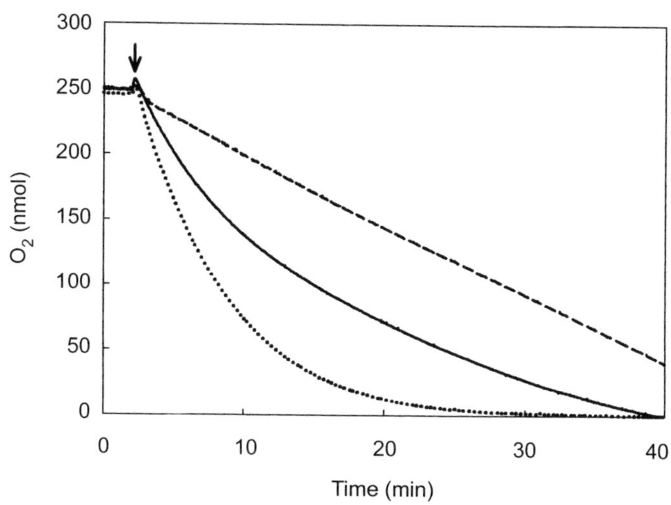

Fig. 27.2. Oxygen consumption by seed fragments during imbibition. Dry seed fragments (100 mg) of approximately 300 μm size were introduced (arrow) into the electrode chamber (Oxytherm, Hansatech) filled with 0.1 M sucrose, 50 mM phosphate (pH 7.5). The electrode chamber was immediately closed to follow the oxygen consumption at 25°C as a function of time. Bold, dashed and dotted lines correspond to control, primed and artificially aged (accelerated ageing for 10 days) seeds, respectively.

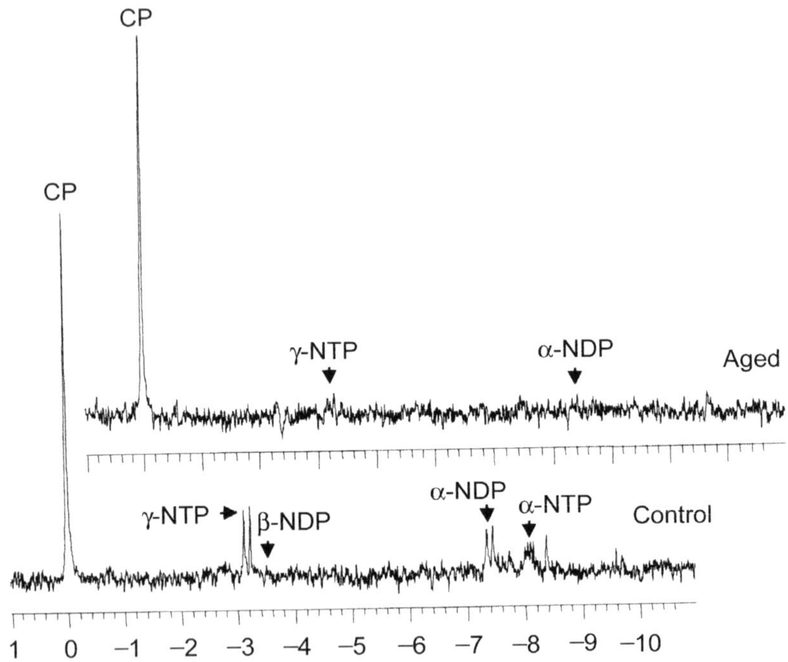

Fig. 27.3. ^{31}P NMR spectra of the nucleotide tri- and diphosphate chemical shift region. The graph shows ^{31}P NMR spectra obtained with perchloric acid extracts of control and aged seed fragments that were allowed to imbibe for 30 min on wet filter paper. The chemical shifts are indicated in ppm, with reference to 10 mM creatine phosphate (CP) that was added to the extracts.

tissue suggests a low coupling of mitochondrial energy transfer, which was indeed confirmed with isolated mitochondria (see below). The lower performance of mitochondria and the increased energy demand in aged seeds were also illustrated by the very low levels of nucleotide di- and triphosphates detected by ^{31}P-NMR in 30 min imbibed fragments (Fig. 27.3).

Oxidative Properties of Pea Seed Mitochondria

The study of mitochondria at the onset of imbibition has remained extremely difficult because of the low water content of seeds, which hampers the isolation of high-quality organelles in aqueous media. We therefore focused our efforts on fully imbibed (12 h after imbibition (HAI)) and ready-to-germinate seeds (22 HAI) from which mitochondria were isolated and purified using Percoll gradients.

While isolated mitochondria oxidized most classical substrates at both times of imbibition (Fig. 27.4), their outer membrane integrity measured with a cytochrome *c* permeation assay, which was around 70% at 12 HAI,

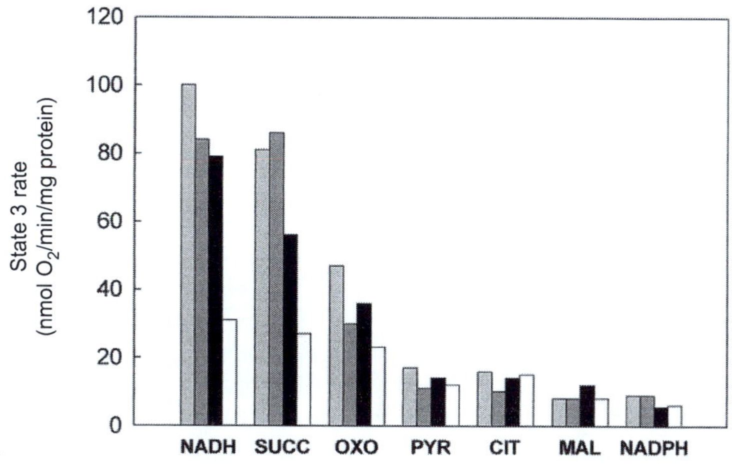

Fig. 27.4. Substrate oxidation profiles of pea seed mitochondria. The graph indicates the state 3 oxidation rate of mitochondria isolated from control seeds after 12 h (light grey bars) or 22 h (black bars) of imbibition, from primed seeds after 12 h of imbibition (dark grey bars) and from 10 days aged seeds after 22 h of imbibition (white bars). SUC, succinate; PYR, pyruvate; MAL, malate; CIT, citrate; OXO, oxoglutarate.

increased to 98% at 22 HAI. A similar improvement was measured for the inner membrane, since the respiratory control measured with several substrates increased on average by 20%. Although the oxidation rates decreased between 12 and 22 HAI, the efficiency of mitochondria in ATP production was thus clearly improved. Isolation injuries might prevent the direct transposition of the results with isolated organelles to the *in vivo* situation, as pointed out by Hoekstra and van Roekel (1983), but we expect isolated mitochondria properties to reflect their original status within the cellular environment. With respiratory controls (RCs) averaging 3.6 with most substrates (22 HAI), seed mitochondria were efficient in generating ATP through the oxidation of the classical Krebs cycle substrates and exogenous NAD(P)H. Although the oxidation rates were found to be lower (two- to threefold) than the usual rates of leaf or tuber mitochondria, particularly for malate, citrate and pyruvate, they were nevertheless consistent with those generally reported for seed mitochondria.

Since external NADH and succinate were the most rapidly oxidized substrates, it is tempting to postulate that they constitute the primary fuel of seed mitochondria during imbibition, thus driving an efficient ATP production without requiring a functional Krebs cycle. Logan *et al.* (2001) suggested that NADH oxidation could drive ATP synthesis in pro-mitochondria from germinating maize embryos deficient in Krebs cycle enzymes, and that would later differentiate into fully active mitochondria. However, in our study the oxidation of the other substrates was already substantial at 12 HAI, and somewhat decreased (state 3 rate) 10 h later, which does not favour a major biogenesis of Krebs cycle machinery during the studied period.

The priming treatment, although quite effective for germination, only slightly improved the mitochondria performance since the organelles isolated at 12 HAI from primed seeds showed oxidation rates (Fig. 27.4), RCs and outer membrane integrity values (not shown) that were intermediate between those of mitochondria from 12 h and 22 h imbibed control seeds.

Several ultrastructural studies of seeds pointed out mitochondria as primary targets in the deterioration of stored seeds (reviewed by Smith and Berjak, 1995) and a study of soybean seed respiration during simulated preharvest deterioration led Amble and Obendorf (1986) to propose mitochondria as a site of primary lesion. The effect of seed deterioration upon mitochondria was investigated by isolating mitochondria at 22 HAI from pea seeds that suffered 10 days of accelerated ageing. The mitochondria appeared to be severely affected in their functions, since their oxidation rates were strongly decreased (e.g. 30% of control rate for NADH, Fig. 27.4) and their RCs appeared much lower (average 2.0). The outer membrane integrity was very low (35% in the cytochrome *c* permeation assay), and these observations were confirmed by transmission electron microscope observations, the outer and inner membranes being clearly damaged in the mitochondria from aged seeds (not shown). Mitochondria from aged seeds appeared clearly hampered in the production of ATP, which is consistent with the low levels of adenylates found in tissue extracts (see Fig. 27.3). It could also explain the higher rates of oxygen consumption recorded for aged seed fragments, which have to face an increased energy demand with less efficient mitochondria.

Conclusion

In this study, mitochondrial integrity and oxidative properties were shown to improve during imbibition of pea seeds, with energy transduction relying mainly on cytosolic NADH and succinate oxidation. A correlation could be established between the germinative capacity (rate, percentage) and mitochondrial performance, with mitochondrial membrane integrity appearing as a primary target of seed deterioration. It will be of keen interest to decipher the mechanisms underlying the maintenance and protection of mitochondria during seed development and germination. In this connection, the recent analysis of mitochondrial proteome during pea development has brought to light dramatic variations in soluble protein composition in different tissues (Bardel *et al.*, 2002). Ongoing work in our laboratory focuses on the identification and functional analysis of seed-specific mitochondrial polypeptides.

Acknowledgements

This work was supported by the Program 'Semences- CER 2000-2003' from the 'Région Pays-de-la-Loire'.

References

Al-Ani, A., Bruzau, F., Raymond, P., Saint-Ges, V., Leblanc, J.-M. and Pradet, A. (1985) Germination, respiration, and adenylate energy charge of seeds at various oxygen partial pressures. *Plant Physiology* 79, 885–890.

Amable, R.A. and Obendorf, R.L. (1986) Soybean seed respiration during simulated preharvest deterioration. *Journal of Experimental Botany* 37, 1364–1375.

Bardel, J., Louwagie, M., Jaquinod, M., Jourdain, A., Luche, S., Rabilloud, T., Macherel, D., Garin, J. and Bourguignon, J. (2002) A survey of plant mitochondria proteome in relation with development. *Proteomics* 2, 880–898.

Bettey, M. and Finch-Savage, W.E. (1996) Respiratory enzyme activities during germination in Brassica seed lots of differing vigour. *Seed Science Research* 6, 165–173.

Bewley, J.D. (1997) Seed germination and dormancy. *The Plant Cell* 9, 1055–1066.

Botha, F.C., Potgieter, G.P. and Botha, A.M. (1992) Respiratory metabolism and gene expression during germination. *Plant Growth Regulation* 11, 211–224.

Ehrenshaft, M. and Brambl, R. (1990) Respiration and mitochondrial biogenesis in germinating embryos of maize. *Plant Physiology* 93, 295–304.

Hoekstra, F.A. and van Roekel, T. (1983) Isolation-inflicted injury to mitochondria from fresh pollen gradually overcome by an active strengthening during germination. *Plant Physiology* 73, 995–1001.

Leopold, A.C. and Musgrave, M.E. (1980) Respiratory pathways in aged soybean seeds. *Physiologia Plantarum* 49, 49–54.

Logan, D.C., Millar, A.H., Sweetlove, L.J., Hill, S.A. and Leaver, C.J. (2001) Mitochondrial biogenesis during germination in maize embryo. *Plant Physiology* 125, 662–672.

Morohashi, Y. (1986) Patterns of mitochondrial development in reserve tissues of germinated seeds: a survey. *Physiologia Plantarum* 66, 653–658.

Parrish, D.J. and Leopold, A.C. (1977) Transient changes during soybean imbibition. *Plant Physiology* 59, 1111–1115.

Smith, M.T. and Berjak, P. (1995) Deteriorative changes associated with the loss of viability of stored desiccation-tolerant and desiccation-sensitive seeds. In: Kigel, J. and Galili, G. (eds) *Seed Development and Germination*. Marcel Dekker, New York, pp. 701–746.

Woodstock, L.W., Furman, K. and Solomos, T. (1984) Changes in respiratory metabolism during aging in seeds and isolated axes of soybean. *Plant and Cell Physiology* 25, 15–26.

28 Gibberellin Regulation of Aleurone Cell Death in Germinating Wheat Seeds

FERNANDO DOMÍNGUEZ[1], JAVIER MORENO[2] AND FRANCISCO J. CEJUDO[1]

[1]*Instituto de Bioquímica Vegetal y Fotosíntesis, Centro de Investigaciones Científicas Isla de la Cartuja, Avda Américo Vespucio s/n, 41092-Sevilla, Spain;* [2]*Departamento de Biología Celular, Facultad de Biología, Universidad de Sevilla, Avda Reina Mercedes s/n, 41012-Sevilla, Spain*

Introduction

The growth and development of complex organisms requires a continuous production of new cells but also the elimination of cells which are no longer needed. In animals, cell death occurs by a process, termed apoptosis, which shows very well defined morphological and biochemical characteristics (Kerr *et al.*, 1995; Raff, 1998; Vaux and Korsmeyer, 1999; Fesik, 2000). In plants, the differentiation of several tissues involves a selective and regulated degeneration of cells, which is termed programmed cell death (PCD) in order to distinguish it from necrosis, a non-regulated degeneration of cells as a consequence of severe external conditions (Greenberg, 1996; Jones and Dangl, 1996; Pennell and Lamb, 1997).

Two phases may be distinguished during the life cycle of the cereal grain: development and germination. At initial stages of development, maternal tissues such as the nucellus and nucellar projection cells suffer degeneration (Morrison *et al.*, 1978, Wang *et al.*, 1994) by PCD (Domínguez *et al.*, 2001). At later stages, 16 days after anthesis (DAA), a process of PCD occurs in the endosperm (Young *et al.*, 1997). Cells located to the centre of the endosperm undergo PCD, which is then extended toward the cap and the base of the kernel. Ethylene was identified as the signal triggering endosperm PCD. In addition, PCD was initiated earlier and was more accelerated in kernels with deficient perception or synthesis of ABA, suggesting that a balance of ethylene and ABA may regulate endosperm PCD in maize (Young and Gallie, 1999).

During germination, gibberellins (GAs) synthesized in the scutellum diffuse to the endosperm and are perceived by the aleurone cells (Appleford

 The Biology of Seeds: Recent Research Advances (eds G. Nicolás, K.J. Bradford, D. Côme and H.W. Pritchard)

and Lenton, 1997). In response to GA, aleurone cells activate the synthesis and secretion of hydrolytic enzymes and then, as a delayed response, the aleurone cell death (Cejudo *et al.*, 2001). Some of the proteases induced by GA in aleurone cells are expressed in other tissues undergoing PCD and therefore may serve as markers of PCD.

GA Defines a Spatio-temporal Pattern of Gene Expression and Starchy Endosperm Acidification in Germinated Wheat Grains

An initial search of GA-regulated genes in wheat aleurone cells identified genes encoding hydrolytic enzymes such as α-amylases and exo- and endo-proteases (Baulcombe and Buffard, 1983). *In situ* hybridization analysis of these genes revealed a characteristic spatio-temporal pattern of expression: transcripts were first detected in aleurone cells proximal to the embryo, by 14 h after imbibition; and then, as germination proceeded, expression progressed towards the distal part of the grain (Domínguez and Cejudo, 1999). This pattern of expression suggested the formation of a gradient of GA from the embryo, where it is synthesized, to the distal part of the grain so that aleurone cells respond to this hormone gradient depending on their location in the grain.

Most of the proteolytic activity occurring in the germinated grain shows an optimum at acidic pH (Domínguez and Cejudo, 1995). Therefore, acidification of the starchy endosperm is a requirement for proteolysis. In barley, endosperm acidification during development was described (Macnicol and Jacobsen, 1992). However, we have shown that the endosperm of the mature wheat grain is neutral and that acidification takes place as a process of the germinative programme. Furthermore, *in situ* pH staining showed that starchy endosperm acidification advanced from the embryo to the distal part of the grain (Domínguez and Cejudo, 1999), i.e. showing the same pattern as gene expression, which suggests that acidification is also regulated by GA. Treatment of de-embryonated half-grains with GA promoted the acidification of the external medium when the treatment was performed at neutral pH but not at acidic pH, thus showing a pH dependence of the response to GA. Similarly, the effect of GA on gene expression in the aleurone layer is highly dependent on the external pH. These results establish that GA not only activates but also coordinates the events that occur in the germinating wheat grain.

Aleurone PCD as Delayed Response to GA

The last response induced by GA in aleurone cells of germinated grains is the cell's own death. The analysis of DNA isolated from aleurone cells of wheat grains at different days after imbibition (DAI) shows the characteristic DNA laddering indicative of PCD. DNA laddering was first observed at 4 DAI and increased progressively up to 7 DAI. Since DNA laddering is a hallmark of PCD, this result suggests an increase in the number of aleurone cells undergoing PCD following germination. The spatio-temporal pattern of aleurone

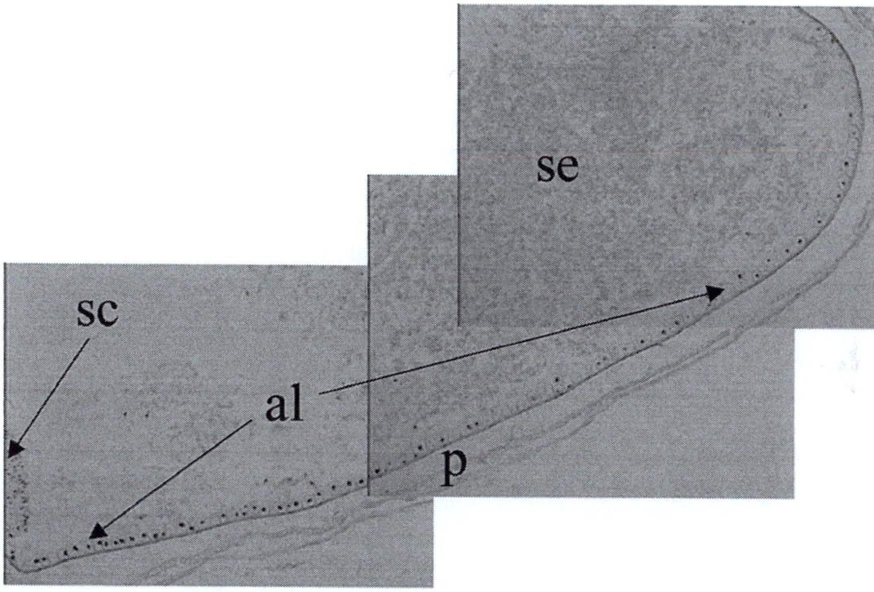

Fig. 28.1. Nuclear DNA fragmentation in the aleurone layer of germinated wheat grains detected by TUNEL labelling (al, aleurone layer; p, pericarp; sc, scutellum; se, starchy endosperm). Wheat grains were allowed to germinate for 7 days and longitudinal sections were prepared and subjected to TUNEL assay.

PCD was established using the TUNEL assay, which allows *in situ* labelling of 3′-OH DNA ends. A gradient of PCD could be established in aleurone cells progressing from the embryo towards the distal part of the grain (Fig. 28.1), in agreement with the increase in DNA fragmentation.

This spatio-temporal pattern suggested the regulation of aleurone PCD by GA, and we therefore tested this possibility as well as the coordination of PCD with other GA-regulated processes in aleurone cells. As mentioned above, external pH exerts a great influence on the response of aleurone cells to GA. When de-embryonated half-grains were incubated in the presence of GA at pH 7.0, a clear activation of cell death was detected. However, the effect of GA on cell death was significantly lower when the treatment was carried out at pH 5.0. The effect of the external pH on GA regulation of PCD was confirmed by the analysis of DNA from aleurone cells. DNA isolated from cells treated with GA at neutral pH showed the characteristic laddering, which was not observed when GA treatment was carried out at acidic pH. DNA fragmentation was observed only after 4 days of treatment with GA, whereas activation of gene expression was evident after 1 day of treatment. Therefore, the response of aleurone cells to GA (gene expression, starchy endosperm acidification and PCD) shows a similar pattern but different timing.

Morphological Changes During Aleurone PCD

Electron transmission microscopy showed that healthy aleurone cells, 1 DAI, present a nucleus with an intact nuclear envelope and abundant lipid bodies surrounding protein bodies, providing a characteristic architecture to the cytoplasm. In cells undergoing PCD this arrangement is lost and the lipid bodies become evenly distributed. Cells become increasingly vacuolated and the plasma membrane is separated from the cell wall. Clear symptoms of degradative processes are noted in the vacuoles, as shown by the presence of electron-dense bodies and membrane fragments. As vacuolization increased in the cytoplasm, the nucleus adopted a highly irregular and invaginated morphology. The nuclear envelope was disrupted; heterochromatin increased and became highly electron-dense. In the cytoplasm, vacuoles lost their integrity so that the cytoplasm contents were degraded. Lastly, a dead cell showed an empty cytoplasm with a few electron-dense bodies. It is remarkable that, though cell contents were degraded, cells did not collapse and that they retained, at least in part, the cell wall.

Nuclease Activity Associated with Aleurone PCD

DNA fragmentation, probably the most characteristic event of cell death, may be produced by one or more nucleases. In order to distinguish PCD-related nucleases from nucleases secreted by the aleurone cells to the starchy endosperm, we have searched GA-regulated nucleases localized into the nucleus of cells undergoing PCD. SDS-PAGE analysis of nuclear extracts of aleurone cells from de-embryonated half-grains treated with GA revealed the presence of a polypeptide with a molecular mass of about 32 kDa, which was absent in nuclear extracts from untreated half-grains. It is likely that this polypeptide corresponded to a PCD-related nuclease, since it showed the same electrophoretical mobility as a nuclease activity detected by in-gel assays of nuclear extracts.

Germination-related Genes as Molecular Markers of PCD

It was assumed that genes encoding hydrolytic enzymes are involved in the mobilization of the reserve compounds of the starchy endosperm. However, two of these genes – a cathepsin B-like and a carboxypeptidase III – are expressed in other tissues of developing and germinating grains. Both genes are expressed in the nucellus, a maternal tissue that degenerates at early stages of grain development (Domínguez and Cejudo, 1998). Further analysis of the nucellus and the nucellar projection cells with the TUNEL assay revealed staining of these cells at early stages of grain development when DNA isolated from these tissues showed the characteristic laddering indicative of PCD (Domínguez *et al.*, 2001).

The electron transmission microscopy analysis of nucellus cells undergoing PCD revealed an increase of heterochromatin, disorganization of the nuclear envelope (which showed vacuolar dilations in the space between the

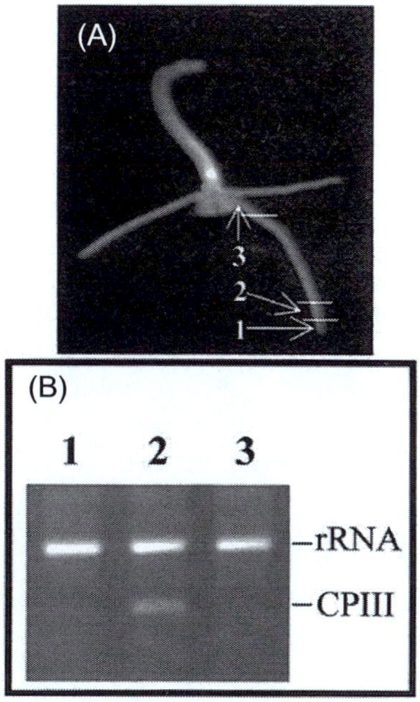

Fig. 28.2. Relative RT-PCR analysis of the accumulation of CPIII mRNA in different parts of wheat roots. (A) Roots from 4-day-old seedlings were dissected into three parts (1, 2, 3) as indicated. (B) DNase I-treated total RNA was isolated from each of these dissected root parts. After retrotranscription and RT-PCR in the presence of gene-specific and 18S rRNA primers, the PCR products were fractionated on agarose gels.

inner and the outer membranes) and progressive degeneration of the cytoplasm contents (Domínguez *et al.*, 2001). Therefore it showed characteristics similar to those observed in aleurone PCD, though nucellar cells did collapse at later stages of grain development. The degeneration of the nucellus and the nucellar projection cells probably serves to feed the endosperm. Indeed, although not well understood, the interaction of both tissues is clear since mutations that affect maternal tissues cause a defective endosperm (Jarvi and Eslick, 1975; Colombo *et al.*, 1997).

The analysis of the carboxypeptidase III gene in the embryo of germinating wheat grains showed expression localized exclusively to tracheary elements of the scutellar vascular tissue (Domínguez *et al.*, 2002). It is known that the scutellum vascular system is not fully differentiated during grain development and appears in the mature grain as provascular tissue (Swift and O'Brien, 1971), so that any passage of compounds from the endosperm is avoided until germination is initiated. Our results show that the differentiation of the scutellar provascular tissue to tracheary elements during germination involves PCD. In addition, carboxypeptidase III transcripts were localized to the midvein provascular tissue in developing shoots and to differentiating tracheary elements of the early metaxylem in developing roots.

The TUNEL assay confirmed that these tissues undergo PCD, demonstrating the association of carboxypeptidase III expression and PCD.

The specific expression of carboxypeptidase III in differentiating tracheary elements allowed the use of this gene as a marker of vascular tissue differentiation in developing wheat seedlings. This was addressed by analysing the distribution of transcripts in the root (Fig. 28.2). To that end, roots of 4-day-old seedlings were dissected in three parts: the root tip (Fig. 28.2A(1)), about 1 cm from the meristem (Fig. 28.2A(2)), where the vascular tissue is being differentiated; and about 1 cm from the grain (Fig. 28.2A(3)), where the root is already differentiated. Accumulation of carboxypeptidase III transcripts was observed exclusively close to the meristem (Fig. 28.2B), i.e. the region where the differentiation of tracheary elements occurs.

References

Appleford, N.E.J. and Lenton, J.R. (1997) Hormonal regulation of α-amylase gene expression in germinating wheat (*Triticum aestivum*) grains. *Physiologia Plantarum* 100, 534–542.

Baulcombe, D.C. and Buffard, D. (1983) Gibberellic-acid-regulated expression of α-amylase and six other genes in wheat aleurone layers. *Planta* 157, 493–501.

Cejudo, F.J., González, M.C., Serrato, A.J., Sánchez, R. and Domínguez, F. (2001) Function and hormonal control of proteases in cereal grains. *Recent Research Development in Plant Physiology* 2, 57–65.

Colombo, L., Franken, J., van der Krol, A.R., Wittich, P.E., Dons, H.J.M. and Angenet, G.C. (1997) Downregulation of ovule-specific MADS box genes from petunia results in maternally controlled defects in seed development. *Plant Cell* 9, 703–715.

Domínguez, F. and Cejudo, F.J. (1995) Pattern of endoproteolysis following wheat grain germination. *Physiologia Plantarum* 95, 253–259.

Domínguez, F. and Cejudo, F.J. (1998) Germination-related genes encoding proteolytic enzymes are expressed in the nucellus of developing wheat grains. *Plant Journal* 15, 569–574.

Domínguez, F. and Cejudo, F.J. (1999) Patterns of starchy endosperm acidification and protease gene expression in wheat grains following germination. *Plant Physiology* 119, 81–87.

Domínguez, F., Moreno, J. and Cejudo, F.J. (2001) The nucellus degenerates by a process of programmed cell death during the early stages of wheat grain development. *Planta* 231, 352–360.

Domínguez, F., González, M.C. and Cejudo, F.J. (2002) A germination-related gene encoding a serine carboxypeptidase is expressed during the differentiation of the vascular tissue in wheat grains and seedlings. *Planta* 215, 727–734.

Fesik, S.W. (2000) Insights into programmed cell death through structural biology. *Cell* 103, 273–282.

Greenberg, J.T. (1996) Programmed cell death – a way of life for plants. *Proceedings of the National Academy of Sciences USA* 93, 12094–12097.

Jarvi, A.J. and Eslick, R.F. (1975) Shrunken endosperm mutants in barley. *Crop Science* 15, 363–366.

Jones, A.M. and Dangl, J.D. (1996) Logjam at the styx: programmed cell death in plants. *Trends in Plant Science* 1, 114–119.

Kerr, J.F.R., Gobé, G.C., Winterford, C.M. and Harmon, B.V. (1995) Anatomical methods in cell death. *Methods in Cell Biology* 46, 1–27.

Macnicol, P.K. and Jacobsen, J.V. (1992) Endosperm acidification and related metabolic changes in the developing barley grain. *Plant Physiology* 98, 1098–1104.

Morrison, I.N., O'Brien, T.P. and Kuo, J. (1978) Initial cellularization and differentiation of the aleurone cells in the ventral region of the developing wheat grain. *Planta* 140, 19–30.

Pennell, R.I. and Lamb, C. (1997) Programmed cell death in plants. *Plant Cell* 9, 1157–1168.

Raff, M. (1998) Cell suicide for beginners. *Nature* 396, 119–122.

Swift, J.G. and O'Brien, T.P. (1971) Vascular differentiation in the wheat embryo. *Australian Journal of Botany* 19, 63-71.

Vaux, D.L. and Korsmeyer, S.J. (1999) Cell death in development. *Cell* 96, 245–254.

Wang, H.L., Offler, C.E. and Patrick, J.W. (1994) Nucellar projection transfer cells in the developing wheat grain. *Protoplasma* 182, 39–52.

Young, T.E. and Gallie, D.R. (1999) Regulation of programmed cell death in maize endosperm by abscisic acid. *Plant Molecular Biology* 42, 397–414.

Young, T.E., Gallie, D.R. and DeMason, D.A. (1997) Ethylene-mediated programmed cell death during maize endosperm development of wild-type and *shrunken2* genotypes. *Plant Physiology* 115, 737–751.

29 What Do We Really Know About Desiccation Tolerance Mechanisms?

Folkert A. Hoekstra[1], Elena A. Golovina[1,2] and Jaap Nijsse[1]

[1]The Graduate School Experimental Plant Sciences, Laboratory of Plant Physiology, Wageningen University, Arboretumlaan 4, 6703 BD Wageningen, The Netherlands; [2]Russian Academy of Sciences, Timiryazev Institute of Plant Physiology, Botanicheskaya 35, Moscow 127276, Russia

Introduction

Desiccation tolerance implies an organism's ability to cope with the stresses of almost complete water loss and of rehydration. In practice, the rehydrated desiccation-tolerant organism is capable of resuming normal metabolism after a previous desiccation to water contents as low as 0.05 g H_2O/g dry matter. Di- and oligosaccharides have been found to accumulate to high levels in desiccation-tolerant (anhydrobiotic) organisms. This has led to an increased interest in the possible role of these sugars. From drying experiments with model systems, it emerged that sugars are capable of preventing phase changes in membranes and conformational changes in proteins (Crowe *et al.*, 1998). Consequently, membranes do not leak and enzyme activities are retained.

The mechanism of these protections is based on hydrogen bonding interactions of the sugar OH with the P=O of the polar headgroup in phospholipids and with the C=O of proteins (Crowe *et al.*, 1998). When the water shell gradually dissipates below a moisture content of 0.3 g H_2O/g dry matter, hydrogen bonding interactions with sugar will replace the original hydrogen bonding interactions with water. This mechanism has been designated as the water replacement hypothesis.

In the case of dehydrating membranes, the sugar interaction prevents an increase in melting temperature (T_m), which would otherwise result from the increased packing density of phospholipids (Fig. 29.1). Often, T_m is considerably lower in model membranes dried in the presence of sugars than in the hydrated controls (Crowe *et al.*, 1996). The spacing between the phospholipid molecules, brought about by the insertion of the sugars, might

 The Biology of Seeds: Recent Research Advances (eds G. Nicolás, K.J. Bradford, D. Côme and H.W. Pritchard)

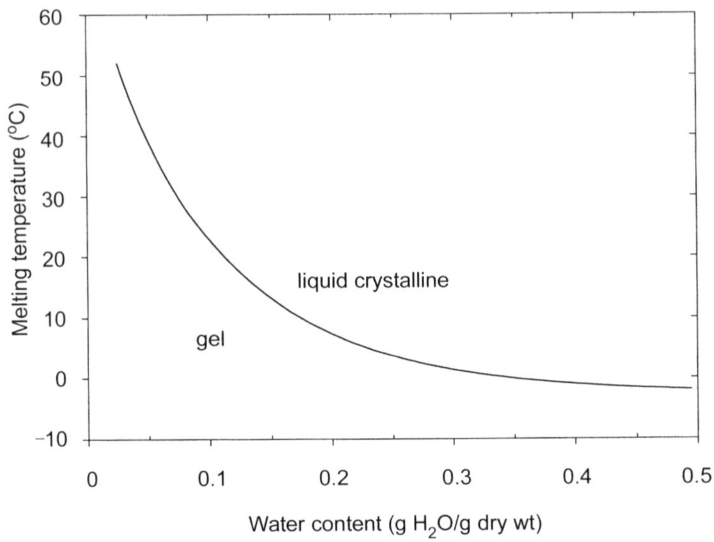

Fig. 29.1. Effect of water content on the melting temperature (phase transition temperature) of the phospholipid, palmitoyl-oleoyl-phosphatidylcholine. Above the curve, POPC is in liquid crystalline phase and below in gel phase.

allow more freedom of motion for the acyl chains than in the case of interactions with water. However, there is evidence that at the same time the headgroups themselves are immobilized in the sugar-protected, desiccated membranes (Lee *et al.*, 1986; Golovina, unpublished work).

In the case of protein secondary structure, the hydrogen bonding interactions between sugars and proteins prevent unfolding and denaturation. These interactions appear to be more effective in preventing structural changes than immobilization of proteins in a glassy matrix. It could thus be found that marginal amounts of glucose prevent unfolding better than similar amounts of sucrose or polymeric dextran (Wolkers *et al.*, 1998b), while glucose is an inferior glass former to sucrose and dextran under the same conditions of temperature and low moisture content. There are extensive pharmaceutical applications of sugars as stabilizers of dried engineered proteins and liposomal drug delivery systems.

A drawback of the early approach to understanding the mechanisms of desiccation tolerance is that mainly model systems have been investigated and that the interaction mainly pertains to the final stages of dehydration. However, the few reports on membrane behaviour *in vivo* in dried anhydrobiotes indicate that, indeed, phase transition temperatures are not considerably increased, which would mean that the liquid crystalline phase is generally maintained (Hoekstra and Golovina, 1999). On the other hand, cooling the dried specimens to far below 0°C often does not harm viability, although membranes have undergone a phase change from the liquid crystalline to the gel phase under these conditions. It is envisaged that immobilization of the membrane surface at low water contents in anhydrobiotes prevents further deteriorative changes in membranes, such as sorting of phospholipids and membrane proteins – a process that is known as phase separation. Also, there is evidence of an extreme *in vivo* stability of protein secondary structure in anhydrobiotic pollen and seed. Even decades after seed death, the native protein structure is still preserved (Golovina *et al.*, 1997), which might depend on the composition of interacting molecules (see 'Glass Formation' below). In contrast, denaturation has been observed in dried seeds/embryos having reduced desiccation tolerance (Wolkers *et al.*, 1998a, 1999).

Are Sugars Sufficient to Confer Desiccation Tolerance?

There are examples showing that manipulations that lead to increased endogenous sugar content also lead to the acquisition of desiccation tolerance (Leslie *et al.*, 1994; de Castro *et al.*, 2000; Guo *et al.*, 2000). However, this may not always hold, because many systems do not acquire tolerance, irrespective of increased sugar contents (Tunnacliffe *et al.*, 2001). Even the human fibroblast cells that had been engineered to accumulate the disaccharide, trehalose, survived the dried state for only a few days (Guo *et al.*, 2000). Living cells are of considerably higher complexity than vesicles filled with proteins. They are actively metabolizing entities containing a wide variety of solutes, which need additional protection against oxidative stress and molecular crowding.

Distinction Between Drought and Desiccation Tolerance

During drought stress, bulk water is preserved in cells, whereas it disappears during desiccation. This difference presumes differences in protective mechanisms between cells being exposed to drought and desiccation stress. There is a similarity with the respective protective mechanisms that are effective during freezing and freeze-drying of model protein and membrane systems (Crowe *et al.*, 1990). With freezing, the water content of the system does not fall below 0.3 g H_2O/g dry matter, whilst with freeze-drying the water is replaced by hydrogen bonding compounds other than water, as in anhydrobiotes.

Problems Encountered When Bulk Water is Still Present

A seed might be confronted by numerous problems during dehydration. These include cell shrinkage, molecular crowding and active metabolism. Due to volume reduction, the cell wall and the cell membrane have to accommodate the shrinking protoplast, the extent of the contraction being dependent on the initial filling of the cells with dry matter. Figure 29.2 shows the results of simple calculations indicating that with an initial water content of 70% (2.3 g H_2O/g dry matter) the cell surface will be reduced by more than half upon full desiccation. Because there is a limit to the compressibility of membranes, it can be expected that folding and/or vesiculation would

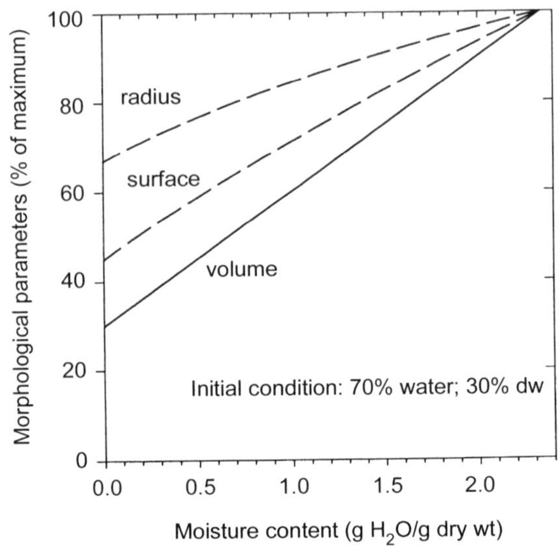

Fig. 29.2. Calculated changes in the dimensions of a hypothetical globular cell during drying. The initial moisture content was set at 70% (on a fresh weight basis). Surface = $4\pi r^2$; volume = $4/3\pi r^3$.

occur. This is exactly what we have noticed in dried seeds and pollen, where both folding and vesiculation appear to occur. Vesiculation has also been observed after osmotic contraction of rye protoplasts (Gordon-Kamm and Steponkus, 1984).

Volume reduction leads to molecular crowding. The changed volume relationship between the aqueous cytoplasmic phase and the lipid phase (oil bodies and membranes) leads to a net flow of endogenous amphiphilic compounds into the lipid fractions. This phenomenon has been demonstrated using an amphiphilic spin probe incorporated in the cytoplasm of seeds and pollen (Golovina *et al.*, 1998; Hoekstra and Golovina, 2002). Electron spin resonance spectroscopy (ESR) revealed the whereabouts of the spin probe during drying; a considerable increase was observed in the cells' membranes. This phenomenon might disturb membrane order and functioning on the one hand, but might provide automatic insertion of amphiphilic antioxidants into membranes on the other hand. Recently, it was found that membrane order becomes gradually disturbed during dehydration of young wheat embryos, most probably by unknown endogenous amphiphiles, and that the occurrence of these amphiphiles coincides with the acquisition of desiccation tolerance (Golovina and Hoekstra, Chapter 37, this volume).

Active metabolism appears to be incompatible with desiccation tolerance. This has been elegantly demonstrated in cucumber seedlings, just after radicle protrusion (Leprince *et al.*, 2000). Desiccation tolerance of the radicle tips can be reinduced by a few days of slow dehydration in a –1.5 MPa polyethylene glycol (PEG) solution. The PEG-induced desiccation-tolerant radicle tips respire considerably less than the untreated desiccation-sensitive radicle tips. This difference in respiration rate is maintained on further drying. There are more indications of reduced respiration rates, for example in carrot somatic embryos. Upon pre-culture in the presence of the plant hormone, abscisic acid (ABA), somatic embryos suspend growth, assume low rates of respiration and reduce the uptake and consumption of sugars (Hoekstra *et al.*, 2001). Subsequent slow drying completes the acquisition of desiccation tolerance. Slow drying without ABA addition is generally insufficient for somatic embryos to acquire tolerance. We assume that metabolism is insufficiently depressed here, which, among others, leads to the exhaustion of protective sugars.

High respiratory activity may lead to the production of potentially dangerous reactive oxygen species (ROS). Disturbance of the electron transport chains (proton leakage) as a result of amphiphile partitioning into mitochondrial membranes might add to the production of ROS. Timely down-regulation of metabolism, thought to be regulated by ABA, reduces the generation of ROS and toxic by-products of metabolism during water loss (Fig. 29.3). Thus, survival of desiccation is secured.

The ROS that nevertheless are produced can be dealt with by free-radical scavengers. The genes encoding for enzymatic oxidants, such as ascorbate peroxidase, glutathione reductase and superoxide dismutase (SOD), are up-regulated during dehydration (Ingram and Bartels, 1996). These enzymatic antioxidant systems can be active only under conditions of sufficient water, before the bulk water is lost. In contrast, molecular antioxidants (e.g. glutathione, ascorbate, polyols, carbohydrates, proteins such as peroxiredoxin, and amphiphilic molecules such as tocopherol, quinones, flavonoids and phenolics) can be active also in the air-dry state (Buitink *et al.*, 2002).

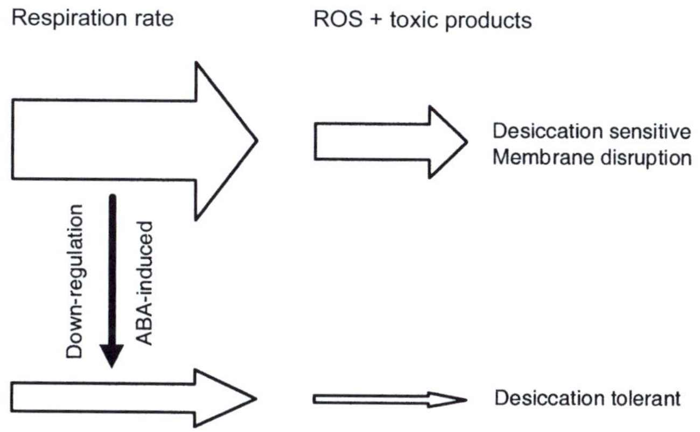

Fig. 29.3. Relationship between respiration rate in the hydrated state and survival of desiccation to the air-dry state. Timely down-regulation of metabolism, thought to be regulated by the plant hormone ABA, reduces the generation of ROS and toxic by-products of metabolism during water loss. Thus, the integrity of membranes is maintained and the survival of desiccation is secured.

Molecular crowding increases the chance of molecular interactions that can cause protein denaturation and membrane fusion. In response to drying, many plants and microorganisms accumulate compatible solutes. Examples of such molecules are proline, glutamate and glycinebetaine and also soluble carbohydrates such as mannitol, sorbitol, fructans, polyols, trehalose, sucrose and oligosaccharides. These solutes are called 'compatible' because they do not interfere with cellular activity, even at elevated concentrations in the cytoplasm. Despite their chemical dissimilarity, these compounds are all preferentially excluded from the surface of proteins, which keeps the proteins preferentially hydrated (Arakawa *et al.*, 1991). Because preferential exclusion is thermodynamically unfavourable, the surface area of a protein will be minimal and the folded conformation will be the most frequent; in the presence of preferentially bound co-solvents, the unfolded denatured state will be the most frequent. In the simultaneous presence of both types of solutes, the net result will be the sum of the effects. As compatible solutes accumulate upon drought stress, preferential exclusion is likely to be the main mechanism of protection of macromolecules in organisms against moderate water loss. Because preferential exclusion is based on a hydrophobic effect, the mechanism of preferential hydration fails to work in the absence of bulk water (below 0.3 g H_2O/g dry matter). Indeed, experiments with model systems show that most of the compatible solutes are unable to protect proteins and membranes against full desiccation (Crowe *et al.*, 1990). Then, only sugars can structurally and functionally preserve proteins and membranes by water replacement, as discussed in the Introduction. Preferential exclusion can therefore be effective only during the drought stage of water loss.

A major scientific effort is presently being made to reveal the function of late embryogenic abundant (LEA) proteins and small heat-shock proteins (HSPs). The accumulation of these proteins to high concentrations coincides with the acquisition of desiccation tolerance. Some of these proteins might act during the drought stage of water loss, particularly because their transcripts have also been detected in recalcitrant (desiccation-sensitive) seeds, and in drought-tolerant tissues submitted to water and/or temperature stress (Buitink *et al.*, 2002). On the basis of the remarkably high number of polar residues within the structure, some LEAs are thought to coat intracellular macromolecules with a cohesive water layer. This mechanism can be interpreted as a sort of preferential hydration. On further dehydration, LEAs would provide a layer of their own hydroxylated residues to interact with the surface groups of other proteins, acting as 'replacement water' (Cuming, 1999).

Small HSPs might act as molecular chaperones during seed dehydration and the first few days of rehydration. Generally, HSPs are able to maintain partner proteins in a folding-competent, folded or unfolded state, to minimize the aggregation of non-native proteins, or to target non-native or aggregated proteins for degradation and removal from the cell. Recently, the LEA-like protein HSP 12 from yeast was observed to be associated with membranes in desiccated yeast cells (Sales *et al.*, 2000), which might point to a function in the protection of membranes.

Problems Encountered with the Loss of the Hydration Shell

When the water shell dissipates from the surface of macromolecular structures below 0.3 g H_2O/g dry matter, intra- and intermolecular interactions become prominent. As argued in the Introduction, this leads to denaturation and aggregation of proteins and fusion and phase changes in membranes. The accumulated sugars then replace the water by hydrogen bonding interactions, which is thought also to hold for dried anhydrobiotic cells. If membranes are insufficiently protected, gel phase formation with drying may be accompanied by phase separations, leading to irreversible membrane damage. Interestingly, we recently observed, using the newest type of high-resolution, low-temperature scanning electron microscope, that radicle tip cells of recalcitrant *Aesculus hippocastanum* seeds (fast-dried in less than a few hours to 3% moisture content) had intact plasma membranes with a homogeneous distribution of intramembrane proteins (Fig. 29.4). This indicates that phase separations had not occurred during the fast dehydration. It might be, however, that slower dehydration would have led to phase separations. On rehydration, membrane integrity decreased gradually over the next hour. This would suggest that recalcitrance in *A. hippocastanum* seeds is not a priori caused by a failure to maintain membrane structure in the dried state.

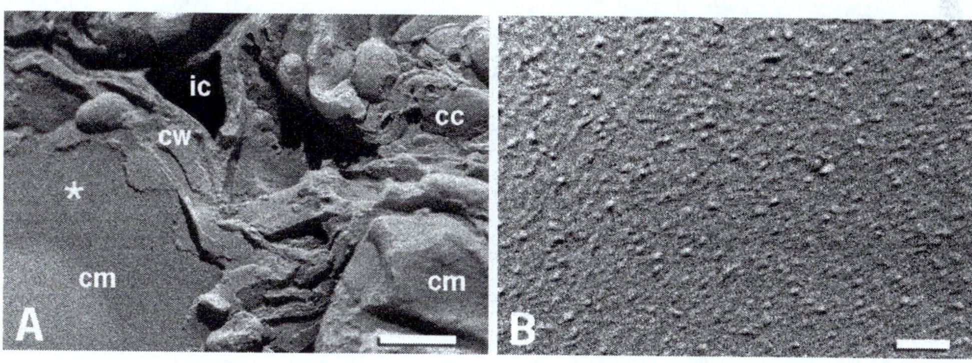

Fig. 29.4. Freeze–fracture micrograph of a fast-dried (3% relative humidity with ventilation) *Aesculus hippocastanum* radicle tip. (A) Overview, bar = 1 μm. (B) Detail from the cytoplasmic membrane at the asterisk, showing numerous homogeneously distributed intramembrane particles (membrane proteins), bar = 100 nm. Abbreviations: cc, cytoplasmic contents; cm, cytoplasmic membrane; cw, cell wall; ic, intracellular space. The high-resolution, low-temperature scanning electron microscope (Hitachi-5200) at the University of Ulm, Germany (Professor Paul Walther) was used.

Glass Formation

On dehydration of organic material, an amorphous glassy state may be formed. A glass is characterized by the lack of crystals, by a glass-to-liquid transition temperature (T_g) that depends on chemical composition and amount of plasticizing water, and by an extremely high viscosity that curtails chemical reaction rates. To appreciate this high viscosity, comparison with the flow rate of water may be informative. While water flows at a rate of 10 m/s, a glass would flow at about 0.3 μm/year.

In orthodox seeds, T_g gradually increases from close to –90°C in the hydrated state to about –60°C at 0.3 g H_2O/g dry matter, after which it rapidly increases to about 60°C at 0.05 g H_2O/g dry matter. During this dehydration the molecular mobility in the cytoplasm decreases by more than five orders of magnitude (Buitink *et al.*, 1999). When such seeds are dried at 20°C, a glass will be formed when the water content has decreased to approximately 0.08 g H_2O/g dry matter, although this value varies somewhat with chemical composition. Desiccation-sensitive seeds and pollen also form glasses when air-dried, although most of them lose viability at water contents far above those at which glasses are formed (Sun *et al.*, 1994; Buitink *et al.*, 1996).

The high amounts of sugars in anhydrobiotic cells have invited comparison between state diagrams of glasses of the prevailing sugar and of the cells under investigation. The similarity in T_g between them has led to the idea that sugars are crucial factors in cellular glasses. However, ESR revealed that there are considerable differences in properties between sugar glasses and cytoplasmic glasses. For example, the molecular mobility of a spin probe rapidly increases above T_g in a sucrose glass, whereas it takes considerable further heating in cytoplasmic glasses to increase mobility to the extent that flow on a practical time scale becomes possible, the so-called critical temperature (Buitink *et al.*, 2000b). This difference in behaviour may be caused by components in the cytoplasmic glass other than sugars – for example, proteins and other biopolymers.

Molecular rearrangements such as protein denaturation are probable only when flow on a practical time scale can occur, i.e. above the critical temperature. In this respect it is no surprise that in dried orthodox seeds protein denaturation usually does not occur on heating until high temperatures (150°C) are reached (Golovina *et al.*, 1997). However, in dried *Arabidopsis thaliana* mutant seeds having reduced ABA responsiveness, the onset temperature of protein denaturation is considerably lower than that of wild-type seeds (Wolkers *et al.*, 1998a; Table 29.1). In the case of the *abi 3-5* mutant seeds and the *aba 1-1 abi 1-3* double mutant seeds, the onset temperature is even close to that of the hydrated wild-type seeds. Seeds with reduced ABA responsiveness to germination also have less LEA protein and reduced desiccation tolerance. These data suggest that high-stability glasses may be a crucial factor in desiccation tolerance. It is tempting to assume that LEA proteins play a role in glass stability, but this has only been demonstrated in a model LEA protein/sucrose system (Wolkers *et al.*, 2001).

Glass stability may pertain to longevity (Sun and Leopold, 1994). It appeared that storage longevity inversely correlated with the molecular mobility of a polar spin probe in the cytoplasm (Buitink *et al.*, 2000a). From

Table 29.1. Impact of ABA defectiveness/insensitivity in mutant seeds of *Arabidopsis thaliana* on half-maximal ABA responsiveness to germination, accumulation of LEA proteins, desiccation tolerance and onset temperature of protein denaturation (air-dry state). Data derived from Wolkers *et al.* (1998a).

Genotype	ABA responsiveness (μM)	LEA	Desiccation tolerance	T_{onset}[a] (°C)
Wild-type	< 1	++	++	> 150
abi 3-1	10	+	++	135
abi 3-7	100	+/–	+	120
abi 3-5	> 1000	–	+/–	70
aba 1-1 abi 3-1	n.a.	–	–	70

[a] T_{onset} of hydrated wild-type seeds was 56°C.
n.a.= not applicable.

such a relationship, longevity can in principle be estimated after measurement of the molecular mobility at any given temperature and moisture content of a seed (Fig. 29.5). At –18°C and relatively high water contents of approximately 10–20%, survival times of several tens of thousands of years have been calculated. Such long survival times might be endorsed by recent findings of partly viable, 30,000-year-old seeds that were encased in the Siberian permafrost (Yashina *et al.*, 2002).

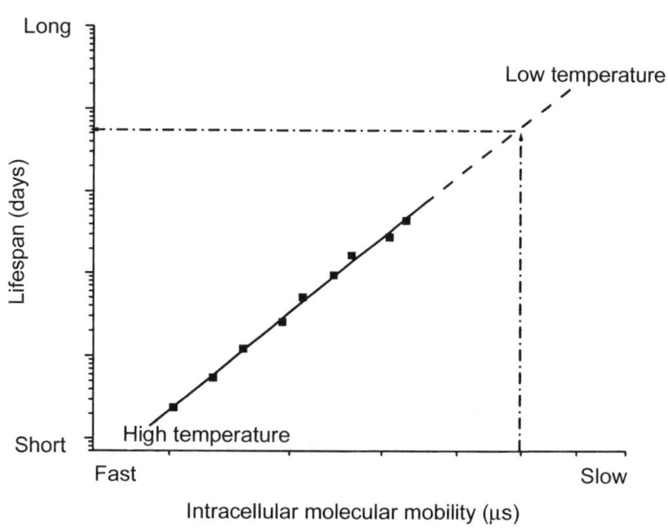

Fig. 29.5. Relationship between lifespan and intracellular molecular mobility in dry anhydrobiotes (modified from Buitink *et al.*, 2000a). Molecular mobility in cells can be estimated by measuring the rotational correlation time using saturation transfer ESR spectroscopy. It is possible, using extrapolation, to predict lifespan after estimating the molecular mobility at a given temperature (dashed lines).

The Danger of Rehydration

Rehydration of dried desiccation-tolerant organisms in water may lead to injury. This injury occurs when the specimens are extremely dry and/or aged, particularly at low temperatures. The injury is linked with a permanent loss of membrane integrity. Scanning electron micrographs show holes in the plasma membranes of pollen within seconds after stressful imbibition (Hoekstra *et al.*, 1999). In complex multicellular systems such as seeds, there is a slower progress of injury with the penetration of water into the tissues. In addition, the proportion of damaged cells in seeds is limited, because the imbibing cells prehumidify the next layers of cells, thus preventing further injury.

The damage can be prevented by warm imbibition or by prehydration of the dried specimens from the vapour phase. These treatments all melt possible gel-phase phospholipids and increase the molecular mobility of the membrane components before the uptake of liquid water (Hoekstra and Golovina, 1999). Thus, a possible membrane phase transition during imbibition is avoided, and the plasma membrane has become sufficiently flexible to accommodate the expanding protoplast. Apparently, rigid membranes rupture at certain sites, and stable holes are subsequently formed. In seeds, the coat or testa has a moderating effect on the extent of imbibitional damage, because it often restricts rapid penetration of water. Seeds with damaged seed coats, caused either mechanically or by ageing, are more sensitive to imbibitional stress (Powell and Matthews, 1979; Priestley, 1986).

Outlook

When focusing on the title of this chapter, which asks how much we really know about desiccation tolerance mechanisms, it appears to us that in the last decade considerable knowledge has been gained on dried model systems, particularly. This means that most biophysical investigations have been focused on phenomena in the dried state such as hydrogen bonding interactions and glass formation. Considering the complexity of molecular interactions in intact cells, the mechanisms of desiccation tolerance should, if at all possible, be studied *in vivo*. Since many desiccation-sensitive organisms die when the water content is still relatively high, future research should preferably be aimed at mechanisms of protection that operate in this particular range of water contents. In this respect it is imperative to resolve metabolic aspects during drying and the functions of dehydration-specific proteins such as LEA proteins and small HSPs.

References

Arakawa, T., Kita, Y. and Carpenter, J.F. (1991) Protein–solvent interactions in pharmaceutical formulations. *Pharmaceutical Research* 8, 285–291.

Buitink, J., Walters-Vertucci, C., Hoekstra, F.A. and Leprince, O. (1996) Calorimetric properties of dehydrating pollen. Analysis of a desiccation-tolerant and an intolerant species. *Plant Physiology* 111, 235–242.

Buitink, J., Hemminga, M.A. and Hoekstra, F.A. (1999) Characterization of molecular mobility in seed tissues: an electron paramagnetic resonance spin probe study. *Biophysical Journal* 76, 3315–3322.

Buitink, J., Leprince, O., Hemminga, M.A. and Hoekstra, F.A. (2000a) Molecular mobility in the cytoplasm: an approach to describe and predict lifespan of dry germplasm. *Proceedings of the National Academy of Sciences USA* 97, 2385–2390.

Buitink, J., van den Dries, I.J., Hoekstra, F.A., Alberda, M. and Hemminga, M.A. (2000b) High critical temperature above Tg may contribute to the stability of biological systems. *Biophysical Journal* 79, 1119–1128.

Buitink, J., Hoekstra, F.A. and Leprince, O. (2002) Biochemistry and biophysics of tolerance systems. In: Black, M. and Pritchard, H.W. (eds) *Desiccation and Survival in Plants: Drying Without Dying*. CAB International, Wallingford, UK, pp. 293–318.

Crowe, J.H., Carpenter, J.F., Crowe, L.M. and Anchordoguy, T.J. (1990) Are freezing and dehydration similar stress vectors? A comparison of modes of interaction of stabilizing solutes with biomolecules. *Cryobiology* 27, 219–231.

Crowe, J.H., Hoekstra, F.A., Nguyen, K.H.N. and Crowe, L.M. (1996) Is vitrification involved in depression of the phase transition temperature in dry phospholipids? *Biochimica et Biophysica Acta* 1280, 187–196.

Crowe, J.H., Carpenter, J.F. and Crowe, L.M. (1998) The role of vitrification in anhydrobiosis. *Annual Review of Physiology* 60, 73–103.

Cuming, A.C. (1999) LEA proteins. In: Shewry, P.R. and Casey, R. (eds) *Seed Proteins.* Kluwer Academic Publishers, Dordrecht, The Netherlands, pp. 753–780.

de Castro, A.G., Bredholt, H., Strom, A.R. and Tunnacliffe, A. (2000) Anhydrobiotic engineering of gram-negative bacteria. *Applied and Environmental Microbiology* 66, 4142–4144.

Golovina, E.A., Wolkers, W.F. and Hoekstra, F.A. (1997) Long-term stability of protein secondary structure in dry seeds. *Comparative Biochemistry and Physiology* 117A, 343–348.

Golovina, E.A., Hoekstra, F.A. and Hemminga, M.A. (1998) Drying increases intracellular partitioning of amphiphilic substances into the lipid phase: impact on membrane permeability and significance for desiccation tolerance. *Plant Physiology* 118, 975–986.

Gordon-Kamm, W.J. and Steponkus, P.L. (1984) The behavior of the plasma membrane following osmotic contraction of isolated protoplasts: implications in freezing injury. *Protoplasma* 123, 83–94.

Guo, N., Puhlev, I., Brown, D.R., Mansbridge, J. and Levine, F. (2000) Trehalose expression confers desiccation tolerance on human cells. *Nature Biotechnology* 18, 168–171.

Hoekstra, F.A. and Golovina, E.A. (1999) Membrane behavior during dehydration: implications for desiccation tolerance. *Russian Journal of Plant Physiology* 46, 295–306.

Hoekstra, F.A. and Golovina, E.A. (2002) The role of amphiphiles. *Comparative Biochemistry and Physiology* 131A, 527–533.

Hoekstra, F.A., Golovina, E.A., van Aelst, A.C. and Hemminga, M.A. (1999) Imbibitional leakage from anhydrobiotes revisited. *Plant, Cell and Environment* 22, 1121–1131.

Hoekstra, F.A., Golovina, E.A., Tetteroo, F.A.A. and Wolkers, W.F. (2001) Induction of desiccation tolerance in plant somatic embryos: how exclusive is the protective role of sugars? *Cryobiology* 43, 140–150.

Ingram, J. and Bartels, D. (1996) The molecular basis of dehydration tolerance in plants. *Annual Review of Plant Physiology and Plant Molecular Biology* 47, 377–403.

Lee, C.W.B., Waugh, J.S. and Griffin, R.G. (1986) Solid-state NMR study of trehalose/1,2-dipalmitoyl-*sn*-phosphatidylcholine interactions. *Biochemistry* 25, 3737–3742.

Leprince, O., Harren, F.J.M., Buitink, J., Alberda, M. and Hoekstra, F.A. (2000) Metabolic dysfunction and unabated respiration precede the loss of membrane integrity during dehydration of germinating radicles. *Plant Physiology* 122, 597–608.

Leslie, S.B., Teter, S.A., Crowe, L.M. and Crowe, J.H. (1994) Trehalose lowers membrane phase transitions in dry yeast cells. *Biochimica et Biophysica Acta* 1192, 7–13.

Powell, A.A. and Matthews, S. (1979) The influence of testa condition on the imbibition and vigour of pea seeds. *Journal of Experimental Botany* 30, 193–197.

Priestley, D.A. (1986) *Seed Aging: Implications for Seed Storage and Persistence in Soil.* Comstock Publishing, Ithaca, New York, pp. 1–304.

Sales, K., Brandt, W., Rumbak, E. and Lindsey, G. (2000) The LEA-like protein HSP 12 in *Saccharomyces cerevisiae* has a plasma membrane location and protects membranes against desiccation and ethanol-induced stress. *Biochimica et Biophysica Acta* 1463, 267–278.

Sun, W.Q. and Leopold, A.C. (1994) Glassy state and seed storage stability: a viability equation analysis. *Annals of Botany* 74, 601–604.

Sun, W.Q., Irving, T.C. and Leopold, A.C. (1994) The role of sugar, vitrification and membrane phase transition in seed desiccation tolerance. *Physiologia Plantarum* 90, 621–628.

Tunnacliffe, A., de Castro, A.G. and Manzanera, M. (2001) Anhydrobiotic engineering of bacterial and mammalian cells: is intracellular trehalose sufficient? *Cryobiology* 43, 124–132.

Wolkers, W.F., Alberda, M., Koornneef, M., Léon-Kloosterziel, K.M. and Hoekstra, F.A. (1998a) Properties of proteins and the glassy matrix in maturation-defective mutant seeds of *Arabidopsis thaliana*. *The Plant Journal* 16, 133–143.

Wolkers, W.F., Van Kilsdonk, M.G. and Hoekstra, F.A. (1998b) Dehydration-induced conformational changes of poly-L-lysine as influenced by drying rate and carbohydrates. *Biochimica et Biophysica Acta* 1425, 127–136.

Wolkers, W.F., Tetteroo, F.A.A., Alberda, M. and Hoekstra, F.A. (1999) Changed properties of the cytoplasmic matrix associated with desiccation tolerance of dried carrot somatic embryos. An *in situ* fourier transform infrared spectroscopic study. *Plant Physiology* 120, 153–163.

Wolkers, W.F., McCready, S., Brandt, W., Lindsey, G.G. and Hoekstra, F.A. (2001) Isolation and characterization of a D-7 LEA protein from pollen that stabilizes glasses *in vitro*. *Biochimica et Biophysica Acta* 1544, 196–206.

Yashina, S.G., Gubin, S.V., Shabaeva, E.V., Egorova, E.F. and Maksimovich, S.V. (2002) Viability of higher plant seeds of late Pleistocene age from permafrost deposits as determined by *in vitro* culturing. *Doklady Academii Nauk* 383, 714–717.

30 Are Sugar-sensing Pathways Involved in Desiccation Tolerance?

O. LEPRINCE, J. BUITINK, B. LY-VU, M.H. MACHEREL AND P. SATOUR

UMR 1191 Physiologie Moléculaire des Semences (INH/INRA/Université d'Angers), Institut National d'Horticulture, 16 bd Lavoisier, 49045 Angers, France

Introduction

Sugars serve as substrates for respiration, synthesis of metabolic intermediates and starch and lipid reserves. Seeds have evolved mechanisms to monitor, respond and adapt their metabolism to intrinsic and external stimuli, including developmental programmes, abiotic stresses, light or nutrient deficiency. A well-known example of this is the sugar-induced repression of some α-amylase genes in germinating cereals (Loreti *et al.*, 2000, 2001). More than 40 sugar-related genes have been identified in plants. In seeds, most of them are related to sucrose metabolism, storage reserves, cell division and fermentation. However, sugars can act, at the millimolar range, as signalling molecules that are sensed by sugar-binding enzymes, proteins and transporters (Gibson, 2000; Smeekens, 2000).

Although the regulatory role of metabolism in sugar responses still prevails, sucrose and hexoses appear to regulate many cellular and biochemical processes controlling embryogenesis, seed maturation, germination, dormancy and seedling establishment in concert with plant hormones. For example, the nature and rate of sugar import into developing embryos of *Arabidopsis* and several legumes influence their development. Upon changes in the sucrose/hexose ratio, embryos can switch from pattern formation or growth to accumulation of storage reserves (Weber *et al.*, 1997). The hypothesis that sugar-induced signal transduction rather than metabolic regulation is involved in the regulation of gene expression and developmental processes in various seed tissues has been further confirmed by several recent discoveries: sugar signalling in *Arabidopsis* seeds can be uncoupled from its metabolism (Pego *et al.*, 1999); hexokinase (HXK) has been identified as the sugar-sensor in seeds (Pego *et al.*, 1999); and phenotypic *Arabidopsis*

 The Biology of Seeds: Recent Research Advances (eds G. Nicolás, K.J. Bradford, D. Côme and H.W. Pritchard)

mutants with disturbed sugar-signalling pathways have been isolated and characterized (Arenas-Huertero *et al.*, 2000; Gibson, 2000; Smeekens, 2000).

Tolerance of desiccation in seeds is a multifactorial trait in which the synthesis of protective substances and the suppression of degradative processes that might be induced during dehydration are equally critical. The role and importance of sucrose and several protective proteins in stabilizing macromolecular structures in the dry state have been demonstrated (Hoekstra *et al.*, 2001; Buitink *et al.*, 2002). Furthermore, the role of additional factors such as concerted down-regulation of metabolism and amphipath partitioning are increasingly recognized (Leprince *et al.*, 2000; Hoekstra *et al.*, 2001; Oliver *et al.*, 2001). However, the signals that trigger the activation and repression of genes leading to desiccation tolerance are largely unknown. ABA plays an important role in inducing desiccation tolerance, as it is responsible for regulating gene expression associated with dehydration, such as LEA genes (Ingram and Bartels, 1996). Interestingly, ABA-signalling pathways have been found to interact with sugar-signalling pathways (Smeekens, 2000). This observation suggests that a web of signal networks controlling various cellular responses is implicated in the induction of desiccation tolerance. Based on the current knowledge on sugar signalling in germinating seeds, we propose that several sugar-sensing mechanisms, in concert with ABA, might be involved in inducing desiccation tolerance.

Hexokinase, a Sugar Sensor Involved in the Reinduction of Desiccation Tolerance in Germinating Radicles

Germination and seedling growth have been found to be amenable physiological model systems to study the signalling functions of sugars and their sensors. Several lines of evidence indirectly indicate that a hexose-sensing mechanism involving hexokinase might be involved in desiccation tolerance. Pego *et al.* (1999) suggested that activation of an HXK-dependent signalling pathway is responsible for the mannose-induced repression of germination in *Arabidopsis*. The loss of desiccation tolerance in germinating seeds is often found to be coincident with radicle emergence (Fig. 30.1A). Treatments that inhibit germination, such as priming, also maintain the tolerance of desiccation. Thus, we hypothesized that activation of an HXK pathway that represses germination could also trigger signalling events leading to desiccation tolerance. In germinated seeds that are desiccation sensitive, desiccation tolerance can conveniently be reinduced by an osmotic stress for several days (Leprince *et al.*, 2000) (Fig. 30.1B). For example, in cucumber, desiccation tolerance is reinduced in desiccation-sensitive radicles by incubating the seeds in a polyethylene glycol (PEG) solution equivalent to a water potential of –1.5 MPa for 3 days (Fig. 30.1B). Similar results were obtained with emerged radicles of *Medicago trunculata* cv. Paraggio (data not shown). This physiological trait was used in this study to test whether hexose sensing is involved in cellular events inducing desiccation tolerance in cucumber and *M. trunculata* radicles, using several non-metabolizable sugar analogues as described in Loreti *et al.* (2001).

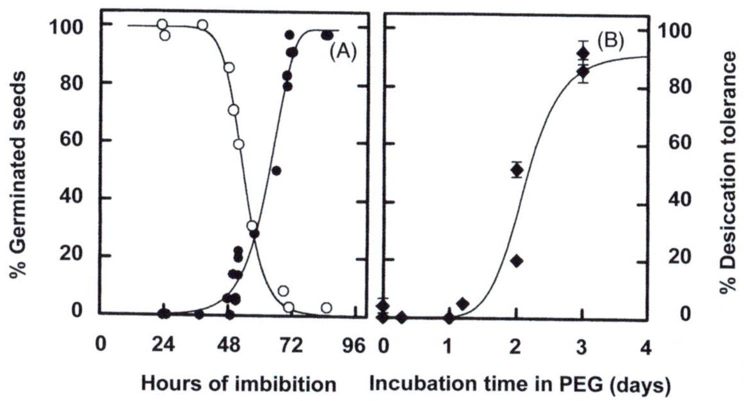

Fig. 30.1. (A) Germination (●) and loss of desiccation tolerance (○) of cucumber radicles during imbibition. Seeds were imbibed in the dark on moist filter paper at 20°C. To test desiccation tolerance, imbibed seeds were dried in a sealed container at 33% relative humidity at 20°C for 3 days. Following drying, seeds were reimbibed on wet filter paper at 25°C. Those exhibiting a growing radicle were scored as desiccation tolerant. (B) Reinduction of desiccation tolerance in sensitive emerged radicles of cucumber. After 72 h imbibition, seeds with emerged radicle of 2 mm long were selected and incubated at 10°C in a PEG 6000 solution with a water potential of −1.5 MPa. At various intervals, seeds were removed from the PEG solution, thoroughly rinsed, dried, then reimbibed on filter paper.

Among the different analogues tested, glucosamine (GAM), an inhibitor of HXK, was found to have an inhibitory effect on the reinduction of desiccation tolerance in both species (Table 30.1). This suggests that activation of a hexose pathway occurs during PEG incubation and is required to induce desiccation tolerance. Control experiments were performed to ascertain that the presence of GAM in dried tissues was not lethal. Seeds imbibed in the presence of 40 mM GAM, then dried again before radicle emergence, remained desiccation tolerant. Table 30.1 also shows that the presence of GAM in PEG does not impede radicle growth when the tissues are transferred from PEG to a moist filter paper without GAM. Glucose and fructose, which are sensed by HXK, had no effect on desiccation tolerance. Mannose also had an inhibitory effect on the reinduction of desiccation tolerance of cucumber radicles but not on those of *M. truncatula*. Since this effect is not common to both species, we suspect that mannose is toxic for cucumber. Cucumber seeds imbibed in the presence of 30 mM mannose germinated but radicle growth ceased after emerging out of the seed coat. Toxic effects of mannose, even at low concentrations, have often been reported (Brouquisse *et al.*, 2001). Phosphorylation of mannose into mannose-6-P may rapidly trap phosphate, thereby preventing rephosphorylation of ADP to ATP and inducing imbalances in the cellular energy status. Thus, testing desiccation tolerance of mannose-treated radicles using resumption of growth as a physiological criterion might be biased in the cucumber system.

Table 30.1. Effectiveness of D-glucose, D-fructose, and D-glucose analogues on radicle growth and induction of desiccation tolerance in emerged radicles of cucumber and Medicago truncatulata.

Compounds added to PEG	Growth after PEG incubation (%)		Desiccation tolerance after PEG incubation (%)	
	Cucumber	Medicago truncatulata	Cucumber	Medicago truncatulata
None (control)	100	100	94 ± 4	91 ± 6
50 mM glucose	100	96	73 ± 10	97 ± 1
50 mM fructose	NT	NT	NT	88 ± 2
10 mM mannose	100	NT	90 ± 7	NT
30 mM mannose	0	NT	0 ± 0	94 ± 10
40 mM glucosamine (GAM)	100	100	0 ± 0	2 ± 2
25 mM 3-O-methylglucose	NT	NT	60	85
100 μM ABA	100	100	100	96
100 μM ABA + 40 mM GAM	NT	NT	NT	0 ± 0

Seeds were germinated in the dark at 20°C on moist filter paper for up to 24 h; then 30–50 seeds having an emerged radicle were selected and incubated for 3 days at 10°C in the dark in PEG 6000 solutions (–1.5 MPa) containing the different compounds. Following incubation, seeds were rinsed and placed on moist filter papers or dried at 40% relative humidity for 48 h. Emerged radicles were scored desiccation tolerant when they resumed growth upon reimbibition.
When indicated by '±', the value corresponds to the average (± SE) of data collected from triplicate independent experiments.
NT, not tested.

Sugar-dependent and -independent Pathways in the Reinduction of Desiccation Tolerance

There is accumulating evidence suggesting that a coordinated down-regulation of metabolism may play an important role in desiccation tolerance of anhydrobiotes (Leprince *et al.*, 2000; Oliver *et al.*, 2001). Upon environmental stress and sugar availability, regulation of metabolism in yeast and plants is controlled by an SNF1 and SnRK1 protein kinase complex, respectively (Halford and Hardie, 1998). The yeast SNF1 is central to the signal transduction pathway which links the sensing of cellular glucose concentrations with the repression and derepression of glucose-repressible genes. By analogy with yeast, the plant homologue SnRK1 that complements the *snf1* yeast mutant is thought to act as a global metabolic regulator (Halford and Hardie, 1998). Considering that hexose sensing may be involved in the reinduction of desiccation tolerance, we tested whether the SNF1 complex participates in the reinduction of desiccation tolerance. For this purpose, we screened the EMBL and GenBank databases for genes encoding the different subunits corresponding to the SnRK1 complex in cucumber. A partial sequence of *SNF1α*, the catalytic unit of the SnRK1 complex, was obtained and its expression was studied during germination and reinduction of desiccation tolerance (Fig. 30.2). RT-PCR analysis revealed that the amount of *SNF1α*

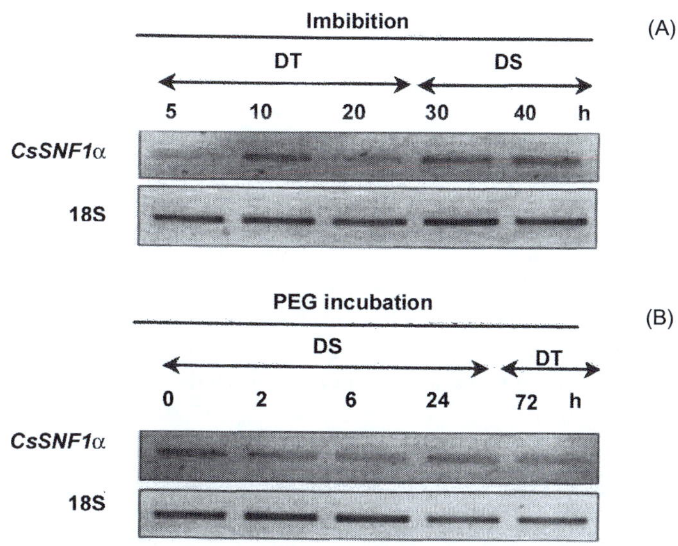

Fig. 30.2. RT-PCR analysis of *SNF1α* gene expression in cucumber radicles during (A) imbibition and (B) PEG incubation. Oligo-(dT)-cDNA synthesis was performed on total RNA and equal aliquots of each reaction were used in PCR with primers specific to the clone Y10036. For positive RT-PCR controls, a combination of primers directed at an 18S RNA was used. Additional controls were performed to ascertain that amplified products were analysed during the exponential phase of the PCR cycles. Desiccation sensitive (DS) and tolerant (DT) stages are indicated.

transcripts did not vary greatly during germination and PEG treatment. These data confirm those previously found with the tomato homologue LeSNF1 during germination by Bradford *et al.* (2000), who showed that the catalytic unit of the SnRK1 complex is constitutively expressed whereas its subunits are regulated at the transcriptional level. We are currently investigating whether the expression of genes encoding the different isoforms of the subunits is correlated with the induction of desiccation tolerance.

Abscisic acid regulates multiple functions in developing seeds, including embryo maturation, dormancy and desiccation tolerance. Some of the sugar-responsive recessive mutants that have recently been isolated harbour a mutation in the coding region of ABA or ABI genes, suggesting that sugar and ABA signals are interacting to modulate various aspects of seed development (Arenas-Huertero *et al.*, 2000; Laby *et al.*, 2000). These observations prompted us to examine whether ABA could repress the effects of GAM. Treatment with ABA was not able to suppress the GAM-induced inhibition of the reinduction of desiccation tolerance in *M. trunculata* seeds (Table 30.1). Further experiments are needed to assess whether the ABA-signalling pathway acts independently or downstream of the HXK pathway.

In conclusion, a hexokinase-dependent pathway and an ABA-dependent pathway appear to be present in germinating seeds of cucumber and *M.*

trunculata. Our data suggest that the activation of both are requisite to reinduce desiccation tolerance in emerged radicles. Further research towards the physiological characterization of sugar-response pathways in seeds is warranted.

Acknowledgements

This work was supported by grants from the Ministère de l'Agriculture, de l'Alimentation, de la Pêche et des Affaires Rurales, the Contrat de Plan Etat-Région 2000-2006, the Conseil Général de Maine-et-Loire and INRA. We thank B. Jettner (Seed-Co Australia Co-Operative Ltd, Hilton, Australia) and T. Bruggink (Syngenta, Enkhuizen, The Netherlands) for the generous gifts of the *M. trunculata* and cucumber seeds, respectively.

References

Arenas-Huertero, F., Arroyo, A., Zhou, L., Sheen, J. and Leon, P. (2000) Analysis of *Arabidopsis* glucose insensitive mutants, gin5 and gin6, reveals a central role of the plant hormone ABA in the regulation of plant vegetative development by sugar. *Genes and Development* 14, 2085–2096.

Bradford, K.J., Chen, F., Cooley, M.B., Dahal, P., Downie, B., Fukunaga, K.K., Gee, O.H., Gurusinghe, S., Mella, R.A., Nonogaki, H., Wu, C.-T., Yang, H. and Yim, K.-O. (2000) Gene expression prior to radicle emergence in imbibed tomato seeds. In: Black, M., Bradford, K.J. and Vazquez-Ramos, J. (eds) *Seed Biology: Advances and Applications*. CAB International, Wallingford, UK, pp. 231–251.

Brouquisse, R., Evrard, A., Rolin, D., Raymond, P. and Roby, C. (2001) Regulation of protein degradation and protease expression by mannose in maize root tips. Pi sequestration by mannose may hinder the study of its signaling properties. *Plant Physiology* 125, 1485–1498.

Buitink, J., Hoekstra, F.A. and Leprince, O. (2002) Biochemistry and biophysics of desiccation tolerance systems. In: Black, M. and Pritchard, H.W. (eds) *Desiccation and Survival in Plants: Drying Without Dying*. CAB International, Wallingford, UK, pp. 293–318.

Gibson, S.L. (2000) Plant sugar-response pathways. Part of a complex regulatory web. *Plant Physiology* 124, 1532–1539.

Halford, N.G. and Hardie, D.G. (1998) SNF1-related protein kinases: global regulators of carbon metabolism in plants? *Plant Molecular Biology* 37, 735–748.

Hoekstra, F.A., Golovina, E.A. and Buitink, J. (2001) Mechanisms of plant desiccation tolerance. *Trends in Plants Science* 5, 431–38.

Ingram, J. and Bartels, D. (1996) The molecular basis of dehydration tolerance in plants. *Annual Review of Plant Physiology and Plant Molecular Biology* 47, 377–403.

Laby, R.J., Kincaid, M.S., Kim, D. and Gibson, S.I. (2000) The *Arabidopsis* sugar-insensitive mutants sis4 and sis5 are defective in abscisic acid synthesis and response. *Plant Journal* 23, 587–596.

Leprince, O., Harren, F.J.M., Buitink, J., Alberda, M. and Hoekstra, F.A. (2000) Metabolic dysfunction and unabated respiration precede the loss of membrane integrity during dehydration of germinating radicles. *Plant Physiology* 122, 597–608.

Loreti, E., Alpi, A. and Perata, P. (2000) Glucose and disaccharide-sensing mechanisms modulate the expression of α-amylase in barley embryos. *Plant Physiology* 123, 939–948.

Loreti, E., De Bellis, L., Alpi, A. and Perata, P. (2001) Why and how do plant cells sense sugars? *Annals of Botany* 88, 803–812.

Oliver, A.E., Leprince, O., Wolkers, W.F., Hincha, D.K., Heyer, A.G. and Crowe, J.H. (2001) Non-disaccharide-based mechanisms of protection during drying. *Cryobiology* 43, 151–167.

Pego, J.V., Weisbeek, P.J. and Smeekens, S.C.M. (1999) Mannose inhibits *Arabidopsis* germination via a hexokinase-mediated step. *Plant Physiology* 119, 1017–1023.

Smeekens, S. (2000) Sugar-induced signal transduction in plants. *Annual Review of Plant Physiology and Plant Molecular Biology* 51, 49–81.

Weber, H., Borisjuk, L. and Wobus, U. (1997) Sugar import and metabolism during seed development. *Trends in Plant Science* 2, 159–174.

31 Gene Expression During Loss and Regaining of Stress Tolerance at Seed Priming and Drying

STEVEN P.C. GROOT, YASUTAKA SOEDA, GEERT STOOPEN, MAURICE C.J.M. KONINGS AND A.H.M. (LONNEKE) VAN DER GEEST

Plant Research International, Wageningen University and Research Centre, PO Box 16, NL–6700 AA Wageningen, The Netherlands

Introduction

Seeds confer a remarkable phase in the life cycle of plants. Most seeds acquire desiccation tolerance during maturation and can be stored for several years without great loss of viability. This phenomenon of storability is diverse among species, among seed lots and even among seeds within a seed lot. In general, storability increases during seed maturation and is gradually lost during seed germination. The latter may create difficulties when seeds are primed. Priming is a process during which seeds are pre-germinated to a certain extent and are re-dried before radicle protrusion to prevent loss of desiccation tolerance. A major advantage of priming is the generally faster and more uniform germination of primed seed, especially under suboptimal conditions. However, this improved germination performance is often accompanied by a reduction in storability, which dictates the need for a careful control of the priming process, in order to avoid progressing the germination process too far. Research performed by Bruggink *et al.* (1999) showed that the loss of storability could be reduced when the drying was performed slowly or at elevated temperatures. Nothing is known about the underlying mechanisms involved in the gaining of storability during seed maturation, its loss during priming and partial restoration during slow drying. Additionally, very little is known about which processes take place during priming that account for the improved germination performance. For seed companies that perform priming treatments, it can be of great help if markers are available to support monitoring of the priming process and optimization of the protocol.

Experimental Model

The aim of our experiments was to set up a model system with seeds of different physiological qualities, including storability, in order to monitor gene expression during seed germination, seed priming and seed drying. Seeds with different levels of stress tolerance were obtained through maturity sorting (by chlorophyll fluorescence according to Jalink *et al.*, 1998), priming and various drying treatments. Seeds from cabbage (*Brassica oleracea* var. *capita*) were used as a model species. To evaluate gene expression we used a cDNA microarray containing cDNA sequences from developing or germinating rape (*Brassica napus*) seeds.

Cabbage seeds were either germinated in water, or primed for 7 days in –1 MPa polyethylene glycol (PEG) at 20°C. After priming, the seeds were briefly washed and the adhering water removed by blotting between paper tissues. Traditional drying (called fast drying) was performed by placing the seeds for 3 days at 32% relative humidity (RH) and 20°C with circulating air. Testing of several drying treatments showed that loss of storability caused by priming could partly be restored by a slow drying treatment (3 days in a desiccator at 75% RH and 30°C with standing air) followed by 3 days under fast drying conditions. Monitoring the rate of water loss during these treatments showed that with fast drying the water loss was about 3%/h during the first 8 h, while with slow drying this figure was about 0.18%/h. Some variation was observed between the different experiments, most likely due to variation in the number of seeds used in the treatments, but in all cases storability was improved by the slow drying procedure. Storability of the seeds was tested in three different ways: by storage on the laboratory bench; by controlled storage at 30°C and 66% RH; or by controlled deterioration. The controlled deterioration (CD) treatment was performed by equilibrating the seeds for 3 days at 85% RH at 20°C, storing the equilibrated seeds in sealed foil bags at 40°C for 5 days, followed by re-drying at 32% RH and 20°C for 3 days. The germination experiments were performed with four replicates of 100 seeds.

Effects on Germination Rate and Storability

Priming of the seeds resulted in an acceleration of germination, as seen by the reduced T_{50} at various temperatures (Fig. 31.1). Although both drying treatments gave similar improvements in germination behaviour, slow-dried primed seeds germinated faster at higher temperatures than fast-dried primed seeds. It is possible that fast drying after priming creates mild damage in the primed seeds, which then need some time to recover, and slows down germination at the suboptimal warmer temperatures. Alternatively, the mild heat stress conferred during the slow drying treatment conditions the seeds and allows them to cope better with the stress encountered during high temperature germination, and CD treatments (see below).

As expected, the priming treatment resulted in a reduction of storability

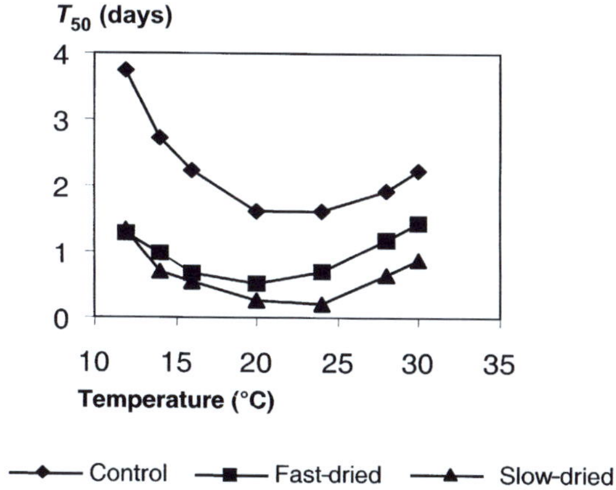

Fig. 31.1. Germination rate of control and primed cabbage seeds after fast or slow drying.

of the cabbage seeds as tested under CD conditions (Fig. 31.2). Slow drying of the primed seeds partly restored the ability to withstand the CD treatment, compared with fast drying. Comparison of storability of the primed and dried seeds under the rather artificial CD conditions, with more natural storage for 4 or 8 months at 30°C and 66% RH, showed a fairly good correlation (Fig. 31.3). The same was observed when comparing with 9 months' open storage in the laboratory, where temperature varied between 20 and 25°C and RH between 30 and 60% (data not shown).

Fast-dried primed seeds, slow-dried primed seeds and control seeds were subjected to an electroconductivity test. Initial electrolyte leakage from the control seeds was higher compared with that from the primed seeds, but

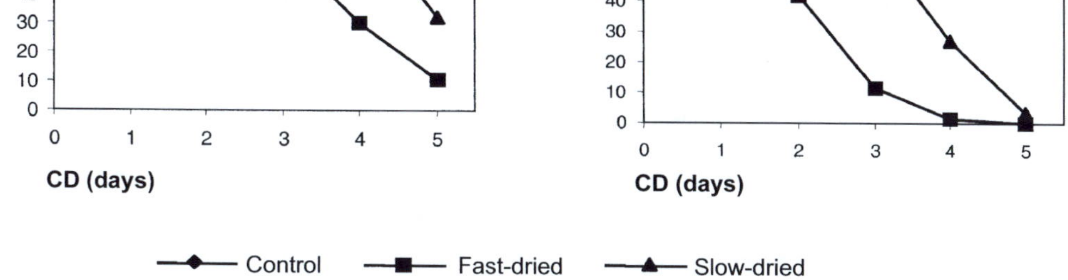

Fig. 31.2. Sensitivity of control (non-primed) and primed cabbage seeds towards a controlled deterioration (CD) treatment of different periods. Primed seeds were either fast- or slow-dried at elevated temperature.

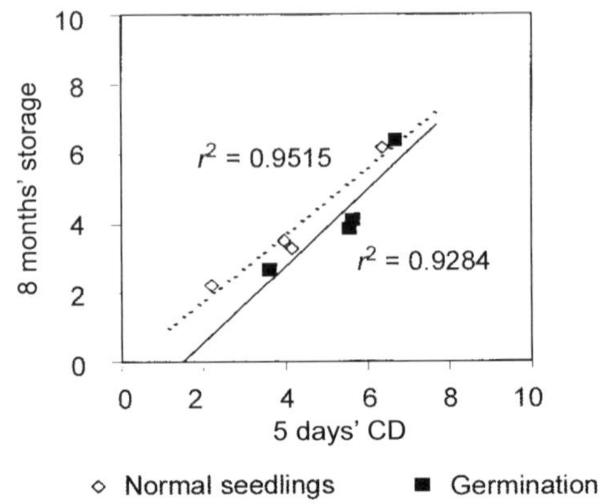

Fig. 31.3. Correlation between survival after storage of seeds under controlled deterioration (CD) (5 days, 85% RH, 40°C) and 'long' storage (8 months, 66% RH, 30°C). Data are presented after probit transformation (a value of 5 on the *x* or *y* axis means 50% germination after 5 days' CD or 8 months' 'long' storage, respectively).

after 4 h it occurred at a similar rate. Fast-dried primed seeds and slow-dried primed seeds did not differ in the level and rate of electrolyte leakage, indicating that their difference in germination performance and storability was not caused by potential differences in cell membrane integrity (data not shown).

Sensitivity to Ultra-drying

Drying of primed seeds at 32% RH results in a moisture content of about 6%, on a fresh weight basis, equal to that of non-primed seeds equilibrated at this RH. In relation to discussions regarding the optimum moisture content for seed storage, which are accompanied by contrasting observations or theories (e.g. Ellis, 1998; Walters, 1998), we were interested to discover whether primed seeds differ in their sensitivity towards drying to very low moisture contents. This was tested by drying samples of control, fast-dried and slow-dried primed seeds (6% moisture content) further, first for 2 weeks and then 4 weeks mixed with freshly regenerated silica gel. Under these conditions the seeds obtained a moisture content as low as 0.6%, on a fresh weight basis. When the seeds were subsequently imbibed with water of different temperatures, it became apparent that ultra-drying hardly affected germination of the control seeds. In contrast, fast-dried primed seeds germinated less efficiently after 6 weeks of ultra-drying and gave rise to a lower number of normal seedlings, especially when imbibed at the lower temperatures (Fig. 31.4). Slow-dried primed seeds were less sensitive to ultra-drying, since an effect was only visible after imbibition at low temperatures. Although under practical conditions imbibition will mostly be performed at room temperatures, this experiment illustrates the differences in the physiological status of the seeds, which can influence the storage behaviour of seeds at dif-

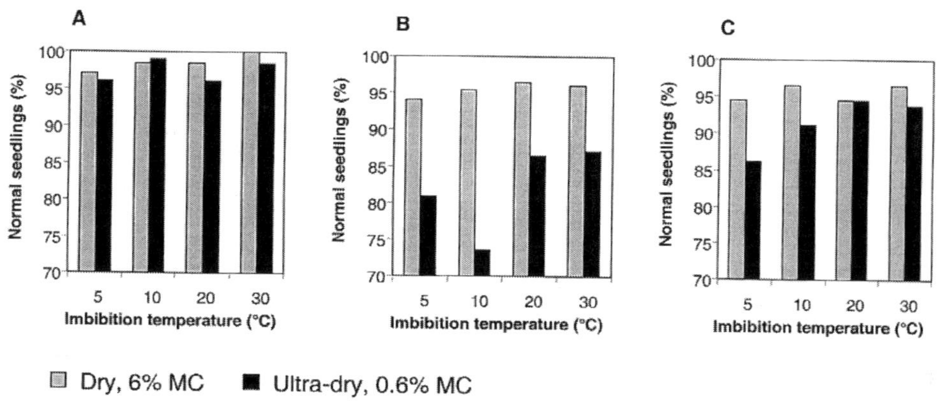

Fig. 31.4. Effect of fast or slow drying after priming on the sensitivity of cabbage seeds to subsequent drying or ultra-drying (0.6% MC) and imbibition at different temperatures. (A) Control non-primed seeds. (B) Seeds dried fast after priming. (C) Seeds dried slowly at elevated temperature after priming.

ferent moisture contents. With several crops some germination processes can start prior to harvest and drying of the seeds when dormancy is lacking (D. Job, C. Job and S.P.C. Groot, unpublished results), resulting in seed lots that contain partly 'primed' seeds. Under moist conditions prior to harvest this may even result in precocious germination (preharvest sprouting). Further research seems appropriate, especially because of the implications that the results may have on genetic conservation programmes performed in gene banks.

Seed Maturity

Since seeds gain in stress tolerance during seed maturation, seeds of different maturity levels were compared at the physiological and molecular level. Making use of the chlorophyll fluorescence sorting technique (Jalink *et al.*, 1998), we separated a cabbage seed sample with a relatively high frequency of less mature seeds into different fractions. The sorting technique makes use of the correlation between seed maturation and chlorophyll degradation. Seeds with low levels of chlorophyll, the most mature ones, were indeed more tolerant to the CD treatment, compared with less mature seeds with medium or high chlorophyll levels (Table 31.1).

Gene Expression Studies using cDNA Microarrays

The invention of DNA microarrays has made it possible to measure the expression levels of thousands of genes simultaneously. The real power of this technology stems from their ability to measure global changes in gene expression, which can be used to gain a molecular understanding of the underlying physiological processes. In order to prepare a cDNA microarray,

Table 31.1. Germination performance of chlorophyll fluorescence (CF) sorted seed fractions prior to and after a controlled deterioration (CD) treatment.

	– CD		+ CD	
CF level	Maximum germination	Normal seedlings	Maximum germination	Normal seedlings
Low	96.0 ± 0.8	95.0 ± 1.0	94.5 ± 2.2	87.5 ± 2.1
Medium	98.5 ± 0.5	96.5 ± 1.0	87.0 ± 2.4	78.5 ± 2.6
High	73.0 ± 2.4	60.0 ± 2.2	42.5 ± 2.4	35.0 ± 2.7

cDNA clones representing individual genes are robotically spotted on to a 1 cm^2 area on a glass slide. Hybridizing these slides with fluorescently labelled cDNAs derived from mRNAs present in various cell or tissue types allows simultaneous detection of changes in gene expression of all the genes that are spotted on the chip. By hybridizing with a mixture of reference and test samples, each labelled with a differently coloured fluorescent dye, relative changes in gene expression compared with the reference sample can be measured accurately (reviewed by Aharoni and Vorst, 2002).

In order to study gene expression during seed maturation, priming and germination prior to radicle protrusion, a cDNA microarray was produced containing 1500 genes from immature (green cotyledon stage) and imbibed (7 and 15 h after addition of water) mature rape seeds. Included on the cDNA microarray were 250 clones of the cotyledon stage seed cDNA library, 600 genes of the 7 h imbibed seed library, and 600 genes of the 15 h imbibed seed library. In addition, about 30 *Arabidopsis* cell cycle or stress-related genes were selected from the literature and spotted on the chip. Foreign clones (luciferase) were included as controls for normalization, and yeast genes that lacked cross-hybridization to plant genes were used as negative controls for subtracting the background. All genes were spotted in duplicate. Approximately 1000 clones were sequenced from the 5′ end and compared with the *Arabidopsis* genome.

Gene expression was studied using the physiologically characterized cabbage seed lots described above. The high level of sequence similarity between cabbage and rape allowed detection of gene expression in cabbage samples using a rape microarray. Swapped dye experiments were performed to estimate experimental variation, resulting in at least four gene expression measurements for each gene in each sample. Figure 31.5 shows the different patterns of gene expression that were observed in the various samples. Several gene classes could be recognized.

Early maturation genes are expressed during seed maturation, and the level of mRNA starts declining in the very last stage of maturation (from high to low chlorophyll level seeds). This group was represented by about 40 genes on the chip and mainly includes various family members of the napin seed storage protein gene family, cruciferins, and seed biotinylated protein. These genes encode proteins that are stored in the seeds in order to provide

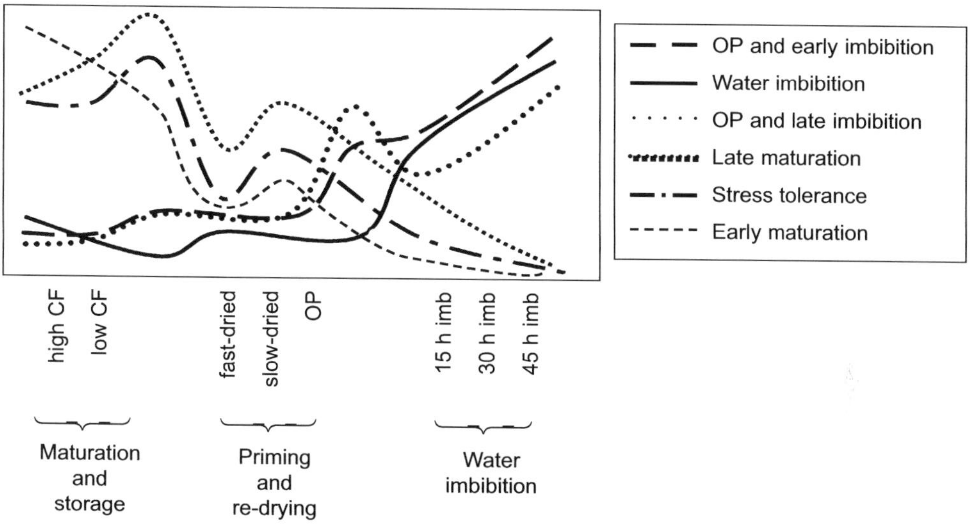

Fig. 31.5. Schematic representation of clusters of genes with specific expression patterns observed during seed maturation, seed priming and subsequent drying and during germination upon imbibition in water. OP, osmoprimed.

nutrition to the germinating seedling. Seed biotinylated protein has been shown to accumulate, particularly during late stages of embryo development, and is hydrolysed early during imbibition (Job *et al.*, 2001).

Late maturation genes reached a maximum level of gene expression during the final drying phase and their mRNA levels are highest in low chlorophyll seeds. This group was represented by about 80 genes on the chip, and included various late embryogenesis abundant (LEA) proteins, such as Em6, heat-shock genes and peroxiredoxin (AtPer homolog; Haslekås *et al.*, 1998). These genes are presumably involved in protecting the seeds during the desiccation phase, where LEA proteins are thought to play a role as chaperones under low water conditions (Kermode, 1997). Although heat-shock genes are predominantly expressed during heat stress, they are known to be expressed under different types of stress as well and may also act as molecular chaperones (Wehmeyer and Vierling, 2000)

Interestingly, a subset of these late maturation genes also differed in expression between slow- and fast-dried osmoprimed seeds, being higher in slow-dried seeds, strongly implicating a role for this subset of genes in acquisition of stress tolerance during both seed maturation and slow drying after priming. One of these genes is homologous to the Em6 gene identified in *Arabidopsis* and rape (Gaubier *et al.*, 1993; Vicent *et al.*, 1998). Em6 expression levels increase during seed maturation and decline during germination, as has been described at both the mRNA and protein level for *Arabidopsis* seeds by Bies *et al.* (1998). Here we found that the gene expression was reinduced during slow drying of the primed seeds, as demonstrated by the chip hybridization experiments and confirmed by Northern analysis (Fig. 31.6).

Water imbibition related genes were only expressed when seeds were

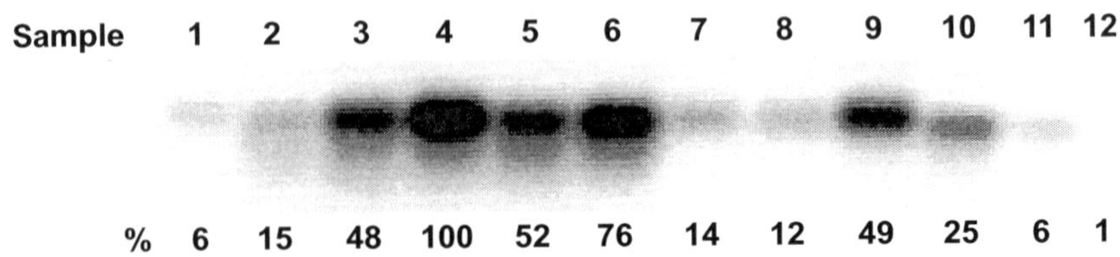

Fig. 31.6. Northern analysis of expression of the Em6 homologue clone during cabbage seed maturation, seed priming, drying and during germination upon imbibition in water. Samples: **(1)** immature seeds; **(2)** near mature seeds (high level of chlorophyll fluorescence); **(3)** dry control seeds (from commercial batch); **(4)** mature seeds (low level of chlorophyll fluorescence); **(5)** control seeds after a controlled deterioration treatment; **(6)** control seeds after 9 months' storage on the laboratory bench; **(7)** osmoprimed seeds before drying; **(8)** osmoprimed seeds after fast drying; **(9)** osmoprimed seeds after slow drying at elevated temperature; **(10)** seeds after 15 h imbibition in water; **(11)** seeds after 30 h imbibition in water; **(12)** seeds after 45 h imbibition in water. The relative expression levels (%) are indicated on the bottom, in reference to the maximum expression observed.

imbibed in water and were not induced during osmopriming. This group was represented by about 300 genes on the chip, and included a large number of housekeeping genes, such as various ribosomal subunits, translational cofactors, ubiquitin, phosphatases and dehydrogenases, but also a number of potentially more interesting genes involved in cell cycle and regulatory functions, such as transcription factors, histone H3, histone deacetylase and hypothetical proteins in the *Arabidopsis* genome for which no function has yet been identified. Onset of cell cycle activity, including DNA replication, occurs in many species, including cabbage, prior to radicle protrusion, as demonstrated by flow cytometry (Górnik *et al.*, 1997). In order to package the newly synthesized DNA into chromatin, histone proteins are required. Newly synthesized histones are acetylated, and deacetylated shortly after their incorporation into nucleosomes. In yeast, these dynamic reactions are essential for proper cell cycle progression, implicating a role for histone deacetylases in cell cycle checkpoint control (Wade *et al.*, 1997). Deacetylated histones in the nucleosomes reduce accessibility of the DNA to transcription factors, and histone deacetylase may therefore also play a role in negatively regulating gene expression. Activation of many of these genes has also been identified by Galardo *et al.* (2001) at the protein level, during *Arabidopsis* seed germination, using a proteomic approach.

The 400 genes that were expressed during osmopriming and imbibition also included housekeeping genes, as above, and additionally genes involved in signal transduction (receptors, MAP kinase), stress response (chaperonins, prefoldin, imbibition protein, cold-induced protein, glutathione transferase), and genes of unknown function. The stress proteins may reflect the stress imposed by the osmotic conditions during priming, and by the fast water uptake during seed imbibition for the array experiments, which was done by floating the seeds in a Petri dish with water.

Acknowledgements

The research presented has been sponsored by the Dutch Ministry of Agriculture, Nature Management and Fisheries.

References

Aharoni, A. and Vorst, O. (2002) DNA microarrays for functional plant genomics. *Plant Molecular Biology* 48, 99–118.

Bies, N., Aspart, L., Carles, C., Gallois, P. and Delseny, M. (1998) Accumulation and degradation of Em proteins in *Arabidopsis thaliana*: evidence for post-transcriptional controls. *Journal of Experimental Botany* 49, 1925–1933.

Bruggink, G.T., Ooms, J.J.J. and Van der Toorn, P. (1999) Induction of longevity in primed seeds. *Seed Science Research* 9, 49–53.

Ellis, R.H. (1998) Longevity of seeds stored hermetically at low moisture contents. *Seed Science Research* 8, Supplement No. 1, 9–10.

Galardo, K., Job, C., Groot, S.P.C., Puype, M., Demol, H., Vandekerckhove, J. and Job, J. (2001) Proteomic analysis of Arabidopsis seed germination and priming. *Plant Physiology* 126, 835–848.

Gaubier, P., Raynal, M., Hull, G., Huestis, G.M., Grellet, F., Arenas, C., Pages, M. and Delseny, M. (1993) Two different Em-like genes are expressed in *Arabidopsis thaliana* seeds during maturation. *Molecular and General Genetics* 238, 409–418.

Górnik, K., de Castro, R.D., Liu, Y., Bino, R.J. and Groot, S.P.C. (1997) Inhibition of cell division during cabbage (*Brassica oleracea* L.) seed germination. *Seed Science Research* 7, 333–340.

Haslekås, C., Stacy, R.A.P., Nygaard, V., Culiáñez-Macià, F.A. and Aalen, R.B. (1998) The expression of a peroxiredoxin antioxidant gene, AtPer1, in *Arabidopsis thaliana* is seed-specific and related to dormancy. *Plant Molecular Biology* 36, 833–845.

Jalink, H., Van der Schoor, R., Frandas, A. and Van Pijlen, J.G. (1998) Chlorophyll fluorescence of *Brassica oleracea* seeds as a non-destructive marker for seed maturity and seed performance. *Seed Science Research* 8, 437–443.

Job, C., Laugel, S., Duval, M., Galardo, K. and Job, D. (2001) Biochemical characterization of atypical biotinylation domains in seed proteins. *Seed Science Research* 11, 149–161.

Kermode, A.R. (1997) Approaches to elucidate the basis of desiccation-tolerance in seeds. *Seed Science Research* 7, 75–95.

Vicent, C.M., Roscoe, T.J. and Delseny, M. (1998) Characterization of an Em-like gene of *Brassica napus*. *Journal of Experimental Botany* 49, 1061–1062.

Wade, P.A., Pruss, D. and Wolffe, A.P. (1997) Histone acetylation: chromatin in action. *Trends in Biochemical Sciences* 22, 128–132.

Walters, C. (1998) Ultra-dry technology: perspective from the National Seed Storage Laboratory, USA. *Seed Science Research* 8, Supplement No. 1, 11–14.

Wehmeyer, N. and Vierling, E. (2000) The expression of small heat shock proteins in seeds responds to discrete developmental signals and suggests a general protective role in desiccation tolerance. *Plant Physiology* 122, 1099–1108.

32 Proteomic Analysis of Maternal Dominance for Cold Germination in Maize

IMAD N. SAAB, KRISHNA P. KOLLIPARA AND EMILY VAN DE MORTEL

Pioneer Hi-Bred International, 7301 NW 62nd Avenue, Johnston, IA 50131, USA

Introduction

The ability of maize seeds to germinate and establish stand under cold conditions is a key trait in regions where the early season is often characterized by cold, wet conditions. This trait is also important in regions where the trend for early planting carries with it increasing risks of cold stress in the early season. The ability of maize to germinate at low temperature has been associated with induction of enzymes that scavenge reactive oxygen species, including catalase, glyoxalase and glutathione S-transferase, as well as organellar properties, including mitochondria, oil and protein bodies, starch grains, membrane integrity, and the structure of chromatin and proteins (e.g. Yacoob and Filion, 1986; Janowiak and Markowski, 1987; Stewart *et al.*, 1990; Schell *et al.*, 1991; Prasad, 1996, 1997; Roxas *et al.*, 1997; van Breusegem *et al.*, 1998; Santis *et al.*, 1999). At the molecular level, much of the cold-stress literature has focused on freezing tolerance in *Arabidopsis*, which is mediated to some extent by the rapid expression of a family of transcription factors (CBF and its homolog DREB1; Jaglo-Ottosen *et al.*, 1998; Kasuga *et al.*, 1999). Over-expression of these transcription factors has led to increased freezing and desiccation tolerance via the coordinated expression of a number of target genes, leading to expression of proteins involved in cellular protection against dehydration and membrane damage (see review by Thomashow, 1999). Seki *et al.* (2001) reported a survey of mRNA expression of ~1300 genes in *Arabidopsis* subjected to drought and cold stress at various stages from germination to maturity. They identified 44 and 19 cDNAs that were inducible by drought and cold stress, respectively. Such detailed molecular studies of cold stress responses have not been reported in maize. In this study we report on wide-scale profiling of protein expression changes associated with germination at low temperature.

There have been several reports where differential phenotypic expression was observed between reciprocal F_1 hybrids in maize, such as seed ger-

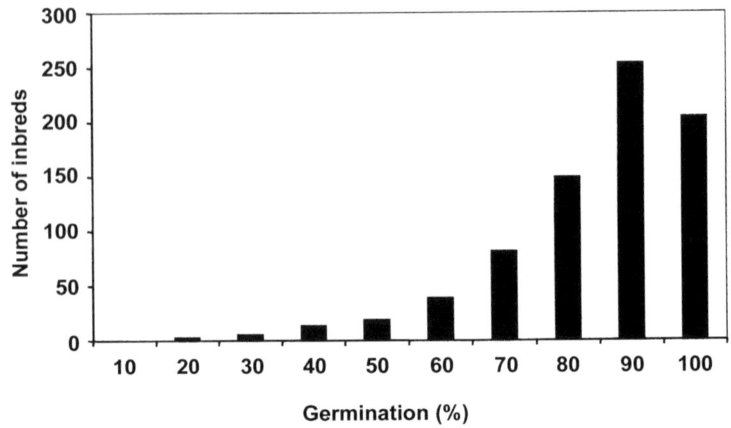

Fig. 32.1. Distribution of cold germination (CG) responses among 771 recombinant inbreds. Inbreds were classified in one of ten germination categories; e.g. inbreds with 85% CG are in the 90% category. Values are means of three replicates of 20 seeds each.

mination at low temperatures (Pinnel, 1949), kernel growth rate (Groszmann and Sprague, 1948) and tolerance to drying injury (Bdliya and Burris, 1988). These differential expressions can be attributed to epigenetic phenomena such as genomic imprinting (parent-of-origin effect) and xenia (effect of pollen on the endosperm phenotype in the same generation), dosage effects (in the case of triploid tissue such as endosperm), and cytoplasmic effects (e.g. mitochondrial and chloroplast genomes). The objective of this study was to identify protein expression patterns related to cold germination (CG), a trait largely influenced by the female parent. We identified recombinant inbred lines (RILs) that are divergent for CG and produced reciprocal F_1 hybrids from them. Reciprocal F_1 hybrids that showed differential phenotypes for CG were profiled for protein expression using an open-ended platform. This approach allowed the profiling of tissues that were genetically identical but phenotypically divergent, thus enhancing the effectiveness of expression profile comparison.

Materials and Methods

Inbred screening and reciprocal hybrid generation

Recombinant inbred populations were generated from a cross between B73 and MO17. The inbred lines were derived by self-pollinating the original F_1 plants (syn0) or sib-mating the original F_1 plants for two (syn2) or four (syn4) cycles, followed by self-pollinating each line of the F_2 progeny for at least 11 generations. Screening was carried out on 771 recombinant inbreds for their ability to germinate and grow under cold conditions. Seeds were germinated in rolls of wet germination paper and incubated at 10°C for 7 days followed by 3 days at 27°C in the dark. Seedlings that had ≥ 2.5 cm shoot and ≥ 2.5 cm primary root lengths were considered germinated (Fig. 32.1). Additional

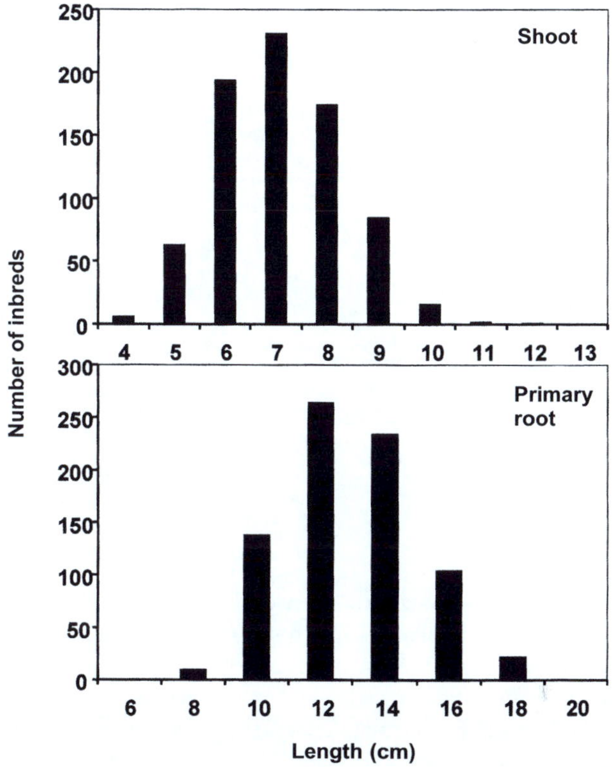

Fig. 32.2. Distribution of shoot and primary-root length (cm) of 771 recombinant inbreds following a 10-day cold germination test. Values represent means of three replicates of 20 seedlings each. Dead seeds were excluded from the measurements.

data were collected on root and shoot growth as well as overall seedling development in the cold (Fig. 32.2).

Five inbreds with high CG (> 90%) and five inbreds with low CG (≤30%) were selected and were self-pollinated or cross-pollinated reciprocally in a diallel experimental design. Seeds from the selected reciprocally crossed ears were grown and self-pollinated in the following season to test for cytoplasmic inheritance of the trait. Seeds from five self- or reciprocally cross-pollinated ears from each line were evaluated for their ability to germinate at a constant low temperature (11 days at 12.5°C in the dark). Reciprocal F_1 hybrids from M0023-syn0 (high CG) and M0081-syn0 (low CG), which displayed significant divergence for CG, were chosen for proteomic analysis along with their inbred parents. Protein samples were extracted from germinated embryonic axes (Fig. 32.3). Three samples were analysed for each genotype, each representing an individual ear, and three gels were run for each sample.

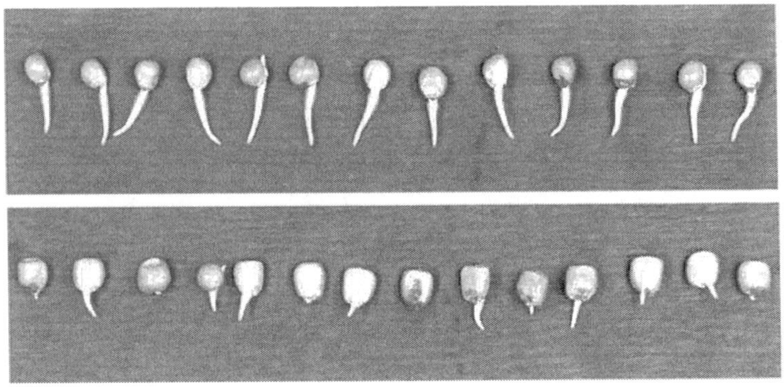

Fig. 32.3. Differential germination of reciprocal hybrids after 11 days at 12.5°C (cold germination). Top: F_1 (M0081-syn0 × M0023-syn0); bottom: F_1 (M0023-syn0 × M0081-syn0).

Protein expression profiling

Proteomic analysis, performed in collaboration with Oxford Glycosciences (Abingdon, UK), involved high-resolution separation on 2D gels, gel image analysis, robotic excision, tandem mass spectral analysis, and protein identification by querying virtual spectra generated from the protein databases (Fig. 32.4). Details for sample extraction and protein analysis were provided

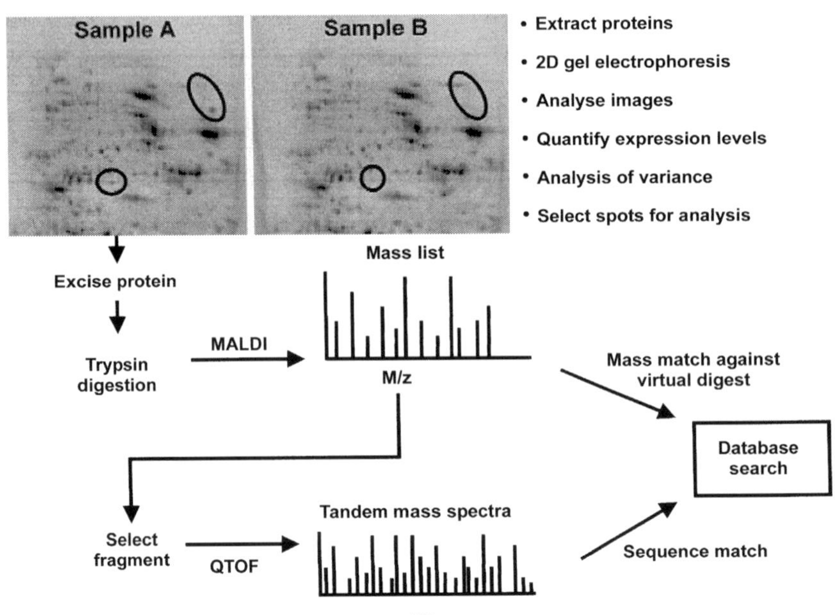

Fig. 32.4. Overview of proteomic analysis of divergence for cold germination (CG). The various steps were performed at Pioneer Hi-Bred International and Oxford Glycosciences as part of a collaboration between the two organizations.

in Kollipara *et al.* (2002). The differences in protein spot intensities among the divergent F_1 genotypes were subjected to statistical analysis of variance and spots that showed a significant difference ($P \leq 0.1$) between the reciprocal F_1s and were clearly visible in the gel images were used to calculate fold-change. Fold-change in expression was derived by taking the ratio of mean intensity of a given spot between the reciprocal F_1 samples. Proteins that showed ≥ 1.5-fold change in expression were excised from the gel by a software-driven robotic cutter and delivered in a 96-well plate for proteolysis and mass spectrometric (MS) analyses. Excised proteins were subjected to trypsin digestion and analysed by a two-stage mass spectrometry process. In the first stage, masses of the trypsin digests were estimated by MALDI-TOF (matrix-assisted laser desorption ionization-time-of-flight) mass spectrometry and the generated mass list was matched with that of the virtual digests of proteins in the databases. In the second stage, tandem mass spectra (residue masses) were generated from individual MALDI masses by a nano-electrospray ionization source and used to query public (GenPept and SWISS-PROT) and proprietary (Pioneer/DuPont and Oxford Glycosciences) databases. Hits were derived by matching the predicted ion series from a complete peptide sequence in the database with the observed fragmentation of peptides in the sample.

Results

Phenotypic analysis of recombinant inbreds and reciprocal hybrids

The recombinant inbred population showed a high degree of divergence for cold germination and seedling traits. Cold germination values ranged between 10 and 100%, with a majority of the inbreds falling in the 80 to 100% range (Fig. 32.1). Germination in the warm, however, was considerably higher for all lines. A surprisingly high number of inbreds showed very poor cold germination, below that of either parent. Low CG values can sometimes be caused by seed source (quality) effects, rather than genetic differences, which can confound investigations into the genetic control of the trait. To address this concern, the recombinant inbred population was re-screened using seeds produced in a subsequent year, and lines that showed consistently high or low CG values with both seed sources were chosen for protein analysis.

The recombinant inbred population also showed high divergence for shoot and primary root lengths at the end of the CG treatment with a two- to threefold length difference between the population tails (Fig. 32.2). In addition, the population was divergent for a number of other seedling traits, including nodal roots, lateral roots and overall seedling vigour (data not shown). The recombinant inbreds were generated from three successive synthetic generations, each with an additional cycle of recombination. Additional generations typically provide added resolution for trait/marker association studies. In this study, based on the phenotypic distribution of inbreds from the three generations, we saw no further divergence among the

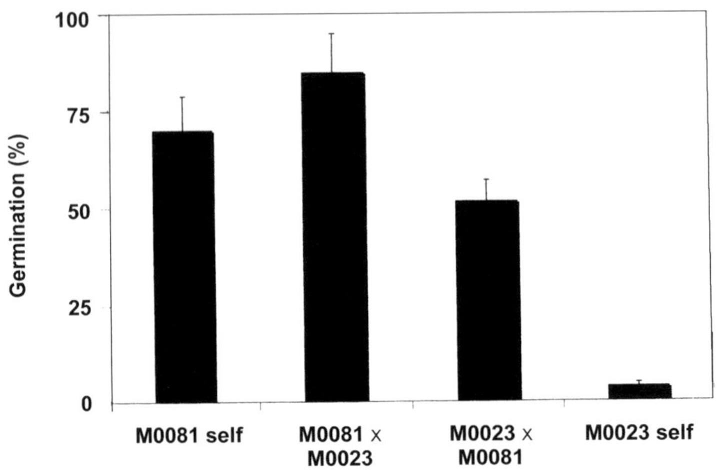

Fig. 32.5. Cold germination response of divergent recombinant inbreds (M0081and M0023) and reciprocal hybrids produced from them. Values represent the mean ± SE of three replicates of 20 seeds each.

inbreds from syn2 or syn4 compared with syn0. Five inbreds with high CG (> 90%) and five inbreds with low CG (≤ 30%) were chosen for generation of reciprocal F_1 hybrids. The hybrids showed varying degrees of CG responses, including transgressive (heterosis) as well as additive responses, compared with their self-pollinated parents. In cases where the reciprocal hybrids showed differential phenotypes, CG of hybrids with the high parent (high CG) as the female was typically higher than that of the F_1 with the low parent (low CG) as the female, indicating maternal dominance.

Of the inbred combinations tested, at least six showed maternal dominance for CG. Due to resource limitations, only one inbred pair (M0023-syn0 and M0081-syn0) and reciprocal hybrids from this pair were selected for expression profiling. The inbreds and reciprocals made from them were highly divergent for CG (Figs 32.3 and 32.5). The divergence of the reciprocal hybrids, however, was reduced compared with that of the parents. This is probably due to hybrid vigour expressed in the F_1 generation, which presumably resulted in higher CG of M0023-syn0 × M0081-syn0 compared with the selfed M0023. When the reciprocal F_1 hybrids were self-pollinated in the summer of 2000, the magnitude of phenotypic differences between them were noticeably reduced (data not shown), suggesting that the phenotypic differences observed in the F_1 generation were controlled largely by nuclear rather than cytoplasmic genes.

Proteomic analysis

Protein samples were obtained from seeds germinated at a constant low temperature (12.5°C) in the dark, a different treatment from that used for screening the inbred population for CG. This treatment was chosen to avoid the impact on cold-related gene expression of the 3-day warm (25°C) period at the end of the standard CG treatment. The phenotypic divergence between the reciprocal hybrids is shown in Fig. 32.3. Divergence between the parents was also maintained in this treatment (data not shown). Protein analysis by 2D gel electrophoresis identified a total of 2641 unique spots (proteins with different isoelectric points and molecular weight (MW) combinations) that were consistently detected above background levels (see Fig. 32.4). Analysis of variance identified 117 spots that were significantly differentially expressed ($P < 0.1$) with at least a 1.5-fold change between the reciprocal hybrids (the number of annotated proteins was smaller due to sequence coverage and/or mass spectral data quality). It was felt that a 1.5-fold threshold change (in combination with rigorous analysis of variance) was sufficiently high to avoid system 'noise' but sufficiently low to identify subtle but meaningful changes in protein levels. Nevertheless, the number of differentially expressed proteins between the divergent reciprocal hybrids was very small, probably as a result of comparing genetically identical material. This comparison increased the likelihood that the differentially expressed proteins are related to the CG trait, rather than to genetic differences between the samples. The identity of differentially expressed proteins, putative functions and expression differences between the parents and reciprocals are listed in Table 32.1. A more complete protein list and companion analysis of RNA expression are presented in Kollipara *et al.* (2002).

Discussion

Our goal was to survey various metabolic changes associated with genetic variability for CG. The recombinant inbred population exhibited wide phenotypic variation for CG, and the variation was maintained in seed from multiple years, indicating a tight genetic control over this trait. Since the recombinant inbred population was derived from a single cross (B73 × MO17 inbreds), only two possible alleles can occur for any gene at a given locus. Furthermore, all inbreds were self-pollinated for at least 11 generations, which resulted in homozygosity at most loci. Thus, the inbred population was ideal for analysis of the genetic control of complex agronomic traits such as CG. We also exploited the phenomenon of maternal dominance for CG, exhibited in divergent reciprocal hybrids, to compare isogenic lines with divergent phenotypes. The comparison of genetically identical tissues probably contributed to the small number of observed protein differences between the reciprocal hybrids and increased the chances of identifying trait-related proteins.

Divergence for CG in reciprocal hybrids has been noted for some time (Pinnell, 1949). This and other types of phenotypic variation related to parent of origin are considered as evidence for genomic imprinting, a phenomenon whose molecular mechanisms are not well understood in plants. For example, in this study, α-tubulin was found to be differentially expressed between the divergent reciprocal hybrids. Lund *et al.* (1995a,b) reported that differential accumulation of α-tubulin RNA was correlated

Table 32.1. Differentially expressed proteins associated with genetic variation for cold germination (CG) in divergent reciprocal hybrids and their inbred parents. Protein names were derived from accession databases and proposed functions were based on published reports. The complete list of annotations including isoforms, accession numbers and amino acid sequence matches and predicted pI/mass data are presented in Kollipara (2002).

Fold difference: High vs. low parent	Fold difference: High vs. low hybrid	Putative function	Protein name
12.74	13.91	Protein processing and degradation	Trypsinogen 7 [Mus musculus]
1.56	3.34	Protein processing and degradation	Cysteine proteinase Mir3 [Zea mays]
−1.18	2.85	Chaperonin	Heat shock protein 70
2.66	2.55	Methionine metabolism	5-methyltetrahydropteroyltriglutamate-homocysteine S-methyltransferase (EC 2.1.1.14)
−1.37	2.49	Chaperonin	Cytosolic chaperonin, delta-subunit
−1.05	2.13	Microtubule-associated protein	Dynactin, 150 kDa isoform
−1.66	2.06	Disease response gene acts as membrane-bound chaperone to stabilize mitochondrial proteins	Prohibitin [Zea mays]
2.42	1.88	Glycolysis / gluconeogenesis	Triosephosphate isomerase 1 [Zea mays] (EC 5.3.1.1)
1.49	1.82	Starch and sucrose metabolism and flavonoids, stilbene and lignin biosynthesis	Beta-glucosidase, chloroplast precusor [Zea mays] (EC 3.2.1.21)
−1.02	1.66	Ubiquitin-mediated protein degradation	Ubiquitin-conjugating enzyme E2 [Catharanthus roseus]
1.81	1.62	Fructose metabolism	Fructokinase
1.78	1.62	Chaperonin	Heat shock protein 82
1.92	1.61	Chaperonin / protein trafficking in ER – stress inducible	78 kDa Glucose-regulated protein precursor (GRP 78)
1.41	1.58	DNA synthesis and replication	Elongation factor 2
1.75	1.55	Microtubule-associated protein	Beta-3 & 6 tubulin [Zea mays]
1.59	1.54	Chaperonin	Heat shock protein 70
1.11	−1.51	Chaperonin	Heat shock 70 kDa protein
−1.31	−1.54	Pyruvate metabolism / cellular detoxification	Glyoxalase II (cytoplasmic isozyme, EC 3.1.2.6)
−2.03	−1.58	Glycolysis / gluconeogenesis and phenylalanine, tyrosine and tryptophan biosynthesis	Enolase 1 (EC 4.2.1.11)
−2.43	−1.63	Methionine metabolism	5-methyltetrahydropteroyltriglutamate-homocysteine S-methyltransferase (EC 2.1.1.14)
−1.74	−1.66	Glycine, serine and threonine metabolism	Hydroxymethyltransferase (EC 2.1.2.1)
1.02	−1.81	Chaperonin	Heat shock protein 70
−1.37	−1.82	Chaperonin	Chaperonin 60 [Zea mays]
1.20	−1.87	Seed storage protein	Globulin-2 precursor
−1.26	−1.94	Cold / stress responsive with unknown function	Late embryogenesis abundant protein in group 3
−1.68	−1.98	Glycolysis / gluconeogenesis and phenylalanine, tyrosine and tryptophan biosynthesis	Enolase 1 (EC 4.2.1.11)
−1.87	−2.19	Chaperone / protein trafficking in ER	Lumenal binding protein cBiPe2 [Zea mays]

Table 32.1. Continued.

Fold difference: High vs. low parent	High vs. low hybrid	Putative function	Protein name
−1.70	−2.21	Pentose and glucuronate interconversions / pyruvate metabolism	Aldose reductase (aldehyde reductase; EC 1.1.1.21)
−1.95	−2.38	Glycolysis / gluconeogenesis	Fructose-bisphosphate aldolase, cytoplasmic isozyme (EC 4.1.2.13)
−4.47	−2.51	Pentose and glucuronate interconversions / pyruvate metabolism	Aldose reductase (aldehyde reductase; EC 1.1.1.21)
−4.00	−2.55	Seed storage protein	Globulin-2 precursor
−2.80	−2.67	Glycolysis / gluconeogenesis	Aldolase [Zea mays] (EC 4.1.2.13)
−1.51	−2.68	Seed storage protein	Globulin-2 precursor
−1.38	−3.02	Cold / stress responsive with unknown function	Late embryogenesis abundant protein (LEA D-34)
−1.75	−3.12	Microtubule-associated protein	Tubulin beta-1 & 5 chain
−1.88	−4.37	Chaperone targeting ubiquitinated proteins to proteosomes for degradation	Valosin-containing protein [Glycine max]
−1.19	−4.84	Cold / stress responsive with unknown function	Late embryogenesis abundant protein (EMB564)
−4.53	−4.95	Pyruvate metabolism / cellular detoxification	Glyoxalase I (EC 4.4.1.5)

with DNA demethylation (DNA methylation is implicated in control of genomic imprinting in mammalian systems). Further, they reported differential DNA demethylation in endosperm tissue between reciprocal crosses of maize inbred lines and that the methylation state of tubulin was maternally transmitted. It is possible that the differential expressions of α-tubulin and the other proteins in our study are a result of genome imprinting for CG; however, other genetic mechanisms can also be invoked to explain these phenomena (Kollipara *et al.*, 2002).

There are few leads in the literature to the molecular mechanisms underlying chilling tolerance and, in particular, the ability of seed to germinate and grow at low temperatures. Much of the published evidence points to a role for reactive-oxygen-species detoxification enzymes in minimizing oxidative damage at chilling temperatures (e.g. Roxas *et al.*, 1997; van Breusegem *et al.*, 1998). Also, acquisition of chilling tolerance has been associated with accumulation of heat-shock proteins (HSPs) (e.g. Cabane *et al.*, 1993; Sabehat *et al.*, 1998). Severe chilling, which involves dehydration damage to cells and organelles, has been shown to be ameliorated by overexpression of transcription factors that in turn modulate biochemical changes associated with dehydration protection, such as accumulation of COR proteins, proline and soluble sugars (e.g. Kasuga *et al.*, 1999; Thomashow, 1999; Gilmour *et al.*, 2000). Proteomic analysis provided insight into the identity of genes associated with CG as well as insights into protein-level biochemical changes associated with CG that are non-transcriptionally regulated. These include post-translational modifications, proteolytic derivation or isoforms of the same protein, which might not be detected by RNA-based expression profiling. For example, enolase 1 was annotated for two different protein features with a mass difference of 680 d and a pI difference of 0.1. This suggests that the proteins are proteolitic derivatives or that they are isoforms of the same enzyme, perhaps localized in different subcellular compartments. In maize, enolase is encoded by two genes (*eno1* and *eno2*) and it has been proposed that the genes undergo post-translational regulation (Lal *et al.*, 1998).

Differentially expressed proteins associated with CG represented various biochemical pathways, including proteolysis, glycolytic/gluconeogenic pathway enzymes, cell wall structure and metabolism, polyamine metabolism, ABA/stress-induced proteins, antioxidants, components of the cytoskeleton, and chaperonins/HSPs. Several of the differentially expressed proteins in this study were previously shown to be associated with stress tolerance. These include ABA/stress-inducible proteins, heat shock proteins, late embryogenesis abundant proteins, and proteases (Yacoob and Filion, 1986; Prasad, 1996; Nieto-Sotelo *et al.*, 1999; White *et al.*, 2000). A discussion of the potential roles of various enzymes/pathways revealed by this analysis is beyond the scope of this chapter but a few observations are worth noting. The two proteins with the widest expression differences between the divergent reciprocal hybrids were trypsinogen and a cysteine protease. Other proteins associated with higher CG include chaperonins involved in protein trafficking. These proteins may play a role in the mobilization of proteins from the scutellum to meristems growing at low temperature. Alternatively, they may enhance degradation of oxidized proteins, thus minimizing cellular damage (Prasad, 1996). Interestingly, the expression of several chaperonins was inversely correlated with higher CG, indicating that the role of chaperonins in stress tolerance is a complex one. Along with chaperonins, levels of cytoskeletal proteins such as tubulin were also modulated in association with higher CG. Chaperonins/HSPs are known to bind to microtubules, influencing their synthesis as well as overall cytoskeletal function (Liang and MacRae, 1997). The role of chaperonins in the cytoskeleton and its responses to cold stress remains to be clarified, however. Glyoxalase I, an enzyme involved in cellular detoxification, accumulated at lower levels in association with CG. This is in contrast to accounts of increased antioxidant levels in maize seedlings acclimated to chilling temperature (Prasad *et al.*, 1994; Prasad, 1996) and also a report by Seki *et al.* (2001) showing increased expression of glyoxalase I in response to cold treatment. In summary, the proteomic analysis provided an overview of various biochemical changes associated with genetic variation for CG, and potential insights into the role of maternal dominance in regulating this trait. Current efforts are directed towards the identification of regulatory pathways for early sensing and acclimation to cold stress as well as mechanisms of genomic imprinting for CG.

References

Bdliya, P.M. and Burris, J.S. (1988) Diallel analysis of tolerance of drying injury in seed corn. *Crop Science* 28, 935–938.

Cabane, M., Calvet, P., Vincens, P. and Boudet, A.M. (1993) Characterization of chilling-acclimation-related proteins in soybean and identification of one as a member of the heat shock protein (HSP 70) family. *Planta* 190, 346–353.

Gilmour, S.J., Sebolt, A.M., Salazar, M.P., Everard, J.D. and Thomashow, M.F. (2000) Overexpression of the Arabidopsis *cbf3* transcriptional activator mimics multiple biochemical changes associated with cold acclimation. *Plant Physiology* 124, 1854–1865.

Groszmann, A. and Sprague, G.F. (1948) Comparative growth rates in a reciprocal maize cross. 1. The kernel and its component parts. *Journal of American Society of Agronomy* 40, 88–98.

Jaglo-Ottosen, K.R., Gilmour, S.J., Zarka, D.G., Schabenberger, O. and Thomashow, M.F. (1998) *Arabidopsis* CBF1 overexpression induces CORgenes and enhances freezing tolerance. *Science* 280, 104–106.

Janowiak, F. and Markowski, A. (1987) Effect of chilling on germination, growth, survival and membrane permeability in seedlings of different breeding forms of maize (*Zea mays* L.). *Acta Physiology Plant* 9, 77–87.

Kasuga, M., Liu, Q., Miura, S., Yamaguchi-Shinozaki, K. and Shinozaki, K. (1999) Improving plant drought, salt, and freezing tolerance by gene transfer of a single stress-inducible transcription factor. *Nature Biotechnology* 17, 287–291.

Kollipara, K.P., Saab, I.N., Wych, R.D., Lauer, M.J. and Singletary, G.W. (2002) Expression profiling of maize hybrids divergent for cold germination and desiccation tolerance. *Plant Physiology* 129, 974–992.

Lal, S.K., Lee, C. and Sachs, M.M. (1998) Differential regulation of enolase during anaerobiosis in maize. *Plant Physiology* 118, 1285–1293.

Liang, P. and McRae, T.H. (1997) Molecular chaperones and the cytoskeleton. *Journal Cell Science* 110, 1431–1440.

Lund, G., Ciceri, P. and Viotti, A. (1995a) Maternal-specific demethylation and expression of specific alleles of zein genes in the endosperm of *Zea mays* L. *Plant Journal* 8, 571—581.

Lund, G., Messing, J. and Viotti, A. (1995b) Endosperm-specific demethylation and activation of specific alleles of α-tubulin genes of *Zea mays* L. *Molecular and General Genetics* 246, 716–722.

Nieto-Sotelo, J., Kannan, K.B., Martinez, L.M. and Segal, C. (1999) Characterization of a maize heat-shock protein 101 gene, *hsp101*, encoding a ClpB/Hsp100 protein homologue. *Gene* 230, 187–195.

Pinnell, E.L. (1949) Genetic and environmental factors affecting maize seed germination at low temperatures. *Agronomy Journal* 41, 562–568.

Prasad, T.K. (1996) Mechanisms of chilling-induced oxidative stress injury and tolerance in developing maize seedlings: changes in antioxidant system, oxidation of proteins and lipids, and protease activities. *Plant Journal* 10, 1017–1026.

Prasad, T.K. (1997) Role of catalase in inducing chilling tolerance in pre-emergent maize seedlings. *Plant Physiology* 114, 1369–1376.

Prasad, T.K., Anderson, M.D., Martin, B.A. and Stewart, C.R. (1994) Evidence for chilling induced oxidative stress in maize seedlings and a regulatory role for hydrogen peroxide. *Plant Cell* 6, 65–74.

Roxas, V.P., Smith, R.K., Alleman, E.R. and Allen, R.D. (1997) Overexpression of glu-

tathione S-transferase/glutathione peroxidase enhances the growth of transgenic tobacco seedlings during stress. *Nature Biotechnology* 15, 988–991.
Sabehat, A., Lurie, S. and Weiss, D. (1998) Expression of small heat-shock proteins at low temperatures. *Plant Physiology* 117, 651–658.
Santis, A.D., Landi, P. and Genchi, G. (1999) Changes of mitochondrial properties in maize seedlings associated with selection for germination at low temperature. Fatty acid composition, cytochrome C oxidase, and adenine nucleotide translocase activities. *Plant Physiology* 119, 743–754.
Schell, L.P., Danehower, D.A., Anderson, J.R. Jr and Patterson, R.P. (1991) Rapid isolation and measurement of adenosine triphosphate levels in corn embryos germinated at suboptimal temperatures. *Crop Science* 31, 425–430.
Seki, M., Narusaka, M., Abe, H., Kasuga, M., Yamaguchi-Shinozaki, K., Carninci, P., Hayashizaki, Y. and Shinozaki, K. (2001) Monitoring the expression pattern of 1300 *Arabidopsis* genes under drought and cold stresses by using a full-length cDNA microarray. *Plant Cell* 13, 61–72.
Stewart, D.R., Martin, B.A., Geding, L. and Cerwick, S. (1990) Respiration and alternative oxidase in corn seedling tissues during germination at different temperatures. *Plant Physiology* 92, 755–760.
Thomashow, M.F. (1999) Plant cold acclimation: freezing tolerance genes and regulatory mechanisms. *Annual Revue of Plant Physiology and Plant Molecular Biology* 50, 571–599.
van Breusegem, F., van Montague, M. and Inze, D. (1998) Engineering stress tolerance in maize. *Outlook on Agronomy* 27, 115–124.
White, C.N., Proebsting, W.M., Hedden, P. and Rivin, C.J. (2000) Gibberellins and seed development in maize. I. Evidence that gibberellin/abscisic acid balance governs germination versus maturation pathways. *Plant Physiology* 122, 1081–1088.
Yacoob, R.K. and Filion, W.G. (1986) The effects of cold-temperature stress on gene expression in maize. *Biochemistry and Cell Biology* 65, 112–119.

33 Seed Lifespan and Telomeres

Daphne J. Osborne[1], Ivan Boubriak[1,3], Hugh W. Pritchard[2] and Roger D. Smith[2]

[1]Oxford Research Unit, Foxcombe Hall, Boars Hill, Oxford OX1 5HR, UK; [2]Millennium Seed Bank Project, Royal Botanic Gardens Kew, Wakehurst Place, Ardingly, West Sussex RH17 6TN, UK; [3]Institute of Cell Biology and Genetic Engineering, Kiev, Ukraine

Introduction

Most of the seeds that are of interest as crops, or as new gene sources to be kept in seed banks for future crop plant improvements, are at their highest germination potential in terms of vigour or viability when they have just matured on the parent plant. Maintained in a cool environment and in a dry state at less than 10% moisture content (MC), the lifespan of the embryo within the seed approaches the maximal; but every percentage increase in moisture during storage, coupled with a rise in temperature, lowers the integrity of the nuclear DNA and the performance of the embryo at germination. For each embryo, the balance for survival is struck between degradative and fragmentation events occurring during the length of time the seed is in storage, and the ability to repair or replace the accumulated damage once the seed is either planted or receives enough water to become metabolically active.

In poorly germinating old seeds, increasing numbers of chromosomal aberrations can be found at the first metaphase or anaphase (Navashin, 1933). These aberrations reflect the overall loss of DNA integrity in the dry seed and the lowered ability to repair DNA on imbibition. It is now known that in almost all eukaryotes there are highly conserved special reiterated tandem short oligonucleotide sequences, the telomeres, that cap and protect the ends of chromosomes. These DNA termini prevent the ends of chromatids from being recognized as DNA breaks and so stop them being repaired and rejoined into aberrant non-linear chromosome abnormalities observed at the first cell divisions of old seeds. In most plants these end telomeres consist of a 5′-$(TTTAGGG)_n$-3′ reiterated sequence.

Bucholc and Buchowicz (1992) were the first to ask the direct questions: are the protective telomere capping sequences lost from chromosomes as part of the fragmentation of DNA in ageing embryos, and is this a critical lesion in the loss of viability or vigour of a seed that is kept too long or poorly stored?

In our model system for rye (*Secale cereale*) we had already shown that the repair of single strand breaks, introduced into the nuclear DNA of embryos by gamma-irradiating them in the dry state, was one of the very earliest events to take place on the imbibition of water. Within 30–90 min, low-molecular-weight fragments are restored to high molecular weight in fresh seed, but repair is progressively less efficient in seed stored for a year or more (Elder and Osborne, 1993) and DNA fragmentation in embryos of dead seed is further exacerbated on imbibition to include the accumulation of nucleosomal multimers (Boubriak *et al.*, 2000).

Telomeres are not directly transcribed as part of chromosomal DNA replication by DNA polymerase. Instead, telomere sequences at 3′ termini of chromosomes are added by a ribonucleoprotein reverse transcriptase enzyme, telomerase, whose RNA subunit serves as a template for the reiterative addition of telomere repeats. To date, assays for telomerase activities in seeds have found them to decrease during seed development and maturation and in barley to be so low as to be undetectable in embryos in the dry state (Kilian *et al.*, 1995).

In common with other eukaryotes, plant telomeres are generally longest in cells with high rates of cell division. In meristems and in tissue cultures, telomerase activity is high but becomes reduced in tissues that are non-dividing. In developing barley embryos, telomeres are recorded as 80 kb but shorten to 30 kb at embryo maturity (Kilian *et al.*, 1995). This loss of 50 kb is considered by McKnight *et al.* (2002) to be too rapid to be accounted for by loss at each cell division.

The question arises, therefore: are the Bucholc and Buchowicz (1992) results for wheat telomeres representative of the ageing and germination of seeds of other species? If so, could telomeres be reliable molecular markers for loss of viability or vigour for seeds stored in the dry state? Further, is the loss of telomeres from chromosomes on storage just a function of general DNA fragmentation to a lower molecular weight, or are there conditions in which telomere-containing sequences are preferentially cleaved from the rest of the nuclear DNA?

We have followed these questions in our model rye system, in which viability is lost after 3–4 years in normal dry storage (< 10% MC, 8–15°C) and under conditions of accelerated ageing at 74% relative humidity and 40°C, when viability is completely lost within 13 days.

Material and Methods

Seed material

High quality seeds of rye *S. cereale* var. Rheidol with 98% germination in 24 h have been yearly gifts from Dr Peter Payne (Unilever and Monsanto) at Plant Breeding International, Cambridge, UK. Normal storage is at room temperature and *c.* 40% relative humidity, providing seed at < 10% MC. All seeds are dead by year 4. For accelerated ageing, seeds of the previous year's harvest are suspended in muslin bags over a saturated solution of NaCl in closed Kilner jars maintained at 40°C. All these seeds attain 14% MC within a day and show germination reduced to 8% by 7 days, with death for all by 13 days. Studies of seeds held in these conditions for beyond the time of seed death are included here. Embryos are excised from the seed by hand just prior to use.

Isolation of DNA

Embryos were homogenized in dry ice; the DNA was then isolated using the Annovis Phytosep DNA Magnetic isolation kit in accordance with the manufacturer's instructions. The recovered DNA was subjected to proteinase K and RNase treatment, then precipitated and washed with ethanol. DNA was quantified either by spectrophotometric analysis at 260 nm or by fluorescent evaluation of double-stranded DNA using Pico Green™ (Molecular Probes, Europe). Concentration adjustments were made to permit equal DNA loading per gel lane for subsequent electrophoresis.

Electrophoresis and membrane transfers

DNA was fractionated on neutral 0.8% agarose (Type 1, Sigma) mini-gels containing 0.0005% ethidium bromide at 45 mA for 1.5 h. Gels were washed in 10 N HCl for 15 min to partially depurinize DNA, followed by *c.* 30 min in high pH denaturing buffer before neutralizing to pH 8.5. DNA was transferred directly from the gel to Hybond N (Amersham) membrane by capillary pressure overnight. Membranes were air dried, the DNA was bound to the membrane by 1 min exposure to a UVC light source and stored preparatory to Southern hybridizations.

DNA quantifications for dot-blot hybridizations

After electrophoretic fractionations, equally DNA-loaded 0.8% agarose gels were divided into segments. The DNA in these segments was eluted using QIAGEN mini-columns as described in the QIAquick Gel Extraction Kit protocol. DNA concentrations were then adjusted to provide equal DNA concentrations in each mobility range segment from the different treatment samples run in the same gel. DNA concentration-adjusted samples (100 µl) at 1, $\frac{1}{10}$, $\frac{1}{100}$ dilutions were then transferred to Hybond N in a BioRad Dotblot module. The membrane was air dried and the DNA bound to the membrane by UVC exposure as before.

Telomere probe labelling and Southern hybridization

Telomere probe PTELC3 5′-(CCCTAAA)$_3$-3′ (Sigma-Genosys) was labelled with Gene-Images 3′-oligolabelling module kit (Amersham-Pharmacia) using fluorescein-11-dUTP and 3′OH-terminal transferase for 1.5 h.

Membranes were prehybridized in 5×SSC buffer containing 0.1% SDS and 0.5% dextran-sulphate (0.25 ml/cm) and Kit Liquid Block for 1 h at 52°C. The fluorescein-labelled probe (10 ng/ml) was added and rotation continued overnight at 52°C. Membranes were washed with first 5×SSC + 0.1% SDS, then 2×SSC + SDS and finally 1×SSC + SDS at 55°C.

Detection of fluorescein probe binding

Gene Images CDP-Star Detection Module (Amersham-Pharmacia) was used as directed by the suppliers. Light production generated by the anti-fluorescein alkaline phosphatase conjugate from the dioxetane substrate was captured on Hyperfilm-MP after exposures for up to 20 h. Film was developed in GBX Developer-Fixer (Sigma-Aldrich) and intensities of probe-labelling assessed with the UVP (UK) Image Analyser and Gelworks™ Advanced Gel Analysis Software.

Nucleosome analysis

Nucleosomes were evaluated in DNA extracts using the Oncogene Research Products Calbiochem, USA, immunological ELISA assay kit for recognition of free histone H_3 DNA termini as described in the kit protocol. Nucleosome relative units are calculated from the amount of nucleosomes produced from UV-irradiated Daudi cells as described in kit standards. Assays are the mean of three replicate samples. Sample replicate differences did not exceed 2%.

Results

We have shown that the DNA of embryos of rye seed stored for both 2 or 7 years at 10% MC (32% and 0% viable, respectively) lose telomere sequences from high-molecular-weight DNA over time, with an accumulation of telomere hybridizing sequences in low-molecular-weight fragments (Fig. 33.1).

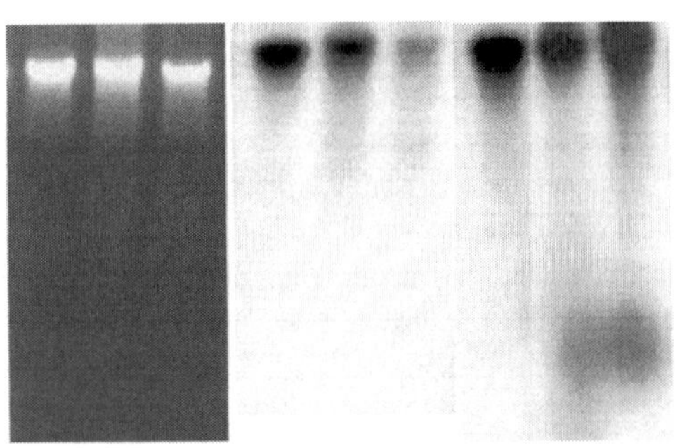
(A) (B) (C)

Fig. 33.1. (A) Gel and (B) Southern blot, telomere probe showing high-molecular-weight DNA binding (0.26 μg DNA/lane); (C) Southern blot, telomere probe showing high- and low-molecular weight DNA binding (0.65 μg DNA/lane). Left to right: viable (98%); low viability (32%); non-viable (0%).

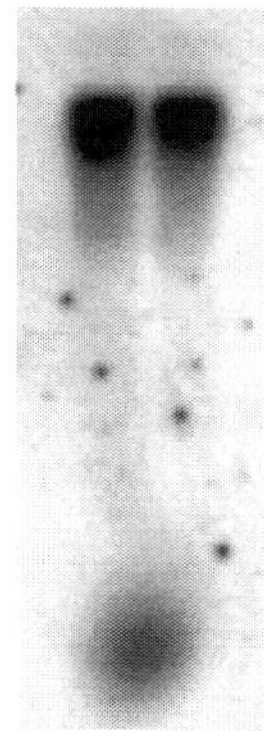

Fig. 33.2. Southern blot with telomere probe binding. Left to right: low viability (78%) dry; low viability (78%) 3 h imbibed in water (0.76 μg DNA/lane).

In a 1-year stored rye, with 78% viability, recovery of low-molecular-weight telomere fragments to high molecular weight DNA can be visualized by comparison of the pattern of probe hybridization with DNA from the dry embryo and with DNA after 3 h imbibition of embryos in water (Fig. 33.2). So essentially what Bucholc and Buchowicz (1992) found for wheat appears to be true also for rye.

Comparison of telomere abundance in DNA from fresh seed, from normally aged dead and from 7-day accelerated aged seed embryos shows some loss of telomeres for the first 7 days under conditions of high temperature and high humidity compared with normal ageing (Fig. 33.3A), but incubation extending to 72 days leads to major loss of telomeres from DNA with almost no accumulation in the low-molecular-weight fractions (Fig. 33.3B).

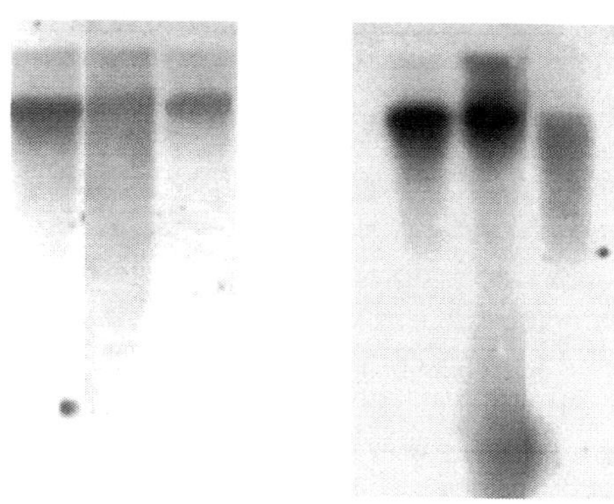

(A) (B)

Fig. 33.3. (A) Southern blot, telomere probe binding showing high-molecular-weight DNA (0.21 μg DNA/lane). Left to right: viable (98%); non-viable (0%); 7-day accelerated aged (8%). (B) Southern blot, telomere probe showing high- and low-molecular-weight DNA (0.69 μg DNA/lane). Left to right: viable (98%); non-viable (0%); 72-day accelerated aged (0%).

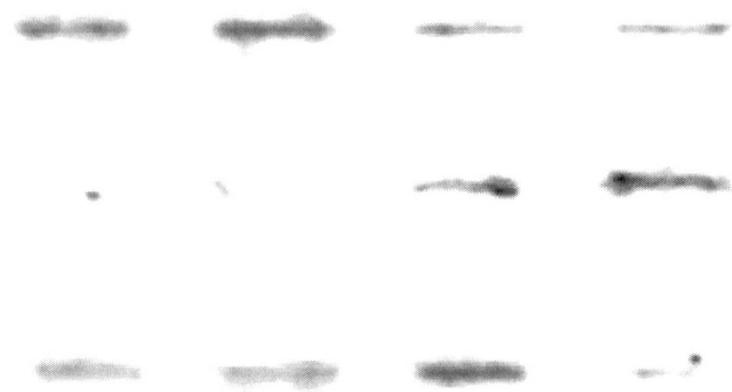

Fig. 33.4. Dot-blot of equal DNA contents for three mobility regions. Left to right: viable (98%); low viability (43%); non-viable (0%); 22-day accelerated aged (0%). Upper row, DNA > 20 kb; middle row, DNA 2–4 kb; lower row, DNA 0.5–1 kb.

Under accelerating ageing conditions, therefore, the overall long-term shift of detectable telomeres in the DNA fragments from high to low molecular weight and their accumulation at low molecular weight, as found for natural ageing, is no longer observed. Instead, there appears to be an overall deletion of telomeres from the chromosomal DNA. This could be due to the loss of very small polynucleotide sequences during the original DNA extraction procedures or to a greater instance of telomere sequence deletion per unit DNA in the 14% compared with 10% MC seed.

Preliminary experiments seeking confirmation of these differences in cell death patterns and telomere abundances at the two levels of storage moisture content have been carried out by dot-blot analysis. Again, loss of telomeres from the high-molecular-weight DNA is evident, but particularly so in the 22-day accelerated aged dead material, where the greatest telomere abundance occurs at intermediate DNA mobilities, not in small DNA fragments – a pattern that appears to be distinct from that found in the normally aged seed embryo DNA, where preferred telomere accumulation is at the lowest mobilities (Fig. 33.4).

Another difference in embryo ageing can be seen in the pattern of nucleosome formation. At < 10% MC, nucleosomes are not formed during fragmentation of DNA, even in embryos of seeds long dead. If held at 14% MC, however, nucleosome multimers increase until after seed death, and if seed is then returned to < 10% MC these multimers remain stable (Boubriak *et al.*, 2000). Table 33.1 shows that if seed continues to be held at high moisture content, nucleosomes will undergo further degradation and by 72 days are essentially eliminated.

Discussion

The loss of telomeres in ageing embryos is shown by the overall reduction in telomere hybridization at high molecular weight (Fig. 33.3). This loss of telomeres on accelerated ageing together with the appearance of nucleo-

Table 33.1. Nucleosome levels in embryos of 98% viable rye seed during accelerated ageing at 40°C, 14% MC. Controls at 40°C, 9% MC.

Duration of incubation	Relative nucleosome units
0 h	0.07
24 h	0.10
3 days	0.25
7 days	0.37
14 days	0.54
72 days	0.00

somes (Table 33.1) may be part of the temporal difference between the two forms of cell death found at the two different moisture contents examined. The overall loss of nucleosomes on continued incubation of dead seed at high temperature and humidity conditions compared with the stability of these nucleosomes in similar seeds when they are transferred back to < 10% MC indicates the differential impact that relatively small changes in hydration levels can have upon the integrity of nuclear DNA (Table 33.1 and Boubriak *et al.*, 2000).

Hybridization of DNA extracted from different mobility fractions following gel electrophoresis, then quantified and analysed by dot-blot hybridizations, confirms the preferential loss of telomeres from the high-molecular-weight DNA of embryos of 22-day accelerated aged seed and their accumulation at the intermediate mobilities. This distribution (Fig. 33.4) and the limited hybridization at low molecular weight suggests a progressive elimination of telomeres by nuclease attrition which may be linked to, or be in parallel with, the eventual loss of nucleosomes. In this respect, the presence of interstitial telomere sequences along the chromosomal DNA may play an important part in a greater nuclease degradation of DNA at 14% MC than at < 10% MC.

The experiments of Bucholc and Buchowicz (1992) with 94% viable wheat and our results with the 78% viable rye show that telomeres lost from high-molecular-weight DNA on storage are restored to chromosomal DNA within 6 h and 3 h imbibition, respectively, that is, before the initiation or completion of S-phase. An interesting question concerns how telomeres are replaced at very early germination when assays for telomerase to date suggest that the enzyme may not be present in the dry seed (Kilian *et al.*, 1995). Increased telomerase activity in other plant tissues has been associated with the S-phase of the cell cycle (Tamura *et al.*, 1999) and so it may be that telomerase is neither reactivated nor newly synthesized until the first S-phase of germination. Extra-chromosomal telomere-containing mini-fragments or mini-circles of DNA may therefore play a critical role in the regaining of telomere multiples at the very early times before S-phase, as considered by Buchowicz (1997) and more recently for yeast by Tomaska *et al.* (2000).

In seeking a molecular marker for loss of viability of seeds in store, it is clear that an assessment of nucleosome multimers would not be useful, for in rye, at least, they are formed only in embryos that die at elevated moisture contents. However, telomere loss does appear to follow the temporal loss of viability, even though the pattern of cell death may be differently determined by the level of hydration that the embryo experiences.

References

Boubriak, I., Naumenko, V., Lyne, L. and Osborne, D.J. (2000) Loss of viability in rye embryos at different levels of hydration: senescence with apoptotic nucleosome cleavage or death with random DNA fragmentation. In: Black, M., Bradford, K.J. and Vazquez-Ramos, J. (eds) *Seed Biology: Advances and Applications*. CAB International, Wallingford, UK, pp. 205–214.

Bucholc, M. and Buchowicz, J. (1992) Synthesis of extrachromosomal DNA and telomere-related sequences in germinating wheat embryos. *Seed Science Research* 2, 141–146.

Buchowicz, J. (1997) Nuclear extrachromosomal DNA of higher plants. *Acta Biochimica Polonica* 44, 13–20.

Elder, R.H. and Osborne, D.J. (1993) Function of DNA synthesis and DNA repair in the survival of embryos during early germination and in dormancy. *Seed Sciences Research* 3, 43–53.

Kilian, A., Stiff, C. and Kleinhofs, A. (1995) Barley telomeres shorten during differentiation but grow in tissue culture. *Proceedings of the National Academy of Sciences USA* 92, 9555–9559.

McKnight, T.D., Riha, K. and Shippen, D.E. (2002) Telomeres, telomerase and stability of the plant genome. *Plant Molecular Biology* 48, 331–337.

Navashin, M.S. (1933) Ageing of seeds is a cause of chromosomal aberrations. *Planta* 20, 233–243.

Tamura, K., Liu, H. and Takahashi, H. (1999) Auxin induction of cell cycle regulated activity of tobacco telomerase. *Journal of Biological Chemistry* 30, 20997–21002.

Tomaska, L., Nosek, J., Makhov, A.M., Pastorakova, A. and Griffith, J.D. (2000) Extra genomic double-stranded DNA circles in yeast with linear mitochondrial genomes: potential involvement in telomere maintenance. *Nucleic Acids Research* 28, 4479–4487.

34 Prediction of Seed Longevity: a Modified Model and Improved Fitting Process

ANDREW MEAD AND DAVID GRAY

Plant Establishment and Vegetation Management Research Team, Horticulture Research International, Wellesbourne, Warwick CV35 9EF, UK

Introduction

Reliable estimates of seed longevity are important both for the management of germplasm collections and for the management of commercial seed production and storage. Working with cereal grains, Roberts and Ellis (Roberts, 1960, 1972, 1973; Ellis and Roberts, 1980a,b) developed equations to predict seed longevity, which are now widely used. The 'improved' model (Ellis and Roberts, 1980a) relates probit percentage viability, v, to storage duration, d (in days), as

$$v = K_i - \left(\frac{1}{\sigma}\right) d \qquad (1)$$

where K_i is the initial probit percentage viability and σ is the standard deviation of the distribution of seed deaths in time.

There are three assumptions behind this model: (i) that the frequency of seed deaths in time under constant storage conditions can be described by a normal distribution; (ii) that the initial viability depends only on species and seed quality; and (iii) that the rate of loss of viability depends only on the storage temperature and moisture content. The modified model described in this chapter addresses the first of these three assumptions, though it impacts also on the second.

The form of equation (1) suggests that viability is 100% at some time prior to storage, so that differences in the initial viability of seed lots are accounted for simply by shifting the viability curve by a fixed amount along the storage duration axis, as shown in Fig. 34.1A. For the curves shown, a reduction in initial viability from 100% to 60% results in a reduction in the time to 50% viability from 450 days to 50 days. This form of the model will be appropriate where the seed lot was originally 100% viable but where deterioration has caused a loss of viability prior to testing and storage.

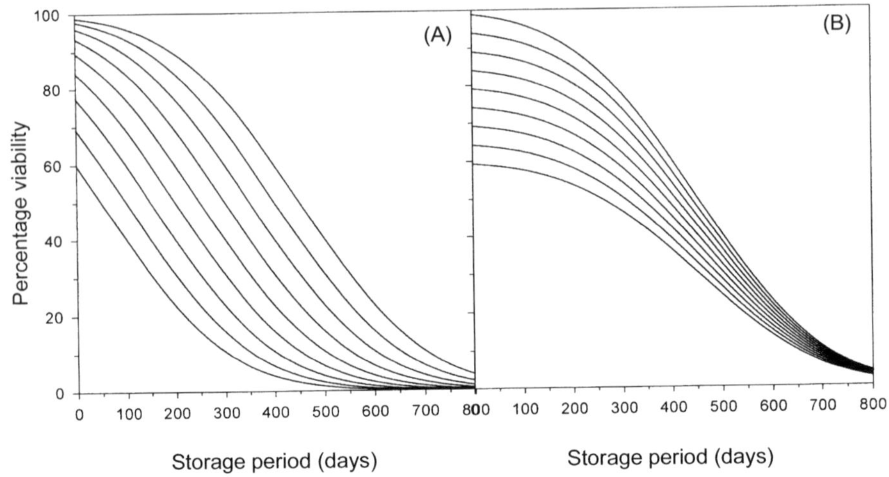

Fig. 34.1. Sets of simulated viability curves constructed following (A) the original Ellis and Roberts model and (B) the modified 'control viability' model. As originally presented in Mead and Gray (1999).

However, the fitting of this model to observed data sets can result in a systematic lack-of-fit, even for seed lots with a relatively high initial viability. This is illustrated later in this chapter in Fig. 34.4, which demonstrates the 'improved' model (Ellis and Roberts, 1980a). Here the model overestimates the viability for storage periods up to about 50 days, and underestimates the viability between 80 and 120 days. Although not shown in the figure, it will also underestimate the viability for storage periods greater than 200 days. This systematic lack-of-fit will occur where the original viability of the seed lot is less than 100%.

The problem with the model is that seeds that are not viable at the start of storage are assumed to be part of the normal distribution of seed deaths in time. However, these initially non-viable seeds can be divided into those that were never viable and those that have lost viability due to the pre-storage conditions. Usually little is known about the events contributing to this non-viability and therefore it is not possible to determine what proportion of the seed lot fall into each group. But the model can be modified to allow for and estimate the initial proportion of the seed lot that was never viable.

One possible solution, suggested by Wilson *et al.* (1989), is to re-scale the observations so that the initial percentage viability is 100%, but this assumes a known, rather than estimated, initial viability. In addition, this solution allows the possibility of observations greater than 100%, which causes problems in fitting the model. Our solution, based on the 'control mortality' probit model used in insecticide bioassays, includes an additional parameter to estimate the original viability of the seed lot. Under this model, the slope and intercept of the response are independent of this original viability, and the time to reach 50% of the original viability is constant for given storage conditions (Fig. 34.1B). Note that this modified model can still allow for a loss of viability due to deterioration prior to testing and storage.

Development of the 'Control Viability' Probit Model

The 'control mortality' probit model for insecticide bioassays was originally proposed by Finney (1971) to allow for the possibility that even when a zero dose of insecticide is applied to a population of insects, some proportion of the population, c_M (referred to as the control mortality), will die. Thus the overall proportion that die at a particular dose is the sum of the proportion, c_M, that die naturally, and a proportion p (given by the probit function) of the remaining proportion $(1 - c_M)$, a result known as Abbott's formula (Abbott, 1925).

The parallel between this model and the seed viability problem is more obvious if we consider the percentage of non-viable seeds rather than the percentage of viable seeds. This is then an increasing response curve, with a percentage of seeds that are non-viable at the start of storage. Algebraically this can be written as

$$\%non\text{-}viability = 100 \times \left\{ c_{NV} + (1 - c_{NV}) \times \Phi \left[- \left(K_i - \frac{d}{\sigma} \right) \right] \right\} \qquad (2)$$

where c_{NV} is the control non-viability, i.e. the original non-viability of the seed lot, $\Phi()$ denotes the cumulative normal function (i.e. the inverse of the probit function), and the other symbols are as defined for equation (1), except that K_i and σ relate to the population of originally viable seeds.

Simple manipulation of equation (2) produces a model for the percentage of viable seeds:

$$\%viability = 100 - 100 \times \left\{ c_{NV} + (1 - c_{NV}) \times \Phi \left[- \left(K_i - \frac{d}{\sigma} \right) \right] \right\}$$

$$= 100 \times (1 - c_{NV}) \times \left\{ 1 - \Phi \left[- \left(K_i - \frac{d}{\sigma} \right) \right] \right\}$$

This can be simplified by writing $c_V = 1 - c_{NV}$, and using the fact that $1 - \Phi(-X) = \Phi(X)$, to give

$$\%viability = 100 \times c_V \times \Phi \left(K_i - \frac{d}{\sigma} \right) \qquad (3)$$

This model is equivalent to the model proposed by Ellis and Roberts (1980a) when $c_V = 1$, i.e. if the original viability is 100%. Equation (1) can then be obtained by dividing the percentage viability by 100 and applying the probit function to the left-hand side of the equation rather than the cumulative normal function to the right-hand side. The additional parameter, c_V, now estimates the original proportion of viable seeds, with K_i still estimating the loss of viability pre-storage.

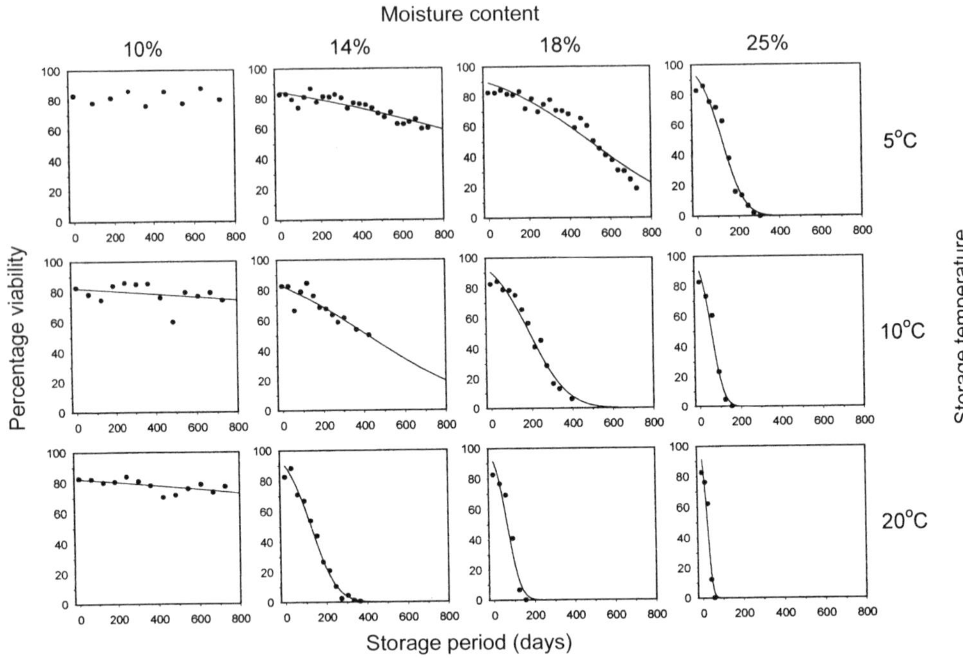

Fig. 34.2. Fitted curves for the Ellis and Roberts model (equation 1). Redrawn from Figure 2 of Mead and Gray (1999).

Comparison of Models

A comparison of the two models was made using data from a storage experiment on carrot seed. This experiment included three replicates of each of 12 storage treatments, comprising all combinations of four moisture contents (10%, 14%, 18% and 25%) and three temperatures (5°C, 10°C and 20°C). The seeds were stored for periods of between 20 weeks and 2 years, with between eight and 24 samples taken for each treatment combination. Full details of the experimental materials and methods are given in Mead and Gray (1999).

The two models were fitted to the data using the generalized linear and non-linear model facilities in GenStat for Windows (Lane and Payne, 2000; Payne, 2000). The fitted curves for the standard model are shown in Fig. 34.2 and those for the modified 'control viability' model are shown in Fig. 34.3. For the standard model the same over- and underestimation of viability can be seen as described earlier, with the estimated initial viability varying from 82% to 92%, the lower values at low moisture content and higher values at high moisture content. In contrast, there is no systematic lack-of-fit for the 'control viability' model, with the estimated initial viability varying from 79% to 84%, with no obvious relationship with temperature or moisture content.

A formal statistical comparison of the models is given in Table 34.1, the mean residual deviance measuring the goodness-of-fit for the whole curve, and the F-test assessing the improvement in fit provided by the inclusion of

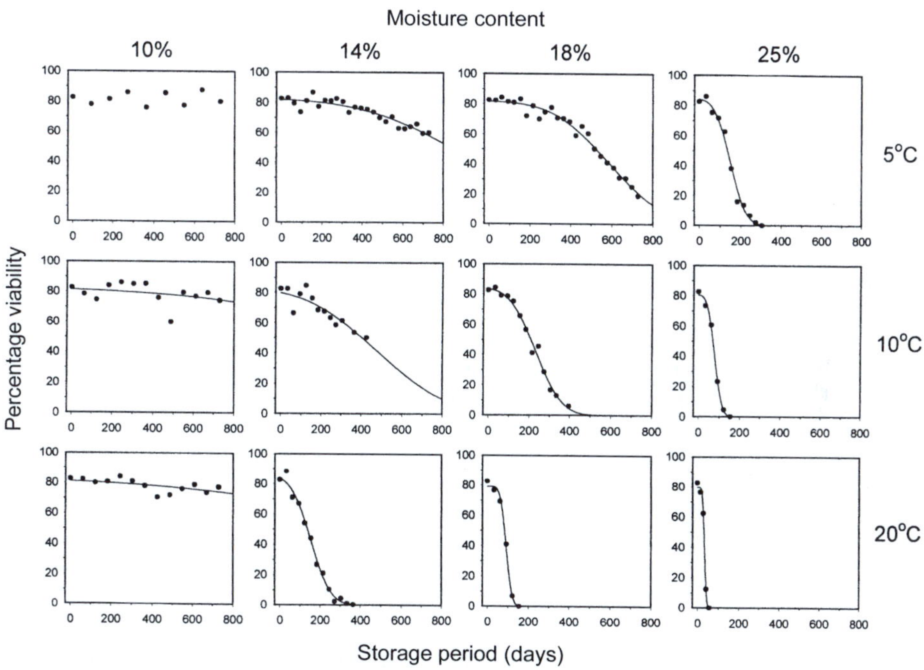

Fig. 34.3. Fitted curves for the modified 'control mortality' model (equation 3). Redrawn from Figure 3 of Mead and Gray (1999).

the extra parameter in the 'control viability' model in comparison with the standard model. In most cases the 'control viability' model gives a significant improvement, and in only one case (20°C, 10%) does this model provide a marginally poorer fit than the standard model.

Modelling the Effects of Storage Conditions

Having developed the modified 'control viability' model to allow for the possibility of an original seed lot viability of less than 100%, the next step is to unify the descriptions provided by this model for a range of storage conditions. As for the standard model developed by Ellis and Roberts (1980a), this can be achieved by modelling the fitted parameters in terms of the storage temperature and moisture content. Following the arguments of Ellis and Roberts (1980a), it still seems reasonable to assume that the intercept parameters are affected only by species, seed lot and pre-storage conditions, whilst the slope parameter, as given by σ, is affected only by the storage conditions.

Taking the additional parameter, c_V, to estimate the original proportion of viable seeds in the population, it would seem reasonable to assume that this is only affected by species and seed lot. The probit intercept parameter, K_i, which in combination with this additional parameter

Table 34.1. Statistical comparison of the modified 'control viability' model with the standard Ellis and Roberts model.

		Mean residual deviance		
Temperature	Moisture content (%)	'Control viability' model	Standard model	F-test for improvement
5°C	14	1.367	1.507	0.001
5°C	18	0.988	2.050	< 0.001
5°C	25	1.586	3.343	< 0.001
10°C	10	4.350	4.250	0.432
10°C	14	2.475	2.414	0.650
10°C	18	1.100	2.113	< 0.001
10°C	25	0.677	3.692	< 0.001
20°C	10	1.900	1.808	N/A
20°C	14	1.635	1.935	0.001
20°C	18	1.066	7.065	< 0.001
20°C	25	0.881	8.348	< 0.001

gives the viability at the start of storage, is then just affected by the pre-storage conditions, and thus represents the loss of viability prior to the start of storage.

Ellis and Roberts (1980a) proposed two possible relationships between the probit slopes and storage conditions:

$$\log(\sigma) = K_L - C_1 \times m - C_2 \times t \tag{4}$$

$$\log(\sigma) = K_E - C_W \times \log(m) - C_H \times t - C_Q \times t^2 \tag{5}$$

where m represents the storage moisture content (% fresh weight) and t the storage temperature. The choice of equation is discussed by Ellis and Roberts (1980a) and it is reasonable to assume that these same relationships will hold for our modified model.

The standard approach to estimating the values of the parameters for either of equations (4) or (5) uses two steps, first estimating values of σ for a range of storage conditions and then regressing these values on the appropriate values of storage moisture content and temperature. As the initial viability must be the same for all storage conditions, the values of the two intercept parameters, c_V and K_i, need to be constant across all storage conditions. Note that the values of these two parameters cannot be estimated simply from the observed initial viability, but that single values for these two parameters must be simultaneously estimated from the combined data set whilst allowing probit slopes (values of σ) to be estimated individually for each storage environment. This can be achieved relatively easily using the generalized non-linear model fitting facilities in GenStat for Windows. Having obtained estimates of the individual probit slopes under this constrained model, either of equations (4) or (5) can then be fitted to the estimated values of σ using simple linear regression.

A major criticism of this two-step approach is that the information about the variability associated with individual estimates of σ is lost or ignored in the linear regression step. The estimates of σ are found to minimize the variability around each fitted probit curve, and the values of the parameters in equation (4) or (5) are found to minimize the variability of these estimates of σ around the appropriate model. This set of parameter estimates will not minimize the overall variability around the combined set of probit curves, and the response curve for a particular treatment might not fit the observed responses particularly well.

A better approach would be to find the combined set of parameter values – both the two intercept parameters and the parameters in equation (4) or (5) – that minimizes the variability of the observed data about the complete set of response curves. Such an approach would also provide sensible standard errors for estimates of σ based on the true variability associated with such estimates for individual storage conditions, rather than the artificial 'between-treatment' variability.

The One-Step Fitting Process

The combined model can be written algebraically as

$$\%\text{viability} = 100 \times c_V \times \Phi\left(K_i - \frac{d}{10^{\,K_E - C_W \times \log(m) - C_H \times t - C_Q \times t^2}}\right) \qquad (6)$$

for the 'control viability' model with the log-linear effect of moisture content and quadratic effect of temperature. Similar equations can be constructed to include the original Ellis and Roberts model or the linear effects of the storage conditions.

Recent advances in statistical methodology, and particularly in statistical computing, mean that such combined models can now be fitted in a single minimization process. This can be achieved using the generalized non-linear modelling facilities in GENSTAT for Windows, using a series of expressions to build up the model, and assuming a binomial error distribution and probit link function (GENSTAT code is available from the first author).

A possible disadvantage of this one-step approach is the need to provide sensible initial values for each of the parameters before starting the minimization process. However, these can usually be obtained by plotting estimates of σ obtained from the first step of the two-step process against the storage temperature or moisture content. Models can then be fitted to subsets of storage conditions, such as particular storage temperatures or storage moisture contents, using these subsets to provide initial values to be used when finally fitting the full model to the combined data set.

Figure 34.4 shows the fitted (predicted) curves for the carrot data from the one-step fitting process, in this case using the 'control viability' model with equation (4) for the effects of moisture content and temperature and having omitted the 10% moisture content treatments from the fitting process. These three treatments were omitted because the limited loss of viability

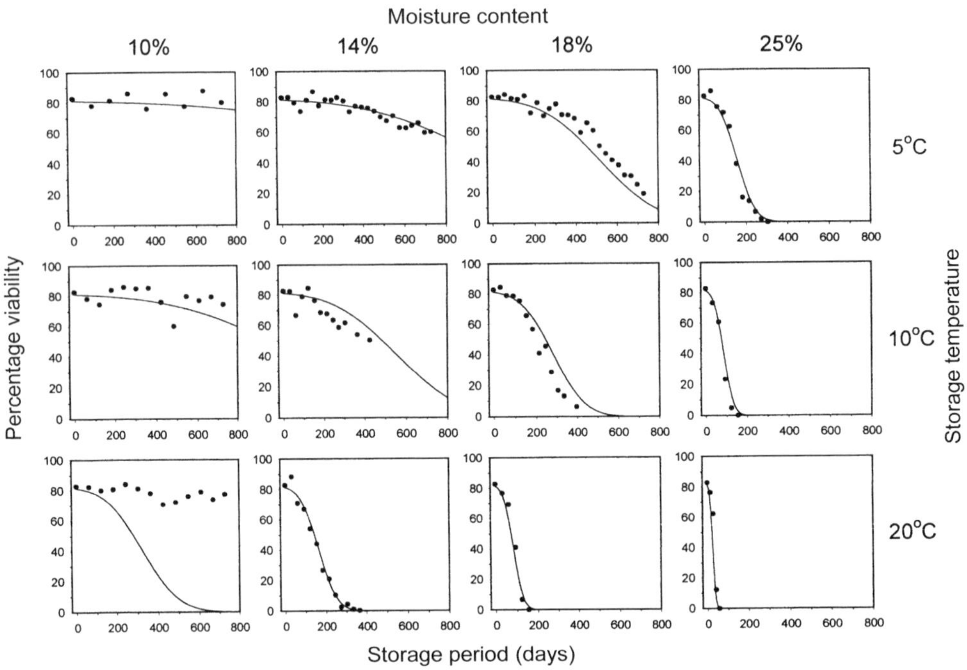

Fig. 34.4. Fitted curves from the one-step analysis of the combined data set, using the modified 'control viability' model with equation (4) for the effects of storage conditions, and fitted omitting the data for the 10% moisture content treatments.

(viability still > 70% after 2 years storage) meant that the values of σ for these treatments were poorly estimated. The simpler model for the effects of storage conditions was chosen because of the limited range of both storage temperature and moisture content assessed in the experiment. The fitted (predicted) curves show reasonable agreement for all storage treatments except 20°C, 10%, for which all models and approaches predict a more rapid loss of viability than that observed.

A further advantage of using the one-step fitting process is that, in addition to providing correct standard errors (confidence intervals) for σ, it is also possible to calculate confidence intervals for the storage duration for viability to drop to some specified level, such as 50%. Interestingly, confidence intervals for σ calculated from the one-step analysis always appear to be smaller than those calculated from the two-step approach.

Conclusions

The standard Ellis and Roberts viability model often shows systematic lack-of-fit, particularly when initial viability is less than 100%. The 'control viability' modification introduces an extra parameter to estimate the original viability of the seed lot, retaining the K_i parameter to estimate the pre-storage loss of viability. The models proposed by Ellis and Roberts for the effect of storage temperature and moisture content on σ are also retained. The modified model provides a better fit to the observed data in this study, and a more general description of the loss of viability during storage.

Advances in statistical methodology and computing now allow the full viability model, with the model for the effect of storage conditions on σ contained within the probit response model, to be fitted in a single minimization step. This provides the advantage over the previously used two-step approach of providing a more robust analysis by simultaneously estimating all model parameters to minimize the overall variability about the fitted response curves.

Together, these two advances improve both the accuracy and precision of predictions of seed longevity during storage.

References

Abbott, W.S. (1925) A method for computing the effectiveness of an insecticide. *Journal of Economic Entomology* 18, 265–267.

Ellis, R.H. and Roberts, E.H. (1980a) Improved equations for the prediction of seed longevity. *Annals of Botany* 45, 13–30.

Ellis, R.H. and Roberts, E.H. (1980b) The influence of temperature and moisture on seed period in barley (*Hordeum distichum* L.). *Annals of Botany* 45, 31–37.

Finney, D.J. (1971) *Probit Analysis*, 3rd edn. Cambridge University Press, London, 333 pp.

Lane, P.W. and Payne, R.W. (2000) Regression analysis. In: Payne, R.W. (ed.) *The Guide to GenStat, Part 2: Statistics*. VSN International, Oxford, UK, pp. 79–234.

Mead, A. and Gray, D. (1999) Prediction of seed longevity: a modification of the shape of the Ellis and Roberts seed survival curves. *Seed Science Research* 9, 63–73.

Payne, R.W. (2000) PROBITANALYSIS. In: Payne, R.W. and Arnold, G.M. (ed.) *Genstat Release 4.2 Reference Manual, Part 3: Procedure Library PL12*. VSN International, Oxford, UK, pp. 260–264.

Roberts, E.H. (1960) The viability of cereal seed in relation to temperature and moisture. *Annals of Botany* 24, 12–31.

Roberts, E.H. (1972) Storage environment and the control of viability. In Roberts, E.H. (ed.) *Viability of Seeds*. Chapman and Hall, London, pp. 14–58.

Roberts, E.H. (1973) Predicting the storage life of seeds. *Seed Science and Technology* 1, 499–514.

Wilson, D.O.J., McDonald, M.B.J. and St Martin, S.K. (1989) A probit planes method for analysing seed deterioration data. *Crop Science* 29, 471–476.

35 Desiccation Rate, Desiccation Response and Damage Accumulation: Can Desiccation Sensitivity Be Quantified?

N.W. Pammenter, Sasha Naidoo and Patricia Berjak

School of Life and Environmental Sciences, University of Natal, Durban, 4041 South Africa

Introduction

Recalcitrant seeds are desiccation sensitive and have a short storage lifespan, even in the hydrated state. It is generally agreed that there are 'degrees' of recalcitrance, but it is not always clear whether this refers to the extent of dehydration that can be tolerated, or to the storage lifespan. Comparisons among species are further complicated by the fact that drying conditions, particularly drying rate, influence the response to dehydration: material that is dried rapidly can survive to lower water contents than that dried slowly. Desiccation damage accumulates in material that is dried slowly, while rapidly dried material reaches low water contents before lethal damage can accumulate (reviewed by Pammenter and Berjak, 1999). (It should be pointed out that rapid drying does not induce desiccation tolerance; such material remains viable for only a short time; Walters *et al.*, 2001). To be able to assess, and so rank, the degree of recalcitrance of seeds, it is necessary to account for the confounding issues of water content and time (stress intensity and stress duration). This chapter proposes and assesses a simple model designed to achieve this.

The model

During drying, the water content of seed tissue will decrease in a non-linear manner (Fig. 35.1A). Damage will accumulate until a point (t_d) at which loss of viability (or some other precise measure of severe damage) occurs. The hypothesis is that the hatched area above the curve (between the curve and the horizontal line corresponding to the initial water content) up to time t_d, integrates the combined effects of water loss and time on the

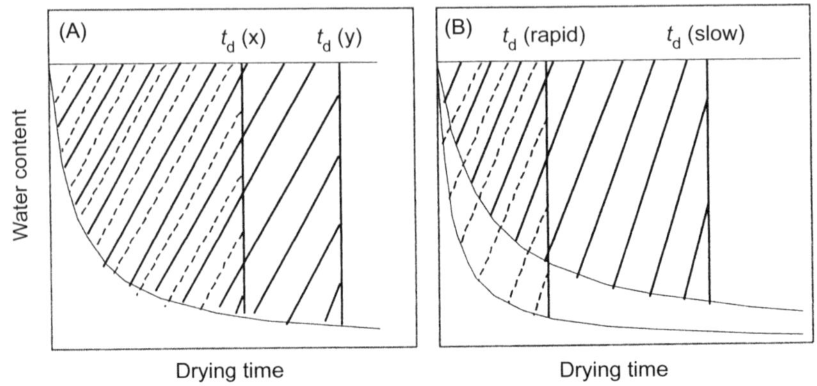

Fig. 35.1. Diagrammatic representation of a typical drying time course. The hatched area above the curve represents the integration of the effects of water loss and time. (A) Species 'x' is more sensitive and so suffers damage at a higher water content and after shorter drying (t_d (x)), than species 'y' (t_d (y)). The area above the curve to t_d(x) is less than that to t_d(y). (B) Slower drying gives rise to damage at higher water contents but after longer drying times, relative to rapid drying. The area above the rapid drying curve is less than that above the slow drying curve.

damage that has accumulated. Thus this area can be used as a measure of recalcitrance. If the area for seeds of species 'x' is less than for seeds of species 'y', then species 'x' is more recalcitrant; the seeds can tolerate less damage and they will die at a higher water content and/or after a shorter drying time.

However, the rate of drying affects the response to desiccation: severe damage accumulation will occur at a higher water content, but later in time, with slower drying rates (Fig. 35.1B). Thus the area above the curve will vary with drying rate, and is invariably greater with slower drying. Thus, to compare between species, it is necessary to consider the differences in the patterns of change in area with drying rate (Fig. 35.2). It is suggested that the

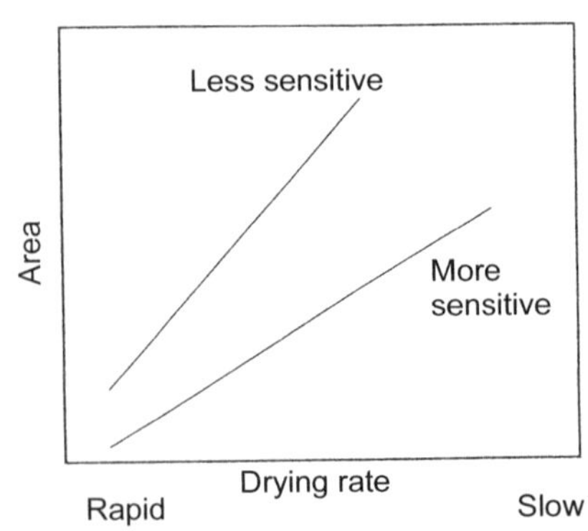

Fig. 35.2. The response of the area above the drying curve to drying rate, for more and less sensitive species, as predicted by the model.

more recalcitrant a seed, the shallower will be the slope of the curve relating drying rate to the integrated area. To construct such a model requires: (i) an objective single descriptor of drying rate, and (ii) an unambiguous definition of the point on the drying curve at which severe damage is deemed to have accumulated.

Single objective descriptor of drying rate

Most recalcitrant seeds are too large to achieve the rapid drying rates required to test the model. Thus excised embryonic axes make suitable experimental material. If a range of drying rates is generated and a suitable function that contains only one constant can be found to describe all the drying rates, the constant of the function can be used as an objective single descriptor of drying rate. Although an exponential function has been found to fit some data sets (Liang and Sun, 2000), it is our experience that in most cases initial drying is considerably faster than described by this function. However, the modified inverse function $wc = a_* b/(b + t)$, where wc is water content (g water/g dry mass), t is time (h) and a and b are constants, was found to fit the data reasonably well in many of our studies; a is the water content at time zero and b is the time required to dry the tissue to half its original water content and describes the curvature of the function. If the data are expressed as a relative water content (RWC), normalized to the initial water content, then a becomes unity and the function simplifies to

$$\mathrm{RWC} = b/(b + t) \qquad (1)$$

This function has only one constant and so b constitutes an objective single descriptor of drying rate. The value of b is determined as the inverse of the slope of a plot of 1/RWC vs. t. (Conventionally, when working with vegetative tissue, RWC is normalized to the water content of fully hydrated tissue. However, when recalcitrant (or orthodox) seeds or excised axes are placed in water to achieve 'full hydration', they continue to absorb water and start germinating. In effect, the concept of 'full hydration' is meaningless for seed material and so RWC in this study was calculated relative to the water content of the tissue at shedding.)

Unambiguous definition of the point of severe damage

The most unambiguous definition of severe damage is loss of viability. However, germinating excised axes frequently requires tissue culture techniques, which are time consuming and cumbersome when relatively large numbers are involved. Instead we measured electrolyte leakage to identify the time of drying (t_d) by which severe damage had accumulated (see below).

Calculation of area above the curve

The area above the curve can be calculated by evaluating the integral of the function up to the point identified for severe damage and subtracting this from the area below the line for RWC = 1. In the case of the function RWC = $b/(b + t)$, this reduces to

$$\text{area} = t_d - b(b + t)^{-1}dt,$$

the integral being evaluated from $t = 0$ to $t = t_d$. This simplifies to:

$$\text{area} = t_d - b\ln(1 + t_d/b) \quad (2)$$

Materials and Methods

The recalcitrant species studied were *Trichilia dregeana* Sond. (*Meliaceae*), *Ekebergia capensis* Sparrm. (*Meliaceae*) and *Quercus robur* L. (*Fagaceae*). *Q. robur* is a temperate species and its seeds are considered to be less recalcitrant than those of the other two species, which are tropical (see Finch-Savage, 1992, for data on *Q. robur*; Pammenter *et al.*, 1998, for *E. capensis*; and Kioko *et al.*, 1998, for *T. dregeana*). Additionally, seeds of the orthodox species *Pisum sativum* L. var. Greenfeast (*Fabaceae*) were imbibed for 6 h, axes excised and further imbibed on filter paper wetted to 2 g water/g dry mass for 48 h to the point at which they were losing desiccation tolerance.

Excised axes of all species were placed on grids in small chambers above silica gel or saturated solutions giving a range of relative humidities. After appropriate drying times the axes were removed, the water content of five was individually determined gravimetrically, and the electrolyte leakage from another five was measured individually after soaking in 2 ml water for 7.5 h in the wells of a multicell conductivity meter.

Results and Discussion

Although the chosen function generally fitted the drying data reasonably well (Fig. 35.3A), this was not always the case. For long drying times at high

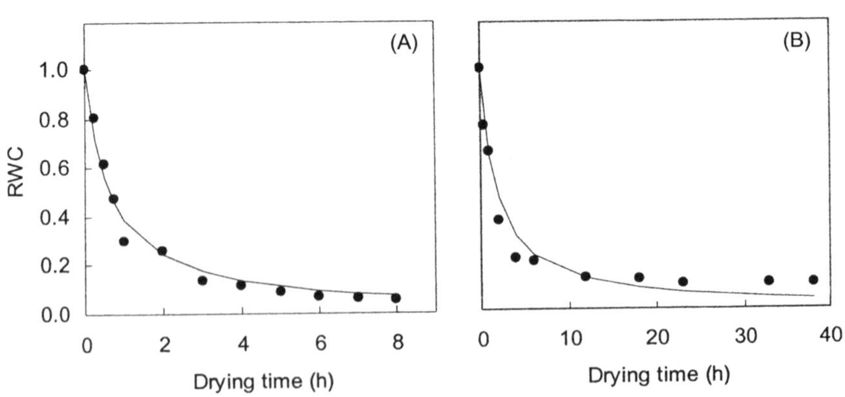

Fig. 35.3. Two examples of drying time courses achieved in this study. (A) The fit of the equation to the data is good for rapid drying (*T. dregeana*, dried at 30% RH). (B) For slower drying (*E. capensis*, dried at 62% RH) the equation overestimates RWC at short drying times and underestimates RWC at longer drying times.

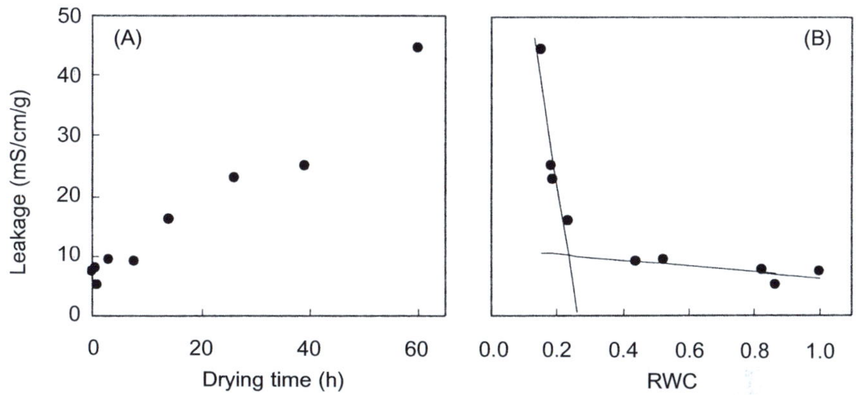

Fig. 35.4. The response of electrolyte leakage to drying (*Q. robur*, dried at 75% RH). (A) When plotted against drying time, the relationship is almost linear. (B) When plotted against RWC, a point at which a marked increase in leakage occurs can be identified.

relative humidity (RH) (e.g. *E. capensis* dried at 62%), where the axes had probably come into equilibrium with chamber RH, the fitted function sometimes underestimated RWC (Fig. 35.3B).

Estimating t_d proved to be difficult. It was hoped that, during the drying time course, a point could be identified where there was a clear marked increase in electrolyte leakage, but leakage tended to be an approximately linear function of time (Fig. 35.4A). To identify t_d, leakage was plotted as a function of RWC; this generally allowed a clear identification of the RWC at which leakage markedly increased (Fig. 35.4B) and t_d was then estimated by substituting this value of RWC into equation 1.

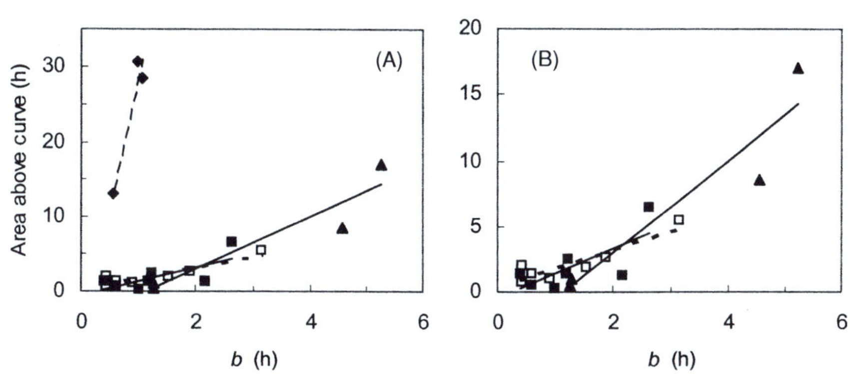

Fig. 35.5. The relationship between the area above the drying time course to severe damage, and the rate of drying as expressed by the constant *b*. (A) The full data set. (B) The data for *Pisum sativum* have been excluded for greater clarity. ◆, *Pisum sativum*; ▲, *Quercus robur*; □, *Trichilia dregeana*; ■, *Ekebergia capensis*.

The results of the model are presented in Fig. 35.5, showing accumulated damage as a function of drying rate. Increasing values of b denote decreasing drying rates (slower drying). Increasing values of 'area' show increasing times of exposure to increasing dehydration up to the point, t_d, at which accumulation of damage becomes severe. For any one species the area increases as b increases. For a 'highly recalcitrant' seed, the area above the drying curve will be small and the increase of this area with increasing b will be shallow. 'Minimally recalcitrant' seeds would have large areas above the drying curve, with steep increases in area above the drying curve with increasing b. The results are consistent with this hypothesis. The lines for *T. dregeana* and *E. capensis* are shallow (slopes of 1.5 and 1.9, respectively), suggesting that these tropical species are highly recalcitrant. The line for *Q. robur*, a temperate species commonly considered to be less recalcitrant, is steeper (slope 3.5). *Pisum sativum*, an orthodox species which had been germinated to the stage of losing desiccation tolerance, had the steepest line (slope 34.1).

Essentially, the slope (a dimensionless ratio) of the plot of the area above the drying curve as a function of b is a quantitative measure of the desiccation sensitivity, or degree of recalcitrance, of an embryonic axis. An advantage of this model is that to compare amongst species, seed lots or various treatments it is not necessary to attain exactly the same drying rates; what is required is a range of drying rates. There are some caveats to be borne in mind when applying the model.

1. Although we have found the modified inverse function used here to be the most widely applicable, at slower drying rates, or times beyond that at which the material is in equilibrium with the humidity of the drying chamber, the fit is not always good. Mathematically this could be accounted for but it would require the introduction of an additional parameter to the equation, conflicting with the requirement for a single descriptor of drying rate. Conceptually, a similar model could be employed whatever function describes the drying rate, as long as a single value can be found that unambiguously describes the drying kinetics. However, the same function must be used across all species and conditions.

2. Some difficulty was experienced identifying t_d, leading to potential errors in the calculation of the area above the drying curve. As t_d had to be estimated by identifying an RWC at which increased leakage occurred, and then calculating from equation (1), estimates of t_d depended upon estimates of b; a poor fit of the equation to the data at long drying times means that estimates of t_d are compromised. This problem can be eased by choosing a value of b that fits the data better at longer drying times (which are more critical in estimating t_d), rather than shorter times.

3. Considerably more data are required than presented in this initial report to determine whether slopes are in fact different. In the data shown in Fig. 35.5, statistically, the slope for *Q. robur* is different from that for *T. dregeana* but overlaps with that for *E. capensis*. However, there are only four data points for *Q. robur*: more data might have separated the overlapping slopes. It is likely, though, that because of the noise invariably associated with this type of data, the model will be able to differentiate between widely different species (temperate vs. tropical, for example) but does not have the resolution to differentiate within these wide groups.

In conclusion, a simple model to assess desiccation sensitivity or degree of recalcitrance semi-quantitatively, accounting for both the intensity and duration of the stress, is proposed. The model shows promise but caveats concerning its application are presented.

References

Finch-Savage, W.E. (1992) Seed water status and survival in the recalcitrant species *Quercus robur* L.: evidence for a critical moisture content. *Journal of Experimental Botany* 43, 671–679.

Kioko, J., Berjak, P., Pammenter, N.W., Watt, M.P. and Wesley-Smith, J. (1998) Desiccation and cryopreservation of embryonic axes of *Trichilia dregeana* Sond. *CryoLetters* 19, 15–26.

Liang, Y. and Sun, W.Q. (2000) Desiccation tolerance of recalcitrant *Theobroma cacao* embryonic axes: the optimal drying rate and its physiological basis. *Journal of Experimental Botany* 51, 1911–1919.

Pammenter, N.W. and Berjak, P. (1999) A review of recalcitrant seed physiology in relation to desiccation-tolerance mechanisms. *Seed Science Research* 9, 13–37.

Pammenter, N.W., Greggains, V., Kioko, J.I., Wesley-Smith, J., Berjak, P. and Finch-Savage, W.E. (1998) Effects of differential drying rate on viability retention of recalcitrant seeds of *Ekebergia capensis*. *Seed Science Research* 8, 463–471.

Walters, C., Pammenter, N.W., Berjak, P. and Crane, J. (2001) Desiccation damage, accelerated aging and respiration in desiccation tolerant and sensitive seeds. *Seed Science Research* 11, 135–148.

36 Is Genetic Fidelity Maintained Following the Cryopreservation of the Seeds of the Endangered African Pepper Bark (*Warburgia salutaris*)?

JOSEPH KIOKO, PATRICIA BERJAK, NORMAN PAMMENTER, JENNY LAMB AND ZANDILE GUMEDE

School of Life and Environmental Sciences, University of Natal, Durban, 4041 South Africa

Introduction

Warburgia salutaris (Bertol. f.) Chiov. (= *W. ugandensis*) is one of the most highly utilized medicinal plants in tropical and subtropical Africa. In pharmacological studies, aqueous suspensions of the bark have been shown to be molluscicidal (Clark and Appleton, 1997), anti-trypanosomal (Kioy *et al.*, 1988) and anti-microbial and to have anti-feedant properties (Hutchings, 1996). Local communities use preparations from the root-bark and stem-bark as an expectorant, for the relief of gastrointestinal disorders and skin complaints, and against malaria (Hutchings, 1996). Herbal medicinal use of *W. salutaris* has created a large demand for the bark and, with the demise of a sustainable traditional gathering ethos, most wild populations of the species are now endangered (Scott-Shaw, 1999).

Since the seeds of this species cannot be stored for appreciable periods using conventional seed storage methods (unpublished observations), this study aims to contribute to the conservation of the germplasm of this species by optimizing the protocol developed by Kioko *et al.* (2000) for the cryopreservation of partially dehydrated seeds, and establishing whether genetic fidelity is maintained following cryopreservation.

Cryopreservation at –196°C been shown to maintain genetic fidelity in plant systems such as shoot-tips of potato (Schäfer-Menuhr *et al.*, 1997) and ginseng (Yoshimatsu *et al.*, 1996), embryogenic cell cultures of barley (Fretz and Lorz, 1994) and nucellar cells of navel orange (Kobayashi *et al.*, 1994). However, genetic aberrations have been reported following the cryopreser-

 The Biology of Seeds: Recent Research Advances (eds G. Nicolás, K.J. Bradford, D. Côme and H.W. Pritchard)

vation of, for example, seeds of mahogany, *Swietenia macrophylla* (Harding *et al.*, 2000).

The objectives of this study were to optimize the seed cryopreservation protocol in order to improve post-thaw recovery, and to assess, using PCR-based random amplified polymorphic DNA (PCR-RAPDs), the effect of cryopreservation on the genetic fidelity of plants regenerated from the cryopreserved seeds.

Materials and Methods

Seed collection and handling

Mature fruits were collected in Lushoto, Tanzania, from more than 30 mother-trees in natural stands, with each tree contributing an equal quantity of fruits in the final bulk collection (H.P. Msanga, Morogoro, Tanzania, 2000, personal communication). The fruits were air-freighted by a commercial courier to South Africa within 2 days and sorted, on arrival, into three categories: green, yellow-green and brown. From yellow-green and brown fruits, seeds were extracted and cleaned immediately; while green fruits, which were still too hard for seeds to be extracted without injury, were kept at 16°C for 4–6 weeks (to soften the pulp) before seed extraction.

Cryopreservation

Cryopreservation of seeds followed the method developed by Kioko *et al.* (2000), which involved the relatively rapid dehydration of the seeds in activated silica gel to a water content of about 0.1 g/g (dry matter basis), before transferral to cryovials and plunging into liquid nitrogen. Five seeds were enclosed in each cryovial, and four cryovials were used for seeds from fruits at each maturity stage. Thawed seeds were sown directly into wet bottom-heated sand beds maintained at 25°C.

Hardening off of seedlings

Germinated seedlings were transplanted into potting soil in pots and maintained in the greenhouse for 8 weeks, under natural light and 25°C/18°C day/night temperature, and watered twice daily. They were then transferred to a shade-house, with 60% light transmittance, watered twice weekly, and fertilized with Multifeed® (Plaaskem, South Africa) once every 10 days. After 6 months, by which time the seedlings had attained heights ranging from 300 to 600 mm, they were transplanted to the field.

Sampling

Following cryopreservation, the surviving seedlings from seeds from the three maturity stages were bulked and then sorted into three experimental populations: seedlings from 'control' (non-dehydrated, non-cryopreserved seeds); seedlings from seeds that were dehydrated only; and seedlings from seeds that were dehydrated and cryopreserved in liquid nitrogen for at least 1 h. Ten individual seedlings were randomly identified from each of three groups and DNA was extracted from their leaves for PCR-RAPD studies.

DNA extraction and quantification

DNA was extracted using the hot CTAB method developed by Doyle and Doyle (1987), with minor modifications, and the purity and concentration of the DNA was determined spectrophotometrically. The purity was assessed by calculating the ratio of the absorbance at 260 nm to that at 280 nm (A_{260}/A_{280}), as a ratio of 1:1.8 corresponds to a pure preparation of DNA (Sambrook *et al.*, 1989). The DNA concentration was then determined based on the relationship that an optical density of 1 at 260 nm corresponds to a concentration of 50 μg/ml of double-stranded DNA. In order to assess the integrity of the extracted DNA, it was subjected to electrophoresis on a 1% agarose gel and stained with ethidium bromide.

Amplification of DNA

Ten arbitrary 10-mer primers (OPA 01, OPA 02, OPA 03, OPA 04, OPA 06, OPA 08, OPA 09, OPA 12, OPA 13 and OPA 18), purchased from Operon Technologies®, were tested for PCR amplification according to the protocol of Bishop (1995). Amplification products were separated by polyacrylamide gel electrophoresis (PAGE) and stained with non-ammoniacal silver.

PCR-RAPD data analysis

Amplified fragments were manually scored: the presence of a specific amplified DNA band was scored as '1', and the absence of such a band was scored as '0'. Two scorers were used, to introduce objectivity into the scoring process. If a band was neither clearly present nor absent, it was scored as '9', indicating that the data were not available. The data were then analysed with the computer program NTSYS-pc (Numerical Taxonomy and Multivariate Analysis System: Rohlf, 1989). The similarity between individuals and populations was measured using the SIMQAL algorithm (similarity for quantitative data), and the similarities clustered using the SAHN cluster analysis algorithm. The resulting relationships were graphically portrayed in the form of a phenogram. Nucleotide diversity within populations and nucleotide divergence between populations were calculated using the RAPDIP computer program.

Results and Discussion

Cryopreservation of seeds

In the present study, seeds from freshly collected green hard fruits were non-germinable, while seeds from the other stages all attained 100% germination,

Table 36.1. Percentage germination of seeds: sown fresh; dehydrated and sown; or dehydrated, cryopreserved in liquid nitrogen, thawed and sown.

	Germination %		
Maturity stage of fruits	Fresh seeds	Dehydrated to *c.* 0.1 g/g	Dehydrated to *c.* 0.1 g/g and cryopreserved
Green hard	0	0	0
Green hard, stored till soft	100	100	55
Yellow-green	100	100	65
Brown	100	100	40

Seeds were sown in sand beds maintained at a temperature of 25°C; germination was scored as seedling emergence.

even after dehydration to about 0.1 g/g. However, after retrieval from cryopreservation, seeds from yellow-green fruits and from the green hard ones that had been stored until soft attained higher germination percentages than seeds from brown fruits (Table 36.1).

The higher post-thaw survival attained in this study, as compared with that reported by Kioko *et al.* (2000), may be attributed primarily to subcellular differences in the seeds used. Embryonic axes of seeds used by Kioko *et al.* had amassed lipid to a lesser extent than axes of seeds used in our study (unpublished results). The latter seeds therefore had a higher lipid-body:organelle volume ratio, accompanied by more cytomatrical compaction, vacuole shrinkage and organelle dedifferentiation, facilitating a higher tolerance to freezing. The recovery medium may also have had an influence on germination: sand (used in our study) afforded the seeds better drainage than vermiculite (used in the earlier cryopreservation studies).

No morphological differences were apparent in seedlings or saplings obtained from seeds that were sown fresh, dehydrated, or cryopreserved. On transplantation to the field, saplings from all three treatments progressed similarly in terms of growth rate and morphology (results not shown).

DNA extraction and quantification

The concentration of DNA extracted from different seedlings, determined by spectrophotometry, ranged from 20 to 1015 ng/µl. This variation was also reflected by the varying brightness of the bands attained on agarose gels after electrophoresis. However, by comparing the brightness of extracted DNA samples with that of *lambda* DNA of known concentration, it was apparent that the results obtained by spectrophotometry overestimated the concentration of high-molecular-weight DNA in the samples, probably as a result of the contribution of contaminants to the sample absorbance.

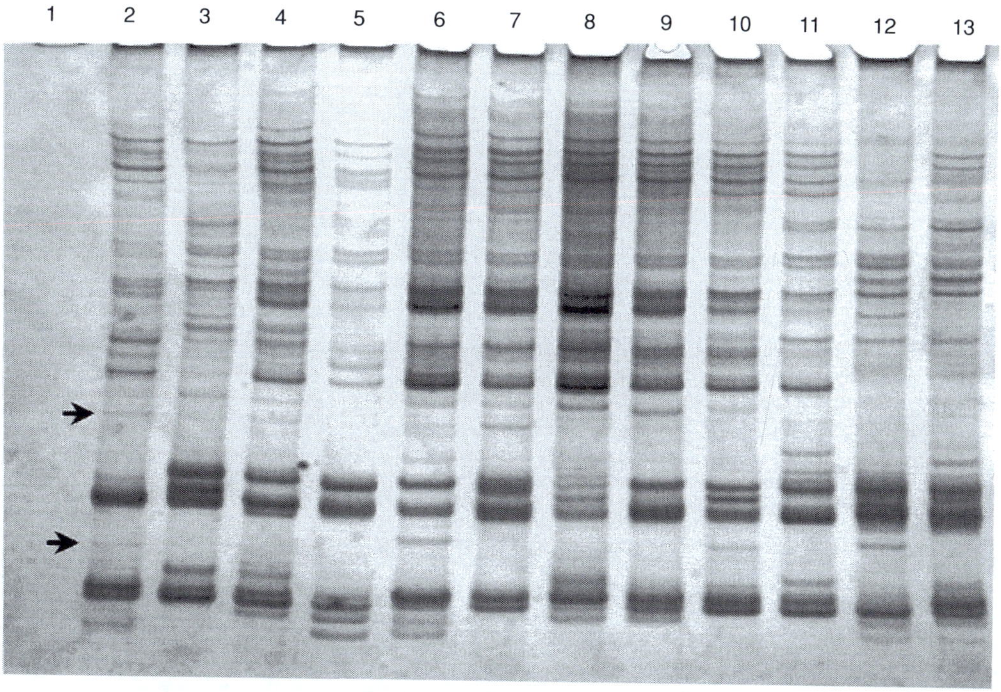

Fig. 36.1. PCR-RAPD products generated using primer OPA09. Lane 1, distilled water (negative control); lanes 2–5, seedlings from freshly harvested untreated seeds; lanes 6–9, seedlings from dehydrated seeds; lanes 10–13, seedlings from cryopreserved seeds. Arrows indicate polymorphic fragments.

Amplification of DNA

Of the ten primers tested, two (OPA 02 and OPA 18) resulted in no amplification, implying that they probably had no homology with *W. salutaris* DNA, or that the reaction simply did not take place. The other eight primers produced a total of 104 fragments, an average of 13 loci per primer. Each primer produced a distinct pattern, but for each primer most samples produced identical banding patterns (Fig. 36.1). The bands were consistent, even over a 100-fold range in DNA concentration. The bands produced by these eight primers were therefore scored. Many workers report scoring of only a subset of primers, particularly those that show the greatest polymorphism (e.g. Connolly *et al.*, 1994; Moukadiri *et al.*, 1999; Edwards, 2000), which may exaggerate the magnitude of the calculated DNA diversity.

Among the bands scored, some polymorphisms were detectable within and between the three populations (treatments) under study: seedlings from freshly collected, non-dried seeds; seedlings from seeds that were only dehydrated; and seedlings from seeds that were cryopreserved in liquid nitrogen. The three populations exhibited varying nucleotide diversities: 0.45 ± 0.08% for seedlings from freshly sown seeds; 0.55 ± 0.09% for those from dehy-

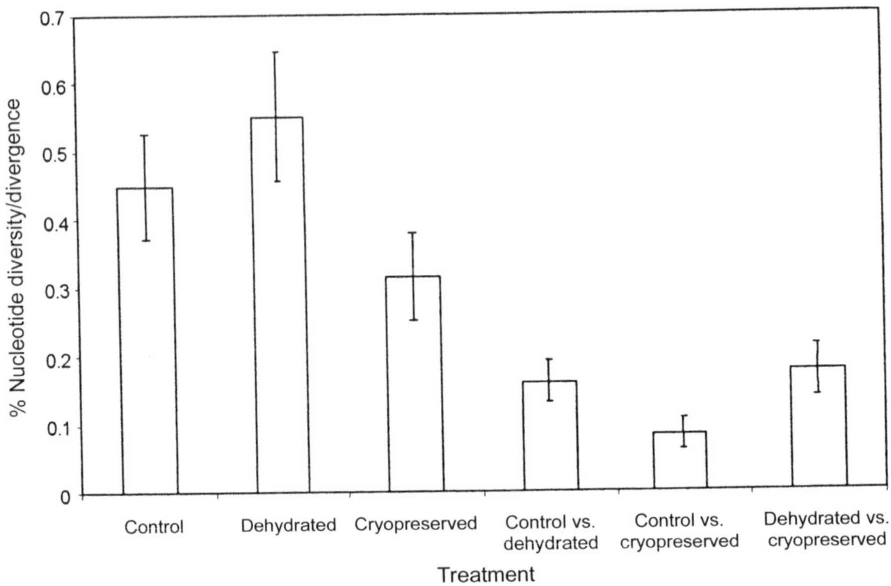

Fig. 36.2. Nucleotide diversity within the populations of seedlings obtained from control (freshly harvested, untreated), dehydrated and cryopreserved seeds (first three bars); and nucleotide divergence between the populations (last three bars).

drated seeds; and 0.32 ± 0.06 % for those from cryopreserved seeds (Fig. 36.2).

These levels of diversity are higher than those in some animal species surveyed (aphids, grasshoppers and leeches), in which the average nucleotide diversity was 0.0135 ± 0.0032% (Borowsky, 2000). This may be because, in plants, the large number of mating systems produces a richer variety of population genetic structures than in animals, with wind-pollinated plants exhibiting the highest levels of heterozygosity, followed by animal-pollinated plants (a category to which *W. salutaris* belongs, based on floral morphology: Dyer, 1975) and then self-pollinated plants (Hamrick and Godt, 1990). However, lower nucleotide diversity has been observed in some cultivated plant populations (e.g. 0.007% in *Arabidopsis thaliana*: Purugganan and Sudith, 1998; and 0.0049% in *Pinus sylvestris*: Dvornyk *et al.*, 2002). The high diversity observed in this study may be a consequence of observing natural populations and may also imply that *W. salutaris* is outcrossing, with partial or complete self-incompatibility (no information is available on the mating system of this species), as such species have higher nucleotide diversity (Liu *et al.*, 1999).

In assessing the effect of cryopreservation on genetic fidelity, it would have been ideal to use tissue from individual seeds before, and after, cryopreservation. This was impossible, as the excision of tissue samples was found to be lethal to the seeds. Hence, nucleotide divergence among the seedlings obtained from the three treatments (control, dried and cryo-

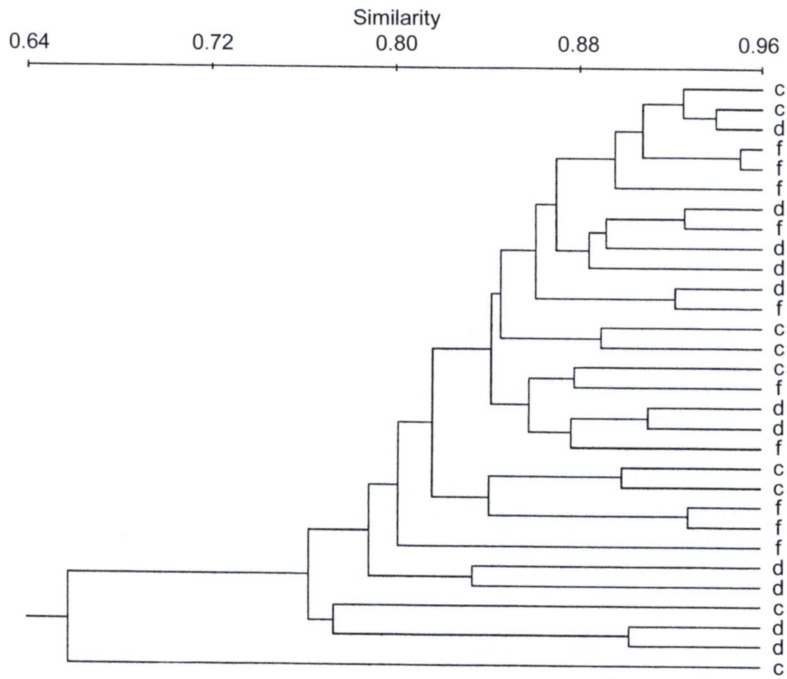

Fig. 36.3. Phenogram showing the absence of clustering according to treatments, in the genetic relationships within the seedlings obtained from three seed treatments: c, control; d, dehydrated; f, dehydrated and frozen.

preserved) was compared with nucleotide divergence within the populations. Significantly, the nucleotide divergence between the populations of seedlings (ranging from 0.08% to 0.17%) was much lower than the nucleotide deversity within them (Fig. 36.2). Thus, there was higher variation within each treatment than between the treatments, showing that neither dehydration nor cryopreservation introduced significant genetic variability within the seeds.

Polymorphic fragments were used to generate a phenogram of all individuals from all three treatments (Fig. 36.3). The phenogram showed that percentage similarities among individuals ranged from about 95% to 65%, and there was variation among individuals from all treatments. This variation was unrelated to either dehydration or cryopreservation and there was no clustering according to treatments. Therefore, neither dehydration nor cryopreservation had an effect on the genetic fidelity of the seeds.

The present results thus indicate that cryopreservation may be used with confidence for the genetic/germplasm preservation of *W. salutaris* and, perhaps, of other endangered species producing non-orthodox seeds.

References

Bishop, K.S. (1995) PCR-RAPD analysis of DNA diversity in selected Natal zebra populations. MSc Thesis, Department of Biology, University of Natal, Durban, South Africa.

Borowsky, R.L. (2000) Estimating nucleotide diversity from random amplified polymorphic DNA and amplified fragment length polymorphism data. *Molecular Phylogenetics and Evolution* 18, 143–148.

Clark, T.E. and Appleton, C.C. (1997) Molluscisidal activity of *Apodytes dimidiata* E. Meyer ex Arn (*Icacinaceae*), *Gardenia thunbergia* L.f. (*Rubiaceae*) and *Warburgia salutaris* (Bertol. F.) Chiov. (*Cannalaceae*), three South African plants. *Journal of Ethnopharmacology* 56, 15–30.

Connolly, A.G., Godwin, I.D., Cooper, M. and DeLacy, I.H. (1994) Interpretation of randomly amplified polymorphic DNA marker data for fingerprinting sweet potato (*Ipomea batatas* L.) genotypes. *Theoretical and Applied Genetics* 88, 332–336.

Doyle, J.J. and Doyle, J.L. (1987) A rapid DNA isolation procedure for small quantities of fresh leaf material. *Phytochemical Bulletin* 19, 11–15.

Dvornyk, V., Sirvio, A., Mikkonen, M. and Savolainen, O. (2002) Low nucleotide diversity at the pal1 locus in the widely distributed *Pinus sylvestris. Molecular Biology and Evolution* 19, 179–188.

Dyer, R.A. (1975) *The Genera of Southern African Flowering Plants*, Vol. 1, *Dicotyledons*. Botanical Research Institute, Pretoria, South Africa, 756 pp.

Edwards, N.R. (2000). Optimization of the randomly amplified polymorphic DNA (RAPD) technique for the characterisation of selected South African maize (*Zea mays* L.) breeding material. MSc Thesis, University of Natal, Pietermarizburg, South Africa.

Fretz, A. and Lorz, H. (1994) Cryopreservation of *in vitro* cultures of barley (*Hordeum vulgare* L. and *H. murinum* L.) and transgenic cells of wheat (*Triticum aestivum* L.). *Journal of Plant Physiology* 146, 489–496.

Hamrick, J.L. and Godt, M.J. (1990) Allozyme diversity in plant species. In: Brown, A.H.D., Clegg, M.T., Kahler, A.L. and Weir, B.S. (eds) *Plant Population Genetics, Breeding and Genetic Resources*. Sinauer Associates, Sunderland, Massachusetts, pp. 43–63.

Harding, K., Marzalina, M., Krishnapillay, B., Nashataul-Zaimah, N.A., Normah, M.N. and Benson, E.E. (2000) Molecular stability assessments of trees regenerated from cryopreserved mahogany (*Swietenia macrophylla*) seed germplasm using non-radioactive techniques to examine the chromatin structure and DNA methylation status of the ribosomal RNA genes. *Journal of Tropical Forest Science* 12, 149–163.

Hutchings, A. (1996) *Zulu Medicinal Plants: an Inventory*. University of Natal Press, Pietermaritzburg, South Africa, 450 pp.

Kioko, J., Berjak, P., Pritchard, H. and Daws, M. (2000) Seeds of the African Pepperbark (*Warburgia salutaris*) can be cryopreserved after rapid dehydration in silica gel. In: Engelmann, F. and Takagi, H. (eds) *Cryopreservation of Tropical Plant Germplasm: Current Research Progress and Application.* Japan International Research Centre for Agricultural Sciences, Tsukuba, Japan / International Plant Genetic Resources Institute, Rome, Italy, pp. 371–377.

Kioy, D.W., Murilla, G., Kofi-Tsekpo, M.W. and Okwara, J. (1998) Anti-trypanosomal effects of some compounds isolated from the extracts of *Warburgia ugandensis*. *African Journal of Health Sciences* 5, 35–37.

Kobayashi, S., Sakai, A., Ohgawara, T. and Nakamura, Y. (1994) Stable maintenance of an integrated gene in nucellar cells of navel orange (*Citrus sinensis* Osb.) under storage in LN_2. *Journal of the Japanese Society for Horticultural Science* 63, 553 – 558.

Liu, F., Charlesworth, D. and Kreitman, M. (1999) The effect of mating system differences on nucleotide diversity at the phosphoglucose isomerase locus in the plant genus *Leavenworthia*. *Genetics* 151, 343–357.

Moukadiri, O., Lopes, C.R. and Cornejo, M.J. (1999) Physiological and genomic variations in rice cells recovered from direct immersion and storage in liquid nitrogen. *Physiologia Plantarum* 105, 442–449.

Purugganan, M.D. and Sudith, J.I. (1998) Molecular population genetics of the *Arabidopsis CAULIFLOWER* regulatory gene: non-neutral evolution and naturally occurring variation in floral homeotic function. *Proceedings of the National Academy of Sciences USA* 95, 8130– 8134.

Rohlf, F.J. (1989) *NTSYS-pc: Numerical Taxonomy and Multivariate Analysis System. Version 1.50*. Exeter Software, New York.

Sambrook, J., Fritsch, E.F. and Maniatis, T. (1989) *Molecular Cloning: a Laboratory Manual*. Cold Spring Harbor Laboratory Press, Cold Spring Harbor, New York, 545 pp.

Schäfer-Menuhr, A., Schumacher, H.M., Mix-Wagner, G. and Altman, A. (1997) Cryopreservation of potato cultivars – design of a method for routine application in genebanks. *Acta Horticulturae* 447, 477–482.

Scott-Shaw, R. (1999) *Rare and Threatened Plants of KwaZulu-Natal and Neighbouring Regions: a Plant Red Data Book*. KwaZulu-Natal Nature Conservation Service, Pietermaritzburg, South Africa, 182 pp.

Yoshimatsu, K., Yamaguchi, H. and Shimomura, K. (1996) Traits of *Panax ginseng* hairy roots after cold storage and cryopreservation. *Plant Cell Reports* 15, 555–560.

37 Structural Changes in Membranes of Developing Wheat Embryos During the Acquisition of Desiccation Tolerance

ELENA A. GOLOVINA[1,2] AND FOLKERT A. HOEKSTRA[2]

[1]Russian Academy of Sciences, Timiryazev Institute of Plant Physiology, Botanicheskaya 35, Moscow 127276, Russia; [2]The Graduate School Experimental Plant Sciences, Laboratory of Plant Physiology, Wageningen University, Arboretumlaan 4, 6703 BD Wageningen, The Netherlands

Introduction

Membranes are considered to be a primary target of desiccation injury. The ability of desiccation-tolerant organisms to avoid membrane damage during a dehydration–rehydration cycle might be related to changes in membrane fluidity. Membrane bilayer structure in dried desiccation-tolerant organisms is thought to be stabilized as a result of interactions of the polar headgroups with sugars and proteins (Crowe *et al.*, 1992, 1998). Such interactions create space between phospholipids and prevent membrane phase changes. Membranes thus remain in the liquid-crystalline phase when the hydration shell is lost.

We have proposed a hypothesis about partitioning of amphiphilic cytoplasmic compounds into membranes, which would occur during drying (Golovina *et al.*, 1998; Hoekstra and Golovina, 2000, 2002; Hoekstra *et al.*, 2001; Golovina and Hoekstra, 2002). The majority of uncharged compounds in the cytoplasm have amphiphilic properties, i.e. they partition between the hydrophilic and hydrophobic phases according to their partition coefficient. A decrease in cytoplasmic volume during water loss causes additional partitioning of amphiphiles into the lipid phase, which leads to a new equilibrium distribution for each amphiphile. Therefore, partitioning of amphiphiles from the cytoplasm into membranes during drying is a physico-chemical phenomenon and an inevitable process in all organisms, which might have a dual effect on cells. On the one hand, such a partitioning could be beneficial, because amphiphilic antioxidants are automatically inserted into membranes. On the other hand, partitioning causes membrane disturbance, which could increase membrane permeability to the extent that cell compartmentation is lost.

The difference between desiccation-sensitive and -tolerant organisms in relation to partitioning might be encompassed in the response of cells to partitioning and/or in the amount and composition of the partitioning molecules. Therefore, both membrane fluidization and immobilization could be expected to occur in dehydrating desiccation-tolerant organisms. Fluidization might originate from the enrichment of cells with amphiphilic antioxidants, the insertion of which into membranes would cause increased disturbance. Immobilization might originate from increased interactions at the membrane surface as a protective mechanism against membrane disturbance.

To elucidate possible structural changes in membranes with drying, also in relation to the acquisition of desiccation tolerance, we used immature wheat embryos in the desiccation-sensitive and -tolerant stages of their development. To correlate the loss/maintenance of membrane integrity during drying with membrane structure, we used *in vivo* electron spin resonance (ESR) spin probe techniques. Because membrane behaviour depends not only on composition, but also on the interaction with the surroundings, *in vivo* studies are imperative for such investigations. The use of different spin probes allows both membrane integrity and membrane fluidity to be studied (Marsh, 1981).

Materials and Methods

Plant material

Plants of spring wheat (*Triticum aestivum* cv. Priokskaya) were grown in pots under greenhouse conditions (16 h light/8 h dark and an average day temperature of 22°C). Ears were tagged at the beginning of anthesis. Embryos were isolated from the kernels during development. Fast drying of the embryos was performed on the laboratory bench (relative humidity approximately 30%), half water loss being reached within 20–30 min of drying, depending on developmental age.

Germination of fresh and air-dried, isolated embryos (50 on average) was carried out in Petri dishes on 0.7% agar in tap water. Air-dry embryos were prehydrated in water vapour-saturated air for 1 h and then transferred to agar medium for germination. Water contents were analysed by weighing the samples before and after heating at 96°C for 24 h.

ESR spin probe study

Membrane integrity was measured with an ESR spin probe technique using perdeuterated oxo-tempo (PDT) (from Professor Igor Grigoriev, Novosibirsk, Russia). The principles of the method have been described elsewhere (Golovina and Tikhonov, 1994; Golovina *et al.*, 1997). Before labelling, the air-dry embryos were prehydrated in humid air for 1 h and then placed on wet filter paper for 3 h for further rehydration. Embryos that were partly dried were immediately placed on wet filter paper (3 h), whilst fresh embryos were directly used for labelling. The hydrated embryos were incubated for 15–20 min in a 1 mM solution of PDT containing 120 mM broadening agent $K_3Fe(CN)_6$. The sample was then loaded into a capillary (2 mm diameter) together with a small amount of the solution. ESR spectra were recorded at room temperature with an X-band ESR spectrometer (Bruker, Rheinstetten, Germany, model 300E). Microwave power was 2 mW and the modulation amplitude was 0.25 gauss.

Membrane fluidity was studied using the membrane spin probe, 5-doxyl-stearic acid (5-DS) (Sigma, St Louis, Missouri, USA). Fresh embryos were placed in a 2 mM solution (final solution containing 2% ethanol) of 5-DS. The solution of 5-DS was freshly prepared from a 0.1 M stock solution in ethanol. Embryos were labelled for 30 min, then carefully blotted with filter paper and placed into a capillary (2 mm diameter) immediately or after air-drying on the surface of a closed Petri dish. The spectra of 5-DS were recorded at a microwave power of 5 mW and 3 gauss modulation amplitude.

Multi-component 5-DS spectra were decomposed by spectral titration, i.e. subtraction after adjusting for position and amplitude of peaks (Berliner, 1976).

Results

Characteristics of wheat embryo development

Up to 15 days after anthesis (DAA), embryos increased in length as a result of cell division, followed by dry weight accumulation, with the main weight gain between 15 and 22 DAA. The rapid decline in moisture content from approximately 3 to 1.5 g H_2O/g dry weight occurs as a result of this dry matter accumulation (Golovina *et al.*, 2000, 2001). The ability of isolated, fresh embryos to germinate was acquired at 15 DAA, but desiccation tolerance, as the ability of embryos to germinate after fast drying and rehydration, was acquired only at 18 DAA (data not shown).

Cellular heterogeneity in the acquisition of desiccation tolerance

To estimate acquisition of desiccation tolerance at the cellular level, plasma membrane integrity was studied during embryo development. Isolated embryos were fast dried on the laboratory bench and then carefully rehydrated, in water vapour first, then on filter paper and followed by incubation in PDT solution with broadening agent.

Briefly, the spin probe method for the estimation of cellular intactness works as follows. The ESR spectrum of only PDT in water is isotropic, i.e. containing three narrow equidistantly separated lines. Application of 120 mM potassium ferricyanide broadens the ESR signal of PDT to invisibility (Eaton and Eaton, 1978). Because ferricyanide does not penetrate intact membranes, whilst PDT does, application of ferricyanide allows only the

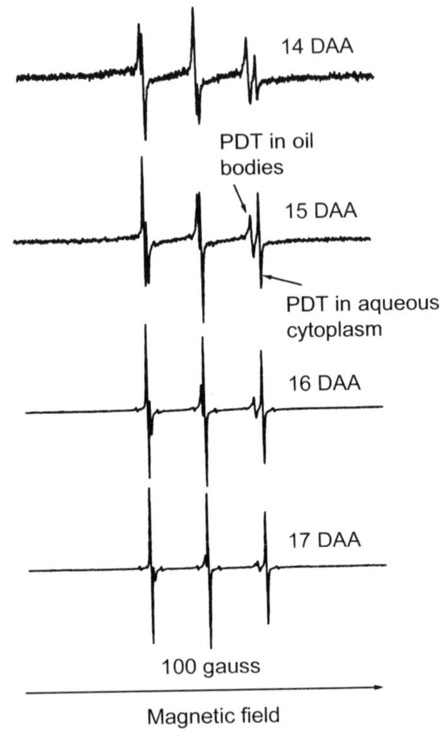

Fig. 37.1. ESR spectra of PDT (perdeuterated oxo-tempo) from wheat embryos of different developmental ages after fast drying and rehydration. The ratio between aqueous and lipid contributions reflects the relative proportion of desiccation-tolerant cells.

ESR signal originating from PDT inside the cells to be observed. Inside cells, PDT is distributed between the hydrophobic (oil bodies) and hydrophilic (aqueous cytoplasm) phases. The ESR signal of PDT in oil bodies has different characteristics and can be resolved from the signal of PDT in aqueous cytoplasm at the right-hand side of the spectrum. If membranes are permeable for ferricyanide ions, then the ESR signal from PDT in cytoplasm also will be broadened. The signal originating from PDT in oil cannot be broadened by ferricyanide because of the inability of these ions to partition into the hydrophobic phase. The proportion of ESR signals originating from aqueous cytoplasm and oil can thus be used to estimate the proportion of cells in the sample having intact membranes.

Figure 37.1 shows PDT spectra from wheat embryos of different developmental ages obtained after a cycle of drying and rehydration. As follows from Fig. 37.1, some desiccation-tolerant cells can be found in wheat embryos as early as at 14 DAA, the proportion of which gradually increased with developmental age up to 17 DAA. At 18 DAA the ESR signals from fresh embryos and from embryos that were dried and rehydrated were similar (data not shown). Also at 18 DAA, germination after drying and rehy-

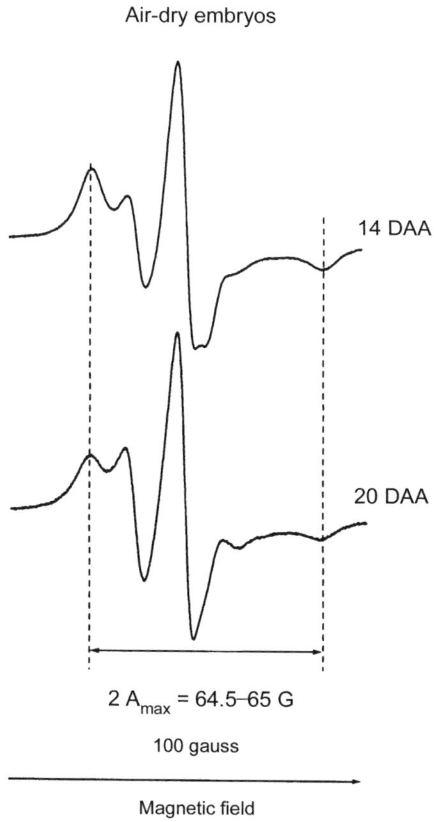

Fig. 37.2. Representative (n = 10) ESR spectra of 5-DS (5-doxyl-stearic acid) from dried desiccation-sensitive (14 DAA) and dried desiccation-tolerant (20 DAA) embryos.

dration was observed for the first time. Apparently, not all cells in wheat embryos acquire desiccation tolerance simultaneously.

Changes in membrane fluidity during acquisition of desiccation tolerance

To correlate acquisition of desiccation tolerance with possible changes in membrane fluidity, we applied the membrane spin label, 5-DS, having the nitroxide radical (doxyl group) at the fifth position along the acyl chain. During incubation of wheat embryos in 5-DS solution, the spin-labelled molecules partition into cellular membranes because of the high affinity of fatty acids for membranes. The shape of the ESR spectra of 5-DS depends on the motional freedom of the labelled part of the acyl chain (Marsh, 1981). In very rigid membranes, the acyl chains are oriented perpendicular to the membrane surface, whereas with fluidization the angular orientation departs from perpendicular. Therefore, an analysis of the spectral shape allows membrane fluidity to be estimated.

Figure 37.2 shows the changes in representative 5-DS spectra obtained from dried wheat embryos at the desiccation-sensitive (14 DAA) and -tolerant (20 DAA) stages of development. These spectra are different from ESR spectra of PDT in Fig. 37.1. Whereas the spectra of PDT in cytoplasm or oil bodies are isotropic, because they originate from rapidly and randomly tumbling nitroxides, those of 5-DS are anisotropic, because they are oriented in membranes (Marsh, 1981). The outermost distance $2A_{max}$ in the 5-DS spectra (Fig. 37.2) reflects the angle of motional freedom. Although $2A_{max}$ was similar for both spectra from 14 and 20 DAA embryos (about 64.5–65.0 gauss), the shapes of these two spectra were different at the left-hand side. Because such spectral shape cannot be described by one mode of motion, these spectra are likely to be the superposition of two spectra with different motion characteristics.

Applying the method of spectral titration (Berliner, 1976), we managed to decompose the original spectra (14 and 20 DAA embryos shown in Fig. 37.2) into two different components (Fig. 37.3). The upper spectrum (mobile component) can be attributed to 5-DS in a weakly ordered environment, with high freedom of motion. The lower spectrum (immobile component) originates from 5-DS molecules in a highly ordered environment with very small angular freedom of motion. Comparison of 5-DS spectra obtained

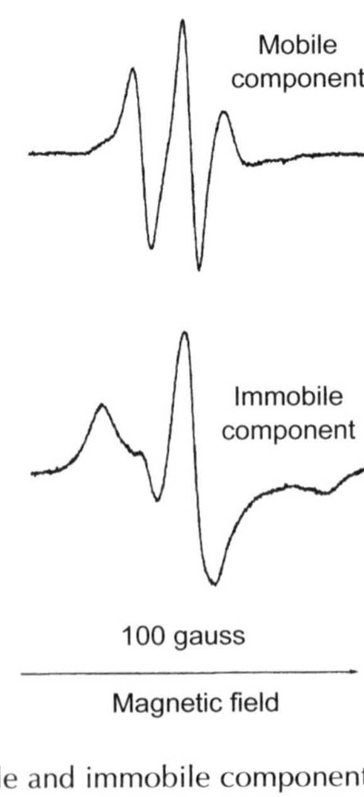

Fig. 37.3. Shape of the mobile and immobile components obtained after decomposition of the 5-DS spectra of Fig. 37.2. The ESR spectra of 5-DS in Fig. 37.2 have both components, but the spectrum from the dried 20 DAA embryo has a higher proportion of the mobile component than that from the dried, 14 DAA embryo. The mobile component corresponds to high angular freedom of motion of the acyl chains (top spectrum); the immobile component corresponds to rigid acyl chains having perpendicular orientation with respect to the membrane surface.

from 14 DAA and 20 DAA embryos (Fig. 37.2) shows that the proportion of the mobile component has increased after the acquisition of desiccation tolerance. This means that membranes are more rigid in dried wheat embryos at the desiccation-sensitive stage than at the desiccation-tolerant stage.

Discussion

We proposed that amphiphilic compounds partition from the cytoplasm into membranes during drying (Golovina *et al.*, 1998; Hoekstra and Golovina, 2000, 2002; Hoekstra *et al.*, 2001; Golovina and Hoekstra, 2002). This suggestion is based on a physico-chemical consideration of equilibrium conditions inside cells during drying. In hydrated cells, amphiphiles are distributed between the aqueous and hydrophobic phases according to their partition coefficient and volume relations between the phases. Drying causes a decrease in aqueous volume and, as a consequence, the establishment of a new distribution of amphiphiles between the two phases. Partitioning of amphiphiles into membranes is known to cause disturbance and even lysis (Maher and Singer, 1984).

The results described above show that drying causes membrane fluidization in desiccation-tolerant embryos. In desiccation-sensitive embryos there was much less fluidization with drying. These results at least show that fluidization with drying is not harmful. On the contrary, membrane fluidization with drying correlates with the acquisition of desiccation tolerance. Although this correlation is not evidence of a causal relationship, we assume that membrane fluidization in wheat embryos reflects the changes in cell composition occurring during the acquisition of desiccation tolerance.

If amphiphile partitioning into membranes would have a causal relationship with the acquisition of desiccation tolerance, two mechanisms can be considered. Firstly, as mentioned above, amphiphile partitioning provides an automatic insertion of amphiphilic antioxidants into membranes. Secondly, a certain level of fluidization might increase membrane elasticity to such an extent that the liquid crystalline phase in the dried state is maintained and that mechanical damage during rehydration (imbibitional damage) is prevented.

References

Berliner, L.J. (1976) *Spin Labeling: Theory and Applications.* Academic Press, New York.

Crowe, J.H., Hoekstra, F.A. and Crowe, L.M. (1992) Anhydrobiosis. *Annual Review of Physiology* 54, 579–599.

Crowe, J.H., Carpenter, J.F. and Crowe, L.M. (1998) The role of vitrification in anhydrobiosis. *Annual Review of Physiology* 60, 73–103.

Eaton, S.S. and Eaton, G.R. (1978) Interaction of spin labels with transition metals. *Coordination Chemistry Reviews* 26, 451–453.

Golovina, E.A. and Tikhonov, A.N. (1994) The structural differences between the embryos of viable and nonviable wheat seeds as studied with the EPR spectroscopy of lipid-soluble spin labels. *Biochimica et Biophysica Acta* 1190, 385–392.

Golovina, E.A. and Hoekstra, F.A. (2002) Membrane behavior as influenced by partitioning of amphiphiles during drying: a comparative study in anhydrobiotic plant systems. *Comparative Biochemistry and Physiology* 131A, 545–558.

Golovina, E.A., Tikhonov, A.N. and Hoekstra, F.A. (1997) An electron paramagnetic resonance spin-probe study of membrane permeability changes with seed aging. *Plant Physiology* 114, 383–389.

Golovina, E.A., Hoekstra, F.A. and Hemminga, M.A. (1998) Drying increases intracellular partitioning of amphiphilic substances into the lipid phase: impact on membrane permeability and significance for desiccation tolerance. *Plant Physiology* 118, 975–986.

Golovina, E.A., Hoekstra, F.A. and van Aelst, A.C. (2000) Programmed cell death or desiccation tolerance: two possible routes for wheat endosperm cells. *Seed Science Research* 10, 365–379.

Golovina, E.A., Hoekstra, F.A. and van Aelst, A.C. (2001) The competence to acquire cellular desiccation tolerance is not dependent on seed morphological development. *Journal of Experimental Botany* 52, 1015–1027.

Hoekstra, F.A. and Golovina, E.A. (2000) Impact of amphiphile partitioning on desiccation tolerance. In: Black, M., Bradford, K.J. and Vasques-Ramos J. (eds) *Seed Biology: Advances and Applications*. CAB International, Wallingford, UK, pp. 43–55.

Hoekstra, F.A. and Golovina, E.A. (2002) The role of amphiphiles. *Comparative Biochemistry and Physiology* 131A, 527–533.

Hoekstra, F.A., Golovina, E.A. and Buitink, J. (2001) Mechanisms of plant desiccation tolerance. *Trends in Plant Science* 6, 431–438.

Maher, P. and Singer, S.J. (1984) Structural changes in membranes produced by the binding of small amphipathic molecules. *Biochemistry* 23, 232–240.

Marsh, D. (1981) Electron spin resonance: spin labels. In: Grell, E. (ed.) *Membrane Spectroscopy. Molecular Biology, Biochemistry and Biophysics*, Vol. 31. Springer Verlag, Berlin, pp. 51–142.

38 Sucrose Pretreatment Increases Desiccation Tolerance in Wampee (*Clausena lansium*) Axes

J.R. Fu, X.M. Huang and S.Q. Song

School of Life Sciences, Sun Yat-Sen University, Guangzhou 510275, PR China

Introduction

Wampee (*Clausena lansium)* is an economically important fruit tree native to South China. Its seeds cannot tolerate dehydration (Fu *et al.*, 1990, 1994) and are termed recalcitrant (Roberts, 1973). The safe seed moisture content of wampee is 45–50% (fresh weight basis). The optimal temperature for seed storage is 15°C (Fu *et al.*, 1989). Viability was less than 50% when seeds were placed under open conditions for 6 days and the water content decreased to 33–35% (Fu *et al.*, 1989). This is the deteriorating seed moisture content region. Being able to preserve recalcitrant seeds by raising desiccation tolerance to a certain extent will be valuable for both agricultural practice and germplasm conservation.

To conserve the genetic resources of species producing recalcitrant seeds, the most promising method is storage of the material (usually the excised embryonic axes) in liquid nitrogen (Roberts *et al.*, 1984) after drying. Embryonic axes of wampee are more tolerant of drying than the whole seeds (Lu and Fu, 1997) and preliminary experiments have shown that axes can be manipulated in culture and tolerate cryopreservation. However, the plumules of excised axes were damaged easily on dehydration, and no normal seedlings were formed. We found that it was better to use sprouting axes instead of fresh embryonic axes (Fu *et al.*, 2000). Preculture of axes on a gradient of sucrose can be used to increase their desiccation tolerance (Huang *et al.*, 2000) and raise their vigour index (Lu and Fu, 1997).

The objectives of the present study were to facilitate increased dehydration of excised wampee axes through the manipulation of culture conditions, leading to increased possibilities of successful cryopreservation.

Materials and Methods

Plant material

Fruits of *C. lansium* (Lour.) Skeels cv. Jixin were hand-harvested when seeds were at physiological maturity. Seeds were removed from the fruits by hand, washed with tap water and mixed with 6% chlorothaloni (a fungicide) containing tetrachloroisophthalonitrile). After slight drying at room temperature, seeds were stored in polythene bags at 15°C until use.

Excision and incubation of embryonic axes

Prior to excision of the embryonic axes, the chlorothaloni was removed. The excised axes were sterilized by a 3 min immersion in 0.1% $HgCl_2$, and washed five times with sterile water. Axes were incubated on woody plant medium (WPM) (McCown and Lloyd, 1981) with 3% sucrose at 25°C with a 12 h photoperiod (800 lux). WPM includes K_2SO_4, NH_4NO_3, KH_2PO_4, $CaCl_2$, $Ca(NO_3)$, $MgSO_4$, $FeSO_4$, Na-EDTA (as macroelements) and $MnSO_4$, $ZnSO_4$, H_3BO_3, $CuSO_4$, Na_2MoO_4 (as microelements). Inositol, nicotinic acid, thiamine, pyridoxine and glycine were also incorporated into the medium prior to adjusting to pH 5.8–5.9.

For all *in vitro* experiments (including cryopreservation), 20 axes were needed per treatment and each treatment was replicated thrice.

The growth of sprouting axes

The excised fresh axes were precultured on WPM for 3 weeks until a 1–2 mm long plumule formed. These sprouting axes (seedlings) had a hypocotyl root 10 mm long. Sprouting axes were precultured on WPM enriched with 27% sucrose for 5–8 days and then dehydrated using activated silica gel and/or high concentrations of sucrose (see below for details).

Induction of adventitious roots from the rootless sprouting axes

The sprouting axes were incubated on WPM for 1 month after drying. Some of them produced normal seedlings but many of them became rootless seedlings. Adventitious root induction was attempted by using either that portion of the sprouting axes remaining after removal of the withered roots or the excised plumule–hypocotyl sections. Such explants were incubated in half-strength WPM containing a combination of the plant growth regulators NAA and IBA.

Cryopreservation

The sprouting axes were precultured successively with 27% (for 7 days), 50% (for 2 days) and 60% (for 2 days) sucrose, and then dried with silica gel, so that the axes reached around 18% water content. The plumule plus around 1–2 mm of hypocotyl was used as experimental material for the cryopreservation experiments.

Table 38.1. The growth of wampee axes (initial length *c.* 3 mm) after desiccation by silica gel.

	Desiccation time (h)				
	0	1	2	3	5
Water content (%)	49.3	31.3	26.7	21.4	15.4
Viability (%)	100.0	100.0	81.3	78.6	27.3
Plantlet formation (%)	88.9	0	0	0	0
Length of hypocotyl (mm)	ND	10.3±1.5	9.4±2.5	7.7±2.8	5.7±1.1
Length of radicle (mm)	ND	8.3±5.0	6.0±2.7	19.1±5.7	3.0±1.7

ND, not detected.

Vitrification solution (PVS2: 15% dimethyl sulphoxide + 15% ethylene glycol + 30% glycerol) was used to pretreat axes before freezing. Ten sprouted axes were put in each cryo-tube with 1 ml 60% vitrification solution and treated for 20 min at room temperature. The loading solution was drained out and replaced with 100% vitrification solution for 10 min in an ice bath. The samples were then directly immersed into liquid nitrogen and kept for 24 h. After the ultra-cool, samples were rapidly warmed in a water bath at 25°C, PVS2 was removed and the axes were washed three times (30 min per cycle) in 1–2 mol/l sucrose. Some of the ultra-cool treated axes were regrown on WPM, whilst the others were used to make slices for ultrastructural observation.

Results

Dehydration of excised fresh axes

When wampee axes were dried by silica gel, their water content decreased from 49.3% to 31.1% (fresh weight basis). The plumules were damaged but the hypocotyls and radicles were still capable of growing well (Table 38.1).

After preculturing with a gradient of sucrose, wampee axes were cultured on WPM at 25°C for 30 days. As the sucrose concentration increased, water content decreased, damage to the plumules was heavier and plantlet formation decreased (Table 38.2). When sucrose concentration was raised to

Table 38.2. Ability of wampee axes to form plantlets following preculturing in increasing sucrose concentrations.[a]

Sucrose preculture	Water content (%)	Viability (%)	Plantlet formation (%)
Medium with 3% sucrose (control)	85.2	100	52.8 ± 21.4
From 3% to 11%, 19% sucrose	57.5	100	28.7 ± 8.8
From 3% to 11%, 19%, 27% sucrose	56.9	100	17.4 ± 6.5
From 3% to 11%, 19%, 27%, 35% sucrose	47.8	100	0[b]

[a]Each sucrose concentration was applied for 2 days.
[b]Only abnormal seedling without plumule.

Table 38.3. Survival of sprouting wampee axes after desiccation with silica gel.

Water content (%)	Desiccation time (h)	Survival (%)	Normal seedling (%)	Seedlings with plumule but no root (%)
82.4	0.0	100	100	0
46.9	23	70	10	60
36.1	29	46	0	30
30.4	32	38	0	28
24.3	35	33	0	25
20.0	37	29	0	11

35%, water content decreased to 47.8% and all axes lost plantlet formation ability, i.e. they had only hypocotyls and radicles, but not plumules and could not form normal seedlings.

Growth and dehydration of sprouting axes

Preculturing wampee axes on WPM for 3 weeks, resulted in the production of seedlings with 10 mm long hypocotyls and roots. The water content of these sprouting axes was 82.4%. They were dehydrated immediately using silica gel, or using a gradient of sucrose and silica gel. Using silica gel to dry sprouting axes to a water content of 46.9% resulted in a 30% loss of viability; 60% exhibited withered roots and only 10% retained a normal appearance. With further dehydration, no normal seedlings survived but most of the remaining seedlings had withered roots (Table 38.3).

When the sprouting axes were precultured for 7 days on WPM containing 27% sucrose prior to dehydration with silica gel, all survived and appeared normal if the water content was 46%. With further dehydration, no normal seedlings survived and most of the seedlings had withered roots (Table 38.4).

When water content was monitored for root tissue separately from the surviving hypocotyl and shoot portion of dehydrated sprouting axes, seri-

Table 38.4. Effect of preculturing on 27% sucrose-enriched WPM on the survival of sprouting axes following silica drying.

Water content (%)	Desiccation time (h)	Survival (%)	Normal seedling (%)	Seedlings with plumule but no root (%)
69.4	0.0	100	100	0
46.6	12	100	100	0
35.7	15	100	0	95
29.7	19	96	0	76
25.6	22	93	0	53
20.9	26	53	0	28

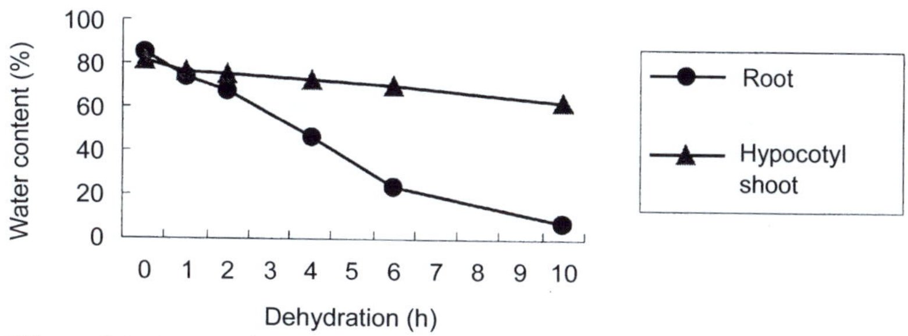

Fig. 38.1. Effect of drying on the water content of wampee root and hypocotyl-shoot tissue.

ous dehydration was observed in roots (Fig. 38.1). In the same situation, the drying rate of the radicle was more rapid than that of the hypocotyl and shoot. These features of the desiccation treatment contributed to the radicle drying earlier than the other parts of the sprouting axes.

Sprouting axes precultured in WPM containing 27% sucrose for 1 week were transferred successively to WPM containing 55%, 60% or 65% sucrose for further drying to 46, 37 and 36% moisture content, respectively (Table 38.5). In such conditions, the sprouting axes grew well after drying to 37% moisture content (60% sucrose), but continued drying (65% sucrose concentration) gave fewer normal seedlings. Using higher sucrose concentration to decrease water content was damaging. An alternative way of drying involved the use of silica gel following preculturing with 55% or 60% sucrose. Gradient-sucrose preculture and silica gel treatment lowered the axes water content to 25.3% (55% sucrose treatment) or 17.6% (60% sucrose treatment), and 64 % or 51% of seedlings, respectively, survived with withered roots. Such pretreated sprouted axes were considered to be more suitable for use in cryopreservation (Table 38.5) (Fig. 38.2).

Table 38.5. The growth of sprouting wampee seedlings monitored after 1 month in culture following desiccation with high concentrations of sucrose and 12 h drying with or without silica gel.

Gradient sucrose preculture	Drying with or without silica gel[a]	Water content (%)	Normal seedling (%)	Seedlings with plumule but no roots (%)
27% (7 days)	–	69.4	100	0
27% (7 days) to 55% (2 days)	–	46.2	100	0
	+	25.3	0	64
27% (7 days) to 60% (2 days)	–	37.4	100	0
	+	17.6	0	51
27% (7 days) to 65% (2 days)	–	35.6	78	0
	+	16.8	0	0

[a]–, without silica gel; + with silica gel.

Fig. 38.2. Normal plantlets growing from sprouting axes without roots after desiccation and culture on WPM. (A) Normal plantlet. (B) Plantlet without plumule. (C) Plantlet without root.

Induction of root growth from the rootless sprouting axes

Adventitious root induction was attemped either using that portion of the sprouted axis remaining after removal of the withered roots or using the excised plumule (connected to the hypocotyl). Explants were cultured for 1 month on WPM containing combinations of NAA and IBA at various concentrations. There was no induction of adventitious roots on the cut surface of the withered root adjacent to the hypocotyl. When dehydrated sprouting axes were excised at the base portion of the shoot, adventitious roots formed from the cut surface of the explant. The combination of 53.7 μmol/l NAA and 49.2 μmol/l IBA was the best for obtaining 69% regeneration of adventitious roots (Table 38.6).

Cryopreservation

Following sucrose (60%) pretreatment, axes were then treated with the vitrification solution PVS2, transferred to liquid nitrogen for 24 h, thawed and incubated in WPM for 1 week. Explants remained green but gradually became brownish. Under the electron microscope, we observed that cells of the hypocotyl were damaged but plumular cells appeared almost normal to subnormal, with the ultrastructure of the mitochondria and plastids evident (Fig. 38.3). Probably, cells of the shoot have a higher tolerance to drying and freezing and also a higher ability to form adventitious roots.

Table 38.6. Effect of plant growth regulators on the regeneration of adventitious roots from shoot-section explants.

Plant growth regulators		Regeneration of adventitious roots (%)
NAA (μmol/l)	IBA (μmol/l)	
2.7	2.5	16.7
10.7	9.8	53.8
53.7	49.2	68.7

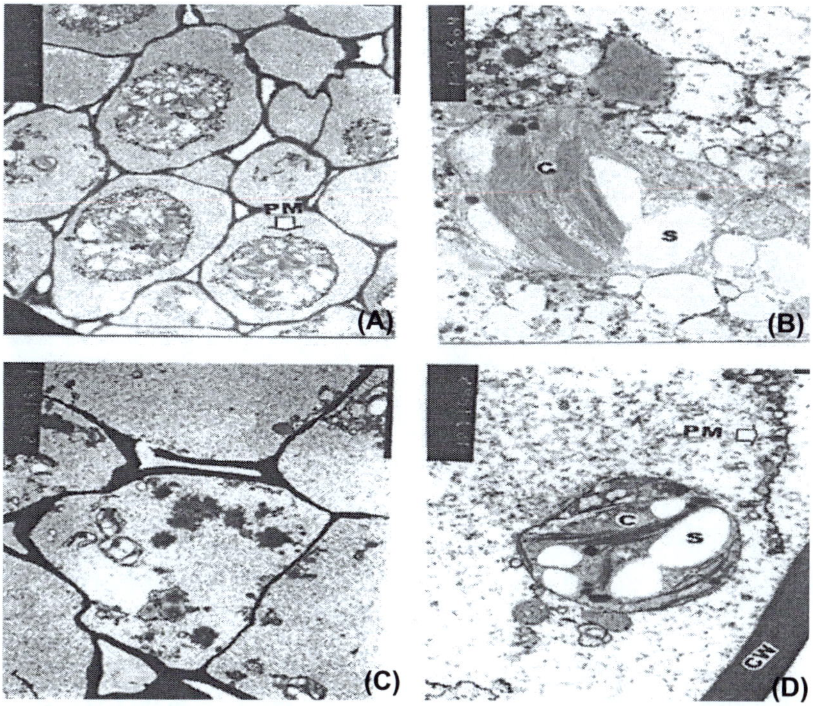

Fig. 38.3. Ultrastructure of wampee axes after vitrification cryopreservation. (A) Plumule, × 2000. (B) Plumule, × 14,000. (C) Hypocotyl, × 1400. (D) Hypocotyl, × 20,000. PM, plasma membrane; C, chloroplast; S, starch grain; CW, cell wall.

Discussion

When wampee axes were isolated, they tolerated dehydration to lower water contents than did whole seeds (Fu *et al.*, 1989), as has been recorded for recalcitrant seeds of other species (Berjak *et al.*, 1990). Therefore, the use of excised axes is preferable for manipulations aimed at increasing tolerance to dehydration. The presence of appropriate levels of soluble sugars has been emphasized in the phenomenon of inherent desiccation tolerance in a variety of organisms and structures, including seed embryos (Crowe *et al.*, 1984; Hoekstra *et al.*, 1989; Leopold *et al.*, 1992). Sucrose pretreatments increased the desiccation tolerance of isolated embryonic axes of wampee and of other species (Dumet and Berjak, 1997; Yap *et al.*, 1998). In fresh (non-sprouted) embryonic axes of wampee seeds, shoot meristems were most sensitive to drying and therefore died first. In contrast, the plumule of sprouting axes endured dehydration down to *c.* 20% axis moisture content, while the root withered rapidly. The present investigation has shown that sprouting axes with a living plumule but withered root retained the capacity for new root formation, when cultured on WPM with the appropriate combination and concentration of plant growth regulators (NAA and IBA).

The vitrification protocol has been applied to a wide range of plant materials (Sakai, 1998). The present experiment showed that the vitrification process increased wampee axes' freezing tolerance. However, the cryopreserved sprouting axes after incubation on WPM remained greenish for only 1 week. Because of this, there is probably cell survival rather than regrowth. As stated by Berjak *et al.* (1999), if cryopreservation procedures are successful, it should merely be a matter of manipulating post-freezing conditions to produce vigorous plantlets. However, real success, measured as the establishment of a high percentage of vigorous plantlets, often proves to be elusive for recalcitrant seeded material.

Acknowledgement

This project was supported by the Natural Sciences Foundation of Guangdong Province (980360, 001224).

References

Berjak, P., Farrant, J.M., Mycock, D.J. and Pammenter, N.W. (1990) Recalcitrant (homoiohydrous) seeds: the enigma of their desiccation-sensitivity. *Seed Science and Technology* 18, 297–310.

Berjak, P., Kioko, J.I., Walker, M., Mycock, D.C., Wesley-Smith, J., Watt, P. and Pammenter, N.M. (1999) Cryopreservation – an elusive goal. In: Marzalina, M., Khoo, K.C., Jayanthi, N., Tsan, F.Y. and Krishnapillay, B. (eds) *IUFRO Seed Symposium 1998 'Recalcitrant Seeds'*. Kuala Lumpur, Malaysia, pp. 96–109.

Crowe, J.H., Crowe, L.M. and Chapman, P. (1984) Preservation of membranes in anhydrobiotic organisms: the role of trehalose. *Science* 223, 701–703.

Dumet, D. and Berjak, P. (1997) Desiccation tolerance and cryopreservation of embryos of recalcitrant species. In: Ellis, R.H., Black, M., Murdock, A.J. and Hong, T.D. (eds) *Basic and Applied Aspects of Seed Biology*. Kluwer Academic Dordrecht, The Netherlands, pp. 771–776.

Fu, J.R., Zhang, B.Z., Wang, X.F., Qiao, Y.Z. and Huang, X.L. (1989) Studies on desiccation and wet storage of four recalcitrant seeds. In: *International Symposium on Horticulural Germplasm of Cultivated and Wild Fruit Trees. Part 1*. International Academic Publishers, Beijing, pp. 121–125.

Fu, J.R., Zhang, B.Z., Wang, X.F., Qiao, Y.Z. and Huang, X.L. (1990) Physiological studies on deciccation, wet storage and cryopreservation of recalcitrant seeds of three fruit species and their excised embryonic axes. *Seed Science and Technology*, 18, 743–754.

Fu, J.R., Jin, J.P., Peng, Y.F. and Xia, Q.H. (1994) Desiccation tolerance in two species with recalcitrant seeds: *Clausena lansium* (Lour.) and *Litchi chinensis* (Sonn.). *Seed Science Research* 4, 257–261.

Fu, J.R., Huang, X.M. and Song, S.Q. (2000) Manipulation of desiccation-sensitive axes of wampee (*Clausena lansium*) to facilitate increased dehydration tolerance. *Seed Science Research* 10, 397–400.

Hoekstra, E.A., Crowe, L.M. and Crowe, J.H. (1989) Differential desiccation sensitivity of corn and *Pennisetum* pollen linked to their sucrose contents. *Plant, Cell and Environment* 12, 83–91.

Huang, X.M., Fu, J.R. and Song, S.Q. (2000) The desiccation-sensitivity of wampee

(Clausena lansium) axes in relation to seedling formation. *Acta Scientiarum Naturalium Universitatis Sunyatseni* 39, 87–90.

Leopold, A.C., Bruni, F. and Williams, R.T. (1992) Water in dry organisms. In: Somero, G.N., Osmond, C.B. and Bolis, C.L. (eds) *Water and Life.* Springer-Verlag, Berlin, pp. 161–173.

Lu, W.J. and Fu, J.R. (1997) The induction of desiccation tolerance in wampee *(Clausena lansium). Acta Scientiarum Naturalium Universitatis Sunyatseni* 36, 118–120.

McCown, B.H. and Lloyd, G. (1981) Wood plant medium (WPM) a mineral nutrient formulation for microculture of woody plant species. *Hortscience* 16, 453.

Roberts, E.H. (1973) Predicting the storage life of seeds. *Seed Science and Technology* 1, 499–514.

Roberts, E.H., King, M.W. and Ellis, R.H. (1984) Recalcitrant seeds: their recognition and storage. In: Holden, J.A. and Williams, J.T. (eds) *Crop Genetic Resources: Conservation and Evaluation.* George Allen & Unwin, London, pp. 38–52.

Sakai, A. (1998) The potential of vitrification technique for conservation of genetic resources of tropical and temperate plant species. In: Marzalina, M., Khoo, K.C., Jayanthi, N.,Tsan, F.Y. and Krishnapillay, B. (eds) *IUFRO Seed Symposium 1998 'Recalcitrant Seeds'*, Kuala Lumpur, Malaysia, pp. 110–118.

Yap, L.V., Hor, Y.L. and Normah, M.N. (1998) Effect of sucrose preculture and subsequent desiccation on cryopreservation of alginate-encapsulated *Hevea brasiliensis* embryo. In: Marzalina, M., Khoo, K.C., Jayanthi, N., Tsan, F.Y. and Krishnapillay, B. (eds) *IUFRO Seed Symposium 1998 'Recalcitrant Seeds'*, Kuala Lumpur, Malaysia, pp. 140–145.

39 Dehydration Damage and its Repair in Imbibed Soybean (*Glycine max*) Seeds

Songquan Song[1,2], Changhua Jiang[2], Runzheng Chen[2], Jiarui Fu[2] and Florent Engelmann[3]

[1]Xishuangbanna Tropical Botanical Garden, The Chinese Academy of Sciences, Menglun, Mengla 666303, Yunnan, China; [2]School of Life Sciences, Sun Yat-Sen University, Guangzhou 510275, China; [3]International Plant Genetic Resources Institute, Via delle Sette Chiese 142, 00145 Rome, Italy

Introduction

Orthodox seeds acquire desiccation tolerance during development and generally undergo substantial drying as the final phase of development. Orthodox seeds may be further dehydrated after they have been shed and will survive in this state for a considerable time.

Imbibition of water by orthodox seeds initiates a series of metabolic steps that lead to germination and, during this sequence of events, desiccation tolerance within the seed is lost. During the early stages of germination and prior to radicle emergence in most species, the seed can be dried to its original moisture content without causing injury. The same degree of drying imposed at progressively later stages of germination reduces seed vigour dramatically and, if imposed after radicle elongation has commenced, usually results in seedling death (McKersie *et al.*, 1988).

Senaratna and McKersie (1983) studied germinating soybean (*Glycine max* L. Merr. cv. Maple Arrow) seeds and found that seeds germinated for 6 h were tolerant of severe drying, while those germinated for 36 h were not. Koster and Leopold (1988) found that germinating soybean (cv. Williams) seeds lost desiccation tolerance between 12 and 18 h of imbibition, while germinating pea (*Pisum sativum* L. cv. Alaska) seeds lost tolerance between 18 and 24 h and maize (*Zea mays* L. cv. Merit) seeds by 48 h imbibition. Koster and Leopold (1988) correlated the loss of desiccation tolerance with changes in embryo sugar composition. Leprince *et al.* (1995) used electrolyte leakage to determine 'critical moisture contents' for germinating bean (*Phaseolus vulgaris* L. cv. Pole Kentucky Wonder) and maize (cv. Kelvedon Glory) seeds.

 The Biology of Seeds: Recent Research Advances (eds G. Nicolás, K.J. Bradford, D. Côme and H.W. Pritchard)

Below these water contents, these seeds were damaged by desiccation, and the injury was hypothesized to be caused by oxygen radicals that resulted from respiratory metabolism. All of the studies described above used whole seeds, but the moisture contents of embryonic axes are usually higher than those of the whole seeds during imbibition. In some studies germinating seeds were dried back to a single water content for the determination of desiccation tolerance. In other studies, only electrolyte leakage was used to assess desiccation tolerance of seed, without survival data. The critera for determining survival also differed among the studies.

Calcium plays an important role as second messenger in animal and plant cell systems. In higher plants, calcium is involved in responses of cells to a number of endogenous or exogenous signals such as hormones, light, mass acceleration or touch (Bush, 1995; Haley *et al.*, 1995). To our knowledge, the effects of calcium on desiccation tolerance of seed or axes, especially imbibed orthodox seeds, are not clear.

Soybean seeds imbibed for different times can provide convenient and reliable experimental material with differing desiccation tolerance for the study of desiccation tolerance and sensitivity. In this study, we attempted to better define the time course of the loss of desiccation tolerance in soybean seeds and to determine how this quantitative trait changes during imbibition/germination. To this end, we looked separately at survival of seeds and axes, at the changes in electrolyte leakage, and at the growth rates of seedlings produced by surviving seeds or axes. The effects of drying at different rates on the desiccation tolerance of axes and the influence of Ca^{2+} on repair of dehydration damage were also investigated. The results will provide a framework for further studies of loss of desiccation tolerance in germinating orthodox seeds and the intolerance of desiccation in recalcitrant seeds.

Materials and Methods

Plant material

Current harvest of soybean (cv. Aijiaozao) seeds were obtained from the Oil Crops Research Institute of the Chinese Academy of Agricultural Sciences (Wuchang, Hubei) and were kept at 15°C until used. The seeds were surface-sterilized in a solution of 1% hypochlorite, rinsed three times in sterile water, and then imbibed by placing the seeds in a shallow layer of distilled water or treatment solution (Ca^{2+}, EDTA, etc.) such that half the seed was immersed. Axes were excised from the seeds that had been imbibed for different times and were then treated as indicated below. All manipulations and germination studies were conducted at room temperature (20°C).

Water content determination

Water contents of 20 individual seeds or axes were determined gravimetrically (80°C for 48 h). Water contents are expressed on a dry mass basis (g H_2O/g dry mass) (g/g).

Desiccation treatment

Desiccation of the differentially pre-imbibed seeds was achieved by burying them for different times in activated silica gel within a closed desiccator. Rapid drying of excised axes was achieved by placing axes over activated silica gel for different times within a closed desiccator in which there was a small fan. Slow drying of excised axes was achieved by placing axes over saturated sodium chloride (relatively humidity 75%) within a closed container. The temperature when axes were dried was 20 ± 1°C.

Assessment of germination and survival

Batches of 30 treated seeds and/or axes were germinated on moist filter paper in Petri dishes for 5 days (20°C, dark). Seeds showing radicle emergence and axes showing a marked increase in length and volume were scored as germinated and thus as having survived the treatment. Although dried seeds and axes take up water during the early stages of re-imbibition, those seeds and axes that were injured by desiccation deteriorated progressively during continued re-imbibition.

Conductivity test

Electrolyte leakage from batches of 15 treated seeds and axes was measured for 3 h in 30 ml distilled water using a DDS-11A conductivity meter (Shanghai, China). The conductivity of the leachates was measured immediately after the drying treatments and leakage rates were expressed as μS/cm/g dry weight/h.

Results

Changes occurring during imbibition of seeds

Germination commences with the uptake of water by the dry seed and is completed when a part of the embryo, usually the radicle, extends to penetrate the structures that surround it. Water content of dry soybean seeds (initially 0.112 g/g) exhibited a triphasic pattern of water uptake with a marked increase during the initial phase of imbibition, and then a slow increase, followed by a second substantial increase as the radicle penetrated its surrounding tissues (Fig. 39.1A). The second phase of water uptake by the soybean axes, however, was shorter than that of whole seeds, and axis water content increased continually after imbibition. The rates of water uptake by whole seeds and axes were about the same until 24 h of imbibition, after which the rate of water uptake by axes was greater than that by the whole seed (Fig. 39.1A). When seeds had imbibed for 24 h, for example, the water contents of seeds and axes were 1.47 and 1.68 g/g, respectively; at 48 h they were 1.84 and 3.37 g/g, respectively. The proportion of germinated seeds increased with imbibition time; the first seeds germinated before about 18 h,

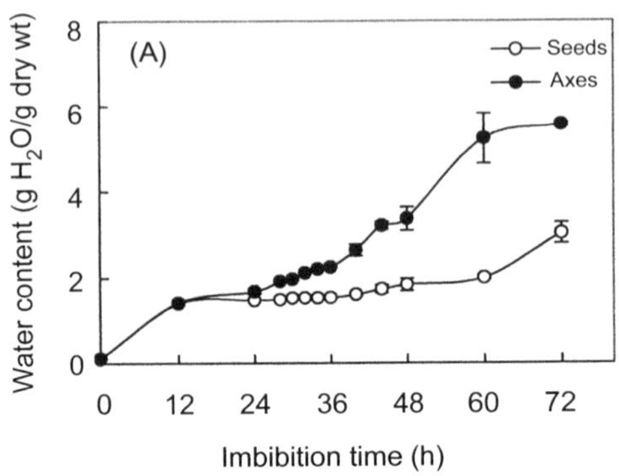

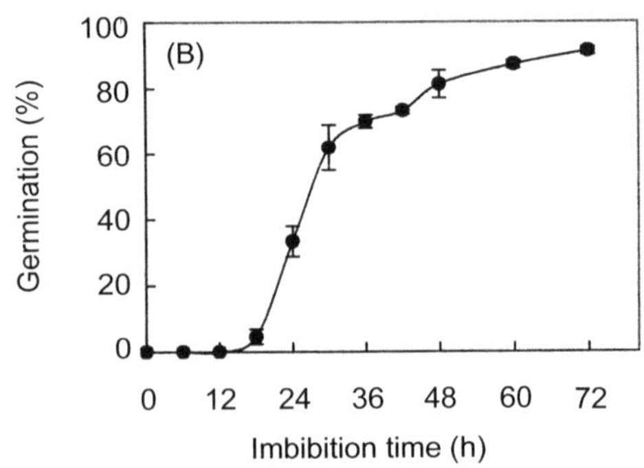

Fig. 39.1. Imbibition and germination of soybean seeds. (A) Time courses of water content in seeds and axes during imbibition at 20°C. Values are means ± SD of three replicates of 20 seeds or 20 axes each and expressed on a dry mass basis. (B) Germination (radicle emergence) of soybean seeds imbibed in water. Values are means ± SD of three replicates of 50 seeds each.

and the time and thermal time required for 50% germination were about 28 h and 23.3°C (Fig. 39.1B).

Effect of pre-imbibition on desiccation tolerance of seeds

Dehydration of seeds pre-imbibed for different times was achieved by burying them in activated silica gel for 24 h to a water content of 0.09–0.1 g/g (water content of dry seeds was 0.11 g/g). Survival of soybean seeds that had been dried after pre-imbibition progressively decreased with pre-imbibition (Fig. 39.2). The pre-imbibition time at which 50% of seeds were killed by subsequent dehydration was about 32 h. The desiccation tolerance of the epicotyl was greater than that of the radicle (data not shown).

The fresh weight of seedlings integrates the desiccation tolerance of the seed and the subsequent metabolism associated with germination and growth. The fresh weight of seedlings produced by surviving seeds declined as the pre-imbibition period increased. The pre-imbibition time at which seedling weight was decreased by 50% was shorter than 24 h (Fig. 39.2).

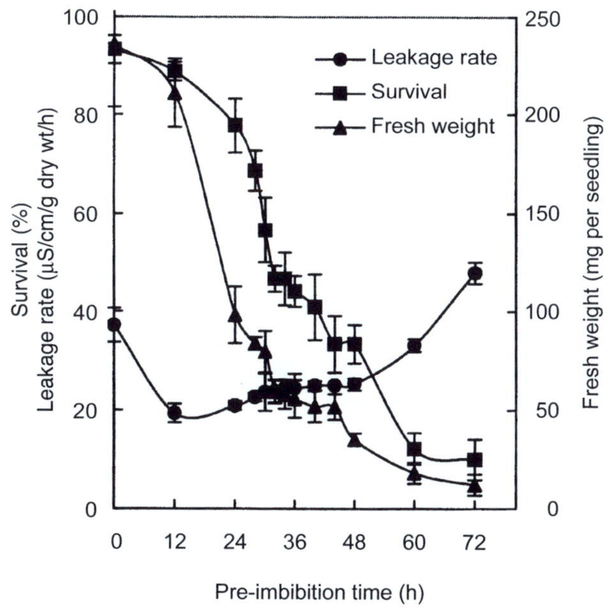

Fig. 39.2. Changes in desiccation tolerance of seeds during pre-imbibition. Seeds were pre-imbibed for various times and then dehydrated (seeds buried in silica gel for 24 h, water content approximately 0.1 g/g) and re-imbibed on moist filter paper for 5 days at 20°C. Seeds showing radicle emergence or showing increase in length and volume of axes were counted as having survived. Fresh weights are of seedlings produced by surviving seeds (does not include cotyledons). Rates of electrolyte leakage were measured by placing seeds in 30 ml distilled water and measuring conductivity of the leachate. All values are means ± SD of three replicates of 30 seeds each for survival and fresh weight and 15 seeds each for electrolyte leakage.

Leakage rates of electrolytes from seeds after dehydration decreased after the initial phase of pre-imbibition, and then increased with increasing pre-imbibition time (Fig. 39.2).

Effects of drying at different rates on desiccation tolerance of soybean axes

To assess and quantify the loss of desiccation tolerance during imbibition and germination, soybean seeds were pre-imbibed for different times; the axes were excised and then dried at different rates for increasing periods.

Water contents of axes from seeds imbibed for 24, 32 and 48 h were 1.70, 1.93 and 3.18 g/g, respectively, and declined rapidly over activated silica gel and slowly over saturated sodium chloride (Fig. 39.3). For all three pre-imbibition times, desiccation tolerance of axes (which was measured by survival, seedling weight produced by surviving axes and leakage rates of electrolytes), gradually decreased with dehydration; however, the longer the pre-imbibition time, the faster the desiccation tolerance of axes was lost (Figs 39.4, 39.5 and 39.6). However, the water content at which 50% of axes

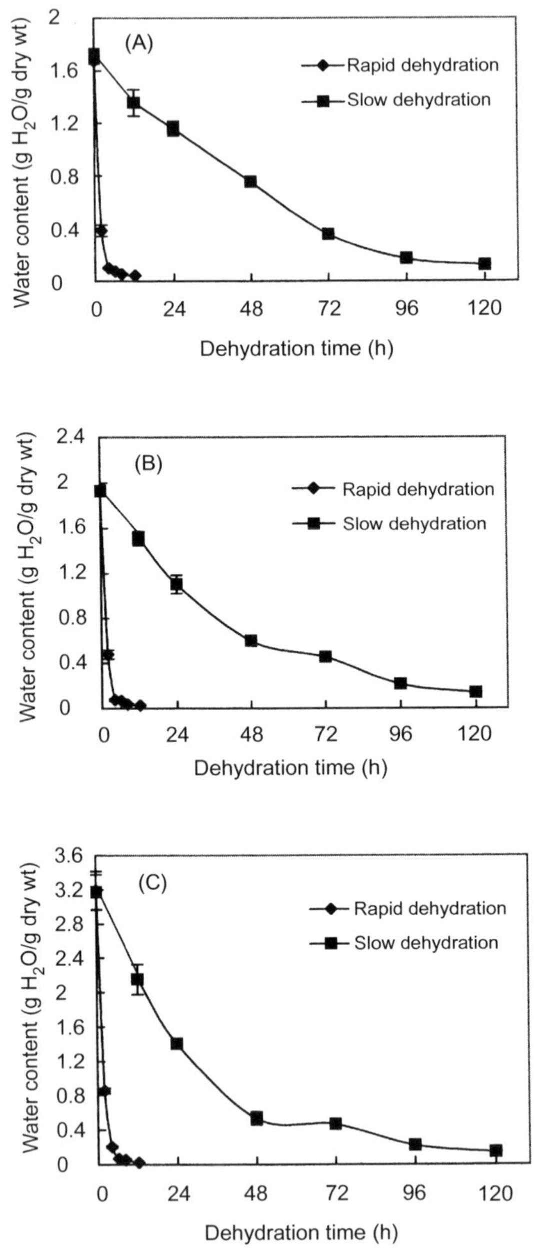

Fig. 39.3. Effects of drying at different rates on water contents of soybean axes. Seeds were imbibed at 20°C for 24, 32 and 48 h, respectively, and axes were excised and dried at different rates for the indicated times at 25°C. Rapid drying of axes was achieved by placing axes over silica gel; slow drying of axes was achieved by placing axes over saturated sodium chloride. Values are means ± SD of three replicates of 20 axes each. (A), (B) and (C) Soybean seeds imbibed for 24, 32 and 48 h, respectively.

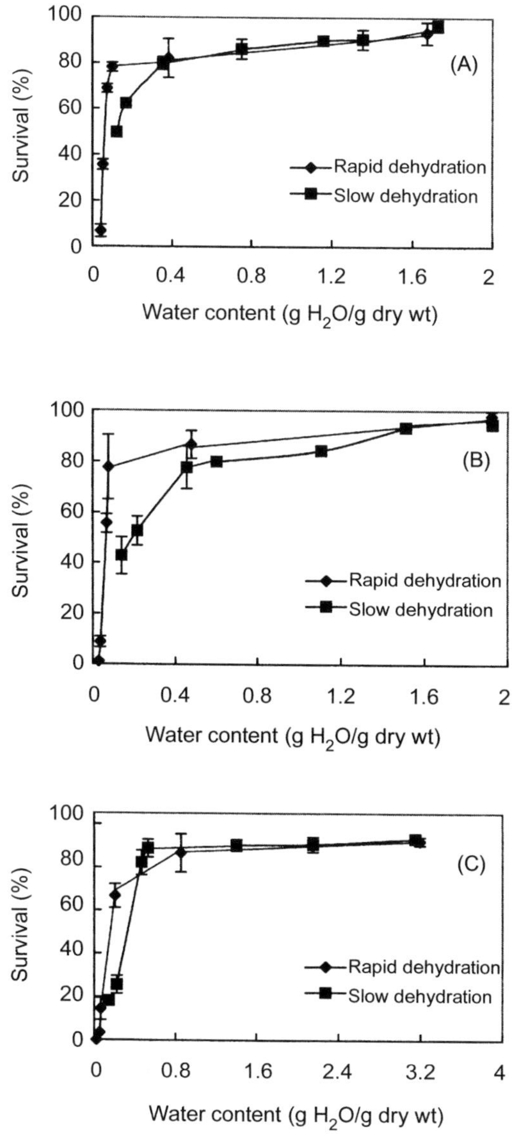

Fig. 39.4. Effects of drying at different rates on survival of soybean axes. Imbibition of seeds, excision and drying of axes were as described in Fig. 39.3. Treated axes were re-imbibed on moist filter paper for 5 days at 20°C. Axes showing an increase in length and volume were counted as surviving. Values are means ± SD of three replicates of 30 axes each. (A), (B) and (C) Soybean seeds imbibed for 24, 32 and 48 h, respectively.

were killed (W_{50}) by rapid and slow dehydration was different. When axes from seeds pre-imbibed for 24, 32 and 48 h, respectively, were rapidly dehydrated, W_{50} was about 0.06, 0.06 and 0.16 g/g, respectively; when slowly dehydrated, these values were about 0.12, 0.18 and 0.32 g/g, respectively

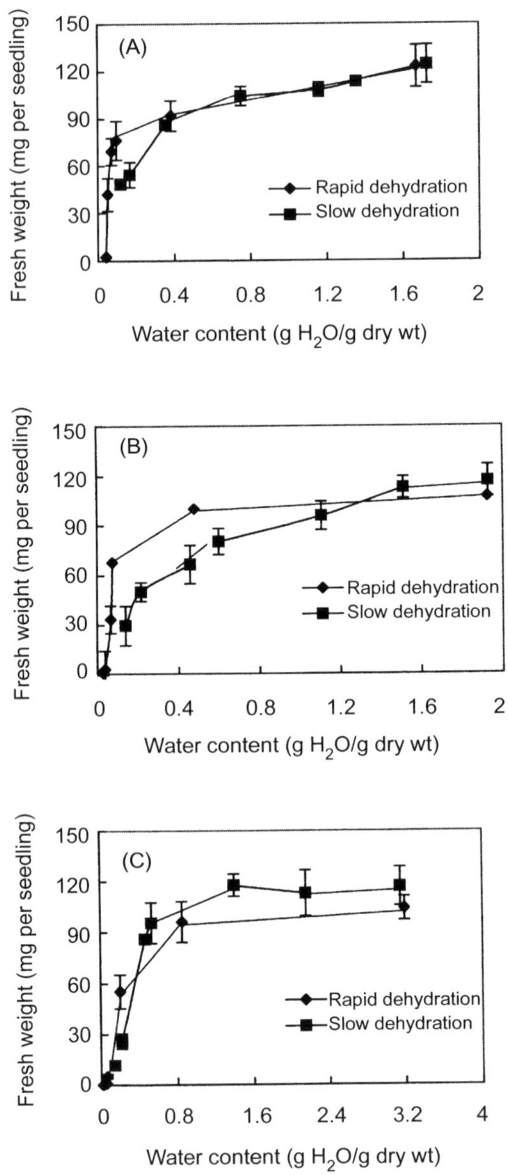

Fig. 39.5. Effects of drying at different rates on seedling weight of soybean axes. Imbibition of seeds, excision and drying and re-imbibition of axes were as described in Fig. 39.4. Seedling fresh weights produced by surviving axes do not include the cotyledons. Values are means ± SD of three replicates of 30 axes each. (A), (B) and (C) Soybean seeds imbibed for 24, 32 and 48 h, respectively.

(Fig. 39.4). Overall, rapid drying improved desiccation tolerance of axes from seeds pre-imbibed for 24 and 32 h relative to slow drying, but this relationship was reversed for seeds pre-imbibed for 48 h (Figs 39.4, 39.5 and 39.6).

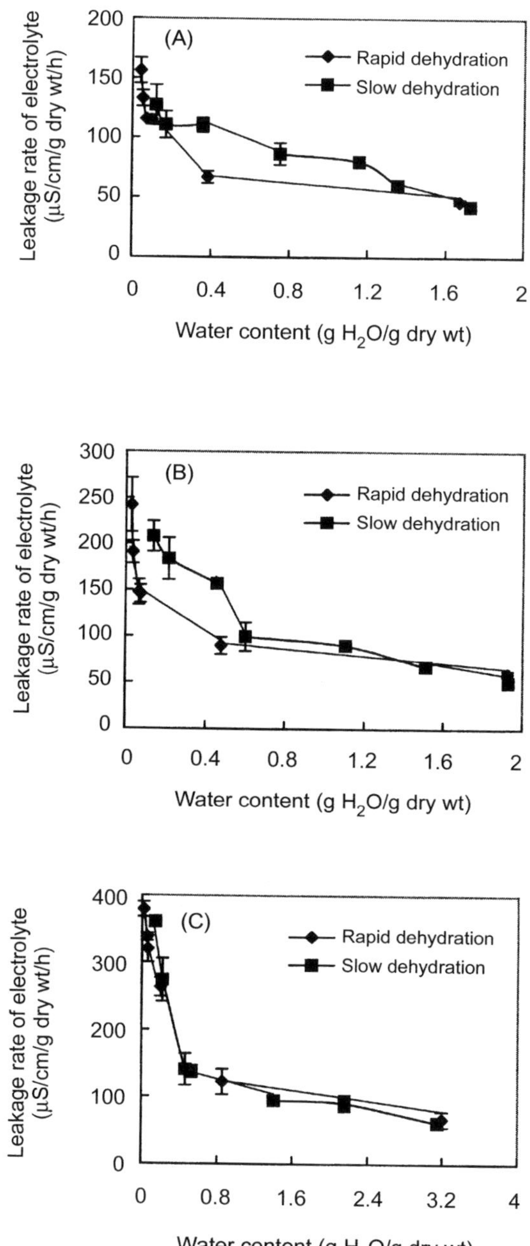

Fig. 39.6. Effects of drying at different rates on the rates of electrolyte leakage from soybean axes. Imbibition of seeds, excision and drying of axes were as described in Fig. 39.3. Treated axes were placed in 30 ml distilled water and conductivity of leachate was measured. Values are means ± SD of three replicates of 15 seeds each. (A), (B) and (C) Soybean seeds imbibed for 24, 32 and 48 h, respectively.

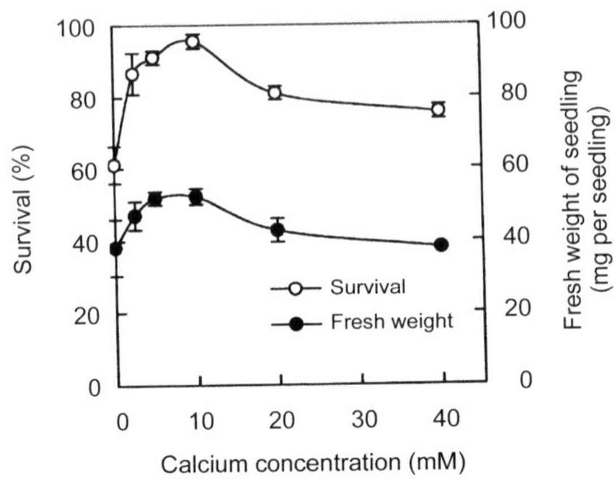

Fig. 39.7. Effects of different concentrations of Ca^{2+} on survival of soybean axes. Axes from soybean seeds which were pre-imbibed for 32 h were dehydrated to a water content of 0.07 g/g over silica gel, and then treated with the indicated concentration of Ca^{2+} for 5 days at 20°C. Axes showing increase in length and volume were counted as surviving. Values are means ± SD of three replicates of 30 axes each.

Effects of calcium on repair of desiccation damage during imbibition of soybean axes

Calcium (0.5–5 mM) supplied during imbibition increased germination percentages and subsequent growth of soybean seeds. The optimal Ca^{2+} concentration was about 1 mM, and germination percentage and seedling growth were inhibited as Ca^{2+} concentration increased over 10 mM (data not shown).

Axes from soybean seeds pre-imbibed for 32 h were dehydrated to a water content of 0.07 g/g, and then treated with different concentrations of Ca^{2+} during imbibition. Ca^{2+} significantly increased survival of axes and seedling weights produced by surviving axes, with an optimal Ca^{2+} concentration of about 10 mM (Fig. 39.7). For example, survival of axes that were treated with 10 mM Ca^{2+} increased by 56.4% relative to that of control axes (treated with 0 mM Ca^{2+}). Similarly, axes from soybean seeds pre-imbibed for 32 h after drying for different times were imbibed in 10 mM Ca^{2+}, which dramatically increased survival of axes and seedling weight, especially those axes that had a lower water content (Fig. 39.8).

EGTA treatment during re-imbibition markedly reduced desiccation tolerance of axes from soybean seeds dried after pre-imbibition in water for 32 h; no axes survived if EGTA concentration was greater than 5 mM in the second imbibition solution (Fig. 39.9). EDTA treatment was even more effective, as survival of axes was zero when EDTA concentration was 2.5 mM (data not shown). Addition of 5 and 10 mM Ca^{2+} could effectively antagonize the effects of EGTA (Fig. 39.9) and EDTA (data not shown) on decreasing desiccation tolerance of soybean axes.

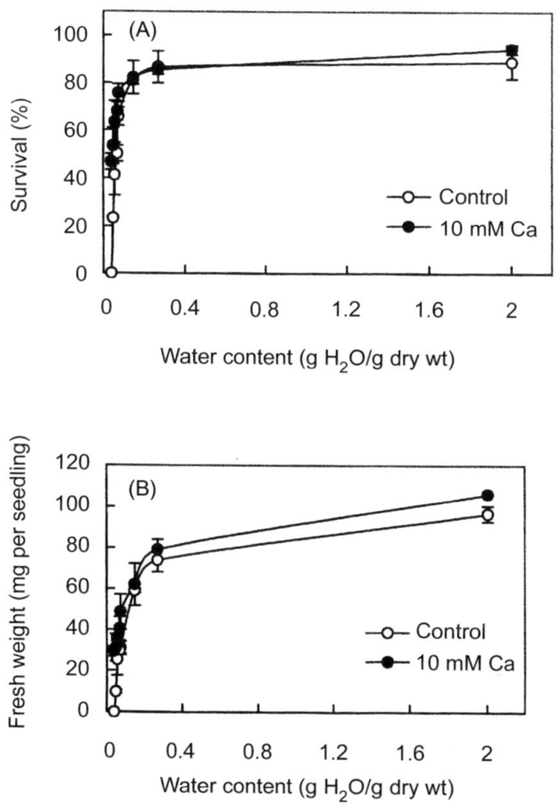

Fig. 39.8. Effects of Ca^{2+} on desiccation tolerance of soybean axes. Axes from soybean seeds which were pre-imbibed for 32 h were dehydrated to the indicated water content over silica gel, and then treated with 10 mM Ca^{2+} for 5 days at 20°C. Axes showing an increase in length and volume were counted as surviving. Seedling fresh weights produced by surviving axes do not include the cotyledons. Values are means ± SD of three replicates of 30 axes each. (A), survival; (B), fresh weight of seedlings.

Discussion

Water uptake by soybean seeds exhibited a typical triphasic pattern of water uptake with a marked increase during the initial phase of imbibition, and then a slower increase, followed by a second substantial increase as the radicle penetrated its surrounding tissues (Fig. 39.1A). The initial rapid hydration phase (phase I) was related to the matric potential of dry seeds; the second slow hydration phase (phase II) corresponded to the period of germination. The length of phase II is affected by imbibition temperature and the water potential of the medium in which the seeds are imbibed (Bradford, 1995; Bewley, 1997). Phase II of water uptake by soybean axes, however, was shorter than that of whole seeds (Fig. 39.1A), and was about 12 h (from 12 to 24 h of imbibition). The soybean axis is located near the surface of the seed and can imbibe water more rapidly than the whole seed.

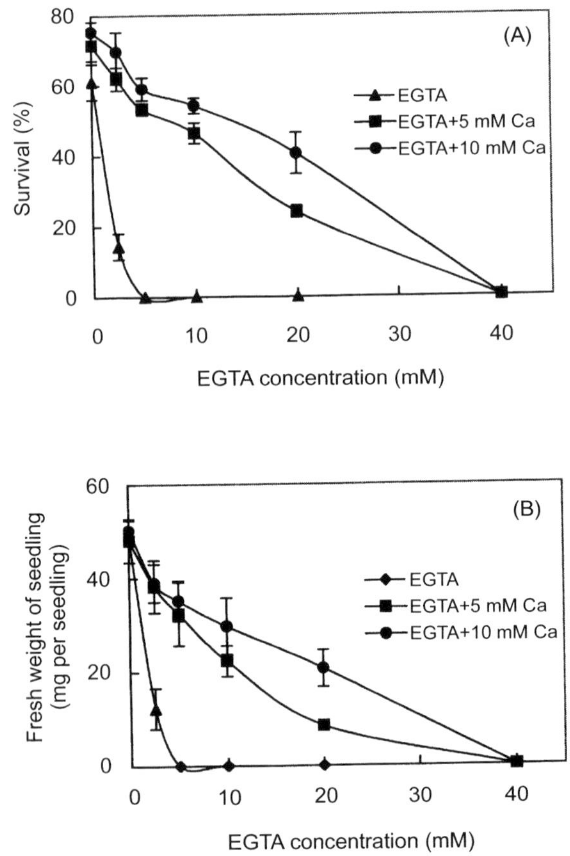

Fig. 39.9. Antagonism of Ca^{2+} to EGTA in decreasing desiccation tolerance of soybean axes. Axes from soybean seeds which were pre-imbibed for 32 h were dehydrated to a water content of 0.092 g/g over silica gel, and then treated with the indicated EGTA concentration, EGTA + 5 mM Ca^{2+}, and EGTA + 10 mM Ca^{2+}, respectively, for 5 days at 20°C. Axes showing an increase in length and volume were counted as surviving. Values are means ± SD of three replicates of 30 axes each. (A), survival; (B), fresh weight of seedlings.

This, together with a different chemical composition, leads to a higher water content of the axis than of whole seed (Fig. 39.1A). By 24 h of imbibition, over 30% of the seeds had germinated (Fig, 39.1B); that is, the cells of some axes had extended, taking up more water, and so phase II of water uptake by soybean axes was shorter. The axes comprised only a small part of the dry mass of the seeds and so the pattern of water uptake by axes did not influence the pattern observed in whole seeds.

Increasing periods of pre-imbibition of seeds prior to the subsequent dehydration of whole seeds led to declining survival and fresh weight of seedlings (see Fig. 39.2), showing that desiccation tolerance was gradually lost in imbibing soybean seeds. The progressive loss of desiccation tolerance during imbibition of soybean seeds is similar to findings for bean and maize by Leprince *et al.* (1995) and for mung bean (*Vigna radiata*) by Sun (1999). The pre-imbibition time at which 50% of seeds were killed by subsequent dehydration was about 32 h, while the pre-imbibition time at which seedling weight was decreased 50% was less than 24 h (Fig. 39.2), showing that seedling weight produced by surviving seeds was a sensitive parameter for assessing desiccation tolerance. The desiccation tolerance of the epicotyl was greater than that of the radicle (data not shown), as observed for mung bean by Qian *et al.* (2000).

Leakage rates of electrolytes from seeds after dehydration decreased at the initial phase of pre-imbibition, and then increased with increasing pre-imbibition time (Fig. 39.2). The influx of water into the cells of dry seeds during the initial phase of pre-imbibition results in temporary structural perturbations, particularly to membranes, which lead to an immediate and rapid leakage of solute and low-molecular-weight metabolites into the surrounding imbibition solution. This is a consequence of the transition of the membrane phospholipid components from the gel phase formed during maturation drying to the hydrated liquid-crystalline state. Within a short time of rehydration, the membranes return to their more stable configuration, at which time solute leakage is curtailed (reviewed by Bewley, 1997). Increasing leakage rates of electrolytes from seeds after drying with increasing time of pre-imbibition was strongly correlated with gradual loss of desiccation tolerance of soybean seeds (Fig. 39.2).

The longer the pre-imbibition time, the faster desiccation tolerance of axes was lost (see Figs 39.4, 39.5 and 39.6). The rate of drying affected desiccation tolerance of axes differently, depending upon the duration of pre-imbibition. For axes from soybean seeds pre-imbibed for 24 and 32 h, rapid drying improved desiccation tolerance of axes more than slow drying (Figs 39.4 and 39.6), similar to results for axes of recalcitrant seeds reviewed by Pammenter and Berjak (1999) and for axes of Chinese wampee by Wu *et al.* (2001). For axes from soybean seeds pre-imbibed for 48 h, however, rapid drying decreased desiccation tolerance (Figs 39.4, 39.5 and 39.6), similar to results for vegetable tissues reviewed by Bray (1997).

One of the earliest symptoms of injury following dehydration is the loss of function or structure of either the plasmalemma or organellar membranes (McKersie *et al.*, 1988). Consequently, sensitivity to desiccation of both orthodox and recalcitrant seed tissues has been quantitatively expressed in terms of a 'critical water content' determined by a leakage assay (Berjak *et al.*, 1993; Vertucci *et al.*, 1993; Leprince *et al.*, 1995). Each of these laboratories quantified electrolyte leakage differently, making direct comparisons difficult. The growth rate of seedlings, reflecting both desiccation tolerance and subsequent germination and/or growth, could also be used as a parameter to assess desiccation tolerance. Additionally, Pammenter *et al.* (1998) pointed out that the response to desiccation can depend on the rate of drying, and consequently they questioned the concept of a 'critical water content'. Therefore, it seems important to include survival, leakage rate and growth rate of seedlings when discussing desiccation tolerance.

We found that Ca^{2+} (< 5 mM) could increase germination percentage and subsequent growth of soybean seeds (data not shown), similar to results for maize seeds by Song *et al.* (1995), who reported that Ca^{2+} can increase activities of α- and β-amylase in the embryo and endosperm, and accelerate mobilization of starch and soluble proteins in the endosperm.

Ca^{2+} dramatically increased survival and seedling weight produced by surviving axes from soybean seeds pre-imbibed for 32 h before drying and re-imbibition (see Figs 39.7 and 39.8). These results were similar to those of Fu *et al.* (1999), who found that pretreatment with 1 mM Ca^{2+} for 48 h could enhance the vigour of wampee seeds following subsequent dehydration. The mechanism by which Ca^{2+} increases desiccation tolerance of soybean axes during imbibition is unclear. Ca^{2+} plays a key role in plant growth and development because changes in cellular Ca^{2+}, acting through Ca^{2+}-modulated proteins and their targets, regulate an astonishing variety of cellular processes (Bush, 1995).

EGTA treatment markedly decreased desiccation tolerance of soybean axes and subsequent growth of surviving axes (see Fig. 39.9). Ca^{2+} (5 and 10 mM) effectively antagonized the negative effects of EGTA on desiccation tolerance of soybean axes (Fig. 39.9). Further research is necessary to determine how Ca^{2+} and Ca^{2+}-chelators improve survival after desiccation.

Soybean seeds progressively lost desiccation tolerance with imbibition; this desiccation tolerance is a quantitative feature which is mainly dependent on the germination activity of the seed or axis. Many characteristics of germinating orthodox seeds are similar to those of recalcitrant seeds. These include desiccation sensitivity, high degree of subcellular development and metabolic activity, the degree of sensitivity being affected by developmental stage and drying environment, and desiccation injury possibly being associated with free radical-mediated oxidative damage (Vertucci and Farrant, 1995; Berjak and Pammenter, 1997; Pammenter and Berjak, 1999; Song and Fu, 1999). Germinating seeds of different desiccation sensitivity and developmental stage can be prepared using the same seed lot, and germination activity and desiccation sensitivity of seeds can be controlled by altering temperature, oxygen and water content during imbibition. Consequently, desiccation-sensitive imbibing orthodox seeds can serve as a model system for studies of seed recalcitrance.

Acknowledgements

We are grateful to the KIP Pilot Project (KSCX2-SW-117) and Hundreds Ta Pent Program of the Chinese Academy of Sciences, to the Natural Science Foundation of Guangdong of China (001224) and to the International Plant Genetic Resources Institute for support.

References

Berjak, P. and Pammenter, N.W. (1997) Progress in the understanding and manipulation of desiccation-sensitive (recalcitrant) seeds. In: Ellis, R.M., Black, M.,

Murdoch, A.J. and Hong, T.D. (eds) *Basic and Applied Aspects of Seed Biology*. Kluwer Academic Publishers, Dordrecht, The Netherlands, pp. 689–703.

Berjak, P., Vertucci, C.W. and Pammenter, N.M. (1993) Effects of developmental status and dehydration rate on characteristics of water and desiccation-sensitivity in recalcitrant seeds of *Camellia sinensis*. *Seed Science Research* 3, 155–166.

Bewley, J.D. (1997) Seed germination and dormancy. *The Plant Cell* 9, 1055–1066.

Bradford, K.J. (1995) Water relations in seed germination. In: Kigel, J. and Galili, G. (eds) *Seed Development and Germination*. Marcel Dekker, New York, pp. 351–396.

Bray, E.A. (1997) Plant response to water deficit. *Trends in Plant Science* 2, 48–54.

Bush, D.S. (1995) Calcium regulation in plant cells and its roles in signaling. *Annual Review of Plant Physiology and Plant Molecular Biology* 46, 95–122.

Fu, J.R., Xiang, X., Song, S.Q., Jin, J.P. and Lu, W.J. (1999) Desiccation tolerance of recalcitrant wampee seed associated with developmental status and its enhancing pretreatments. In: Marzalina, M., Khoo, K.C., Jayanthi, N., Tsan, F.Y. and Krishnapillay, B. (eds) *IUFRO Seed Symposium 1998, 'Recalcitrant Seeds', Proceedings of the Conference*, Kuala Lumpur, Malaysia, pp. 1–13.

Haley, A., Russell, A.J., Wood, N., Allan, A.C., Knight, M., Cambell, K. and Trewavas, A.T. (1995) Effects of mechanical signaling on plant cell cytosolic calcium. *Proceedings of the National Academy of Sciences USA* 92, 4124–4128.

Koster, K.L. and Leopold, A.C. (1988) Sugars and desiccation tolerance in seeds. *Plant Physiology* 88, 829–832.

Leprince, O., Vertucci, C.W., Hendry, G.A.F. and Atherton, N.M. (1995) The expression of desiccation-induced damage in orthodox seeds is a function of oxygen and temperature. *Physiologia Plantarum* 94, 233–240.

McKersie, B.D., Senaratna, T., Walker, M.A., Kendall, E.J. and Hetherington, P.R. (1988) Deterioration of membranes during aging in plants: evidence for free radical mediation. In: Nooden, L.D. and Leopold, A.C. (eds) *Senescence and Aging in Plants*. Academic Press, San Diego, California, pp. 441–464.

Pammenter, N.W. and Berjak, P. (1999) A review of recalcitrant seed physiology in relation to desiccation-tolerance mechanisms. *Seed Science Research* 9, 13–37.

Pammenter, N.W., Greggains, V., Kioko, J., Wesley-Smith, J. and Finch-Savage, W.E. (1998) Effects of differential drying rates on viability retention of recalcitrant seeds of *Ekebergia capensis*. *Seed Science Research* 8, 463–471.

Qian, C.M., Song, S.Q., Wu, X.J. and Fu, J.R. (2000) Temporal pattern of changes in desiccation tolerance during imbibition of *Phaseolus radiatus* L. seeds. *Acta Scientiarum Naturalium Universitatis Sunyatseni*, 39 (Supplement), 104–111.

Senaratna, T. and McKersie, B.D. (1983) Dehydration injury in germinating soybean (*Glycine max* L. Merr.) seeds. *Plant Physiology* 72, 620–624.

Song, S.Q. and Fu, J.R. (1999) Desiccation sensitivity and peroxidation of membrane lipids in lychee seeds. *Tropical Sciences* 39, 102–106.

Song, S.Q., Chen, J. and Fu, J.R. (1995) Seed vigor of *Zea mays* L. improved by calcium. *Journal of Tropical and Subtropical Botany* 3, 56–60.

Sun, W.Q. (1999) Desiccation sensitivity of recalcitrant seeds and germinated orthodox seeds: can germinated orthodox seeds serve as a model system for studies of recalcitrance? In: Marzalina, M., Khoo, K.C., Jayanthi, N., Tsan, F.Y. and Krishnapillay, B. (eds) *IUFRO Seed Symposium 1998, 'Recalcitrant Seeds', Proceedings of the Conference*, Kuala Lumpur, Malaysia, pp.29–42.

Vertucci, C.W. and Farrant, J.M. (1995) Acquisition and loss of desiccation tolerance. In: Kigel, J. and Galili, G. (eds) *Seed Development and Germination*. Marcel Dekker New York, pp. 237–271.

Vertucci, C.W., Farrant, J.M. and Crane, J. (1993) The status of and requirement for water in developing bean seeds. In: Close, T.J. and Bray, E.A. (eds) *Plant Response to Cellular Dehydration During Environmental Stress.* American Society of Plant Physiologists, Rockville, Maryland, pp. 259–260.

Wu, X.J., Song, S.Q., Qian, C.M., Cheng, L. and Fu, J.R. (2001) Effects of drying at different rates on desiccation sensitivity and membrane lipid peroxidation in Chinese wampee [*Clausena lansium* (Lour.) Skeels] axes. *Acta Phytophysiologica Sinica* 27, 407–412.

40 New Approaches to the Study of the Evolution of Physical and Physiological Dormancy, the Two Most Common Classes of Seed Dormancy on Earth

JERRY M. BASKIN[1] AND CAROL C. BASKIN[1,2]

[1]*Department of Biology, University of Kentucky, Lexington, KY 40506-0225, USA;* [2]*Department of Agronomy, University of Kentucky, Lexington, KY 40546-0091, USA*

Introduction

At maturity, the dormancy status of seeds can be classified broadly as dormant or non-dormant. A dormant seed cannot germinate under any set of normal environmental conditions, while a non-dormant seed can germinate over the widest range of conditions possible for the species. Five classes of seed dormancy can be distinguished (Table 40.1). In addition, some of the classes can be subdivided into levels of dormancy, and seeds with the non-deep level of physiological dormancy (PD) into five types (Baskin and Baskin, 1998). Based on a sample of 5250 species representing all major taxonomic groups of seed plants and vegetation regions of the world, 30.4% had non-dormant seeds and 69.6% dormant seeds. Of the 3653 species with dormant seeds (Table 40.1), PD is by far the most important of the five classes of dormancy, with physical dormancy (PY) being second.

The fact that about 70% of extant seed plants produce dormant seeds is a strong indication that this trait is adaptive. In short, dormancy is an adaptation for survival during periods when the abiotic and/or biotic factors of the environment are unfavourable for seedling establishment and/or growth of plants to maturity. An interesting aspect of seed dormancy is the various anatomical and physiological ways in which it is achieved. However, very little is known about the evolutionary origins and relationships of the various classes, levels or types of seed dormancy, but see Baskin and Baskin (1998), Nikolaeva (1999) and Baskin *et al.* (2000).

This chapter discusses two approaches that could be used to increase knowledge of the evolutionary aspects of seed dormancy. The first part of

(eds G. Nicolás, K.J. Bradford, D. Côme and H.W. Pritchard)

Table 40.1. Classes and cause(s) of seed dormancy (modified from Nikolaeva, 1977, and Baskin and Baskin, 1998) and proportion of species in each class. Of seeds of 5250 species for which information on dormancy is available, 1597 (30.4%) are non-dormant and 3653 (69.6%) are dormant. The first percentage after the name of each dormancy class is based on non-dormant + dormant seeds (5250) and the second on dormant seeds (3653) only.

Class of dormancy	Cause(s) of dormancy
A. Physiological, PD (45.1%, 64.8%)	Low growth potential of embryo, which cannot overcome mechanical restraint of covering layer(s)
B. Morphological, MD (1.5%, 2.1%)	Small differentiated (but underdeveloped), or small undifferentiated, embryo that simply needs time to grow (or grow/differentiate) before seed germinates, i.e. growth (or growth/differentiation) period = period of dormancy
C. Morphophysiological, MPD (8.1%, 11.6%)	Combination of underdeveloped (or undifferentiated) and physiologically dormant embryo
D. Physical, PY (14.5%, 20.8%)	Water-impermeable layer(s) of palisade or palisade-like cells in seed (or fruit) coat
E. Combinational, (PY + PD) (0.5%, 0.7%)	Water-impermeable seed (or fruit) coat and physiologically dormant embryo

the chapter shows how seed (or fruit) anatomy, phylogenetic schemes and fossil evidence can be used to contribute to knowledge of the evolution of PY. In the second part, we suggest that detailed information on world biogeography of the five types of non-deep PD could be used to help to resolve their origins and evolutionary relationships, using Type 2 as an example to illustrate the approach.

Physical Dormancy

Taxonomic occurrence

Physical dormancy is known in nine orders and 15 families of angiosperms; seven of the families also have PD, i.e. (PY+PD). The orders and families (*sensu* APG, 1998) are shown in Table 40.2.

Comparative ecological anatomy

Physical dormancy is caused by a water-impermeable layer (or layers) in the seed (or fruit) coat. Within the water-impermeable layer(s) is an anatomically specialized water-restriction region or 'water gap', through which water enters the seed after PY is broken. In nature, dormancy break involves physical disruption or dislodgement (opening) of this water gap by changes in physical environmental factors, e.g. fire or high daily fluctuations of soil surface temperatures. These environmental changes are likely to be followed by conditions suitable for germination, establishment and survival to matu-

Table 40.2. Orders and families in which physical dormancy is known (Baskin et al., 2000).

Order	Family	Class of dormancy
Cucurbitales	Cucurbitaceae	PY and (PY + PD)
Fabales	Fabaceae	PY and (PY + PD)
Geraniales	Geraniaceae	PY and (PY + PD)
Malvales	Bixaceae	PY
	Cistaceae	PY
	Cochlospermaceae	PY
	Dipterocarpaceae (subfamilies Monotoideae and Pakaraimoideae)	PY
	Malvaceae	PY and (PY + PD)
	Sarcolaenaceae	PY
Proteales	Nelumbonaceae	PY
Rosales	Rhamnaceae	PY and (PY + PD)
Sapindales	Anacardiaceae	PY and (PY + PD)
	Sapindaceae	PY and (PY + PD)
Solanales	Convolvulaceae	PY
Zingiberales	Cannaceae	PY

rity. Thus, a water gap in the seed (or fruit) coat serves as an environmental signal detector of high adaptive significance (Baskin and Baskin, 2000).

The seed coat develops from integuments, chalazal region of the ovule and raphe, all of maternal origin. Seed coat structure in its chalaza-, integument- and raphe-derived portions is continuous and similar (Boesewinkel and Bouman, 1984), except in the region where the water gap is formed. In bitegmic ovules, the inner integument gives rise to the tegmen and the outer integument to the testa. Seeds of taxa in which most of the seed coat develops from the chalaza are called pachychalazal (Boesewinkel and Bouman, 1995).

The water-impermeable palisade layer (or layers) originates in the ovule (or ovule + raphe) in 13 of the 15 families of angiosperms and in the pericarp in two families. Developmental origin of the water-impermeable layer of the integuments may be exotestal, (testal)-hypodermal (*Convolvulaceae* and *Cucurbitaceae*: Corner, 1976), endotestal or exotegmic (Baskin *et al*., 2000). Although in *Geraniaceae* the inner layer of the testa and the outer layer of the tegmen each form a mechanical layer of the seed coat (Boesewinkel and Been, 1979), the origin of the water-impermeable layer in seeds of *Geraniaceae* is endotestal (Meisert *et al*., 1999). In the pachychalazal seeds of *Canna*, about 90% of the seed coat is derived from the chalaza (Grootjen and Bouman, 1988). In *Anacardiaceae*, the three innermost layers of the four-layered endocarp, which develops from the inner epidermis of the ovary wall, are water-impermeable (Li *et al*., 1999); the seed coat does not have a mechanical layer (Corner, 1976). In *Nelumbonaceae*, the water-impermeable layer is formed in the subdermal region of the pericarp (Baskin *et al*., 2000).

Not surprisingly, then, there is also quite a bit of variation in development and anatomy of the water gap. The kind of water gap or 'plug', for example, ranges from the lens in *Fabaceae* to the anatomically complex bixoid chalazal plug (formed from the hypostase in *Bixaceae* and other families of the superorder Malvanae: Takhtajan, 1997) to the suberized 'stopper' (from nucellus) in *Geraniaceae* to the imbibition lid (from raphe) in *Canna* to the carpellary micropyle (in endocarp) in *Anacardiaceae* (Baskin *et al.*, 2000).

Phylogeny and fossils

As noted above, PY seems to be restricted to monocots and eudicots (*sensu* Friedman and Floyd, 2001). Interestingly, in monocots PY occurs only in *Cannaceae*, an evolutionarily advanced family in Zingiberales (Kress, 1990), with a fossil record back to the Cretaceous (Rodríguez-de la Rosa and Cevallos-Ferriz, 1994). Following the APG (1998) ordinal phylogeny scheme (Fig. 40.1), within the eudicots 11 of the 14 families (in five orders) with PY occur in the rosids, with one family each in *Geraniales* (*Geraniaceae*), sister group to the rosids, *Proteales* (*Nelumbonaceae*) and asterids (*Solanales*, *Convolvulaceae*). Six of the 15 families with PY occur in *Malvales* and two in its sister group, *Sapindales*.

When occurrence of PY, (PY + PD), non-dormancy (ND) and PD are plotted on a phylogenetic diagram for *Anacardiaceae* based on wood anatomy (Martínez Millán, 2000), it appears that PY, and also (PY+PD), has evolved only in the *Rhus* complex (*sensu* Miller *et al.*, 2001). Further, within the *Rhus* complex, PY, or PY and (PY+PD), occurs in *Rhus sensu stricto* (*s.s.*). as well as in several genera in the three non-*Rhus* clades in the complex, e.g. *Malosma*, *Cotinus* and *Toxicodendron*. (PY+PD) has also been documented in all four of these genera except *Malosma*. However, although *Rhus s.s.* is monophyletic, the *Rhus* complex is not (Miller *et al.*, 2001). PY, or (PY + PD), does not occur in *Pistacia*, used as the outgroup in the phylogenetic analysis of the *Rhus* complex, or in *Schinus*, a member of the *Schinus–Searsia–Toxicodendron* clade (*sensu* Miller *et al.*, 2001). These data suggest that PY evolved more than once within the *Rhus* complex.

The fossil record strongly suggests that PY was present in seeds (true seed + endocarp) of the genus *Rhus* by the middle Eocene. Endocarp anatomy of the extinct middle Eocene *Rhus rooseae* is nearly identical to that of extant *Rhus* species (Baskin *et al.*, 2000), and all extant species of *Rhus sensu lato* investigated have PY or (PY+PD). Based on the fossil record of the families, and assuming that dormancy in extinct species was the same as that in their extant close relatives, it is very likely that PY in *Anacardiaceae* and in other families had evolved by the late Cretaceous or early Tertiary.

In summary, evidence from seed coat anatomy and phylogeny suggests that although PY is the most phylogenetically constrained class of seed dormancy (Baskin *et al.*, 2000), it originated several times during evolution of the angiosperms. A polyphyletic origin for PY is supported by: (i) the considerable variation among taxa in developmental origin of the water-impermeable layer and of the origin and anatomy of the water gap; and (ii)

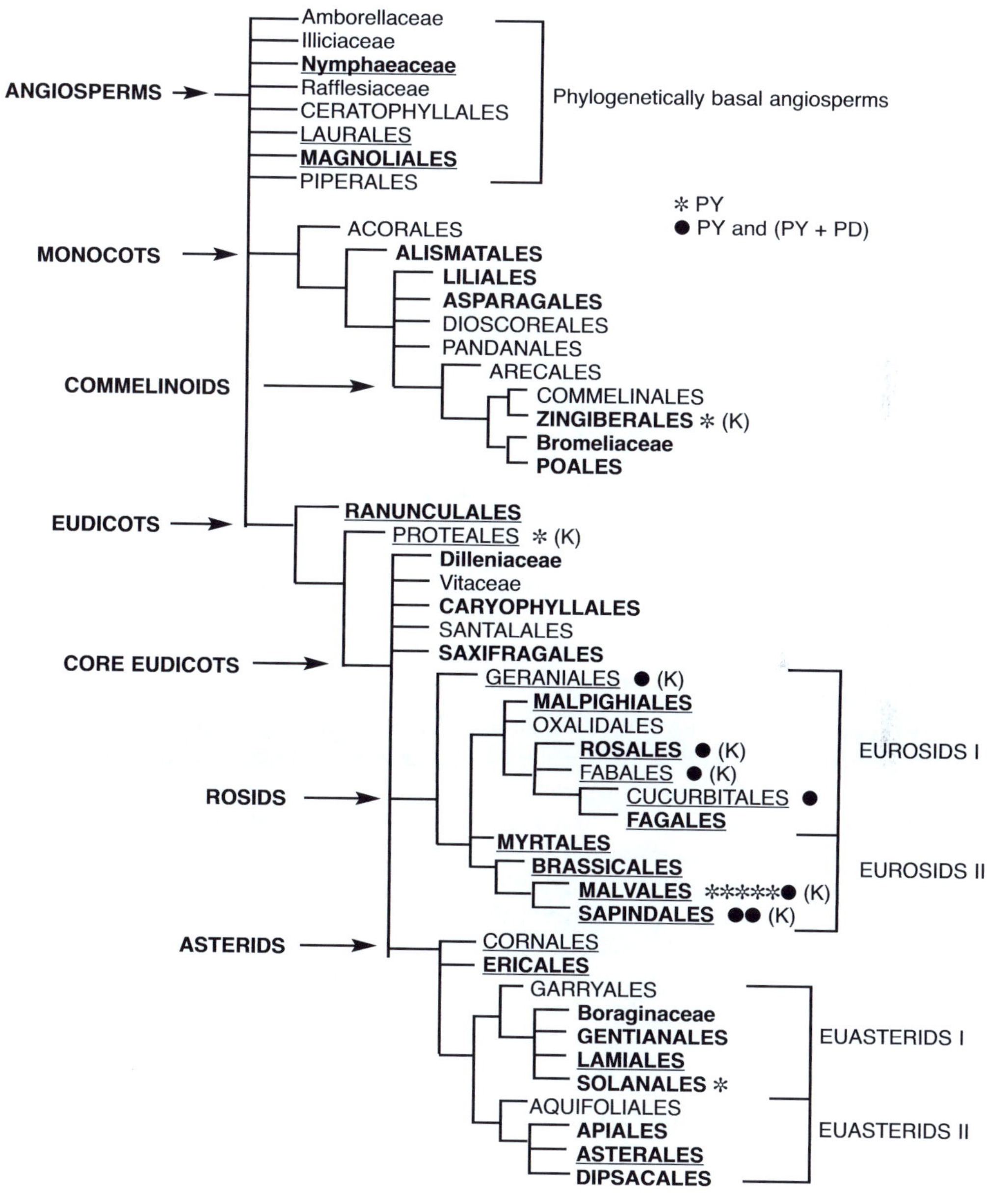

Fig. 40.1. Ordinal phylogenetic tree of the angiosperms, as modified by Bremer *et al.*, 1999, from tree published by the Angiosperm Phylogeny Group (APG, 1998). Each asterisk represents one family with PY, and each solid circle represents one family with PY and (PY+PD). (K) indicates a Cretaceous record for that order (Wing and Boucher, 1998). Recalcitrance occurs in orders that are underlined (von Teichman and van Wyk, 1994). Taxa in bold type are those discussed by Bremer *et al.* (1999) in the text of their publication; it has no special meaning in the diagram as presented here. (Figure is modified from Baskin *et al.*, 2000.)

its phylogenetic occurrence in distantly related clades and in advanced, but not in primitive, members of a particular taxonomic group. Thus, although water gaps have a similar function, they differ in origin. Within superorder Malvanae, however, the great similarity among families in the anatomically complex bixoid chalazal plug supports a monophyletic origin of PY for *Cistaceae* and closely related families in this group (Nandi, 1998). Finally, PY almost certainly had evolved by the middle Eocene, and it had probably done so by the late Cretaceous (Baskin *et al.*, 2000). Seven of the nine orders that contain taxa with PY (see Fig. 40.1) have a fossil record extending back to the Cretaceous (Wing and Boucher, 1998).

Physiological Dormancy

Types of non-deep PD

Physiological dormancy is caused by low growth potential of the embryo that prevents emergence of the radicle through covering layers (Table 40.1). As PD is broken, growth potential of the embryo increases and the role of covering structures in preventing germination decreases greatly. Three levels of PD have been distinguished: deep, intermediate and non-deep (Nikolaeva, 1969). In addition, five types of non-deep PD have been identified, based on the pattern of changes during dormancy break in temperature requirements for germination (Vegis, 1964; Baskin and Baskin, 1998).

- **Type 1.** At the beginning of dormancy break, seeds germinate only at low temperatures, and as dormancy break continues the maximum temperature for germination increases.
- **Type 2.** Seeds germinate first at high temperatures only and as dormancy break continues the minimum temperature for germination decreases.
- **Type 3.** Seeds germinate first at intermediate temperatures only and as dormancy break continues the minimum and the maximum temperature at which seeds will germinate decreases and increases, respectively.
- **Type 4.** During dormancy break, seeds gain the ability to germinate at high temperatures only.
- **Type 5.** During dormancy break, seeds gain the ability to germinate at low temperatures only.

The type of non-deep PD has been determined for one or more species in 51 angiosperm families (Baskin and Baskin, 1998 and unpublished), all of which occur in the portion of southeastern USA with a Cfa-type climate (Cf = temperate rainy climate, moist in all seasons; a = hot summers) (Müller, 1982).

Relationships between the types of non-deep PD

Although the five types of non-deep PD seem to represent a fine-tuning of species to their environment, little is known about their evolutionary origins and relationships (Baskin and Baskin, 1998). One reason for thinking that there might be close evolutionary relationships between the various types of non-deep PD is that more than one type has been found in several families (Baskin and Baskin, 1998 and unpublished). Thus, detailed phylogenetic relationships in families known to have two or more types of non-deep PD (e.g. *Asteraceae*, *Poaceae*) would allow us to address questions of relationships between the types. This approach is not possible at present, due to lack of phylogenetic trees for these families. Data for the types of non-deep PD have been 'plotted' on the phylogenetic tree published by Magallón *et al.* (1999) and on the more expanded version by Soltis *et al.* (1999). On both trees (with the exception of Type 4, where only two species are known), the different types of non-deep PD are in both basal and derived families.

Ecology and evolution of Type 2

Species whose seeds have Type 2 are well suited for growth in a climate with hot summers and cold winters. Dormancy break occurs in winter and seeds germinate in spring, at the beginning of the growing season. One hypothesis for the origin of Type 2 is that it is derived from Type 4. In seeds with Type 4, dormancy break occurs during exposure to high temperatures and non-dormant seeds germinate at high temperatures. If the year-round warm climate favourable for dormancy break and germination of seeds with Type 4 changed into a climate with warm summers alternating with cool winters, a lowering of the minimum temperature for germination would mean that germination could occur in early spring. It seems feasible that increased fitness resulting from seeds germinating early in the growing season could have resulted in development of Type 2. Was something new (ability to germinate at low temperatures) added to something old (ability to germinate at high temperatures) in the development of Type 2 (Baskin *et al.*, 1993)?

One of the difficulties in testing the hypothesis that Type 4 gave rise to Type 2 is that very little is known about the phylogenetic and biogeographical occurrence of Type 4. Although seed dormancy has been studied in several hundred species from the Cfa climatic zone in southeastern USA (Baskin and Baskin, 1988), Type 4 is known only in seeds of *Callicarpa americana* L. (Baskin and Baskin, unpublished). Also, Type 4 may be present in a species of *Poaceae* growing in the Cf climatic zone of New South Wales, Australia (Lodge and Whalley, 1981). If Type 4 gave rise to Type 2, it seems reasonable that Type 4 should be found in many plant families.

Questions about Type 2

Did Type 2 originate before angiosperms diversified?

If Type 2 evolved in the early stages of angiosperm diversification and before they spread geographically, then it should be present in many families and in various climatic zones throughout the world. However, most information on Type 2 is for regions of the northern hemisphere with temperate rainy climates. One exception is *Arctotheca calendula* L., a South African *Asteraceae* that has become a weed in Australia. Its achenes come out of dormancy during dry storage in summer and have Type 2 non-deep PD (Chaharsoghi and Jacobs, 1998). Although *A. calendula* is the first species from the southern hemisphere reported to have Type 2, this is not the first report of Type 2 being broken by high temperatures. Five species in the Cfa region of southeastern USA also have Type 2 and come out of dormancy during summer (Baskin and Baskin, 1988 and unpublished).

Consequently, it seems reasonable that families with Type 2 in regions with temperate rainy climates could also have Type 2 in other regions of the world that have a climate with an unfavourable (e.g. dry) season for seed germination and seedling establishment. Presence of Type 2 in members of the same family living in different climatic zones would lend support to the idea that Type 2 could have originated during the early divergence of angiosperms, in which case occurrence of Type 2 in a habitat is determined by presence of members of families with Type 2 and not by environmental conditions *per se*.

Did Type 2 originate after angiosperms diversified, and if so did it originate in regions with temperate rainy (Cfa and/or Cf) climate(s)?

Southeastern USA and eastern Mexico are the only places on earth with a Cfa climate. All other regions with a temperate rainy climate are mapped as Cf (Müller, 1982). There is a fairly good picture of the types of non-deep PD in the southeastern portion of the USA with a Cfa climate and to some extent the Cf region of Europe. However, very little is known about the types of non-deep PD in Cf climate regions in southeast Asia, South America, South Africa and Australia, or in the Cfa region of Mexico.

Before it can be concluded (or not) that Type 2 is more common in regions with temperate rainy climates than in those with other climates, it is necessary to determine what types of non-deep PD occur in these regions. More specifically, is Type 2 present in these regions? Does Type 2 occur in the same families in regions with temperate rainy climates and in those with warm climates and a dry season? If Type 2 is rare outside regions with a temperate rainy climate, this might suggest that occurrence of Type 2 in a habitat depends on whether environmental factors are appropriate for development of Type 2 and not on presence of members of particular families *per se*.

In summary, if Type 2 originated before angiosperms diversified and spread geographically, it should be present not only in many plant families but also in members of these families growing in a diversity of climates. In contrast, if Type 2 originated *in situ* as temperate rainy climates developed, it should be present in members of many families growing in regions with temperate rainy climates but not in members of these same families growing in regions with other kinds of climates. Thus, information on world biogeography of the types of non-deep PD would contribute significantly to the understanding not only of the adaptations of species to their habitats but also of the evolutionary origins and relationships of the various types of non-deep PD.

References

APG (Angiosperm Phylogeny Group) (1998) An ordinal classification for families of flowering plants. *Annals of the Missouri Botanical Garden* 85, 531–553.

Baskin, C.C. and Baskin, J.M. (1988) Germination ecophysiology of herbaceous plant species in a temperate region. *American Journal of Botany* 75, 286–305.

Baskin, C.C. and Baskin, J.M. (1998) *Seeds: Ecology, Biogeography, and Evolution of Dormancy and Germination*. Academic Press, San Diego, California, xiv + 666 pp.

Baskin, C.C., Baskin, J.M. and Leck, M.A. (1993) After-ripening pattern during cold stratification of achenes of ten perennial Asteraceae from eastern North America, and evolutionary implication. *Plant Species Biology* 8, 61–65.

Baskin, J.M. and Baskin, C.C. (2000) Evolutionary considerations of claims for physical dormancy-break by microbial action and abrasion by soil particles. *Seed Science Research* 10, 409–413.

Baskin, J.M., Baskin, C.C. and Li, X. (2000) Taxonomy, anatomy and evolution of physical dormancy in seeds. *Plant Species Biology* 15, 139–152.

Boesewinkel, F.D. and Been, W. (1979) Development of the testa of *Geranium pratense* L. and some other representatives of the Geraniaceae. *Acta Botanica Neerlandica* 37, 111–120.

Boesewinkel, F.D. and Bouman, F. (1984) The seed: structure. In: Johri, B.M. (ed.) *Embryology of Angiosperms*. Springer-Verlag, Berlin, pp. 567–610.

Boesewinkel, F.D. and Bouman, F. (1995) The seed: structure and function. In: Kigel, J. and Galili, G. (eds) *Seed Development and Germination*. Marcel Dekker, New York, pp. 1–24.

Bremer, K., Bremer, B. and Thulin, M. (1999) *Introduction to Phylogeny and Systematics of Flowering Plants*, 5th edn. Department of Systematic Botany, Evolutionary Biology Center, Uppsala University, Sweden, 102 pp. + 8-page appendix.

Chaharsoghi, A.T. and Jacobs, B. (1998) Manipulating dormancy of cape weed (*Arctotheca calendula* L.). *Seed Science Research* 8, 139–146.

Corner, E.J.H. (1976) *The Seeds of Dicotyledons*, 2 volumes. Cambridge University Press, Cambridge, UK, ix + 311 pp. and vi + 552 pp.

Friedman, W.E. and Floyd, S.K. (2001) Perspective: the origin of flowering plants and their reproductive biology – a tale of two phylogenies. *Evolution* 55, 217–231.

Grootjen, C.J. and Bouman, F. (1988) Seed structure in Cannaceae: taxonomic and ecological implications. *Annals of Botany* 61, 363–371.

Kress, W.J. (1990) The phylogeny and classification of Zingiberales. *Annals of the Missouri Botanical Garden* 77, 698–721.

Li, X., Baskin, J.M. and Baskin, C.C. (1999) Anatomy of two mechanisms of breaking physical dormancy by experimental treatments in seeds of two North American *Rhus* species (Anacardiaceae). *American Journal of Botany* 86, 1505–1511.

Lodge, G.M. and Whalley, R.D.B. (1981) Establishment of warm- and cool-season perennial grasses on the north-west slopes of New South Wales. I. Dormancy and germination. *Australian Journal of Botany* 29, 111–119.

Magallón, S., Crane, P.R. and Herendeen, P.S. (1999) Phylogenetic pattern, diversity, and diversification of eudicots. *Annals of the Missouri Botanical Garden* 86, 297–372.

Martínez Millán, M. (2000) Biogeografía histórica (Terciario y Cuaternario) de *Anacardiaceae* con base en caracteres anatómicos de la Madera. Tesis que para obtener el titulo de Biología, Universidad Nacional Autonoma de México, México DF, México, iv + 72 pp.

Meisert, A., Schulz, D. and Lehmann, H. (1999) Structural features underlying hardseededness in Geraniaceae. *Plant Biology* 1, 311–314.

Miller, A.J., Young, D.A. and Wen, J. (2001) Phylogeny and biogeography of *Rhus* (Anacardiaceae) based on its ITS sequence data. *International Journal of Plant Sciences* 162, 1401–1407.

Müller, M.J. (1982) *Selected Climatic Data for a Global Set of Standard Stations for Vegetation Science.* Dr W. Junk Publishers, The Hague, xxviii + 306 pp.

Nandi, O.I. (1998) Ovule and seed anatomy of Cistaceae and related Malvanae. *Plant Systematics and Evolution* 209, 239–264.

Nikolaeva, M.G. (1969) *Physiology of Deep Dormancy in Seeds.* Izdatel'stvo 'Nauka', Leningrad, iv + 220 pp. (Translated from Russian by Z. Shapiro, National Science Foundation, Washington, DC.)

Nikolaeva, M.G. (1977) Factors controlling the seed dormancy pattern. In: Khan, A.A. (ed.) *The Physiology and Biochemistry of Seed Dormancy and Germination.* North-Holland, Amsterdam, pp. 51–75.

Nikolaeva, M.G. (1999) Patterns of seed dormancy and germination as related to plant phylogeny and ecological and geographical conditions of their habitats. *Russian Journal of Plant Physiology* 46, 369–373.

Rodríguez-de la Rosa, R.A. and Cevallos-Ferriz, S.R.S. (1994) Upper Cretaceous Zingiberalean fruits with *in situ* seeds from southeastern Coahuila, Mexico. *International Journal of Plant Sciences* 155, 786–805.

Soltis, P.S., Soltis, D.E. and Chase, M.W. (1999) Angiosperm phylogeny inferred from multiple genes as a tool for comparative biology. *Nature* 402, 402–404.

Takhtajan, A. (1997) *Diversity and Classification of Flowering Plants.* Columbia University Press, New York, x + 643 pp.

Vegis, A. (1964) Dormancy in higher plants. *Annual Review of Plant Physiology* 15, 185–224.

von Teichman, I. and van Wyk, A. (1994) Structural aspects and trends in the evolution of recalcitrant seeds in dicotyledons. *Seed Science Research* 4, 225–239.

Wing, S.L. and Boucher, L.D. (1998) Ecological aspects of the Cretaceous flowering plant radiation. *Annual Review of Earth and Planetary Sciences* 26, 379–421.

41 Germination Deferment Strategies

ENRIQUE JURADO[1] AND ANGELA MOLES[2]

[1]*Facultad de Ciencias Forestales, UANL, AP 41 Linares, NL 67700, México;*
[2]*Department of Biological Sciences, Macquarie University, 2109 NSW, Australia*

Introduction

Angevine and Chabot (1979) described germination as a process in which a plant bets its life as it can usually not return to a quiescent stage. Because germination poses such great risks, selection should favour strategies that prevent germination during unfavourable conditions or at times when conditions are unlikely to remain favourable for long enough for seedling establishment. Such adaptations are often referred to as seed dormancy (Rees and Long, 1992). However, the term described in the literature as dormancy has at least two main aspects. They both refer to a deferment of germination to improve the chances of seedling establishment, through: (i) matching germination time with suitable environmental conditions to maximize establishment success, and (ii) risk-spreading, which results in some seeds germinating while others 'wait' in case conditions are not good enough. In order to avoid confusion, we have used the term 'germination deferment' throughout.

Safe time for germination and establishment might be influenced by environmental variables such as water and temperature and seasonal variation of dispersal agents, predators and pathogens of seeds and seedlings. Duration and frequency of favourable times for germination and establishment vary broadly across environments. In this chapter, environments have been divided into three main categories: (i) those that are constantly favourable for germination and seedling establishment; (ii) those in which favourable conditions occur at predictable times; and (iii) those in which favourable times for germination are unpredictably distributed. Matching germination time with suitable environmental conditions to maximize establishment success may vary according to the predictability of the environment.

 The Biology of Seeds: Recent Research Advances
(eds G. Nicolás, K.J. Bradford, D. Côme and H.W. Pritchard)

Constantly favourable environments

In an environment with adequate conditions for seeds to germinate all year (perhaps deep lakes in non-freezing environments and tropical rainforests near the equator) seeds could be produced any time of the year with similar establishment chances. In this type of environment, germination deferment would not increase fitness, and most seeds would be expected to germinate immediately.

Predictably favourable/unfavourable environments

Plants in environments with predictable periods of favourable and unfavourable conditions (e.g. seasonal environments) could either shed mature seeds at a time coincident with suitable conditions for seedling establishment or produce seeds with an inbuilt mechanism to defer germination until suitable conditions occur. These mechanisms could include an after-ripening period, so that seeds actually mature after shedding. Also chilling requirements for germination might help a seed to germinate in spring after a cold winter, rather than in autumn with the risk of imminent frosts. Alternatively, seeds can have a temperature requirement condition for germination, such as a temperature that only occurs in spring when the risk of frosts is over.

Unpredictably favourable environments

In some environments favourable conditions for germination and establishment occur sporadically, generally in response to rare events that alter resource availability – for example, increases in water availability in arid environments resulting from heavy rainfall events, increases in light availability resulting from treefall, or increases in available nutrients and space resulting from fires or floods.

In these environments it seems unlikely that plants could produce their seeds to coincide with favourable time for germination. Hence, species that depend on such rare and unpredictable events for seedling establishment are expected to have some strategy for germination deferment. Often, the cue used for germination will be associated with the event, such as rising temperatures after a period of cold (Densmore and Zasada, 1983), rainfall after a drought (Beatley, 1974; Gulmon, 1992; Veenendaal *et al.*, 1996; Clauss and Venable, 2000), changes in light quality and availability following windthrow of forest trees (Vázques-Yanes and Orozco-Segovia, 1993) and the presence of smoke or charred wood following the heat of a fire (Keeley, 1991).

Although the physical environment is expected to have a major influence on germination deferment strategies, other factors such as pathogens, seedling competition, seed predators, pollinators and dispersers could also have an effect. Thus germination deferment strategies across environments should not be expected to be a perfect fit, but rather a general trend.

Risk spreading

Because a plant cannot always accurately predict the favourability of the coming growth season, and because density-dependent mortality may affect the chances of seedling establishment, it is not always advantageous for a plant to risk all of its seeds on one germination event. Risk spreading could be achieved through germination deferment strategies such as the presence of leachable germination inhibitors in seed coats, or the requirement for abrasion of a thick seed coat. Alternatively, dimorphic seeds may be produced, some of which readily germinate while others remain dormant until the next suitable condition occurs.

Evidence for a relationship between density-dependent mortality and germination deferment has been found by Hyatt and Evans (1998), who showed that germination fraction in *Lesquerella fendleri* (*Brassicaceae*) decreased with increasing sibling competition, and by Hacker and Ratcliffe (1989), who showed that seeds of *Cenchrus ciliaris* from fascicles with more than one seed were more dormant than those that matured in one-seeded fascicles. Lundberg *et al.* (1996) also concluded that the mean and variance of the ecologically stable strategy (ESS) dormancy period increase with increasing sibling competition. Thus, the effect of density-dependent mortality is to increase variation in germination time. However, density-dependent mortality can act within any environment. Therefore this line of theory does not result in a clear hypothesis relating the frequency of risk-spreading strategies to different environments. On the other hand, the selective process for a trait that matches germination with favourable germination times might decrease with increasing environmental heterogeneity. Therefore we predict higher levels of risk-spreading germination deferment in unpredictable environments.

There is a massive body of work regarding germination responses to different environmental conditions. This work has been extensively collated (Baskin and Baskin, 1998). There is also a well-developed body of theoretical literature surrounding the subject of regeneration strategies in different environmental conditions (Rees, 1994). In this study we brought these two areas of literature together and tested the predictions of the models with empirical data.

In order to test the hypothesis that the proportion of species showing no germination deferment would be greater in more constant environments, data were obtained from aquatic environments, evergreen tropical rainforest, semi-evergreen tropical rainforest, tropical deciduous forest and dry woodland.

To test the hypothesis that risk spreading is more common in unpredictable environments, we compared data from central Australia (an environment with extremely unpredictable rainfall) (Southwood, 1977) and northeastern Mexico (a harsh but seasonal environment).

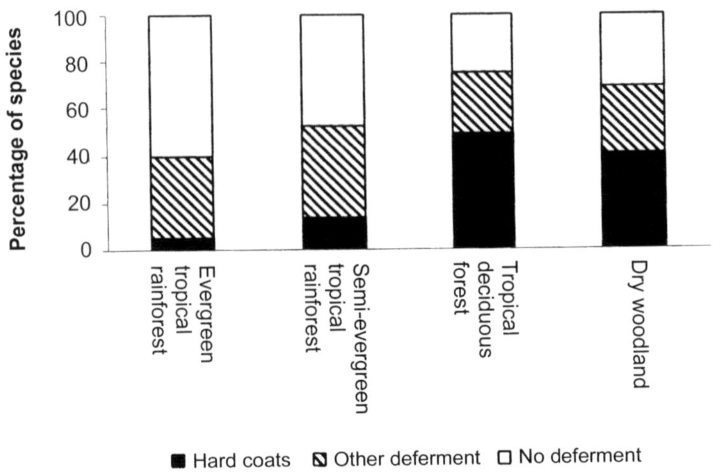

Fig. 41.1. Germination deferment for trees in habitats with contrasting moisture availability. Evergreen tropical rainforest (458 species), semi-evergreen tropical rainforest (97), tropical deciduous forest (79) and dry woodland (71). Data from Baskin and Baskin (1998).

Germination Deferment for Selected Habitats

Germination deferment in relation to environmental heterogeneity

The ecosystems with least within-year variation in terms of rainfall and temperature are likely to be evergreen tropical rainforests, followed by semi-evergreen tropical rainforests, tropical deciduous forests and dry woodlands. Most trees from evergreen tropical rainforests (60%) did not show dormancy in their seeds (Fig. 41.1). For trees in a semi-evergreen tropical rainforest this percentage was somewhat smaller (50%), and more so for trees growing in tropical deciduous forests (20%) and dry woodlands (22%). Seeds with physical dormancy (hard teguments), interpreted as risk spreading, were more common in drier environments than in tropical rainforests (Fig. 41.1), suggesting that risk spreading may be more selected for where rainfall is a limiting factor. The overall pattern of germination deferment vs. no deferment also holds when comparing herbaceous species of the same communities (Fig. 41.2). However, germination deferment strategies other than 'hard seeds' were more common in drier environments.

It is interesting to note that many forms of dormancy were found in each environment. This could reflect the different life history strategies of the many species in each habitat, or could result from phylogenetic constraints; that is, deferment strategies of some species could result from adaptations to previous environments. More research should focus on phylogeny vs. environment in germination deferment.

A second test of the prediction that germination deferment strategies might be more common in heterogeneous environments was made using data for aquatic species. Some aquatic plants grow fully underwater in

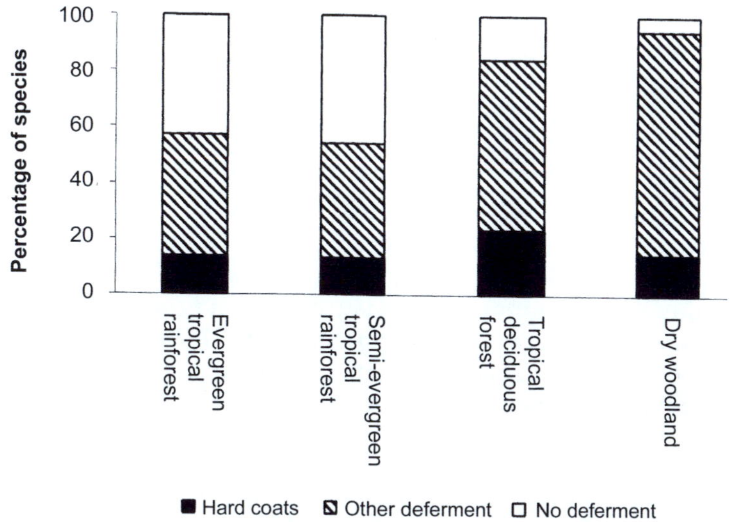

Fig. 41.2. Germination deferment for herbaceous plants in habitats with contrasting moisture availability. Evergreen tropical rainforest (35 species), semi-evergreen tropical rainforest (37), tropical deciduous forest (59) and dry woodland (124). Data from Baskin and Baskin (1998).

places with few temperature and water level changes throughout the year, and thus experience very little seasonal variation. We compared data for these submerged species with data for emergent aquatic plants, which generally grow in shallower water, where water level might be expected to fluctuate more dramatically. When looking at freshwater plant germination traits from a variety of places, we found more species showing no germination deferment in submerged species than in emergent species (Fig. 41.3).

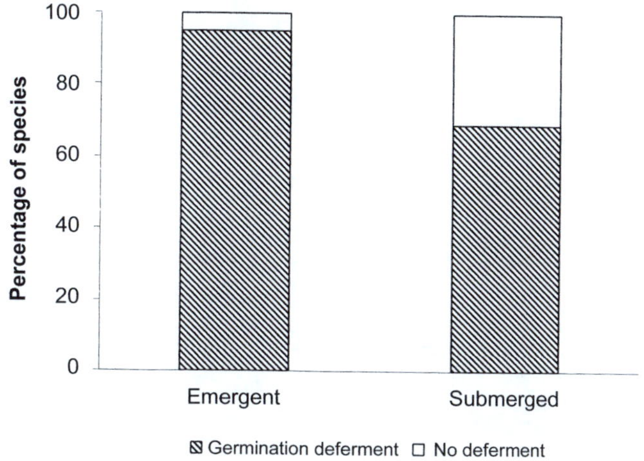

Fig. 41.3. Germination deferment for freshwater emergent (164 species) and submerged plants (36 species) across environments. Data from Baskin and Baskin (1998).

Thus our prediction that a lower proportion of species would have germination deferment strategies in more constant environments was supported, both in terrestrial and aquatic plants.

Germination deferment according to rainfall predictability

We contrasted two environments with different rainfalls and also different rainfall predictability. The study area in central Australia has a mean annual rainfall of 263 mm (mostly in summer), while northeastern Mexico has a rainfall of 700 mm falling in late spring and (mostly) in early autumn, with a dry period in between and a dry winter. In central Australia the main risk that a seed faces is germinating without sufficient follow-up rainfall for seedling establishment, whereas in northeastern Mexico the main risk is germinating with late spring rains and facing the summer drought as a small seedling.

Few species (*c.* 20%) in either central Australia or northeastern Mexico had no germination deferment strategy (Fig. 41.4). Species that lack any germination deferment strategy are likely to be perennial species that produce seeds over more than one season. Hard seeds were common in central Australia (> 70%) and rare in northeastern Mexico (*c.* 30%) (Fig. 41.4). Other germination deferment strategies were common for Mexico (*c.* 60%) and rare in central Australia (*c.* 10%). In an environment with sufficient rainfall occurring infrequently and unreliably, it seems more advantageous to have seeds that do not all germinate at the same time and hence spread the risk of seedling mortality. In northeastern Mexico, on the other hand, mechanisms to defer germination until the suitable once-a-year optimum germination time occurs are more common.

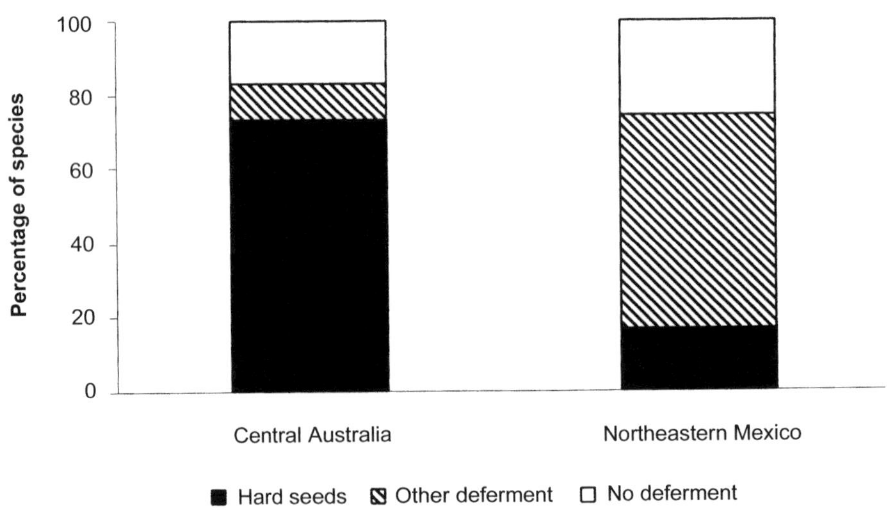

Fig. 41.4. Germination deferment for 106 plant species in central Australia (Jurado and Westby, 1992) and 66 in northeastern Mexico (Jurado *et al.*, 2000).

The greater incidence of risk-spreading dormancy in the less predictable environment is in line with the predictions made by the theoretical literature on this subject (Rees, 1994). However, this result should be interpreted with some caution, as it is based on a single contrast between environments, and the observed differences could be due to some other difference between central Australia and northeastern Mexico.

Conclusions

The main findings of this work were: (i) that germination deferment strategies became increasingly common as we moved from more constant environments, such as evergreen tropical rainforests, through to more heterogeneous environments, such as deciduous forests and dry woodland; and (ii) that risk-spreading germination deferment strategies were more common in an environment with unpredictable rainfall than in an environment with seasonal rainfall events. These findings were entirely in line with the predictions of the theoretical literature on germination deferment strategies.

Acknowledgements

This research was sponsored by IFS D/2256-1&2 and PAICYT CN-515 01-02.

References

Angevine, M.W. and Chabot, B.F. (1979) Seed germination syndromes in higher plants. In: Solbrig, O.T., Jain, S., Johnson, G.B. and Raven, P.H. (eds) *Topics in Plant Population Biology*. Columbia University Press, New York, pp. 188–206.

Baskin, C.C. and Baskin, J.M. (1998) *Seeds, Ecology, Biogeography, and Evolution of Dormancy and Germination*. Academic Press, San Diego, California, 666 pp.

Beatley, J.C. (1974) Phenological events and their environmental triggers in Mojave desert ecosystems. *Ecology* 55, 856–863.

Clauss, M.J. and Venable, D.L. (2000) Seed germination in desert annuals, an empirical test of adaptive bet hedging. *American Naturalist* 155, 168–186.

Densmore, R. and Zasada, J. (1983) Seed dispersal and dormancy patterns in northern willows, ecological and evolutionary significance. *Canadian Journal of Botany* 61, 3207–3216.

Gulmon, S.L. (1992) Patterns of seed germination in Californian serpentine grassland species. *Oecologia* 89, 27–31.

Hacker, J.B. and Ratcliffe, D. (1989) Seed dormancy and factors controlling dormancy breakdown in buffel grass accessions from contrasting provenances. *Journal of Applied Ecology* 26, 201–212.

Hyatt, L.A. and Evans, A.S. (1998) Is decreased germination fraction associated with risk of sibling competition? *Oikos* 83, 29–35.

Jurado, E. and Westoby, M. (1992) Germination biology of selected central Australian plants. *Australian Journal of Ecology* 17, 341–348.

Jurado, E., Aguirre, O., Flores, J., Navar, J., Villalon, H. and Wester, D. (2000) Germination in Tamaulipan thornscrub of Northeastern Mexico. *Journal of Arid Environments* 46, 413–424.

Keeley, J.E. (1991) Seed germination and life history syndromes in the Californian chaparral. *The Botanical Review* 57, 81–116.

Lundberg, S., Nilsson, P. and Fagerström, T. (1996) Seed dormancy and frequency dependent selection due to sib competition, the effect of age specific gene expression. *Journal of Theoretical Biology* 183, 9–17.

Rees, M. (1994) Delayed germination of seeds – a look at the effects of adult longevity, the timing of reproduction, and population age/stage structure. *American Naturalist* 144, 43–64.

Rees, M. and Long, M.J. (1992) Germination biology and the ecology of annual plants. *American Naturalist* 139, 484–508.

Southwood, T.R.E. (1977) Habitat, the templet for ecological strategies? *Journal of Animal Ecology* 46, 337–365.

Vázques-Yanes, C. and Orozco-Segovia, A. (1993) Patterns of seed longevity and germination in the tropical rainforest. *Annual Review of Ecology and Systematics* 23, 69–87.

Veenendaal, E.M., Ernst, W.H.O. and Modise, G.S. (1996) Effect of seasonal rainfall pattern on seedling emergence and establishment of grasses in a savanna in south-eastern Botswana. *Journal of Arid Environments* 32, 305–317.

42 Dormancy and Germination Ecology of Annual Ryegrass (*Lolium rigidum* Gaud.)

AMANDA J. ELLERY[1*], ROBERT S. GALLAGHER[2] AND SIMONE V. DUDLEY[1]

[1]*CSIRO, Plant Industry, Private Bag 5, Wembley, WA 6913, Australia;*
[2]*Department of Crop and Soil Science, Washington State University, Pullman, WA 99164-6420, USA*

Introduction

Annual ryegrass (*Lolium rigidum*) is a significant weed of continuous cereal-cropping systems in southern Australia. Although originally introduced into pasture–cropping rotations as a desirable forage species, its high fecundity and similar phenology to cereal crops have rendered it a weed of the cropping system. The weedy characteristics of annual ryegrass have been compounded by the development of herbicide-resistant biotypes; resistance to most herbicide groups has been reported (Gill, 1995). Nevertheless, crops can be grown effectively without in-crop herbicides when annual ryegrass seed banks are reduced to low levels (Revell and Hudson, 2001; Roy, 2001). The goal of this chapter is to explore the factors that regulate germination of annual ryegrass and to consider some possibilities for the management and depletion of annual ryegrass seed banks.

Seed Bank Persistence

Annual ryegrass is a winter annual, originating in the Mediterranean region. Seeds mature in spring and after-ripen during summer and autumn, before germinating in late autumn or early winter. Although seed bank persistence of annual ryegrass is relatively short-lived compared with many other weedy species, seeds persist long enough to be a management impediment in continuous cropping systems. In the absence of reliable control measures, annual ryegrass seed banks increase rapidly.

*See list of contributors for new address.

Seed bank longevity estimates of 1 year (McGowan, 1970) to 4 years (Peltzer and Matson, 2002) have been reported. Revell and Hudson (2001) found that small numbers of annual ryegrass seeds were still present after 4 years of seed set control, but that autumn tillage accelerated seed bank decline. Seed burial may promote rapid seed death in higher rainfall regions (R.S. Gallagher, Perth, 2002, personal communication). Thus regional climatic conditions, seed production management and seed microsite can all affect seed bank longevity.

Seed Dormancy

Role of dormancy

Primary seed dormancy serves at least three adaptive functions in Mediterranean climates. First, it prevents germination after summer rainfall. Southern Australian summers are generally dry but sporadic summer rainfall occurs in most years (Chapman and Asseng, 2001). Summer rainfall is often sufficient to promote germination but rarely sustains plant growth. Secondly, seed dormancy distributes seedling emergence through the growing season. Although a majority of emergence occurs in late autumn (Peltzer and Matson, 2002), cohorts can continue to emerge throughout winter (McGowan, 1970), enabling plants to escape early-season weed control. Finally, we expect dormancy to facilitate seed persistence from year to year. Since a large portion of the seed rain tends to emerge in the following year, persistent dormancy may only occur in a small proportion of a seed population.

Dynamics of dormancy release

Loss of primary dormancy is generally curvilinear over time and can be affected by regional and microclimatic factors. Gramshaw (1972) showed that germination under optimal conditions was low at seed harvest, but rose to near 80% after 12–15 weeks of after-ripening. Dormancy release was similar in seeds after-ripened under a range of field environments and at room temperature, except for some evidence that temperatures in excess of 60°C delayed after-ripening, suggesting that dormancy release was not strongly affected by temperature. Chapman *et al.* (1999) also showed that field temperatures that often exceeded 50°C delayed dormancy loss, supporting the suggestion that high temperature inhibits after-ripening.

Between 6 and 35°C, the rate of dormancy loss increases with after-ripening temperature (Steadman, 2002). A minimum temperature for dormancy loss of 5.4°C was derived (Steadman *et al.*, 2003), suggesting that dormancy release during winter might be slow or negligible. Further evidence for temperature-dependent dormancy release was obtained from seeds placed at locations with contrasting climates (R.S. Gallagher, Perth, 2002, personal communication). Dormancy release of seeds on the soil surface was slightly enhanced by higher temperature but when seeds were buried, dormancy release was faster at the site with higher summer rainfall. These results, along with those of Lush *et al.* (1981), indicate that imbibition in the dark can hasten the release of dormancy.

Genetic regulation of dormancy release

Annual ryegrass is distributed across a range of climatic zones in southern Australia. Adaptation of dormancy strategies to local environments may be needed to ensure persistence of individual populations. For example, it might be expected that regions with frequent summer rainfall would have populations with more intense dormancy than regions with predictably dry summers. In eight seed populations collected throughout Western Australia but grown in common environments, there were no differences in germination potential in light of after-ripened seed, but differences of up to 15% in the dark (Gramshaw, 1976). We found differences in germination potential among seed populations collected at locations ranging from the low to high rainfall zones of Western Australia but grown at a single location (Table 42.1). These differences reflected the differences in germination potential of the seeds collected originally (Ellery and Chapman, 2001). Although genetics clearly plays a role in regulating seed dormancy, the specific link between the local environment and seed dormancy has not been made. More research in this area is needed if ecologically sound weed management strategies are to be applied to climatically diverse regions.

Effects of seed maturation environment

In agronomic situations, annual ryegrass seeds develop and mature under a range of environmental conditions. For example, late-emerging plants are likely to experience greater competition from the crop for light, water and nutrients than early cohorts. Stress during plant and seed development can

Table 42.1. Variation in germination at seed maturity of annual ryegrass seeds originating from six locations throughout Western Australia and grown in a common garden during 2000. Seeds were germinated under a diurnal temperature fluctuation of 25/17°C with a 12 h photoperiod. Values followed by the same letter do not differ significantly at $P < 0.05$. Values for germination were averaged over four replicates of each of 12 germination treatments ($n = 48$).

Population	Annual rainfall (mm)	Maturity type	Initial germination (%)
Mullewa	328	Early	69.8[a]
Geraldton	465	Early	61.9[b]
Northam	432	Intermediate	48.0[c]
Merredin	328	Intermediate	42.7[d]
Wongan Hills	390	Intermediate	36.6[e]
Mt Barker	738	Late	58.3[b]

reduce resource allocation to root, stem and seeds; for example, annual ryegrass grown in 55% of ambient light produced similar biomass and number of seeds per plant compared with plants grown in full sun, but seeds were smaller and less dormant. Thus, resource-limiting conditions that do not affect plant biomass or seed production can still influence seed quality and fitness.

Environmental Germination Cues

Seed dormancy ranges from highly dormant, where seeds do not germinate, to relatively non-dormant, where seeds germinate in response to a wide variety of environmental cues (Baskin and Baskin, 1998). Temperature, light and soil chemicals, such as ethylene and nitrate, are among the environmental cues that can trigger germination. Germination in response to environmental cues is probably an adaptive strategy to coordinate germination with respect to season, position in the soil and disturbance.

Temperature

Annual ryegrass germinates at temperatures between 8 and 35°C (Gramshaw, 1976; Turner *et al.*, 2001). Response to germination temperature can depend on the level of seed dormancy and the light environment during germination. For example, freshly harvested seeds germinated better at 12°C than at 24°C but after 18 weeks of after-ripening there was little response to germination temperature (Gramshaw, 1972). In contrast, we found that germination of fresh seeds was significantly higher at 25/10°C (12 h photo-thermal period) than at 15/10°C although germination of after-ripened seeds was unaffected by temperature (R.S. Gallagher, Perth, 2002, personal communication). Germination potential is enhanced by daily temperature fluctuations of at least 5°C (Gramshaw, 1972, 1976; Cocks and Donald, 1973), which may be an adaptive mechanism to sense the depth of seed burial.

Light

Germination of conditionally dormant annual ryegrass seeds tends to be stimulated by light (Gramshaw, 1972, 1976; Gramshaw and Stern, 1977). Once after-ripening is complete, most seeds will germinate in the dark but up to 20% of seeds may remain dark dormant. These seeds germinate readily when exposed to continuous light but not to short flashes of light characteristic of low- or very low-fluence responses (R.S. Gallagher, Perth, 2002, personal communication). Likewise, germination is largely inhibited by far-red light.

Soil chemistry

The effects of soil chemistry on germination of annual ryegrass seeds have not been widely considered. Nitrate, in association with light, promotes ger-

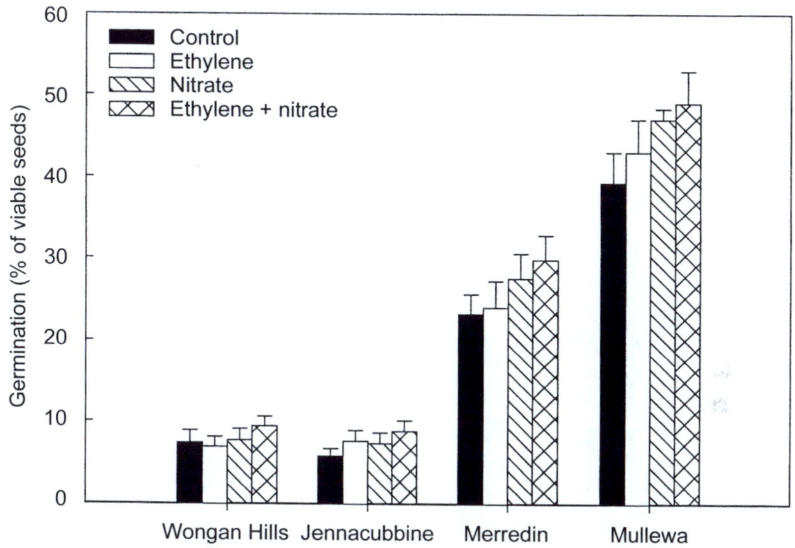

Fig. 42.1. Germination response of annual ryegrass seeds from four independent populations to ethylene (0.4 mM) and nitrate (0.8 mM). Error bars represent standard error of germination averaged over four replicates, two germination temperatures (15 and 20°C) and five sampling times.

mination in many plant species (Hilhorst and Karssen, 2000). Ethylene can also stimulate germination, although amongst weedy species grasses seem to be less sensitive than broadleaf weeds (Taylorson, 1979). Germination of annual ryegrass seeds was slightly stimulated by the presence of biologically active concentrations of nitrate (0.8 mM) and ethylene (0.4 mM) but, as observed in *Avena fatua* (Saini *et al.*, 1985), the two compounds did not interact to increase germination, rather the effects were additive (Fig. 42.1).

Maintenance of Seed Vigour

Factors that affect viability and vigour of seeds in the soil affect germination and emergence of seedlings and can be important considerations when estimating the impact of weeds on crop productivity. Generally, low temperature and moisture content increase seed longevity (Ellis and Roberts, 1980). Annual ryegrass seeds, when maintained at moisture contents above 15% (fresh weight basis) at 20°C in small laminated foil envelopes, decayed rapidly (Fig. 42.2), presumably because the available oxygen was rapidly consumed, thus inhibiting cellular repair mechanisms. Loss of viability was reduced at lower temperatures, probably due to a reduction in the rate of decay processes. Annual ryegrass seeds therefore seem susceptible to damage at high moisture contents and moderate temperatures. Management strategies that retain moisture near the soil surface, such as mulching or cover cropping, may offer the opportunity to accelerate depletion of the seed bank by increasing seed mortality.

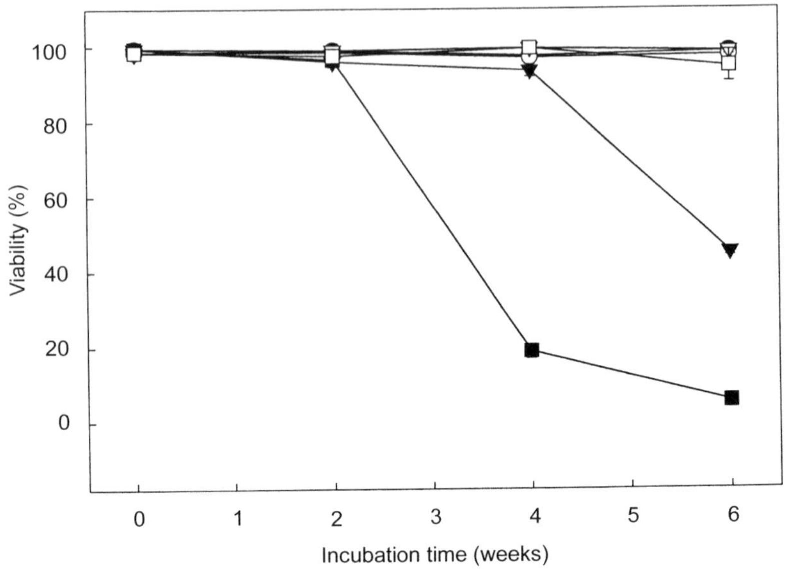

Fig. 42.2. Loss of viability, determined by tetrazolium test, in annual ryegrass maintained at seed moisture contents (fresh weight basis) of 10% (open symbols) and 33% (closed symbols), at temperatures of 6°C (circles), 15°C (triangles) and 20°C (squares). Error bars (where larger than symbol) represent standard error of germination at 25/17°C, averaged over four replicates.

Management of Annual Ryegrass Seed Banks

Recent research has shown that annual ryegrass can be managed without the use of in-crop herbicides. Key features of these systems are the removal of paddocks from cereal cropping for several years, promotion of ryegrass germination by soil disturbance and effective seed set control.

Seed set control

Controlling seed set is a critical aspect of managing annual ryegrass, which is a highly fecund species with seed production rates of around 3000 seeds per plant. In an undisturbed system, Revell and Hudson (2001) found that including seed set control in weed management operations halved the number of ryegrass plants establishing in the subsequent wheat crop. Roy (2001) also showed that cropping systems including 1 or 2 years of seed set control yielded the greatest reductions in ryegrass plant numbers.

Tillage to promote germination

Soil tillage promotes rapid seed bank decline in many weed species (Roberts and Feast, 1973; Mulugeta and Stoltenberg, 1997). In Australia, autumn tillage prior to annual ryegrass emergence has been found to promote seedling recruitment and seed bank decline (Revell and Hudson, 2001). Disturbance probably enhances emergence by redistributing dark-dormant seeds to the soil surface, exposing them to light and more extreme soil temperature fluctuations, although there may also be some contribution from nitrate mineralized after disturbance. Thus, timely soil disturbance may improve annual ryegrass seed bank management efforts.

Cover crops to change soil chemistry

We have shown that germination of annual ryegrass seeds may be enhanced by the presence of biologically active levels of nitrate and ethylene (Fig. 42.1). Depleting the seed bank by modifying soil chemistry may therefore be a useful component in an integrated weed management strategy. Soil nitrate and ethylene concentrations may be increased by the use of cover or green manure crops (Arshad and Frankenberger, 2002), particularly those with a low carbon to nitrogen ratio, as decomposition of these residues generally results in net nitrogen mineralization (Paul and Clark, 1996). Inclusion of green manure phases in cropping systems has been associated with reduced numbers of in-crop ryegrass plants as a result of seed set control, so this practice may provide the dual benefits of stimulating germination and controlling seed set.

Conclusion

Seeds of annual ryegrass can persist in soils for a number of years, facilitating a rapid accumulation of plants in the event of herbicide failure. By exploiting the germination sensitivity of this species to light, nitrate, ethylene and temperature fluctuation, it has been possible to reduce weed seed banks to extremely low levels. Attention to seed set control ensures that seed banks are not replenished, and crops may be grown successfully without the use of in-crop herbicides.

References

Arshad, M. and Frankenberger, W.T. (2002) *Ethylene: Agricultural Sources and Applications.* Kluwer Academic/Plenum Publishers, New York, 342 pp.

Baskin, C.C. and Baskin, J.M. (1998) *Seeds: Ecology, Biogeography, and Evolution of Dormancy and Germination*. Academic Press, San Diego, California, 666 pp.

Chapman, R. and Asseng, S. (2001) An analysis of the frequency and timing of false break events in the Mediterranean region of Western Australia. *Australian Journal of Agricultural Research* 52, 367–376.

Chapman, R., Bolger, T.P. and Le Coultre, I.F. (1999) Seed dormancy release in three common pasture grasses from a Mediterranean-type environment under contrasting conditions. *Australian Journal of Experimental Agriculture* 39, 143–147.

Cocks, P.S. and Donald, C.M. (1973) The germination and establishment of two annual pasture grasses (*Hordeum leporinum* Link and *Lolium rigidum* Gaud.). *Australian Journal of Agricultural Research* 24, 1–10.

Ellery, A.J. and Chapman, R. (2001) Variation in seed dormancy and management of annual ryegrass. In: Stewart, V. (ed.) *Agribusiness Crop Updates*. Department of Agriculture Western Australia, Perth, pp. 29–30.

Ellis, R.H. and Roberts, E.H. (1980) Improved equations for the prediction of seed longevity. *Annals of Botany* 45, 13–30.

Gill, G.S. (1995) Development of herbicide resistance in annual ryegrass populations (*Lolium rigidum* Gaud.) in the cropping belt of Western Australia. *Australian Journal of Experimental Agriculture* 35, 67–71.

Gramshaw, D. (1972) Germination of annual ryegrass seeds (*Lolium rigidum* Gaud.) as influenced by temperature, light, storage environment and age. *Australian Journal of Agricultural Research* 23, 779–787.

Gramshaw, D. (1976) Temperature/light interactions and the effect of seed source on germination of annual ryegrass (*Lolium rigidum* Gaud.) seeds. *Australian Journal of Agricultural Research* 27, 779–786.

Gramshaw, D. and Stern, W.R. (1977) Survival of annual ryegrass (*Lolium rigidum* Gaud.) seed in a Mediterranean type environment. II. Effects of short-term burial on persistence of viable seed. *Australian Journal of Agricultural Research* 28, 93–101.

Hilhorst, H.W.M. and Karssen, C.M. (2000) Effect of chemical environment on seed germination. In: Fenner, M. (ed.) *Seeds. The Ecology of Regeneration in Plant Communities.* CAB International, Wallingford, UK, pp. 293–310.

Lush, W.M., Groves, R.H. and Kaye, P.E. (1981) Presowing hydration–dehydration treatments in relation to seed germination and early seedling growth of wheat and ryegrass. *Australian Journal of Plant Physiology* 8, 409–425.

McGowan, A.A. (1970) Comparative germination patterns of annual grasses in north-eastern Victoria. *Australian Journal of Experimental Agriculture and Animal Husbandry* 10, 401–404.

Mulugeta, D. and Stoltenberg, D.E. (1997) Increased weed emergence and seed bank depletion by soil disturbance in a no-tillage system. *Weed Science* 45, 234–241.

Paul, E.A. and Clark, F.E. (1996) *Soil Microbiology and Biochemistry,* 2nd edn. Academic Press, San Diego, California, 340 pp.

Peltzer, S. and Matson, P. (2002) Understanding the weed seed bank life of important agricultural weeds. In: Stewart, V. (ed.) *Agribusiness Crop Updates.* Department of Agriculture Western Australia, Perth, pp. 13–14.

Revell, C.K. and Hudson, C. (2001) Short term pasture phases for weed control. In: Stewart, V. (ed.) *Agribusiness Crop Updates.* Department of Agriculture Western Australia, Perth, pp. 21–22.

Roberts, H.A. and Feast, P.M. (1973) Changes in the number of viable weed seeds in soil under different regimes. *Weed Research* 13, 298–303.

Roy, W. (2001) A systems approach to managing herbicide resistant ryegrass. In: Stewart, V. (ed.) *Agribusiness Crop Updates.* Department of Agriculture, Western Australia, Perth, pp. 9–11.

Saini, H.S., Bassi, P.K. and Spencer, M.S. (1985) Interactions among ethephon, nitrate and after-ripening in the release of dormancy of wild oat (*Avena fatua*) seed. *Weed Science* 34, 43–47.

Steadman, K.J. (2002) A summary of dormancy in annual ryegrass (*Lolium rigidum*) seeds: dry after-ripening versus imbibition in the dark. In: Spafford Jacob, H., Dodd, J. and Moore, J.H. (eds) *Proceedings of the 13th Australian Weeds Conference.* PPS/CAWSS, Perth, Western Australia, pp. 468–470.

Steadman, K.J., Crawford, A.D. and Gallagher, R.S. (2003) Dormancy release in *Lolium rigidum* seeds is a function of thermal after-ripening time and seed water content. *Functional Plant Biology* 30, 345–352.

Taylorson, R.B. (1979) Response of weed seeds to ethylene and related hydrocarbons. *Weed Science* 27, 7–10.

Turner, N.C., Thomson, C.J. and Rawson, H.M. (2001) Effect of temperature on germination and early growth of subterranean clover, capeweed and Wimmera ryegrass. *Grass and Forage Science* 56, 97–104.

43 Phenotypic Plasticity of Seed Germination as a Component of the Complementary Sets of Adaptations and Survival Strategies During the Life Cycle of Ephemerals Occurring in the Negev Desert of Israel

YITZCHAK GUTTERMAN

Wyler Department of Drylands Agriculture, Jacob Blaustein Institute for Desert Research and Department of Life Sciences, Ben-Gurion University of the Negev, Sede Boker Campus 84990, Israel

Introduction

Many of the ephemeral plant species with unrelated taxa occurring in the Negev Desert (Zohary, 1966, 1972; Feinbrun-Dothan, 1978, 1986) have been found to have developed ecologically equivalent or possibly even convergent (Cloudsley-Thompson, 2001) sets of complementary adaptations and survival strategies throughout the different stages of their life cycles (Gutterman, 1993, 2000, 2002a,b). These include phenotypic plasticity of seed germination (PPSG) influenced during seed development, maturation and storage, seed dispersal and germination (Fig. 43.1).

Post-maturation environmental factors affecting PPSG in Australia were found by Quinlivan (1966) and in North America by Went (1953). Seed position influences on PPSG were found to affect the shape and mucilaginous layer of seeds in the Namaqualand desert of South Africa (Beneke, 1991; Beneke *et al.*, 1992a,b).

The Effects on PPSG During Seed Development, Maturation, Dry Storage and Germination

The PPSG is a component of the complementary sets of ecological equivalents of adaptations and strategies. These include close physiological and/or

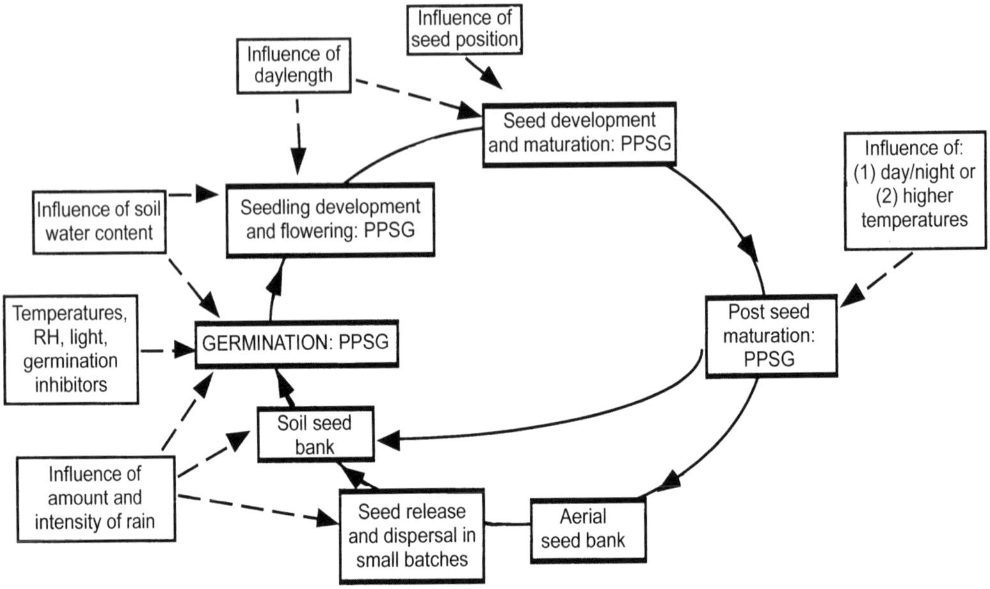

Fig. 43.1. The influences of environmental and maternal factors on phenotypic plasticity of seed germination (PPSG) during the different stages of the annual life cycle of ephemerals in the Negev Desert of Israel (Gutterman, 2002a).

morphological and anatomical resemblance among seeds of different plant species. Differing germinability may occur in seeds of plants of the same species, on one plant, or even according to the position of the seeds in a single capsule. Seed germinability may also be affected by environmental factors during the history of each seed, from the time of maturation, seed dispersal, storage and germination (Gutterman, 1993, 2001, 2002a,b) (Fig. 43.1).

The amount of water during the process of germination, as well as temperature, relative humidity, light conditions, depth of dormancy and germination inhibitors may affect the percentage of germination (Baskin and Baskin, 1998; Gutterman, 1993, 2002a,b).

In many plant species, and in particular those with escape strategies of seed dispersal, the PPSG is one of the most important survival strategies in more extreme deserts. The PPSG may regulate the number of seeds in the soil seed bank that are 'ready to germinate' after a suitable rain, which in turn affects the spread over time of the seeds that join the seed bank (Gutterman 1993, 2000, 2002a,b).

PPSG affected by daylength

In plant species occurring in the Negev Desert, daylength is one of the most important environmental factors that may affect seed germinability during the time of seed development and maturation on the mother plant.

Daylength effects on seed coat permeability to water

In some plant species of the *Fabaceae*, the seeds that matured under short days were found to have less developed seed coat structures than those that matured under long days. The latter seeds, with well-developed seed coat structures, did not germinate, because the seed coats were impermeable to water. Their embryos are ready to germinate and germination quickly follows wetting after scarification of the seed coat. Such a daylength effect has been found in seeds of *Ononis sicula* Guss. and *Trigonella arabica* Delile (Gutterman, 1993, 2002a).

Daylength effects on embryo germinability

In some plant species of different taxa, the daylength affects the germinability of the embryo. Examples are *Schismus arabicus* Nees, *Polypogon monspeliensis* (L.) Desf. (*Poaceae*), *Carrichtera annua* (L.) DC. (*Brassicaceae*), *Spergularia diandra* (Guss.) Heldr. et Sart. (*Caryophyllaceae*), *Portulaca oleracea* L. (*Portulaceae*), *Plantago coronopus* L. subsp. *commutata* (Guss.) Pilger (*Plantaginaceae*) and *Cucumis prophetarum* L. (*Cucurbitaceae*) (Gutterman, 2000, 2002a,b).

Critical time for daylength effect on seed germination

In some plants, the critical time when daylength affects seed germinability is during the last 5–15 days of seed maturation. This was observed in *T. arabica* *C. annua* and *P. oleracea* (Gutterman, 1993).

PPSG influenced by seed position

Differences in seed size, morphology and germinability have been found between aerial seeds and subterranean seeds of amphicarpic plants such as *Gymnarrhena micrantha* Desf. (*Asteraceae*) (Koller and Roth, 1964) and *Emex spinosa* (L.) Campd. (*Polygonaceae*) (Evenari *et al.*, 1977).

Different PPSG was found according to the seed position in the multi-seeded lignified dispersal units of species such as *Pteranthus dichotomus* Forssk. (*Caryophylaceae*) (Evenari, 1963) and *Medicago laciniata* (L.) Miller (*Fabaceae*) (Koller, 1972). In the following plant species differences in germinability, size, colour and hairiness were found according to the position of the caryopses in the spikelets of the spike, which acts as the dispersal unit: *Aegilops geniculata* Roth (= *A. ovata* L.) (Datta *et al.*, 1970, 1972) and *A. kotschyi* Boiss. (*Poaceae*) (Würzburger and Koller, 1976).

Different PPSG was found between the achenes (seeds) according to their position in their capitulum, as in *Asteriscus hierochunticus* (Michon) Wikl. (= *A. pygmaeus*) (*Asteraceae*) (Gutterman, 1993).

There are differences in the period of wetting required for seeds to be separated and dispersed in *Mesembryanthemum nodiflorum* L. (*Aizoceae*) between the groups of about 20 seeds located at the upper, middle or lower part of the capsule, which contains about 60 seeds. Differences in PPSG between these three groups of seeds were found even 30 years after maturation (Gutterman, 1980/81, 2002a,b).

Environmental factors affecting PPSG in post-matured seeds

At the end of the season with rains, and before the long, dry and hot summer, the seeds of the winter annuals of the Negev will have matured. Different plant species have developed various adaptations and strategies to prevent germination of the freshly matured seeds after a late rain when there is no chance for the plants to develop and produce seeds.

Strategies of primary, secondary or skotodormancy of seeds

Strategies such as primary dormancy (Evenari, 1965; Evenari *et al.*, 1982; Gutterman, 1990, 2002a,b), thermodormancy (Small and Gutterman, 1991, 1992a) and skotodormancy (Small and Gutterman, 1992b), which may prevent seed germination after a late rain near the beginning of the long, hot and dry summer, are important survival strategies found in plants occurring in the Negev.

Primary dormancy was found in winter annual plants such as *Schismus arabicus* (Gutterman, 1996) and *Hordeum spontaneum* C. Koch (*Poaceae*) (Gutterman and Nevo, 1994; Gutterman *et al.*, 1996; Gutterman, 2002a,b), *Ammochloa palaestina* Boiss. (*Asteraceae*), *Stipa capensis* Thunb. (*Poaceae*) (Gutterman, 2002a,b) and *Plantago coronopus* (Gutterman *et al.*, 1998).

Similar strategies were found in 21 populations of the winter annual *Bromus tectorum* (*Poaceae*) collected in a wide range of habitats in western North America, including the Mojave Desert (Meyer *et al.*, 1997; Meyer and Allen, 1999).

Post-maturation strategies of hard seeds

Various mechanisms have been found in different plants that enable water to penetrate the seeds after exposure to particular environmental conditions.

GRADUAL INCREASE IN AIR RELATIVE HUMIDITY AND GERMINATION OF HARD SEEDS

The higher the relative humidity (RH) under which 'hard' seeds of *Lathyrus hierosolymita* Boiss. (*Fabaceae*), occurring in the Mediterranean, Judean and Negev deserts, are stored, the faster was the swelling of the seeds. When seeds of *Medicago laciniata*, a common winter annual of the Negev Desert highlands, were stored for 2 months in RH increasing from 13% to 85%, germination was progressively higher, up to 55% (Koller, 1972). This mechanism was first found in seeds of some of the *Fabaceae* by Hyde (1954). During a gradual increase of RH there is an increase in the volume of the embryo. As a result of this increase, the hilum valve opens, enabling free water to enter the seed when they are immersed in water.

TEMPERATURE FLUCTUATIONS AFFECTING GERMINATION OF HARD SEEDS Fluctuating day/night temperatures may affect seed germinability according to the depth at which the seeds are located in the soil and whether the soil is bare, or if it is covered by litter. This may cause phenotypic differences in the seed bank population.

Quinlivan (1961, 1965, 1966, 1968) found this phenomenon in pasture areas east of Perth, in the arid region of Western Australia. In seeds of Geraldton subterranean clover (*Trifolium subterraneum* L.), which mature below the soil surface and germinate *in situ,* as well as in seeds of *Lupinus pilasus* (= *L. varius* L.) (blue lupin), the greater the difference between day and night temperatures during the hot, dry and long summer, the higher is the percentage of seeds that may germinate after wetting. The fluctuation in temperatures causes a split to occur in the strophioler zone of the seed coat through which water can penetrate. A similar mechanism of water uptake through the strophiolum was found by Manning and van Staden (1987) in *Sesbania punicea* (Cav.) Benth., collected near Pietermaritzburg, South Africa.

Impaction and scarification of hard seeds of plants such as *Dalea spinosa, Olneya tesota* and *Cercidium aculeatum* by flood water, gravel and stones at the bottoms of wadis in the Mojave and Colorado deserts of North America, were found to increase their germination (Went, 1953).

Bet-hedging Germination by Seed Release and Dispersal

Bet-hedging germination (Philippi and Seger, 1989) is affected by the PPSG as well as the synaptospermic (Zohary, 1962) or serotinous (Lamont *et al.,* 1991) strategies of seed release, dispersal and germination. At one particular rain event, only small batches of the seeds of the protected aerial species' seed bank are released and dispersed, and then germinate or are collected by seed eaters. Another small batch may join the species' soil seed bank (Fig. 43.1) (Gutterman, 2002a).

Rapid germination after seed dispersal and wetting

Seeds in aerial seed banks of many plant species of the Negev Desert may remain there for many years until they are released year by year in small batches by rain and then dispersed. The dispersed seeds may adhere to the soil surface by their mucilaginous layer, and germinate within a very short time. Examples are *Blepharis* spp. (*Acanthaceae*) (Gutterman *et al.,* 1967) and *Anastatica hierochuntica* L. (*Brassicaceae*) (Gutterman, 1993, 2002a,b).

Germination a few days or weeks after dispersal

In plants such as *Plantago coronopus,* only a few of the dispersed seeds germinate shortly after dispersal and the rest are buried near the soil surface, from where they may germinate after a suitable rainfall a few weeks later. In other plant species, such as *Asteriscus hierochunticus,* a few of the achenes are released each year by a suitable rain event. Later these achenes are dispersed by wind and some of them germinate (Gutterman, 1993, 2002a,b).

Germination some years after dispersal

In other plant species, such as *Mesembryanthemum nodiflorum*, the seeds are released by rain from their capsules during the winter following maturation and remain for many years in the soil seed bank. They germinate from there in small batches over the years according to their original position in their capsules during maturation (Gutterman, 2002a,b).

Conclusion

The history and location of each seed, from the time of seed development and maturation to the time of seed wetting and germination, may affect the PPSG. The many plant species tested were found to have developed their own pathways to PPSG adaptations and strategies at the different stages of their annual life cycle (Fig. 43.1). These pathways may contribute to the survival of the plant species by germination of small batches of seeds under the unpredictable distribution of the small amounts of rain in deserts.

References

Baskin, C.C. and Baskin, J.M. (1998) *Seeds: Ecology, Biogeography, and Evolution of Dormancy and Germination.* Academic Press, San Diego, California, 666 pp.

Beneke, K. (1991) Fruit polymorphism in ephemeral species of Namaqualand. MSc Thesis, University of Pretoria, Pretoria, South Africa.

Beneke, K., von Teichman, I., van Rooyen, M.W. and Theron, G.K. (1992a) Fruit polymorphism in ephemeral species of Namaqualand. I. Anatomical differences between polymorphic diaspores of two *Dimorphotheca* species. *South African Journal of Botany* 58, 448–455.

Beneke, K., von Teichman, I., van Rooyen, M.W. and Theron, G.K. (1992b) Fruit polymorphism in ephemeral species of Namaqualand. II. Anatomical differences between polymorphic diaspores of *Arctotis fastuosa* and *Ursinia cakilefolia. South African Journal of Botany* 58, 456–460.

Cloudsley-Thompson, J.L. (2001) Ecological analogues and convergence among desert organisms. *Biogeographica* 77, 51–63.

Datta, S.C., Evenari, M. and Gutterman, Y. (1970) The heteroblasty of *Aegilops ovata* L. *Israel Journal of Botany* 19, 463–483.

Datta, S.C., Gutterman, Y. and Evenari, M. (1972) The influence of the origin of the mother plant on yield and germination of their caryopses in *Aegilops ovata* L. *Planta* 105, 155–164.

Evenari, M. (1963) Zur Keimungsökologie zweier Wüstenpflanzen. *Mitteilungen der Floristisch-Soziologischen Arbeitsgemeinschaft* 10, 70–81.

Evenari, M. (1965) Physiology of seed dormancy, after-ripening and germination. *Proceedings of the International Seed Testing Association* 30, 49–71.

Evenari, M., Kadouri, A. and Gutterman, Y. (1977) Eco-physiological investigations on the amphicarpy of *Emex spinosa* (L.) Campd. *Flora* 166, 223–238.

Evenari, M., Shanan, L. and Tadmor, N. (1982) *The Negev. The Challenge of a Desert*, 2nd edn. Harvard University Press, Cambridge, Massachusetts, 438 pp.

Feinbrun-Dothan, N. (1978) *Flora Palaestina,* Part Three – Text. Israel Academy of Sciences and Humanities, Jerusalem, 485 pp.

Feinbrun-Dothan, N. (1986) *Flora Palaestina,* Part Four – Text. Israel Academy of Sciences and Humanities, Jerusalem, 468 pp.

Gutterman, Y. (1980/81) Annual rhythm and position effect in the germinability of *Mesembryanthemum nodiflorum. Israel Journal of Botany* 29, 93–97.

Gutterman, Y. (1990) Do germination mechanisms differ in plants originating in deserts receiving winter or summer rain? *Israel Journal of Botany* 39, 355–372.

Gutterman, Y. (1993) *Seed Germination in Desert Plants. Adaptations of Desert Organisms.* Springer, Berlin, 253 pp.

Gutterman, Y. (1996) Temperatures during storage, light and wetting, affecting caryopses germinability of *Schismus arabicus,* a common desert annual grass. *Journal of Arid Environments* 33, 73–85.

Gutterman, Y. (2000) Environmental factors and survival strategies of annual plant species in the Negev Desert, Israel. *Plant Species Biology* 15, 113–125.

Gutterman, Y. (2001) *Regeneration of Plants in Arid Ecosystems Resulting from Patch Disturbance.* Geobotany 27. Kluwer Academic Publishers, Dordrecht, The Netherlands, 260 pp.

Gutterman, Y. (2002a) *Survival Strategies of Annual Desert Plants. Adaptations of Desert Organisms.* Springer, Berlin, 348 pp.

Gutterman, Y. (2002b) Mini review: survival adaptations and strategies of annuals occurring in the Judean and Negev Deserts of Israel. *Israel Journal of Plant Sciences* 50, 165–175.

Gutterman, Y. and Nevo, E. (1994) Temperatures and ecological-genetic differentiation affecting the germination of *Hordeum spontaneum* caryopses harvested from three populations: the Negev Desert and opposing slopes on Mediterranean Mount Carmel. *Israel Journal of Plant Sciences* 42, 183–195.

Gutterman, Y., Witztum, A. and Evenari, M. (1967) Seed dispersal and germination in *Blepharis persica* (Burm.) Kuntze. *Israel Journal of Botany* 16, 213–234.

Gutterman, Y., Corbineau, F. and Côme, D. (1996) Dormancy of *Hordeum spontaneum* caryopses from a population on the Negev Desert highlands. *Journal of Arid Environments* 33, 337–345.

Gutterman, Y., Shem-Tov, S. and Gozlan, S. (1998) The effect of post-maturation temperatures and duration on seed germinability of *Plantago coronopus* occurring in natural populations in the Negev Desert highlands. *Journal of Arid Environments* 38, 451–463.

Hyde, E.O.C. (1954) The function of the hilum in some Papilionaceae in relation to the ripening of the seed and the permeability of the testa. *Annals of Botany* 18, 241–257.

Koller, D. (1972) Environmental control of seed germination. In: Kozlowski, T.T. (ed.) *Seed Biology,* II. Academic Press, London, pp. 2–102.

Koller, D. and Roth, N. (1964) Studies on the ecological and physiological significance of amphicary in *Gymnarrhena micrantha* (Compositae). *American Journal of Botany* 51, 26–35.

Lamont, B.B., Le Maitre, D.C., Cowling, R.M. and Enright, N.J. (1991) Canopy seed storage in woody plants. *Botanical Review* 57, 277–317.

Manning, J.C. and van Staden, J. (1987) The role of the lens in seed imbibition and seedling vigour of *Sesbania punicea* (Cav.) Benth. (Leguminosae: Papilionoideae). *Annals of Botany* 59, 705–713.

Meyer, S.E. and Allen, P.S. (1999) Ecological genetics of seed germination regulation in *Bromus tectorum* L. I. Phenotypic variance among and within populations. *Oecologia* 120, 27–24.

Meyer, S.E., Allen, P.S. and Beckstead, J. (1997) Seed germination regulation in *Bromus tectorum* (Poaceae) and its ecological significance. *Oikos* 78, 475–485.

Philippi, T. and Seger, J. (1989) Hedging one's evolutionary bets, revisited. *Trends of Ecological Evolution* 4, 41–44.

Quinlivan, B.J. (1961) The effect of constant and fluctuating temperatures on the permeability of the hard seeds of some legume species. *Australian Journal of Agricultural Research* 12, 1009–1022.

Quinlivan, B.J. (1965) The influence of the growing season and the following dry season on the hardseededness of subterranean clover in different environments. *Australian Journal of Agricultural Research* 16, 277–291.

Quinlivan, B.J. (1966) The relationship between temperature fluctuations and the softening of hard seeds of some legume species. *Australian Journal of Agricultural Research* 17, 625–631.

Quinlivan, B.J. (1968) The softening of hard seeds of sand-plain lupin (*Lupinus varius* L.). *Australian Journal of Agricultural Research* 19, 507–515.

Small, J.G.C. and Gutterman, Y. (1991) Evidence for inhibitor involvement in thermodormancy of Grand Rapids lettuce seeds. *Seed Science Research* 1, 263–267.

Small, J.G.C. and Gutterman, Y. (1992a) Effect of sodium chloride on prevention of thermodormancy, ethylene and protein synthesis and respiration in Grand Rapids lettuce seeds. *Physiologia Plantarum* 84, 35–40.

Small, J.G.C. and Gutterman, Y. (1992b) A comparison of thermo- and skotodormancy in seeds of *Lactuca serriola* in terms of induction, alleviation, respiration, ethylene and protein synthesis. *Plant Growth Regulation* 11, 301–310.

Went, F.W. (1953) The effects of rain and temperature on plant distribution in the desert. In: *Proceedings International Symposium on Desert Research, May 1952*. Research Council of Israel, Special Publication 2, pp. 230–240.

Würzburger, J. and Koller, D. (1976) Differential effects of the parental photothermal environment on development of dormancy in caryopses of *Aegilops lotschyi*. *Journal of Experimental Botany* 27, 43–48.

Zohary, M. (1962) *Plant Life of Palestine*. Ronald, New York, 262 pp.

Zohary, M. (1966) *Flora Palaestina*, Part One – Text. Israel Academy of Sciences and Humanities, Jerusalem, 364 pp.

Zohary, M. (1972) *Flora Palaestina*, Part Two – Text. Israel Academy of Sciences and Humanities, Jerusalem, 489 pp.

44 Cold-induced Germination Promotion in *Hemerocallis dumortieri* var. *esculenta* and *H. fulva* var. *littorea* Seeds

Kojiro Suzuki, Hiroko Nakajima and Chikayasu Hamano

Faculty of Regional Environment Science, Tokyo University of Agriculture, 1-1-1 Sakuragaoka, Setagaya-ku, Tokyo 156-8502, Japan

Introduction

Hemerocallis L. (day lilies), a genus of the *Liliaceae* according to Engler's classification (Melchior, 1964), includes several wild species distributed in Japan, such as *H. dumortieri* Morr. var. *esculenta* (Koidz.) Kitam. and *H. fulva* L. var. *littorea* (Makino) M.Hotta. *H. dumortieri* var. *esculenta* plants are predominantly distributed in wet grassland, mountainous or subalpine zones with high snowfall, ranging from the mainland (Honshu) to the northern region (Hokkaido) of Japan. *H. fulva* var. *littorea* plants grow in the area from the west of the mainland to the southern region (Kyushu), favouring craggy and dry coastal slopes.

H. dumortieri var. *esculenta* seeds are thought to show characteristics of cold stratification, because of their distribution pattern in high snowfall areas, and many seeds are dispersed under snow during winter; therefore germination promotion progresses under cold wet conditions. *H. fulva* var. *littorea* seeds may not respond similarly, due to their different habitat profile. We investigated cold response of the seeds of these two species that belong to the same genus but grow under entirely different conditions.

Material and Methods

Sources of plant materials

Mature seeds were collected from several geographical locations. Details of the locations and dates of seed collection are described in Table 44.1.

Table 44.1. Collection sites and dates of seed lots for the two *Hemerocallis* species. Figures in parentheses are altitudes.

Species	Date	Location
Hemerocallis dumortieri var. *esculenta*	6 October 1995	Mt Makihata (~1750–1800 m), Niigata Prefecture, northern Japan
	4 September 1998 27 September 1999 27 August 2000	Mt Kurumayama (~1800 m), Nagano Prefecture, central Japan
	2 September 2001	Shiga Height (~1700–2000 m), Nagano Prefecture, central Japan
Hemerocallis fulva var. *littorea*	22 August 2001	Jyogasaki coast, Shizuoka Prefecture, central Japan
	5 October 2001	Jyogashima, Kanagawa Prefecture, central Japan

Seed pretreatment

To determine when and how seed dormancy loss is achieved in seed lots of *H. dumortieri* var. *esculenta* and *H. fulva* var. *littorea*, we conducted two experiments, comprising germination tests and tests to determine the ratio of embryo to albumen/endosperm. A portion of the collected seeds was subjected to germination testing immediately after harvest; the remaining seeds were stored under two sets of conditions: (i) cold, dry conditions in a desiccator at 5°C (9.2% seed moisture content); or (ii) cold, wet conditions in a hermetically sealed polyethylene bag at 5°C (46.6% seed moisture content). The seed lots were subjected to germination tests monthly and the ratio of embryo to albumen was tested every 2 weeks.

Germination

Fifty seeds were placed in 9-cm diameter glass dishes in which two sheets of filter paper soaked in 4 ml tap water had been placed. Dishes were placed in individual chambers (NK System Co.) maintained at 10, 15, 20, 25 or 30°C and illuminated by a fluorescent lamp under 12 h light cycle conditions (28.7 μmol/m^2/s). Three dishes were placed in each chamber at one temperature. Germinated seeds were recorded and removed from the dish every week and the germination percentage was calculated.

Ratio of embryo to albumen

Embryos were sectioned, using a razor blade, immediately after harvest or after storage for certain periods under dry or wet conditions at intervals of 2 weeks, and the sections were photographed using a digital camera attached to a stereoscopic microscope (Nikon SMZ800). The areas of embryo and albumen were measured using a planimeter. The percentage (embryo area/albumen area $\times$ 100) was calculated.

Table 44.2. Germination percentage of *Hemerocallis dumortieri* var. *esculenta* seeds collected on (A) 6 October 1995, on Mt Makihata, Niigata Prefecture and (B) 4 September 1998, at Kurumayama, Nagano Prefecture.

	(A)		(B)	
Temperature (°C)	MGP ± SD	WRM	MGP ± SD	WRM
10	88.3 ± 7.1	6	0.7 ± 1.6	19
15	76.7 ± 9.4	5	3.3 ± 3.7	22
20	76.7 ± 14.1	6	22.3 ± 7.9	21
25	60.0 ± 9.4	4	41.0 ± 19.5	9
30	30.0 ± 9.4	3	10.0 ± 7.7	7

MGP, mean of germination percentage; WRM, weeks to reach maximum observed.
Germination tests: (A) 30 November 1995, after storage in polythene bag at 5°C for about 2 months; (B) soon after seed harvesting. All data are the average of three dishes.

Results

Cold stratification of *H. dumortieri* var. *esculenta* seeds

The results for seed lots collected on 6 October, 1995, at Mt Makihata and on 4 September, 1998, at Mt Kurumayama are presented in Table 44.2.

The germination percentage at 15°C was 76% for the October 1995 seed lot, but the germination percentage at 25°C of those collected in September 1998 at Mt Kurumayama was only 41% (Table 44.2). These percentages diverge more than would be expected. The difference may be partly attributable to the 1-month time difference in the collection date and the fact that the seeds collected in 1995 were stored at low temperature under wet conditions. Different cold stratification treatment thus may have resulted in different germination levels.

We then compared the germination percentages of seed lots collected on 27 September 1999, immediately after harvest and after storage for set periods under dry or wet conditions. The results showed that the seed lots stored under wet conditions had a higher germination percentage (Fig. 44.1B) than those immediately after harvest (Fig. 44.1A). Those stored under dry conditions showed a low germination percentage (Fig. 44.1B).

We also observed cross-sections of the embryos of seeds that had been stored for 6 months under wet or dry conditions at 5°C. The results showed that the embryos had enlarged after storage under wet conditions, for which the ratio of the cross-section of embryo to albumen was changed from 40 to 56%, but the embryos that had been stored under dry conditions had shrunk and detached from the albumen, with the ratio of the cross-section of embryo to albumen decreased to 33% (Fig. 44.2).

Seed lots collected on 27 August 2001 were stored under cold, wet conditions, and were subjected to germination testing at 15°C every month for a period of 6 months. The ratio of embryo to albumen cross-section was also determined (Fig. 44.1C). The results showed that the germination percentage increased progressively over 4 months, as did the embryo size.

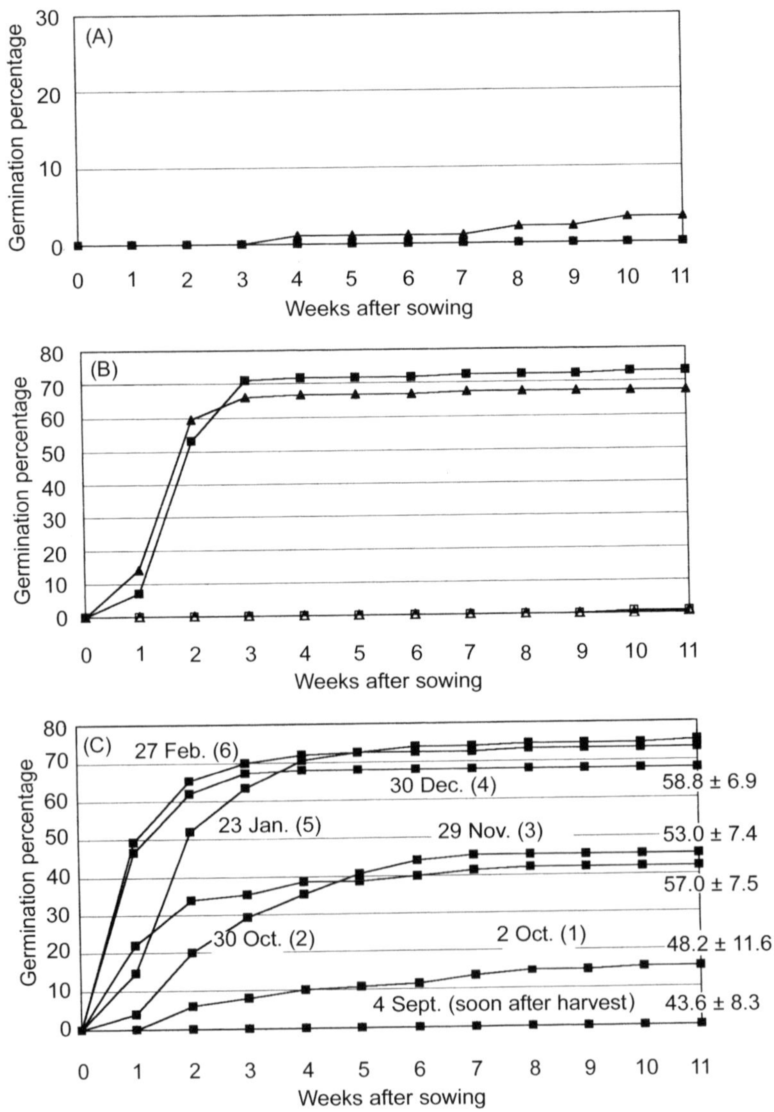

Fig. 44.1. Germination percentage of *H. dumortieri* var. *esculenta* seeds stored under different conditions. The seed lots used were collected on 27 September 1999 (A and B) and on 27 August 2002 (C), at Mt Kurumayama, Nagano Prefecture. The seed germination tests were done soon after seed harvest (A) and 5 months (3 March 2000) after seed harvest or storage (B). The numbers in parentheses on graph (C) represent months after seed harvest: values at the end of the lines represent the ratio of embryo to albumen (% ± SD). The seed lots used were stored under wet conditions (solid line and black symbol) or dry conditions (white symbol). The seed lots from storage under dry conditions were not able to germinate in (B). The seed germination tests were done at 15°C (■ or □) and at 20°C (▲ or △). Storage temperature was 5°C in all conditions. All data are the average of three dishes.

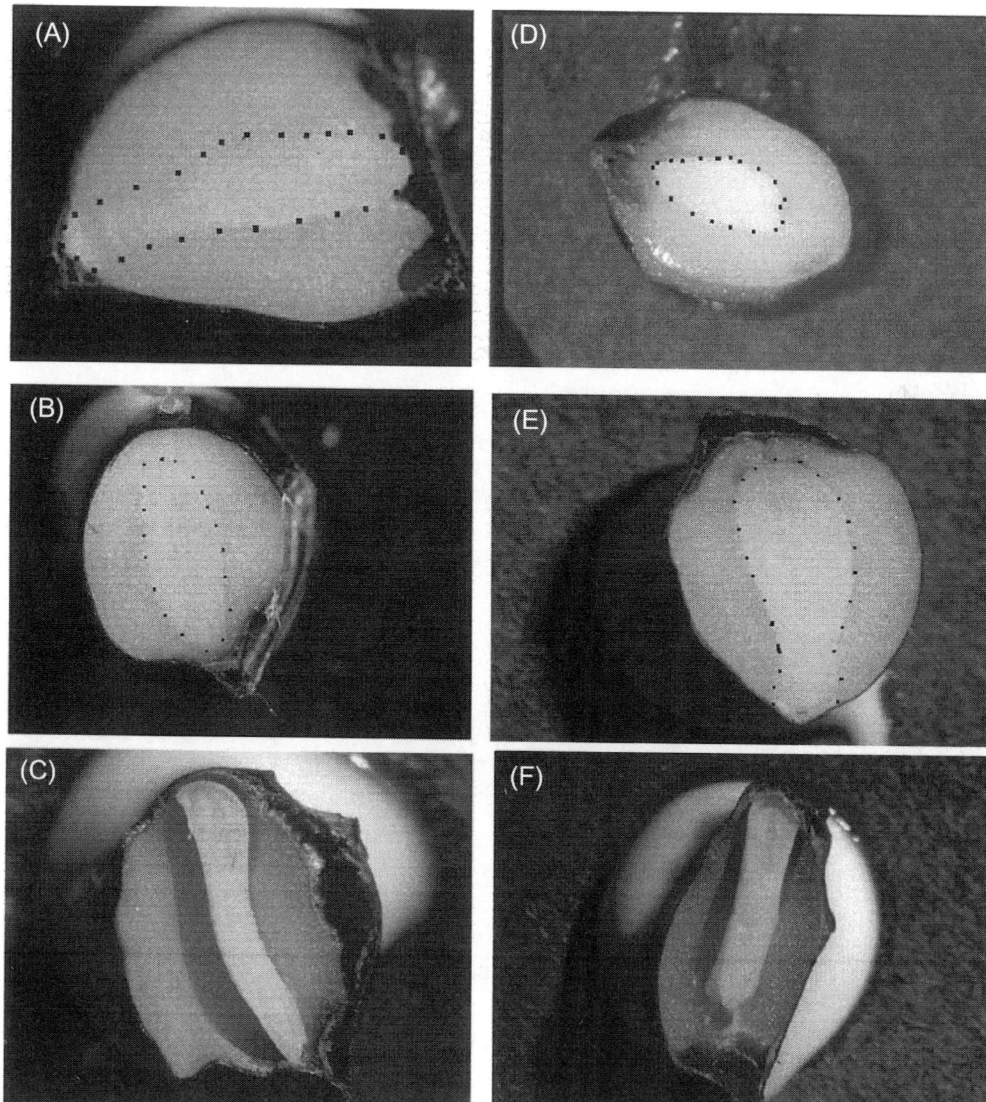

Fig. 44.2. Changes in ratio of cross-section of embryo to albumen of *H. dumortieri* var. *esculenta* (A, B, C) and *H. fulva* var. *littorea* (D, E, F). (A) Soon after seed harvest on 11 September 2001; (B) 3 months after storage under wet condition at 5°C; (C) 3 months after storage under dry condition at 5°C; (D) soon after seed harvest on 22 August 2001; (E) 4 months after storage under wet condition at 5°C; (F) 4 months after storage under dry condition at 5°C.

Seeds collected on 2 September 2001 at Shiga Heights were tested for germination at 10, 15, 20, 25 and 30°C after storage for 3 months under wet or dry conditions. The results showed that seed lots stored under dry conditions did not show improved germination percentages in comparison with the percentage immediately after harvest, while those stored under wet con-

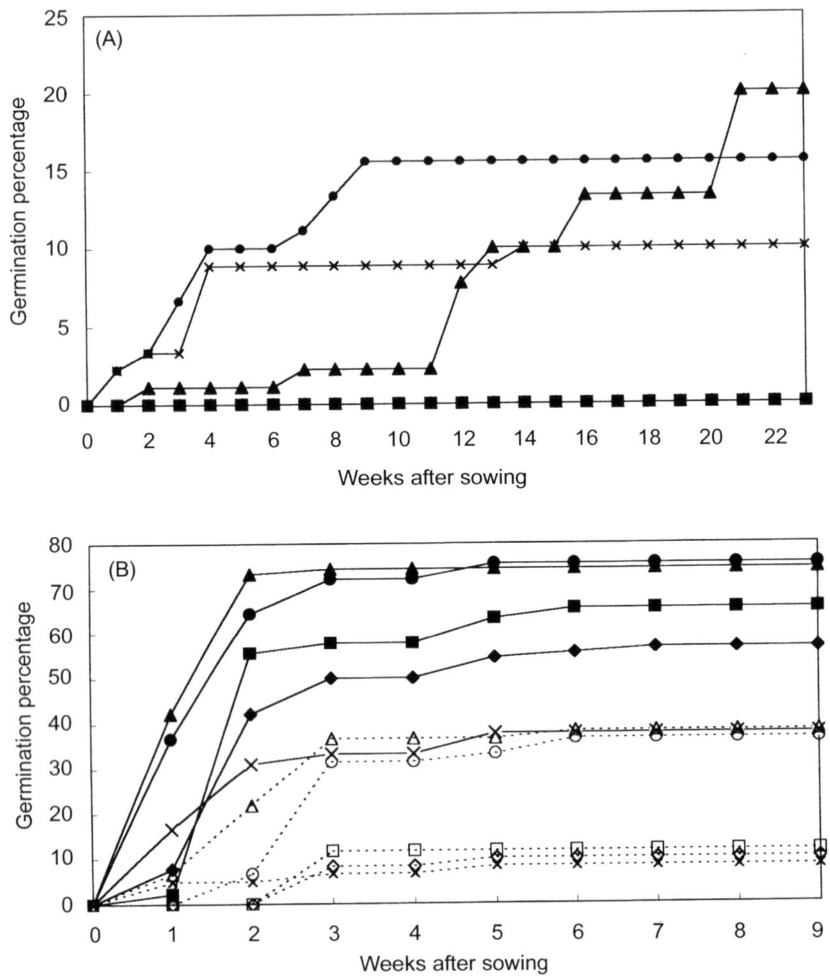

Fig. 44.3. Germination percentage of *H. dumortieri* var. *esculenta* seeds collected in 2001 and stored under different conditions. All seed lots used were collected on 2 September 2001, in Shiga, Nagano Prefecture. Germination tests: (A) soon after harvest; (B) 3 months after seed harvest. The seed lots used were stored under wet conditions (solid lines and black symbols) or dry conditions (dotted lines and white symbols) at 5°C. The seed germination tests were done at 10°C (◆, ◇), 15°C (■, □), 20°C (▲, △), 25°C (●, ○), 30°C (×). All data are the average of three dishes.

ditions showed a higher germination percentage than when tested immediately after seed harvest (Fig. 44.3).

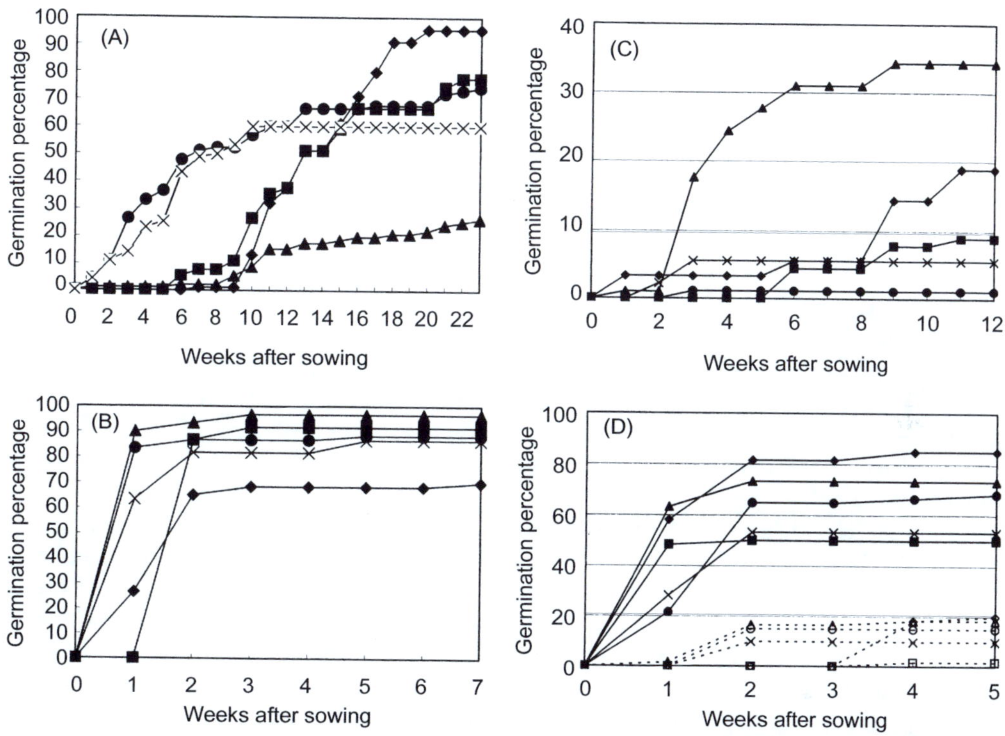

Fig. 44.4. Germination of *H. fulva* var. *littorea* seeds stored under different conditions. The seed lots used were collected on 22 August 2001, on Jogasaki coast, Shizuoka Prefecture (A, B), and collected on 5 October 2001, in Jogashima, Kanagawa Prefecture (C, D). Germination tests: (A, C) soon after seed harvest; (B) 3 months after seed harvest and storage (on 5 December 2001); (D) 4 months after seed harvest and storage (on 5 February 2002). The seed lots used were stored under wet conditions (solid lines and black symbols) or dry conditions (dotted lines and white symbols) at 5°C. The seed germination tests were done at 10°C (◆, ◇), 15°C (■, □), 20°C (▲, △), 25°C (●, ○), 30°C (×). All data are the average of three dishes.

Cold treatment of *H. fulva* var. *littorea* seeds

H. dumortieri var. *esculenta* plants are distributed in wet grasslands located from northern to central Japan, especially in alpine or subalpine zones, while *H. fulva* var. *littorea* plants are distributed from central to southern Japan, especially in sunny places along the coast. We investigated the cold response of *H. fulva* var. *littorea* seeds that had grown in different places.

The seed germination percentage at 25°C and 30°C of seed lots collected at Jyogasaki on 22 August 2001, was about 50% immediately after the seed harvest at 9 weeks after sowing (Fig. 44.4A). However, those stored under wet conditions for 3 months showed a 95% or higher germination percentage (Fig. 44.4B). Concerning the seed lots collected at Jyogashima on 5 October 2001, the highest seed germination percentage was 35% at 20°C immediately after harvest (Fig. 44.4C). The seed germination percentage increased to 50% or higher after storage for 4 months under wet conditions, and to 80% or higher at 10°C. The seed germination percentage was only 20% after storage under dry conditions (Fig. 44.4D).

Discussion

Both *H. dumortieri* var. *esculenta* and *H. fulva* var. *littorea* seeds are able to respond positively to storage under wet conditions at 5°C. However, the germination percentage of seed lots stored under dry conditions fell in both species, suggesting that *H. dumortieri* var. *esculenta* and *H. fulva* var. *littorea* seeds can both be classified as recalcitrant seeds or that there is a complex reaction between desiccation and depth of dormancy.

Since *H. dumortieri* var. *esculenta* plants are distributed in regions with heavy snowfall and wet soil, we conclude that the seeds are covered with snow after seed dispersal, allowing the promotion of seed germination. On the other hand, *H. fulva* var. *littorea* is distributed along the coast in areas with a dry climate, but its seeds also respond to cold wet treatment.

The different germination behaviour of *H. dumortieri* var. *esculenta* and *H. fulva* var. *littorea* seeds may be due to genetic factors, since these plants grow in areas with very different germination conditions. Studies of other *Hemerocallis* species have shown that, for example, *H. minor* also responds to wet cold stratification for 3 weeks at 3.5°C (Griesbach and Voth, 1957). It therefore remains possible that seed lots collected from areas with different, even contrary, germination conditions (such as wet areas and dry areas) may be induced to show the same germination habits.

Several questions remain, such as under what conditions *H. fulva* var. *littorea* seeds are dispersed, how dormancy is lost and how the seeds germinate in the field. These issues will be solved in future studies. We also plan to study the germination behaviour of other *Hemerocallis* species distributed in specific regions in Japan.

Acknowledgements

We thank Dr Megumi Aso, Mr Kazuya Kurita and Miss Jyunko Abe, Faculty of Regional Environmental Science, Tokyo University of Agriculture, for collecting the fruits and seeds of *H. dumortieri* var. *esculenta* and *H. fulva* var. *littorea* for use in the present study. Thanks also to Dr Shunichiro Nakamura for his comments and Mrs Junko Muto for her help. This research was funded by the Grants-in-Aid for Scientific Research in 2001–2002.

References

Griesbach, R.A. and Voth, P.D. (1957) On dormancy and seed germination in *Hemerocallis*. *Botanical Gazette* 118, 223–237.

Melchior, H. (ed.) (1964) *A. Engler's Syllabus der Pflanzen familien*, Vol. 2. Borntrager, Berlin-Nikolasse, pp. 512–542.

45 Germination Stimulation of Weed Species by Smoke

Steve W. Adkins[1], Nick C.B. Peters[2], Mary F. Paterson[1] and Sheldon C. Navie[1]

[1] Tropical and Subtropical Weed Research Unit, The University of Queensland, St Lucia Campus, Brisbane, Qld 4072, Australia; [2] Institute of Arable Crops Research, Long Ashton Research Station, Department of Agricultural Science, University of Bristol, Bristol BS41 9AF, UK

Introduction

Compounds produced by the combustion or charring of plant material stimulate germination in a number of species (De Lange and Boucher, 1993; Baxter *et al.*, 1994; Dixon *et al.*, 1995; Brown and van Staden, 1997; Keeley and Fotheringham, 2000). The compound or compounds involved in this stimulation of germination have not been definitively identified (Baldwin *et al.*, 1994) but they are produced by a variety of wood types, are water soluble and are derived from the hemicellulose fraction of the combustion materials (Baldwin *et al.*, 1994). It is not known how the active component acts on seed, though Egerton-Warburton (1998) found that aerosol smoke acted on the seed coat of *Emmenanthe penduliflora* Bench. in a way similar to scarification, a process in which the passage of water and oxygen into the dormant embryo can be promoted.

Smoke is effective on species from a wide range of families that vary in ecology, reproductive strategy, seed size and morphology (Dixon and Roche, 1995). To date, germination enhancement has been shown in more than 170 species from 37 families (Roche *et al.*, 1997). Most studies have focused upon species native to fire-prone areas but germination can also be stimulated in lettuce (*Lactuca sativa* L.: Drewes *et al.*, 1995), celery (*Apium graveolens* L.: Thomas and van Staden, 1995), red rice (*Oryza sativa* L.: Doherty and Cohn, 2000) and a number of arable weed species (Adkins and Peters, 2001).

Several products containing smoke in solution are on the market, including 'Regen 2000', 'Seed Starter, Australian Smoky Water' and 'Kirstenbosch Instant Smoke Plus Seed Primer'. These smoke-water products have applications in the rehabilitation of disturbed areas, horticultural industries, ecological management and crop production, particularly in

 The Biology of Seeds: Recent Research Advances (eds G. Nicolás, K.J. Bradford, D. Côme and H.W. Pritchard)

organic farming systems. Such applications are primarily for species that are difficult to germinate under normal circumstances. Investigation of smoke for rehabilitation has predominantly been carried out for fire-prone environments (Keeley and Fotheringham, 2000). Very little examination of the effect of smoke on germination of introduced weedy species has been carried out (Doherty and Cohn, 2000). One feature of weeds is their ability to persist in the soil seed bank for several years, due to dormancy. Should smoke prove to be an effective agent in stimulating the germination of these species, this would have implications for rehabilitation of sites with native species.

The aims of the present study are: (i) to develop a better understanding of the effect of smoke water upon the germination of a wide range of native and introduced weedy species under laboratory conditions; and (ii) to investigate the effect of aerosol smoke as a possible tool for the regeneration of a degraded grassland community containing native and introduced weedy species.

Materials and Methods

Seed materials and germination

Three replicates of 50 seeds (see Table 45.1) were placed in 9-cm diameter Petri dishes lined with four layers of 9 cm Whatman No. 2 filter paper moistened with 8 ml of distilled water (control) or smoke-water solution. All dishes were incubated in darkness at a constant temperature of 15 ± 1°C for species of temperate origin and 25 ± 1°C for all other species. Germination was recorded periodically for a period of 35 days with any seedlings or dead seeds removed. The smoke-water solution used (unless otherwise stated) was diluted from a stock solution of 'Seed Starter®'(Kings Park and Botanical Gardens, Perth, Australia) to a concentration of 10%. In one experiment three other smoke-water solutions were used: Regen 2000 (Tecnica Propriety Ltd, Bayswater, Victoria, Australia) and two solutions prepared at the Institute of Arable Crops Research (IACR), Long Ashton Research Station, Bristol, UK – one from the combustion of wheat (*Triticum aestivum* L.) straw, and one prepared from charred willow (*Salix* sp.) wood chips (Thornton *et al.*, 1999).

Physiological measurements

The rate of water uptake induced by smoke water was monitored over time by soaking intact seeds of kangaroo grass (*Themeda triandra* Forssk.) and *Bothriochloa pertusa* (L.) A. Camus in a 1:4 CsCl solution. Following the desired imbibition time the seeds were dried, embedded in an apoxyresin and the blocks sanded and polished using a series of carbon papers and diamond pastes. The samples were viewed using a backscatter detection technique while placed in a JEOL 6400 scanning electron microscope (SEM). The areas of brightness seen in the scans were indicative of water presence and time-course observations illustrated the way in which water was taken up by the seeds over time. To evaluate further the possibility that smoke acts on seeds similarly to surfactants by scarifying the seed coat surface, the external seed husk structures were examined on *T. triandra* prior to and after a 15 min aerosol smoke treatment, using a 6200 SEM.

The field site

The grassland site chosen in northern Queensland had been used for beef pasture for several decades prior to the commencement of this research but had recently been declared a national park. *Parthenium hysterophorus* L., a serious weed of this area, had been present for many years. The introduced pasture grass *Cenchrus ciliaris* L. and several native species of *Aristida* dominated the extant vegetation. A random collection of soil samples (7 cm in diameter and 6 cm deep) was undertaken over an area of 10 × 15 m. In the glasshouse the soil samples were spread thinly over sterilized soil held in shallow trays. Smoke from black speargrass (*Heteropogon contortus* (L.) Beauv. ex Roem & Schult.) was trapped in a tent and the trays were exposed for either 0, 10 or 60 min. The trays were then gently misted with distilled water to settle the smoke and randomly distributed on a bench in a glasshouse and kept watered to field capacity. Emerging seedlings were marked and identified. Differences between the numbers of seed germinating from different treatments were analysed with a non-parametric Friedman's test.

Results

Germination studies on a range of species

All of the monocotyledonous species tested either germinated well in water (non-dormant, e.g. *Bromus diandrus* Roth) or were strongly stimulated by smoke water (eight species; Table 45.1). The dicotyledonous species, on the other hand, were either strongly stimulated (e.g. *Malva neglecta* Wallr.), moderately stimulated (three species), slightly stimulated (seven species), unaffected (eight species) or inhibited (*Lamium purpureum* L.). Most of the monocotyledonous species studied originated in habitats that are prone to fires (Table 45.1) but the origins of the dicotyledonous species were much more varied and only about half of these species originated in fire-prone habitats (Holm *et al.*, 1991). Only about half of the smoke-responsive species could be tracked back to originating in a fire-prone habitat (Table 45.1). When smoke-water solutions prepared from four different organic sources were tested on two representative species (*Avena fatua* L. and *Polygonum persicaria* L.), three solutions were active in promoting the germination of both species, while the fourth (charred wood solution) was only active on *A. fatua*. All solutions proved to be inhibitory to germination at the highest concentration tried.

Physiological responses to smoke treatment

Examination of water uptake in non-dormant seed of *Themeda triandra*, a typical smoke-responsive grass species, showed that there were distinctive

Table 45.1. The effect of smoke water on the germination of several native and weed species. Life history: whether the species is an Australian native plant, an introduced pasture species or a weed; Smoke effect: whether the germination of the species is strongly stimulated (+++), moderately stimulated (++), slightly stimulated (+), unaffected (=) or inhibited (–) by smoke water; Geographic origin: location at which the species originated; Fire-prone habitat: whether the species originated in a location prone to fires (+) or not (–).

Species	Life history	Smoke effect	Geographic origin	Fire-prone habitat
Monocotyledonous				
Alopecurus myosuroides Huds.	Weed	+++	Europe and Mediterranean	+
Aristida ramosa R.Br.	Native	+++	Australia?	+
Avena fatua L.	Weed	+++	Mediterranean	+
Bromus diandrus Roth.	Weed	=	Mediterranean	+
Cenchrus ciliaris L.	Weed/pasture	+++	Africa and Middle East	+
Heteropogon contortus (L.) Beauv.	Native	+++	S. Asia, Polynesia and USA?	+
Phalaris paradoxa L.	Weed	+++	Mediterranean	+
Sorghum halepense (L.) Pers.	Weed	++	Mediterranean	+
Themeda triandra Forssk.	Native	++	Southern Asia	+
Dicotyledonous				
Angelica sylvestris L.	Weed	=	Europe	–
Boronia fastigiata Bartl.	Native	=	Australia	+
Boronia rosmarinifolia A.Cunn.	Native	=	Australia	+
Boronia safrolifera Cheel.	Native	+	Australia	+
Chamaescilla corymbosa (L.) A.Love	Native	+	Australia	+
Fallopia convolvulus L.	Weed	+	Europe	±
Galium aparine L.	Weed	++	Europe and Northern Asia?	–
Heracleum sphondylium L.	Weed	=	Europe and Central Asia	–
Hibbertia amplexicaulis Steud.	Native	+	Australia	+
Hibbertia commutata Steud.	Native	++	Australia	+
Lamium purpureum L.	Weed	–	Eurasia	–
Malva neglecta Wallr.	Weed	+++	Eurasia and North Africa	–
Mercurialis annua L.	Weed	=	Europe, N. Africa & S.W. Asia	–?
Polygonum aviculare L.	Weed	=	Europe and Asia	–?
Polygonum pennsylvanicum L.	Weed	+	?	–
Polygonum persicaria L.	Weed	+	Europe	–
Sinapis arvensis L.	Weed	=	Eurasia (Mediterranean)	+
Veronica hederifolia L.	Weed	=	Britain, Eurasia	+
Veronica persica Poir.	Weed	++	Western Asia	+
Zieria laxiflora (Benth) Domin	Native	+	Australia	+

points of entry throughout the length of the seed coat. In a softer-coated species such as *Bothriochloa pertusa*, the major areas for water uptake were found to be through the callus and through the join of the glumes, with limited and slower water uptake through the seed coat. Upon examination of the external seed coat structure of dormant *T. triandra*, it was discovered that there was a very distinctive pore system that was likely to be responsible for the observed pattern of water uptake. However, in dormant seed, materials that resembled cuticular waxes plugged these pores. In response to a 15 min aerosol smoke treatment, the plugging materials formed into spherical structures, which allowed for the opening of the pore apertures.

Smoke-stimulated seedling emergence

The total germinable soil seed bank varied between 20,396 seeds/m^2 in the control treatment and 25,124 seeds/m^2 in the smoke-60 treatment (Table 45.2). The introduced weed *Parthenium hysterophorus* was the most abundant species present and accounted for between 79 and 88% of the seed bank, depending on the germination treatment. Grasses were also common in the seed bank (9–17% of the total), particularly the introduced *C. ciliaris*, which represented the vast majority of the germinable grass seed bank

Table 45.2. Seedling emergence from the soil seed bank (seedlings/m^2) after two different smoke treatments were applied for either 10 or 60 min. Figures with the same letter after are not significantly different at the $P = 0.05$ level.

Species	Origin	Control	Smoke (10)	Smoke (60)
Parthenium hysterophorus	Introduced	17786*a*	21685*b*	20713*b*
Cenchrus ciliaris	Introduced	1678*a*	1883*b*	3285*c*
Brachiaria sp.	Native	133	102	154
Eleusine indica	Introduced		10	
Other Poaceae	–	215	256	225
Argemone ochroleuca	Introduced	61	133	72
Chenopodium cristatum	Native	31	20	41
Conyza sp.	Introduced		20	10
Einadia trigonos	Native		10	
Gnaphalium spp.	Introduced	10	10	10
Malva parviflora	Introduced			10
Oxalis corniculata	Native		10	
Physalis minima	Native	10		
Phyllanthus sp.	Native	10	10	
Portulaca filifolia	Native		10	
Portulaca oleracea	Native	10	10	41
Sida sp.	?	41	72	92
Sonchus oleraceus	Native			10
Verbena officinalis	Introduced		10	10
Wahlenbergia tumidifructa	Native	10		10
Other dicotyledons	–	82	41	30
Cyperus gracilis	Native	31	20	20
Cyperus rotundus	Introduced	51	41	61
Other Cyperaceae	–			
Unknown	–	235	266	327
TOTAL	–	20396*a*	24622*b*	25124*b*

(1678–3920 seeds/m^2). There were significantly greater numbers of *Parthenium hysterophorus* seeds germinating from the two smoke treatments as compared with the control. A significant increase in seedling production was also found in the total number of seeds from all other species germinating after the two smoke treatments. The smoke-60 treatment produced significantly more seedlings than the other smoke treatment and the control.

Discussion

When a number of introduced weed and native species were tested for their germination response to smoke water, a wide range of responses was obtained (Table 45.1). Some species were strongly stimulated, while others less stimulated or unaffected. Of the two types of species tested, the monocotyledonous (as compared with the dicotyledonous) were by far the most responsive. This may indicate that smoke water is more easily able to stimulate those species that, in the main, have a simple permeability-based, coat-imposed dormancy mechanism, and is less effective on those species with a more complex mechanical-based, coat-imposed or embryo dormancy mechanism. It was interesting to note that amongst the dicotyledonous species there was a poor relationship between origin from a fire-prone habitat and a positive response to smoke. About half of the smoke-responsive species could be tracked back to smoke-prone habitats. Further work now needs to be undertaken to establish the significance of this trend.

The identity of the compounds in smoke responsible for dormancy release remains unknown. Past studies have shown that a range of combusting plant materials can produce the active components and this view is again supported in the present study. Both species were stimulated by smoke water generated from burnt Australian vegetation and burnt wheat straw, as well as one species being stimulated by a solution washed over the charred remains following pyrolysis of willow wood chips. Egerton-Warburton (1998), based on transmission and scanning electron microscopic analysis, reported a surfactant-like effect exerted by smoke water, in which two seed coat layers were destabilized/re-organized by components of smoke. In the present series of experiments a similar sort of action is suspected in the weaker seed-coated species studied (e.g. the grasses). The scarification effect of smoke was demonstrated in *T. triandra* where the wax plugs were partially removed from the pores. This presumably allowed for greater water and oxygen uptake to the dormant embryo.

The smoke treatment of soil samples taken from a degraded pasture indicated that both undesirable weeds and desirable native grasses could be stimulated to emerge from the soil seed bank (Table 45.2). As grasses (i.e. *C. ciliaris*) were one of the dominant components of the seed bank, it is likely that the smoke stimulation of germination was by a similar mechanism to that described above for *T. triandra*. These findings suggest that burning of vegetation in degraded rangeland sites that generate significant volumes of smoke may have a desirable effect of clearing the land of unproductive weed species and promoting the germination of desirable pasture grasses.

However, the increase in the number of *P. hysterophorus* seedlings, promoted into germination in response to smoke, nullifies any advantage gained. Also, there are many other aspects of the fire that should be considered before its value as a management tool is promoted. For example, burning of existing vegetation, leaving bare areas in the pasture, will create ideal sites for *P. hysterophorus* to grow and establish.

Acknowledgements

We gratefully acknowledge seed samples from Dr M. Parham, Herbiseed, UK, Dr T.H. Thomas and M.A. Thornton, IARC-Long Ashton Research Station, for the wheat straw and charred willow wood smoke-water solutions, and Wendy Armstrong, University of Queensland for help with the electron microscope imaging. The seed bank study was part of a larger project supported by the Rural Industries Research and Development Corporation and administered by Wane Vogler of the Queensland Department of Natural Resources and Mines, QDN-7A.

References

Adkins, S.W. and Peters, N.C.B. (2001) Smoke derived from burnt vegetation stimulates germination of arable weeds. *Seed Science Research* 11, 213–222.

Baldwin, I.T., Staszak-Kozinski, L. and Davidson, R. (1994) Up in smoke. I. Smoke-derived germination cues for postfire annual, *Nicotiana attenuata* Torr. ex. Watson. *Journal of Chemical Ecology* 20, 2345–2371.

Baxter, B.J.M., van Staden, J., Granger, J.E. and Brown, N.A.C. (1994) Plant-derived smoke and smoke extracts stimulate seed germination of the fire-climax grass *Themeda triandra*. *Environmental and Experimental Botany* 34, 217–223.

Brown, N.A.C. and van Staden, J. (1997) Smoke as a germination cue: a review. *Plant Growth Regulation* 22, 115–124.

De Lange, J.H. and Boucher, C. (1993) Autecological studies on *Audouinia capitata* (Bruniaceae). VIII. Role of fire in regeneration. *South African Journal of Botany* 59, 188–202.

Dixon, K.W. and Roche, S. (1995) The role of combustion products (smoke) in stimulating *ex situ* and *in situ* germination of Western Australian plants. *Proceedings of the International Plant Propagation Society* 45, 53–56.

Dixon, K.W., Roche, S. and Pate, J.S. (1995) The promotive effect of smoke derived from burnt native vegetation on seed germination of Western Australian plants. *Oecologia* 101, 185–192.

Doherty, L.C. and Cohn, M.A. (2000) Seed dormancy in red rice (*Oryza sativa*). XI. Commercial liquid smoke elicits germination. *Seed Science Research* 10, 415–421.

Drewes, F.E., Smith, M.T. and van Staden, J. (1995) The effect of a plant-derived smoke extract on the germination of light sensitive lettuce seed. *Plant Growth Regulation* 16, 205–209.

Egerton-Warburton, L.M. (1998) A smoke-induced alteration of the sub-testa cuticle in seeds of the post-fire recruiter, *Emmenanthe penduliflora* Bench. (Hydrophyllaceae). *Journal of Experimental Botany* 49, 1317–1327.

Holm, L.G., Plucknett, D.L., Pancho, J.V. and Herberger, J.P. (1991) *The World's Worst Weeds: Distribution and Biology*. Krieger Publications Co., Malabar, Florida, 609 pp.

Keeley, J.E. and Fotheringham, C.J. (2000) Role of fire in regeneration from seed. In: Fenner, M. (ed.) *Seeds. The Ecology of Regeneration in Plant Communities*, 2nd edn. CAB International, Wallingford, UK, pp. 311–330.

Roche, S., Dixon, K.W. and Pate, J.S. (1997) Seed ageing and smoke: partner cues in the amelioration of seed dormancy in selected Australian native species. *Australian Journal of Botany* 45, 783–815.

Thomas, T.H. and van Staden, J. (1995) Dormancy break of celery (*Apium graveolens* L.) seeds by plant-derived smoke extract. *Plant Growth Regulation* 17, 195–198.

Thornton, M.A., Thomas, T.H. and Peters, N.C.B. (1999) The promotive effect of combustion products from plant vegetation on the release of seeds from dormancy. *Plant Growth Regulation* 28, 129–132.

46 Genetic Analysis and QTL Mapping for Cytosolic Glutamine Synthetase (GS) and Germination Traits in Maize Grain

Anis M. Limami[1], Clothilde Rouillon[1], Gaëlle Glevarec[1], André Gallais[2] and Bertrand Hirel[3]

[1]*UMR Physiologie Moléculaire des Semences, Université d'Angers, 2 bd Lavoisier, 49045 Angers cedex, France;* [2]*Station de Génétique Végétale du Moulon, INRA-UPS-INAPG, Ferme du moulon, 91190 Gif/Yvette, France;* [3]*Unité de Nutrition Azotée des Plantes, INRA, route de St Cyr, 78026 Versailles cedex, France*

Introduction

At the end of the development stage, maturation of seeds is characterized by dehydration that leads to the loss of more than 90% of water content and a dramatic decrease in metabolic activity. The induction of germination by rehydrating the seeds leads to an increase in respiration and metabolic activity that allows for mobilization of seed C and N reserves. The mobilization of carbon reserves has been well studied with respect to the regulation of α-amylase in barley in relation to programmed cell death in aleurone cells of germinating grain under the control of the phytohormones gibberellin and abscisic acid (Bethke and Jones, 2001; Fath *et al.*, 2000; Gomez-Cadenaz *et al.*, 2001). Mobilization of nitrogen reserves has received less attention although several earlier works proposed that amino acid (AA) catabolism, as a source of energy and nutrients to supply expanding new tissue, would be involved in the control of germination (Below *et al.*, 2000). There is also a considerable interconversion of AA in germinating seeds, because the types and amounts of AA delivered by storage protein hydrolysis do not match the composition of cytoplasmic proteins, nor do they meet the requirements of the transport system (Lea and Joy, 1983).

The overall mobilization of nitrogen reserves – protein hydrolysis as well as oxidative deamination of amino acids – leads to a release of ammonium. The latter is recycled in germinating seeds by transfer on to glutamate to generate glutamine; this reaction is catalysed by glutamine synthetase (GS), a key enzyme in nitrogen metabolism in plants. In several species, GS

 The Biology of Seeds: Recent Research Advances (eds G. Nicolás, K.J. Bradford, D. Côme and H.W. Pritchard)

isoforms specific to seed tissue or specifically expressed during germination have been reported (Swarup *et al.*, 1990; Stanford *et al.*, 1993; Watanabe *et al.*, 1994). The induction during early stages of germination of seed-specific GS isoforms is probably due to specific needs of nitrogen metabolism in germinating seed and the crucial role of this enzyme in the germination process.

We have studied the role of GS in maize kernel germination at both physiological and genetic levels on two maize genotypes Io and F2 and a population of 140 F_6 recombinant inbred lines (RILs) (Limami *et al.*, 2002). The physiological role of GS was assessed throughout the germination of the parental lines Io and F2 by the determination of AA and ammonium contents and activities of GS and the related enzymes glutamate synthase (NADH/Fd GOGAT), glutamate dehydrogenase (GDH) and phosphoenol pyruvate (PEPc) involved in ammonium assimilation. The efficiency of germination was estimated by the determination of the T50 that represents the imbibition time at which 50% of seeds germinate. Thousand-kernel weight (TKW) was determined as an easy and accurate estimation of kernel size. Then the role of GS in germination was assessed at the genetic level in an experiment carried out on the 140 F_6 RILs. The aim of this study was the determination of correlation among agronomic traits, T50, TKW and GS activity as well as quantitative trait locus (QTL) determination for the studied traits.

Germination Properties of Two Maize Genotypes Io and F2

Germination properties of two maize parental genotypes, Io and F2, were determined at two temperatures, 21 and 25°C. Io and F2 are the two parental lines that have been crossed for the production of a population of the 140 F_6 RILs. Compared with F2, the parental line Io was characterized by kernels of both smaller size and weight that germinated faster at both temperatures, 21°C and 25°C. As shown in Fig. 46.1, the T50 of F2 was greater by 11 and 12 h when the seeds were germinated at 25°C and 21°C, respectively. Decreasing the temperature from 25°C to 21°C slowed down germination speed in the same manner in both genotypes, resulting in a delay of the T50 by about 28 h. For both genotypes, 100% of the seeds germinated regardless of the temperature conditions, indicating probably that differences in T50 between the two genotypes are physiologically and/or genetically relevant and were not due to any damage or ageing side effects (Fig. 46.1).

Physiological Characterization of Nitrogen Metabolism During Maize Kernel Germination

Changes in both nitrogen metabolite concentrations (Fig. 46.2) and enzyme activities related to N assimilation and recycling (Fig. 46.3) illustrate the induction of *de novo* metabolic activity very early after the onset of imbibition. These metabolic changes were characterized by an increase in total AA, particularly Gln and ammonium (data not shown), despite an increasing GS activity. Moreover, we observed a shift in the composition of the AA pool

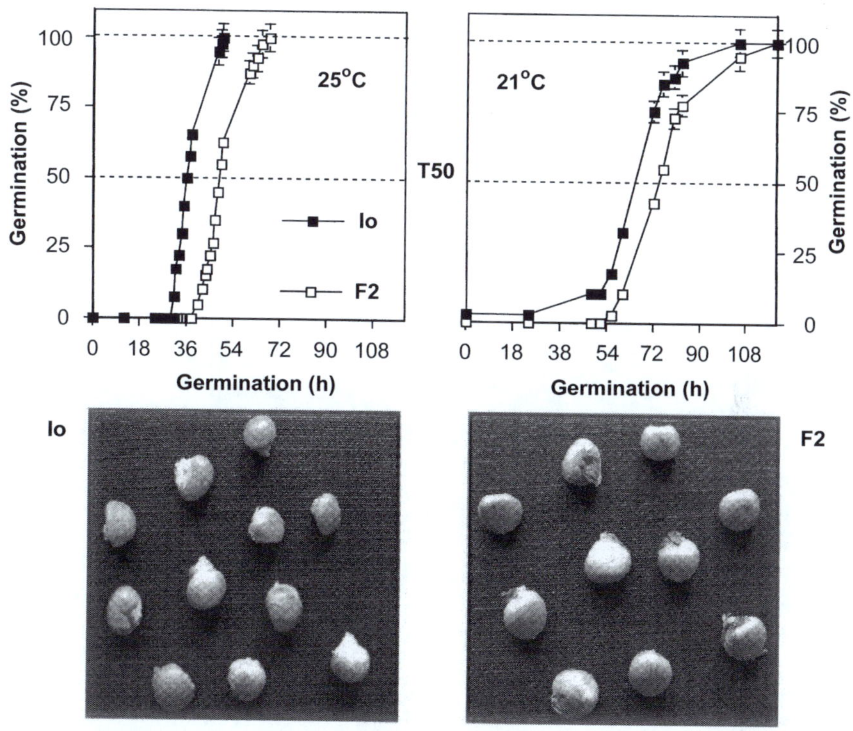

Fig. 46.1. Germination efficiency of two maize lines, Io and F2, determined as percentage of germinated kernels at various times after the beginning of imbibition at two temperatures, 21 and 25°C.

from a storage profile in which Asn was the most abundant amino acid, to an anabolic activation profile in which Gln, Glu and Pro became predominant (Fig. 46.2). In particular, our results highlight the central role of Gln during germination of maize kernels, probably acting as a sink for ammonium released during both storage protein degradation and AA deamination. Furthermore, we found that the increase in Gln occurred earlier in Io compared with F2, a result consistent with its faster germinating capacity revealed by the T50 of the two parental lines. Because Io was the genotype that germinated earlier, when germination was delayed (T50 increased) by a decrease in germination temperature from 25°C to 21°C, the changes above mentioned in the AA pool were delayed accordingly. Although the increase in Pro was somehow surprising, this amino acid is directly linked to glutamate metabolism and may be a source of reducing power as well as substrate for major constituents of cell wall structural proteins in young metabolically active tissues (Nakashima *et al.*, 1998). Protein hydrolysis along with AA oxidative deamination is probably the source of ammonium in the germinating maize kernel. Ammonium recycling is therefore likely to occur through the reaction catalysed by the enzyme of GS consistent with the increase in both Gln content concomitant to the increase in the enzyme activ-

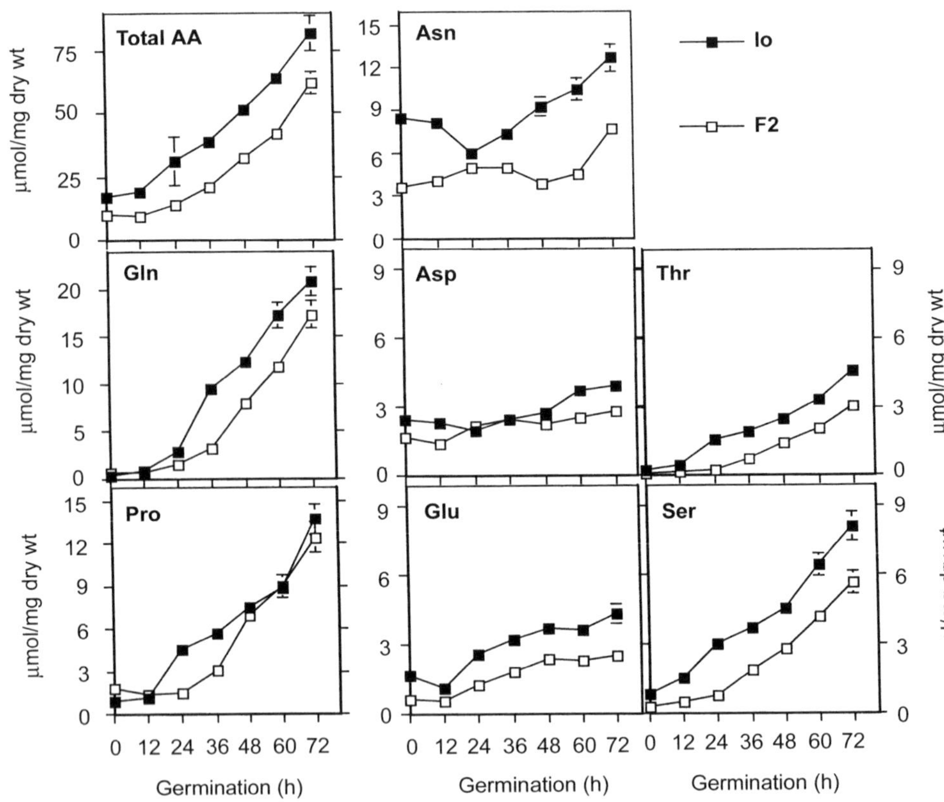

Fig. 46.2. Variation in kernel amino acid contents of two maize lines, lo and F2, germinating at 25°C.

ity (Fig. 46.3). Since NADH-dependent GOGAT activity (Fig. 46.3) increases during germination and is largely predominant compared with its Fd-dependent counterpart (data not shown), it can be hypothesized that the enzyme regenerates Glu for the reaction catalysed by GS. An increase in the GDH aminating activity was also observed, which represents a typical response to increased ammonium concentration in plant organs or tissues (Fig. 46.3) (Melo-Oliveira *et al.*, 1996). However, the precise role of GDH during N assimilation and recycling still remains to be clearly assessed, taking into account both its cellular and subcellular localization. During germination PEPc activity remains practically constant. This characteristic is probably in relation to its role as an enzyme of carbon metabolism that interacts with nitrogen metabolism in providing (through its anapleurotic function) carbon skeletons to the tricarboxylic acid cycle, thus allowing the synthesis of 2-oxoglutarate for the GS/GOGAT pathway and fuelling up the respiratory chain for the production of energy.

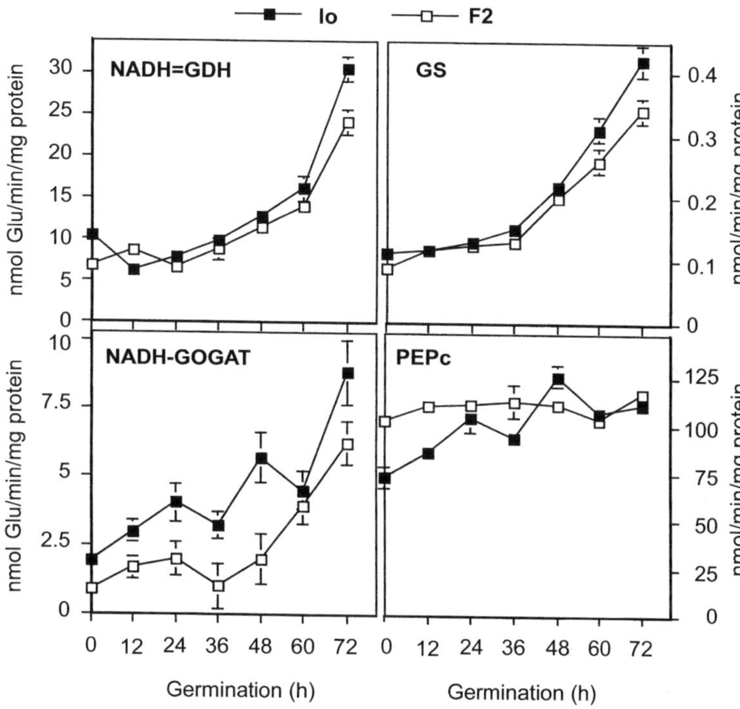

Fig. 46.3. Variation in kernel enzyme activities throughout the germination process of two maize lines, Io and F2, at 25°C. GS, glutamine synthetase; NADH-GDH, glutamate dehydrogenase tested for its oxoglutarate-aminating activity; NADH-GOGAT, glutamate synthase; PEPc, phosphoenol pyruvate carboxylase.

Genetic Analysis and QTL Mapping of Nitrogen Metabolism in Relation to Maize Kernel Germination

Since the enhancement of Gln synthesis was one of the most important events occurring in nitrogen metabolism during germination, this observation led to the logical conclusion that the reaction catalysed by GS may play a major role in the determination of the germination process. A quantitative genetic approach was therefore undertaken to support this hypothesis by associating germination traits (T50 and TKW) and GS activity to molecular markers on a genetic map of maize. The characters TKW, T50 and GS activity exhibited a normal distribution and transgression in both directions, indicating that they are rather under multiple gene control, which is a characteristic of polygenic traits (Fig. 46.4). QTLs detected for germination traits and biochemical trait GS activity are reported in Table 46.1. The position of different QTLs on the maize RFLP map is shown in Fig. 46.5. In germinating maize kernels, three QTLs for GS activity were detected. Two corresponded to GS activity at the early stages of germination (48 h) and one corresponded to GS activity at later stages of germination (72 h). The two

Table 46.1. QTLs for glutamine synthetase (GS) activity at early stages of germination (48 h) and late stages of germination (72 h) and traits related to grain germination efficiency (T50) and grain size/weight (TKW). QTLs detected by simple interval mapping.

Trait	R^2	Location				Confidence interval	LOD	Additive effect	Favourable allele from parental line
		Chr.	Marker	+ cM	Distance (cM)				
Early GS activity	17.7	4	SC419	+ 0	130	122 – 134	3.11	- 0.009	F2
		5	SC258A	+ 13	182	166 – 214	2.64	+ 0.012	Io
Late GS activity	7.3	1	SC145	+ 17	104	88 – 138	2.08	+ 0.022	Io
T50	21.9	2	SC199	+ 2	84	78 – 94	2.53	+ 5.728	Io
		3	SC224A	+1	58	50 – 70	1.84	- 4.989	F2
		4	SC59C	+ 10	114	104 – 126	2.17	- 5.271	F2
TKW	17.0	4	UMC133	+ 27	206	178 – 234	2.40	- 38.298	F2
		5	SC343B	+ 1	106	98 – 118	2.93	+ 18.320	Io
		5	SC258A	+2	170	156 – 186	2.80	+ 18.986	Io

R^2, Percentage of phenotypic variance explained by the QTLs.
Chr, Chromosome number; confidence interval is for LOD – 1.

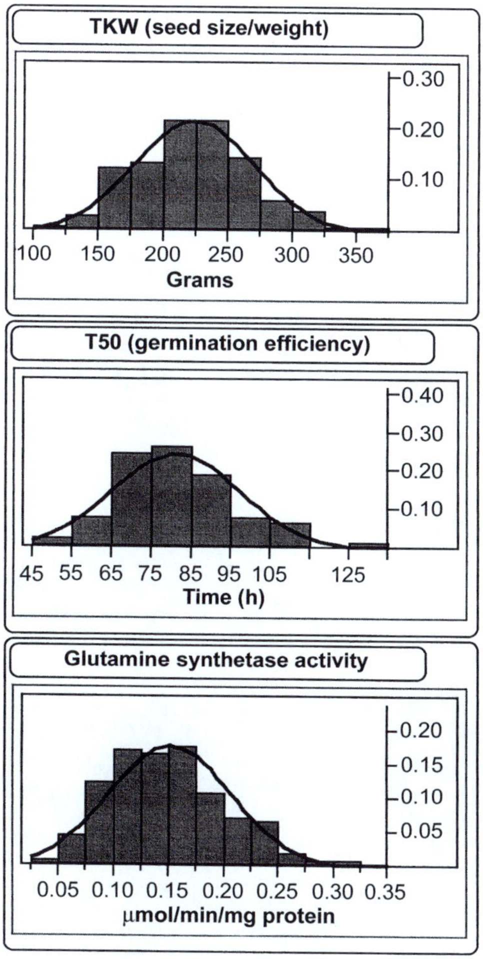

Fig. 46.4. Frequency distribution of germination trait T50, kernel size/weight trait TKW and biochemical trait kernel GS (glutamine synthetase) activity of 140 recombinant inbred lines (RILs).

QTLs for GS activity at early stages of germination co-localized with two of the structural genes encoding for cytosolic GS, *gln3* on chromosome 4 and *gln4* on chromosome 5 (Hirel *et al.*, 2001). A QTL for GS activity at a later stage of germination was also detected on chromosome 1. Northern blot analysis in the parental line Io showed that of the five genes encoding cytosolic GS, only *gln3* and *gln4* were expressed during germination (Fig. 46.6); *gln3* was transiently expressed, exhibiting a maximum between 24 and 36 h, whereas *gln4* was constitutively expressed within 72 h of germination. This result is consistent with the detection of two QTLs for GS activity on two different chromosomes each co-localizing with the *gln3* and *gln4* loci.

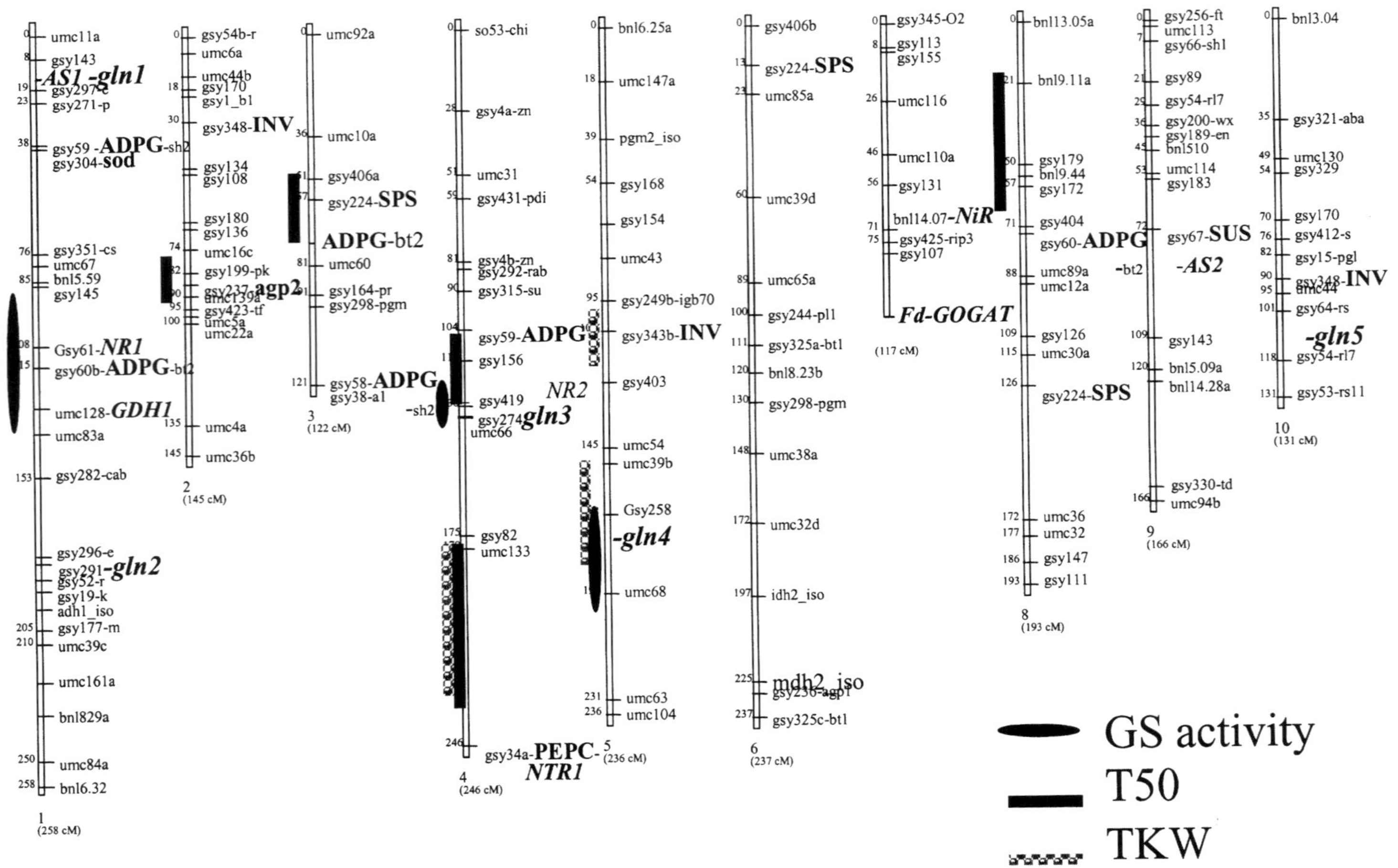

Fig. 46.5. Location of the QTLs of germination trait T50, kernel size/weight trait TKW and biochemical trait kernel GS (glutamine synthetase) activity on the maize RFLP genetic map. The positions of the loci for genes encoding enzymes involved in nitrogen and carbon assimilation are indicated in bold italics.

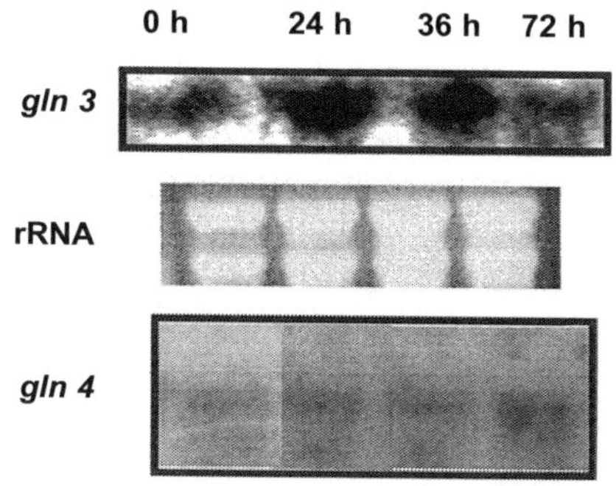

Fig. 46.6. Expression of genes encoding two maize cytosolic Gs isoforms *gln3* and *gln4* throughout germination. Northern blots were performed on total RNA extracted from kernels of the line Io germinating at 25°C.

The identification of a QTL for GS activity co-localizing with both *gln3* structural gene and a QTL for T50 on chromosome 4 (both with the favourable allele from the parental line, Io), suggests that its transient expression may be of major importance in the control of germination efficiency. This hypothesis is further supported by the finding that a highly significant negative correlation was observed between GS activity and T50 ($P < 0.001$, $r = -0.30$). This means that kernels exhibiting a faster germination rate (low T50) have a higher GS activity.

On chromosome 5, the triple co-localization between *gln4* (another structural gene encoding GS1) and two QTLs for GS activity and TKW is also an interesting finding which may help to unravel the physiological function of the corresponding translation product. In a previous study a co-localization between leaf GS activity in young developing maize plants and the same GS1 structural gene was identified (Hirel *et al.*, 2001). Moreover, in this region of chromosome 5 a QTL for TKW reflecting the agronomic performances of the hybrids derived from the RILs population was identified in the same position (Fig. 46.5). This result shows that RILs and derived hybrids share common QTLs for TKW both with the favourable allele from the parental line Io. Both of these observations confirm that *gln4* is a good candidate gene that may influence grain yield and as a consequence grain size, which may be one of the main components controlling the efficiency of germination.

Genetic Analysis of Maize Kernel Germination in Relation to its Size/Weight

The observation in the parental lines Io and F2 that the smaller the kernel size the higher is their germination speed suggests that kernel size interferes with its germination speed. The hypothesis of a genetic basis of the association between both traits was put forward, since we have shown that differences in germination speed of Io and F2 did not rely on differences in rehydration speed. This hypothesis was confirmed by the overlapping on chromosome 4 near the marker umc133 of two QTLs, one for germination speed (T50) and the other for grain size/weight (TKW), both with the favourable allele from the parental line F2. The results indicate that both traits are controlled by the same gene(s) or by tightly clustered gene(s) in this chromosomal region. Statistical data further support this hypothesis, since both traits show a highly significant ($P < 0.001$) positive correlation ($r = 0.29$). The comparison of the position of this QTL with several maize genetic maps (INRA, France, and University of Missouri, Columbia, USA) allowed us to identify a candidate gene because of its involvement in basic development. The gene is *knox7* (knotted related homeobox7) that belongs to the class 2 kn1-like homeobox gene family encoding for transcription factor homeodomain proteins (Kerstetter *et al.*, 1994). Unlike class1 kn1-like homeobox genes that are well known to control shoot apical meristem development, little is known about the role of class 2 kn1-like homeobox genes (Bharathan *et al.*, 1999). Nevertheless, expression of *KNAT3*, a class 2 kn1-like homeobox gene in *Arabidopsis thaliana*, has been shown to be confined to early organ development, leaves, buds and pedicels at and near junctions between organs, including hypocotyl–root boundary in young seedlings. The validation of this candidate gene will be undertaken by first studying its expression during kernel development and germination (Serikawa *et al.*, 1996, 1997).

Conclusion

Study of inheritance of quantitative traits became more powerful with the advent of molecular markers that allowed for construction of genetic maps and detection of QTLs. Even highly complex traits such as crop yield can be under the control of a small number of QTLs. QTL mapping has been largely used for agronomical and morphological traits with the aim of performing marker-assisted selection that would result in gathering favourable alleles and breaking their linkage with unfavourable alleles in elite genotypes. Recently QTL mapping has proved to be an instrumental method for studying quantitative physiological traits. In this case the goal is the search for a relationship between polymorphism of genes of known functions (candidate genes) and the variation of the studied trait. The method enables the evaluation of the genetic control exerted by a single gene on a quantitative trait (for reviews, see Prioul *et al.*, 1997; Limami and de Vienne, 2001). In the present work, germination was for the first time studied as a complex developmental trait in parallel with a physiological trait, GS activity, a major enzyme of amino acid metabolism during kernel germination. The approach allowed the identification of GS as a candidate gene in the control of germination, which was strongly supported by the co-localization of QTLs of germination (T50) and GS activity at *gln3* locus on chromosome 4 and the transient expression of *gln3* during the germination process.

References

Below, F.E., Cazetta, J.O. and Seebauer, J.R. (2000) Carbon/nitrogen interactions during ear and kernel development of maize. In: *Physiology and Modeling Kernel Set in Maize*. Crop Sciences Society of America, Society of Agronomy, Special Publication No. 29, pp. 15–24.

Bethke, P.C. and Jones, R. (2001) Cell death of barley aleurone protoplasts is mediated by reactive oxygen species. *Plant Journal* 25(1), 19–29.

Bharathan, G., Janssen, B.-J., Kellogg, E.A. and Sinha, N. (1999) Phylogenetic relationship and evolution of KNOTTED class of plant homeodomain proteins. *Molecular Biology Evolution* 16(4), 553–563.

Fath, A., Bethke, P., Lonsdale, J., Meza-Romero, R. and Jones, R. (2000) Programmed cell death in cereal aleurone. *Plant Molecular Biology* 44(3), 255–266.

Gomez-Cadenas, A., Zentella, R., Walker-Simmons, M.K. and Ho, T.H.D. (2001) Gibberellin/abscisic acid antagonism in barley cells: site of action of the protein kinase PKABA1 in relation to gibberellin signaling molecules. *Plant Cell* 13(3), 667–679.

Hirel, B., Bertin, P., Quilleré, I., Bourdoncle, W., Attagnant, C., Dellay, C., Gouy, A., Cadiou, S., Retaillau, C., Falque, M. and Gallais, A. (2001) Towards a better understanding of the genetic and physiological basis for nitrogen use efficiency in maize. *Plant Physiology* 125, 1258–1270.

Kerstetter, R., Vollbrecht, E., Lowe, B., Veit, B., Yamagushi, J. and Hake, S. (1994) Sequence analysis and expression patterns divide knotted1-like homeobox genes into two classes. *Plant Cell* 6(12), 1877–1887.

Lea, P.L. and Joy, K.W. (1983) Amino acids interconversion in germinating seeds. In: Nozillo, C., Lea, P.J. and Loewus, F.A. (eds) *Mobilization of Reserves in Germination. Recent Advances in Phytochemistry*, Vol. 17. Plenum Press, New York, pp. 77–109.

Limami, A.M. and de Vienne, D. (2001) Natural genetic variability in nitrogen metabolism. In: Lea, P. and Morot-Gaudry, J.-F. (eds) *Nitrogen and Plants*. Springer-Verlag, Heidelberg, Germany, pp. 369–378.

Limami, A.M., Rouillon, C., Glevarec, G., Gallais, A. and Hirel, B. (2002) Genetic and physiological analysis of germination efficiency in maize in relation to nitrogen metabolism reveals the importance of cytosolic glutamine synthetase. *Plant Physiology* 130, 1860–1870.

Melo-Oliveira, R., Oliveira, I.C. and Corruzzi, G.M. (1996) Arabidopsis mutant analysis and gene regulation define a nonredundant role for glutamate dehydrogenase in nitrogen assimilation. *Proceedings of the National Academy of Sciences USA* 93, 4718–4723.

Nakashima, K., Satoh, R., Kiyosue, T., Yamagoshi-Shinozaki, K. and Shinozaki, K. (1998) A gene encoding proline dehydrogenase is not only induced by proline and hypoosmolarity, but is also developmentally regulated in the reproductive organs of Arabidopsis. *Plant Physiology* 118, 1233–1241.

Prioul, J.-L., Quarrie, S., Causse, M. and de Vienne, D. (1997) Dissecting complex physiological functions through the use of molecular quantitative genetics. *Plant Physiology* 48(311), 1151–1163.

Serikawa, K.A., Martinez-Laborda, A., Kim, H.S. and Zambrisky, P.C. (1996) Three knotted1-like homeobox genes in Arabidopsis. *Plant Molecular Biology* 32(4), 673–683.

Serikawa, K.A., Martinez-Laborda, A., Kim, H.S. and Zambrisky, P.C. (1997)

Localization of expression of KNAT3, a class 2 knotted1-like gene. *Plant Journal* 11(4), 853–861.

Stanford, A.C., Larsen, K., Baker, D.G. and Cullimore, J.V. (1993) Differential expression within the glutamine synthetase gene family of the model legume *Medicago truncatula*. *Plant Physiology* 103, 73–81.

Swarup, R., Bennett, M.J. and Cullimore, J.V. (1990) Expression of glutamine synthetase genes in cotyledons of *Phaseolus vulgaris* L. *Planta* 183, 51–56.

Watanabe, A., Hamada, K., Yokoi, H. and Watanabe, A. (1994) Biphasic and differential expression of cytosolic glutamine synthetase genes of radish during seed germination and senescence of cotyledons. *Plant Molecular Biology* 26(6), 1807–1817.

47 Physico-chemical Factors Influence Beet (*Beta vulgaris* L.) Seed Germination

A.G. Taylor[1], M.C. Goffinet[1], S.A. Pikuz[2], T.A. Shelkovenko[2], M.D. Mitchell[2], K.M. Chandler[2] and D.A. Hammer[2]

[1]*Department of Horticultural Sciences, New York State Agricultural Experiment Station, Cornell University, Geneva, NY 14456, USA;*
[2]*Laboratory of Plasma Studies, Cornell University, Ithaca, NY 14853, USA*

Introduction

Uniformity in harvest maturity and product size of vegetable crops is directly related to uniformity of seedling emergence. Therefore, germination synchrony is the first essential step to achieve uniform plant growth and development (Taylor, 1997). Table beet (*Beta vulgaris* L.) does not germinate uniformly, and factors responsible for this lack of uniformity are the focus of this investigation.

A table beet seed is botanically a fruit that is generically termed a 'seedball' (Hayward, 1938), since most cultivars are multi-germed. The seedball is a tightly compressed assemblage of one to several fruits, each containing one seed with an embryo. Conventional varieties most commonly have two embryos in each seedball, while some propagules may contain three or four embryos. An X-ray image is illustrated with four embryos oriented in different planes (Fig. 47.1). Sugarbeet (also *Beta vulgaris* L.) and certain table beet varieties are monogerm but factors that influence germination rate are similar for both table and sugarbeets.

The germination environment influences beet germination potential and seeds are most sensitive under wet conditions, which limits oxygen availability (Perry and Harrison, 1974). However, table beet seed lots have non-synchronous germination, even under ideal environmental conditions. There are three major seed factors that negatively influence beet germination: (i) a mucilaginous layer that can surround the seedball; (ii) the ovary cap tenacity; and (iii) the presence of phenolic chemical inhibitors. A mucilaginous layer has been observed on sugarbeet seeds and those cultivars with greater incidence of mucilage had lower germination potential

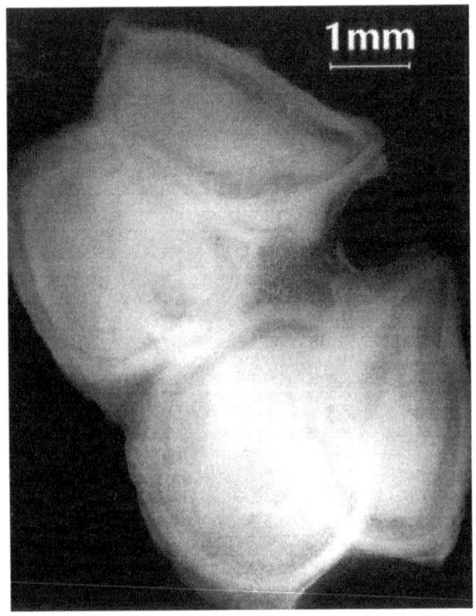

Fig. 47.1. X-ray image of a table beetseed ball with four embryos.

(Duan and Burris, 1997). The ovary cap or operculum is a morphological dome that covers the embryo. The operculum is composed of the non-embedded portion of the ovary and remnants of the stigma-style. Ovary caps act as barriers to oxygen diffusion (Heydecker *et al.*, 1971; Coumans *et al.*, 1976) and varietal differences exist in ovary cap tenacity (Morris *et al.*, 1985). Finally, a number of inhibitors are found in the seed coverings of beet seeds and several phenolic compounds have been isolated and identified (Chiji *et al.*, 1980). Collectively, these physico-chemical barriers interact to reduce the germination potential of beet seeds.

The objective of this study was to elucidate both the effect of mucilage and the effect of the ovary cap tenacity on table beet germination. Light microscopy studies were conducted on the ovary cap and X-ray imaging of the mucilaginous layer. Time course studies were conducted to quantify phenolic compound leakage from seeds with and without a mucilaginous layer.

Materials and Methods

Seed lot and germination studies

A seed lot of the non-hybrid, multi-germed cultivar 'Ruby Queen' was provided by Chris Seeds, Mount Vernon, Washington, USA. The lot was not commercially treated with chemical or biological treatments. Germination tests were performed on rolled towels at a constant 20°C in the dark, and visible germination was recorded daily. There were 25 propagules (seeds) per replication, with four replications per treatment. Seeds with two embryos were selected for all studies. The presence or absence of mucilage was observed under a stereo microscope at 10×. Seeds could be separated by appearance in the dry stage but these differences became more apparent after soaking seeds for 10 min. To facilitate separating batches of seeds, seeds were soaked for 10 min and then separated by density in water. Seed in the floating fraction had a mucilaginous layer, which was due to air entrapped in the mucilage matrix. Seeds without a mucilaginous layer sank in the water.

Light microscopy

A seed development study was conducted to observe the formation of the ovary cap. In late autumn of 2000, mature roots were removed from the field plots at the NYSAES Vegetable Research Farm and kept at 5°C until removed to a greenhouse bench in January 2001 for regrowth and flowering. At first flowering, newly opened individual flowers and inflorescence segments were tagged with a thread and hand pollinated. On a daily basis, newly opening flowers were tagged with a different colour thread and pollinated, thus providing a population of flowers at known post-bloom age. Florets and developing fruits were harvested 21 days post-bloom, and these were killed/fixed in an aqueous solution of formalin–acetic acid–ethanol (10%, 5%, 48% by volume). Specimens were later removed, embedded in paraffin-based medium, sectioned at 9 μm, stained in safranin–orange gold–tannic acid–iron alum (Sharmon, 1943) and analysed for tissue development via microscopy.

X-ray imaging

An X-pinch X-ray source (Pikuz *et al*., 2001) was used for phase contrast X-ray radiography of beet seeds. The X-pinch is a high-temperature plasma X-ray source that is produced by passing a 0.1 μs, 200–400 kA current pulse through two or more fine metal wires (such as 25 μm diameter titanium) that are configured in the form of an X, touching only at the central cross point. A very bright X-ray source, the spectrum of which depends upon the wire material used, is generated at the cross point. When the radiation below 3 keV is eliminated using a thin foil filter, the X-ray source is very small (< 1 μm) and the pulse duration is short (< 1 ns), but enough X-ray energy is emitted to produce radiographs in a single pulse using ordinary Kodak TMAX film. For the present application, the X-ray energy was predominantly in the 3–6 keV band and was generated using either 25 μm molybdenum or 38 μm titanium wires for the X-pinch. The low X-ray photon energy and small size enable point-projection imaging of low absorption biological objects with good spatial resolution. Because the source is so small, coherent addition of X-rays that are refracted to the same point on the film from the two sides of a boundary enhances the contrast at the boundary in the image. The images presented here were of a seedball with four embryos (Fig. 47.1) and of the mucilage (Fig. 47.2), which were placed about 7 cm from the X-ray source in a small chamber. The magnification on to the film was 4 : 1 and the film was scanned with a 1200 dpi AGFA scanner.

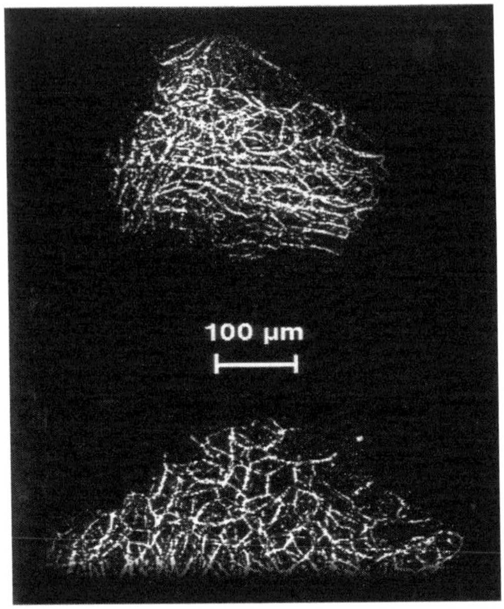

Fig. 47.2. X-ray image of two sections of the mucilaginous layer. The fibrous appearance reveals cell wall debris in the layer.

Phenolic compound leakage

Many organic and inorganic compounds are present in beet seeds and some of these constituents are inhibitory to germination (Santos and Pereira, 1989). Absorbance of the seed leachate at 265 nm was demonstrated to quantify phenolic compounds in sugarbeets (Duan and Burris, 1997). The seed leachate from table beet seeds was measured in time course studies at 25°C. The concentration of phenolics was determined by using 9300 as the molar extinction coefficient of the major phenolic present (Chiji *et al.*, 1980) and expressed as nmol per seed.

Results and Discussion

Mucilage

X-ray imaging revealed that the mucilaginous layer was cellular in nature and appeared as crushed cell walls (Fig. 47.2). The mucilage cannot simply be washed off the seeds. The beet mucilage is really a well-hydrated insoluble polysaccharide mixture, probably with internal cross-linking to give it strength and physical structure. The amazing feature is its ability to reversibly hydrate and dehydrate and retain structure. Dr B. Lewis, in the Division of Nutritional Sciences, Cornell University, conducted carbohydrate characterization of the mucilage. Preliminary investigations revealed the presence of pectins, hemicellulose and cellulose (Lewis, unpublished).

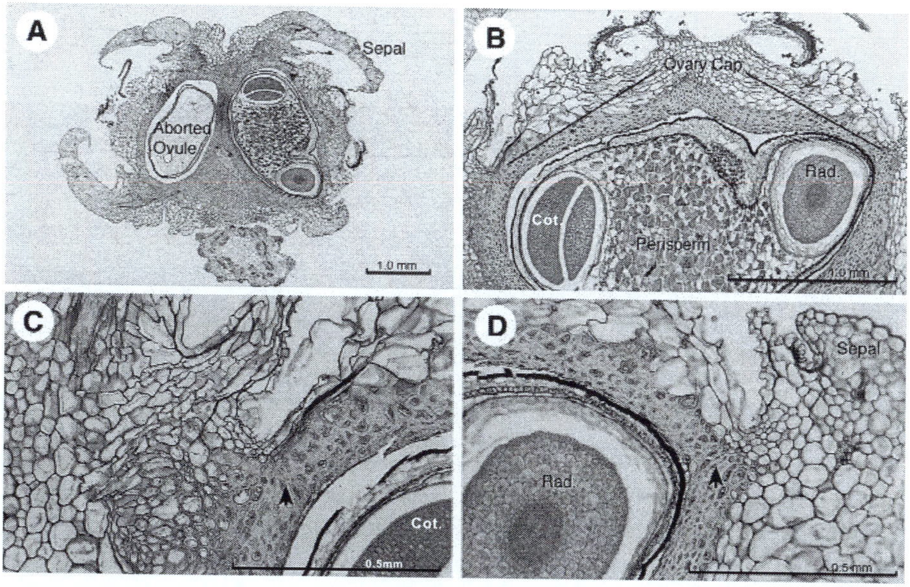

Fig. 47.3. Cross-section of 21-day post-pollination seedball with two embedded ovules. (A) Cross-section showing one aborted and one developing ovule. (B) Close-up of (A) showing ovary cap as a dome above developing ovule. (C), (D) Ovary cap junction adjacent to cotyledon (C) and radicle (D), indicated by arrowheads.

Ovary cap

The upper ovary wall (cap) must separate from the lower fruit wall to allow radicle emergence in germinating seeds. One factor responsible for poor germination in beets is the failure of cells to separate at the cap periphery due to tight bonding between lignified cell walls in this region. Light microscopy studies were conducted during seed development and cross-sections of the ovary with ovules at 21 days after pollination are shown in Fig. 47.3. Two ovules are shown, with the aborted ovule on the left (Fig. 47.3A); the structure at the bottom is a cross-section through the stem of one partial inflorescence of the mother plant. A close-up of the developing ovule is shown in Fig. 47.3B, revealing the cotyledon, perisperm, radicle and ovary cap. Further magnification revealed a junction in cells separating the ovary cap from the ovary wall at both the cotyledon (Fig. 47.3C) and radicle ends (Fig. 47.3D) of the nearby embryo. The cap juncture cells tended to be oriented in the radial direction in the ovary wall (indicated by arrowheads), while adjacent cells of the ovary wall were larger and more randomly oriented.

Phenolic compound leakage

Leakage of phenolic compounds from seed balls within a 24 h period followed first-order reaction kinetics (Fig. 47.4) such that a significant rela-

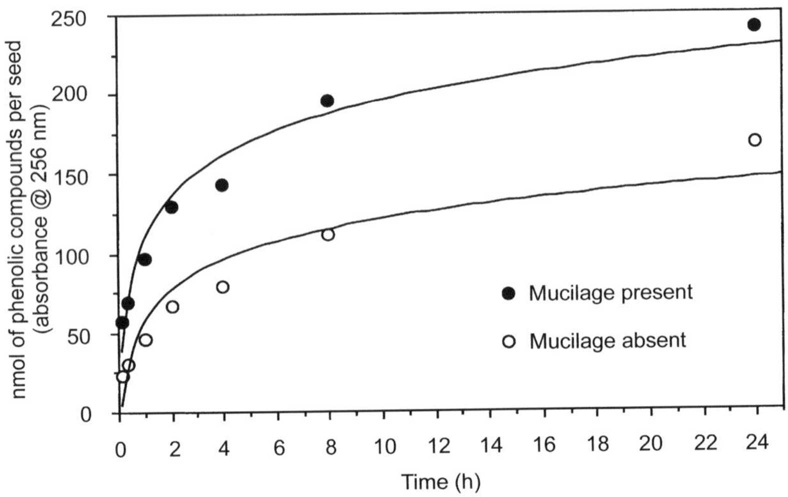

Fig. 47.4. Phenolic compound leakage from table beet seeds sorted for the presence or absence of mucilage from the same seed lot.

tionship existed between log nmol phenolics per seed and soak time ($R^2 = 0.92$, $P = 0.01$). Seeds with mucilage leaked more compounds than those without mucilage, which was attributed to a greater concentration of phenolics present in seeds with a mucilaginous layer.

Influence of mucilage and ovary cap on germination

Seeds could be sorted by density separation after a 10 min soak, thus providing a simple method of separating seeds. Using this method, about 75% of the seed lot was categorized as having significant amounts of mucilage. The presence of a mucilaginous layer had a profound influence on reducing both the rate and final germination in comparison with seeds without a mucilaginous layer (Fig. 47.5).

The ovary cap can be surgically manipulated by soaking for 60 min to soften the seeds. The ovary cap was carefully lifted but remained in place (cap lifted), or could be completely removed (decapped). Lifting or decapping seeds enhanced the germination rate of seeds with and without a mucilaginous layer (Fig. 47.6), while the greatest improvement was recorded in those seeds with a mucilaginous layer.

Conclusions

Germination of beet seeds may be impaired by at least three seed-related factors: (i) a mucilaginous layer that surrounds the propagule; (ii) the ovary cap tenacity; and (iii) the presence of phenolic chemical inhibitors. In this particular seed lot, the mucilaginous layer had the primary deleterious effect while ovary cap tenacity and phenolics were secondary factors. Collectively,

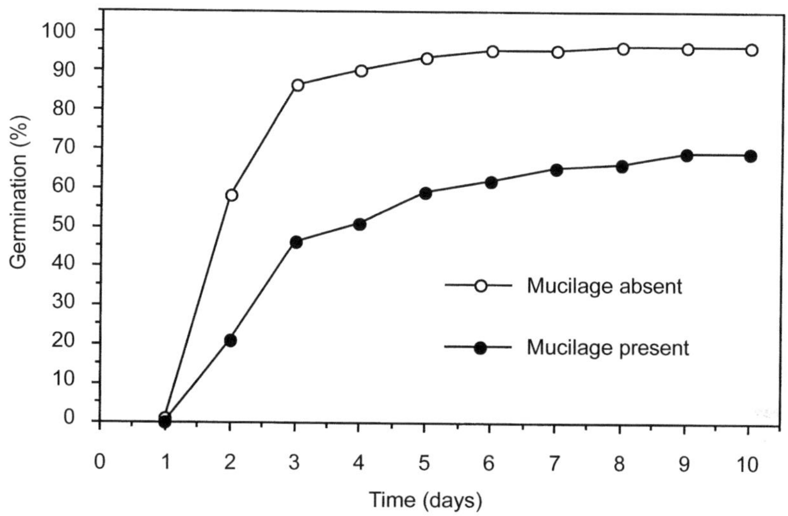

Fig. 47.5. Germination rate of table beet seeds sorted for the presence or absence of mucilage from the same seed lot.

these three limiting factors must be considered when developing pre-sowing treatments for this crop. Understanding the science of seeds provides the foundation for seed technology and supports further efforts on seed enhancements.

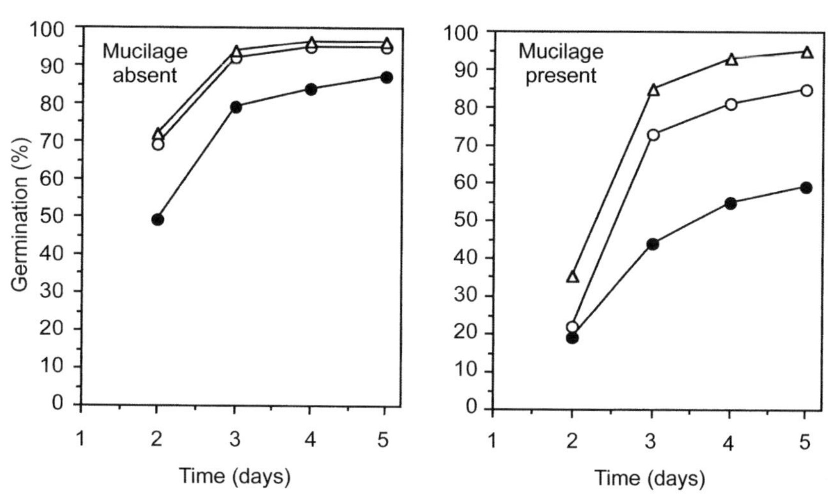

Fig. 47.6. The effect of raising and removal of the ovary cap on germination rate of table beet seeds sorted for the presence or absence of mucilage from the same seed lot: ●, ovary cap intact; ○, cap raised; △, decapped.

Acknowledgements

The authors wish to thank C. Huftalen, N. Suzuki, D. Paine and M.J. Welser for technical assistance. This work was partially supported by the New York Vegetable Research Association/Council. X-ray imaging was partially supported by a DOE grant DE-FG02-98ER54496.

References

Chiji, H., Tanaka, S. and Izawa, M. (1980) Phenolic germination inhibitors in the seed balls of red beet. *Agricultural Biological Chemistry* 44, 205–207.

Coumans, M., Côme, D. and Gaspar, T. (1976) Stabilized dormancy in sugarbeet fruits. I. Seed coats as a physicochemical barrier to oxygen. *Botanical Gazette* 137, 274–278.

Duan, X. and Burris, J.S. (1997) Film coating impairs leaching of germination inhibitors in sugar beet seed. *Crop Science* 37, 515–520.

Hayward, H.E. (1938) *The Structure of Economic Plants*. Macmillan, New York, 674 pp.

Heydecker, W., Chetram, R.S. and Heydecker, J.C. (1971) Water relations of beetroot seed germination. II. Effects of the ovary cap and of the endogenous inhibitors. *Annals of Botany* 35, 31–42.

Morris, P.C., Grierson, D.G. and Whittington, W.J. (1985) The influence of the fruit structure on germination of sugar beet (*Beta vulgaris*). *Seed Science and Technology* 13, 41–51.

Perry, D.A. and Harrison, J.G. (1974) Studies on the sensitivity of monogerm sugar beet germination to water. *Annals of Applied Biology* 77, 51–60.

Pikuz, S.A., Shelkovenko, T.A., Sinars, D.B., Chandler, K.M. and Hammer, D.A. (2001) Phase-contrast x-ray radiography using X pinch radiation. In: Kyrala, G.A. and Gauthier, J.-C.J. (eds) *Applications of X-rays Generated from Lasers and Other Bright Sources II*. SPIE, Bellingham, Washington, pp. 234–239.

Santos, D.S.B. and Pereira, M.F.A. (1989) Restrictions of the tegument to the germination of *Beta vulgaris* seeds. *Seed Science and Technology* 17, 601–611.

Sharmon, B.C. (1943) Tannic acid and iron alum with safranin and orange gold in studies of the shoot apex. *Stain Technology* 18, 105–111.

Taylor, A.G. (1997) Seed storage, germination and quality. In: Wien, H.C. (ed.) *The Physiology of Vegetable Crops*. CAB International, Wallingford, UK, pp 1–36.

48 Ageing in Tomato Reduces the Capacity of Seeds to Produce Ethylene, While Priming Increases Ethylene Evolution During Germination

GUNCHING SIRIWITAYAWAN, MANJUL DUTT, SHARON KESTER, BRUCE DOWNIE AND ROBERT GENEVE

Department of Horticulture, University of Kentucky, Lexington, KY 40546, USA

Introduction

During germination, there is often a rise in ethylene production following imbibition that peaks just after radicle emergence (Matilla, 2000). Increased ethylene production during the early stages of germination has been documented for numerous species (Lalonde and Saini, 1992; Petruzzelli *et al.*, 1993; Khan, 1994). Although this increased ethylene production appears to be common among diverse species, studies using ethylene inhibitors or ethylene-insensitive mutants demonstrate that non-dormant seeds do not generally require ethylene production to complete radicle emergence (Lalonde and Saini, 1992; Hua and Meyerowitz, 1998).

The ability of a seed lot to produce ethylene following imbibition has consistently been associated with seed viability and is negatively correlated with seed ageing (Samimy and Taylor, 1983; Gorecki *et al.*, 1991; Khan, 1994; Chojnowski *et al.*, 1997). In all cases, aged seed lots produced ethylene later after imbibition and produced less ethylene overall than non-aged seed lots. One possible mechanism reducing ethylene production in aged seeds is a reduction in 1-aminocyclopropane-1-carboxylic acid (ACC) oxidase activity or synthesis. In general, seeds exposed to saturating concentrations of exogenous ACC amplify subsequent ethylene production, but the magnitude of this increased production is considerably greater in non-aged vs. aged seeds (Khan, 1994), suggesting reduced ACC-oxidase activity in the latter. Siriwitayawan (2002) showed reduced ACC-oxidase mRNA abundance in aged tomato seeds.

Seed priming reduces the time to radicle protrusion and often ameliorates the deleterious effects of seed ageing on germination (McDonald, 2000). The mechanisms for improved germination in primed seeds are not completely understood, but maintaining seeds in the lag phase of germination is thought to promote cellular repair and initiate metabolic processes required for germination. There are several studies that show an increased ability for primed seeds to produce ethylene during germination (Fu *et al.*, 1988; Chojnowski *et al.*, 1997; Cantliffe *et al.*, 2000; Habdas *et al.*, 2000). However, it is not clear whether ethylene production is integral to obtaining a priming effect in seeds or whether it is simply the result of higher vigour displayed by primed seeds.

The objective of the present study was to compare ethylene production in a tomato seed lot after moderate ageing. A seed lot aged for 18 months was chosen, because it showed reduced seed vigour but maintained high standard germination. Therefore, differences in ethylene production would be due to vigour loss without being confounded with reduced viability. In addition, the relationship between seed priming and ethylene production in aged and non-aged seeds was compared with the intent of using ethylene-insensitive mutants in tomato to study whether there is a causal relationship between ethylene and the priming response in seeds.

Materials and Methods

Tomato seeds (*Lycopersicon esculentum* L. Mill. cv. Moneymaker) were originally obtained from the C.M. Rick Tomato Genetics Resource Center (TGRC) (University of California at Davis) and subsequently grown at the University of Kentucky Horticultural Research Farm. Seeds were collected from mature fruits and placed in sealed glass bottles and stored at 4°C or room temperature (~23°C) for 18 months. Standard germination was in Petri dishes on moist blotters incubated at alternate 8 h at 30°C in light (20 μmol/s/m^2) and 16 h at 30°C in dark (AOSA, 1992). Four replications of 50 seeds were tested for standard germination.

Seeds were osmotically primed by placing 2000 seeds in 500 ml of an aerated 3% KNO_3 solution for 7 days d at 20°C in the dark and subsequently dried at room temperature for 24 h to a moisture content of 8.2 ± 1.1% (fresh weight basis).

In separate Petri dishes, seeds were imbibed on water or ACC (5 mM) for various times (12, 24, 36, 48 and 60 h). Petri dishes were placed in germination conditions as previously described. Ethylene evolution was quantified by moving 50 seeds from the pool of ACC-treated seeds to dry 25 ml Erlenmeyer flasks. Flasks were sealed with serum stoppers. After 3 h of incubation, a syringe was used to withdraw a 1 ml gas sample for ethylene evaluation. A Buck Scientific gas chromatograph with flame ionization detector (155°C) and alumina column (125°C) with a nitrogen flow rate of 1 ml/min was used to determine ethylene concentration using a standard curve of dilutions of pure ethylene in air.

Four replications of 50 seeds were frozen in liquid nitrogen and placed in a mortar and pestle and ground in 2 ml of 80% ethanol. The slurry was transferred into a test tube and incubated at 70°C for 30 min. Following centrifugation, the supernatant was removed and evaporated to dryness *in vacuo*. Subsequently, chloroform (1 ml) plus water (1 ml) was used to resuspend and separate recovered ACC into the aqueous phase, which was assayed for ACC-derived ethylene production according to methods from McKeon *et al.* (1982) and Lizada and Yang (1979). Internal standards for ACC indicated extraction efficiency for the assay of 88.0 ± 1.9%.

Results and Discussion

Tomato seeds aged for 18 months showed only an 8% reduction in final germination percentage (data not shown) but a significant reduction in seed vigour as indicated by radicle emergence after 60 h (Fig. 48.1) and accelerated ageing test (data not shown). Tomato seeds germinating on water evolved less ethylene compared with seeds imbibed in the presence of 5 mM ACC (Fig. 48.1). ACC content in seeds imbibed on 5 mM ACC increased from less than 5 to approximately 500 and 1500 pmol per seed after 12 and

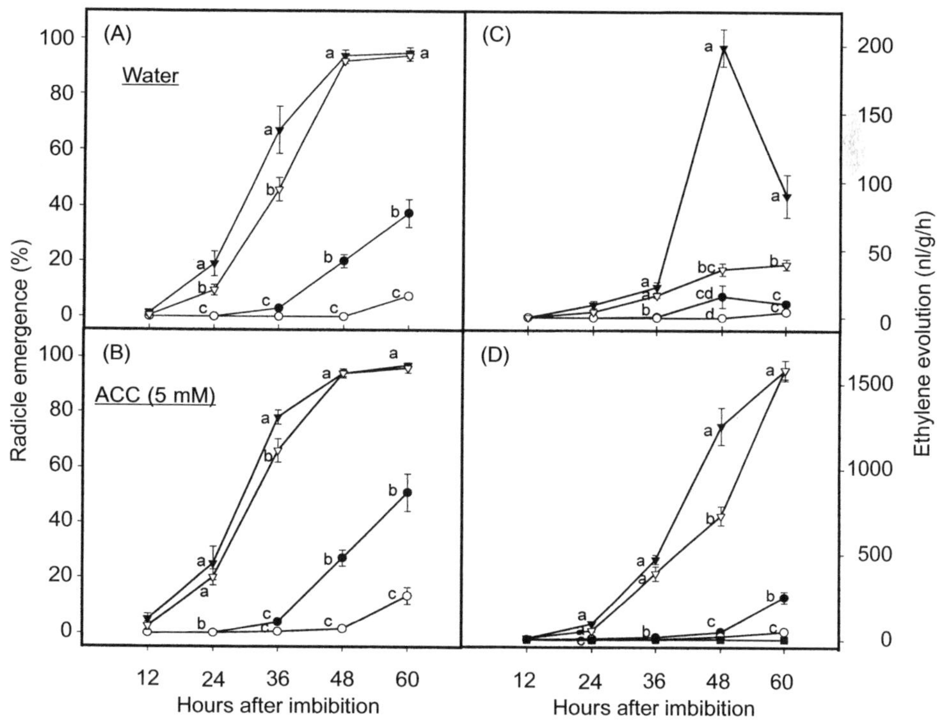

Fig. 48.1. Time to radicle protrusion and ethylene evolution from tomato seeds aged for 18 months with or without priming and imbibed on water (A, C) or 5 mM ACC (B, D). Closed symbols non-aged, open symbols aged, circles non-primed, triangles primed. Values are means ± standard error. Means followed by the same letter were not different using Tukey's test ($P \leq 0.05$).

24 h, respectively. Aged seeds showed a reduced capacity to evolve ethylene and this was more apparent in ACC-treated seeds (Fig. 48.1). Non-primed aged seeds began to evolve ethylene 24 h later than non-primed non-aged seeds. Khan (1994) has suggested that ethylene produced during germination in the presence of exogenous ACC would be a sensitive biochemical test for seed vigour. Several other studies have come to the same conclusion that ethylene evolution is diagnostic for seed vigour (Samimy and Taylor, 1983; Gorecki *et al.*, 1991; Chojnowski *et al.*, 1997). However, in these studies, accelerated ageing was used to deteriorate seeds, resulting in a significant reduction in viability as well as vigour. Seed vigour begins to be reduced prior to loss in viability. The current study indicates that even between seed lots with comparable viability, reduced vigour results in a reduced capacity to produce ethylene (Fig. 48.1).

Non-primed tomato seeds attained 50% radicle protrusion after 77 h, while primed seeds required only approximately 36 h (Fig. 48.1). Priming was accompanied by a significant increase in ethylene production during radicle protrusion compared with non-primed seeds. At 48 h, primed seeds produced ten times the ethylene of non-primed seeds (Fig. 48.1). Primed aged seeds recovered most of the vigour lost during storage. Primed aged and non-aged seeds had similar radicle emergence percentages after 48 h. However, ethylene production, though enhanced, was significantly lower in primed aged seeds compared with primed non-aged seeds (Fig. 48.1C). Primed seeds imbibed on ACC produced an order of magnitude more ethylene compared with non-primed seeds (Fig. 48.1D). The ethylene production of primed aged and primed non-aged seeds germinated on 5 mM ACC was comparable, only differing at 48 h.

A correlation between priming and ethylene evolution during germination has been shown in a number of species (Esahsi *et al.*, 1990; Chojnowski *et al.*, 1997; Habdas *et al.*, 2000). Priming appears to reduce the time required to initiate synthesis and/or activity of both ACC-synthase and ACC-oxidase. Fu *et al.* (1988) showed that ACC content and ACC-synthase activity were increased in groundnut (*Arachis hypogea* L.) seeds by osmotic priming. In the current study, endogenous ACC content in tomato seeds following priming was 4.0 ± 0.9 pmol per seed compared with 0.69 ± 0.08 and 1.1 ± 0.06 pmol per seed in non-primed non-aged and non-primed aged seeds, respectively. Based on the ability to produce ethylene during germination, it appears that endogenous ACC may not increase in primed aged seeds to the same degree as in primed non-aged seeds. Ethylene evolution in primed aged seeds was one-fourth that of primed non-aged seeds after 48 h when both sets of seeds had similar percentages of radicle protrusion (Fig. 48.1). ACC-oxidase activity, as indicated by the ability to convert saturating levels of ACC to ethylene, was greatly increased by priming (Fig. 48.1B). There was less difference between primed aged and primed non-aged seeds when imbibed on ACC relative to on water, but primed non-aged seeds still produced twice the ethylene at 48 h compared with primed aged seeds.

The significance of ethylene production to the priming process has received some attention. Cantliffe *et al.* (2000), investigating the interaction between priming and ethylene production in thermoinhibited lettuce (*Lactuca sativa* L.) seeds, determined that ethylene promoted endo-β-mannanase activity, which was partly responsible for inducing germination at elevated temperatures. A second interaction between ethylene production and seed priming may involve ethylene-mediated osmoregulation (Esahsi *et al.*, 1990). Primed cocklebur (*Xanthium pennsylvanicum* Wallr.) as well as ethylene-treated seeds showed increased osmolarity in their cell sap, enabling them to complete germination on mannitol solutions with reduced water potential.

The ethylene-insensitive *Nr* mutation in tomato was used to investigate further the interaction between ethylene and seed priming (Fig. 48.2). The time to radicle protrusion in *Nr* seeds was significantly reduced compared with wild type seeds (T_{50} for wild type was 75 h, *Nr* was 52 h). This has been shown to be more related to the impact of reduced ethylene perception during seed development, rather than a direct effect of ethylene perception on germination speed (Siriwitayawan, 2002). Seeds from both *Nr* and wild-type plants responded in a similar way to priming by significantly reducing the time to radicle emergence compared with non-primed seeds (Fig. 48.2). This suggests that ethylene perception is not required for seeds to show the benefits of priming. However, there are multiple receptors in tomato for ethylene and it is possible that additional receptors (i.e. *LeETR2* or *LeETR4*) act independently from *Nr* during seed germination to allow ethylene perception. Given this limitation, the maintenance of a seed priming effect in *Nr*

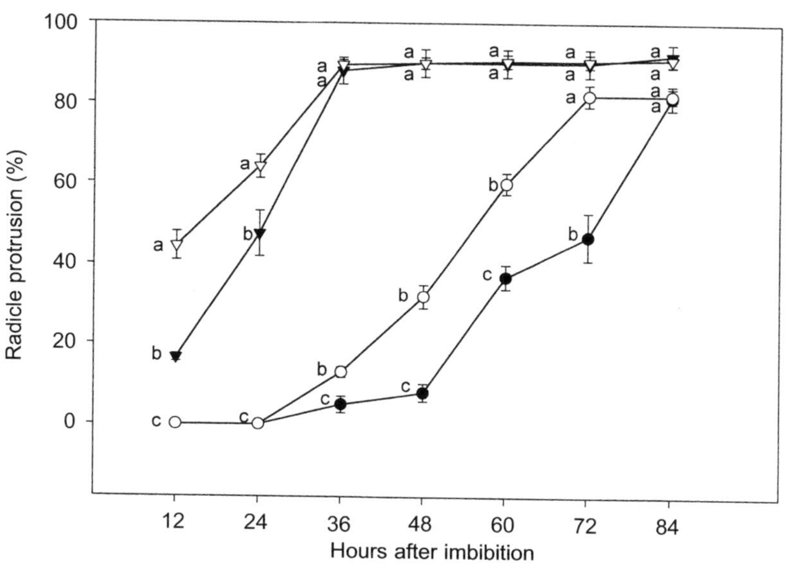

Fig. 48.2. Time to radicle protrusion from wild type (wt) or *never ripe* (*Nr*) tomato seeds before and after priming. Closed symbols wild type, open symbols *Nr*, circles untreated, triangles primed. Values are means ± standard error. Means followed by the same letter were not different using Tukey's test ($P \leq 0.05$).

tomato seeds is the best evidence to date to indicate that ethylene perception is not a requirement for reduced time to radicle protrusion in primed seeds.

References

AOSA (1992) *Seedling Evaluation Handbook*, No. 35. Association of Official Seed Analysts, Las Cruces, New Mexico, 101 pp.

Cantliffe, D.J., Nascimento, W.M., Sung, Y. and Huber, D.J. (2000) Lettuce endosperm weakening: a role for endo-β-mannanase in seed germination at high temperature. In: Black, M., Bradford, K.J. and Vazquez-Ramos, J. (eds) *Seed Biology: Advances and Applications*. CAB International, Wallingford, UK, pp. 277–285.

Chojnowski, M., Corbineau, F. and Côme, D. (1997) Physiological and biochemical changes induced in sunflower seeds by osmopriming and subsequent drying, storage and aging. *Seed Science Research* 7, 323–331.

Esashi, Y., Matsuyama, S., Hoshina, M., Ashino, H. and Ishizawa, K. (1990) Mechanism of action of ethylene in promoting the germination of cocklebur seeds. I. Osmoregulation. *Australian Journal of Plant Physiology* 17, 537–550.

Fu, J.R., Lu, X.H., Chen, R.Z., Zhang, B.Z., Liu, Z.S., Li, Z.S. and Cai, D.Y. (1988) Osmoconditioning of peanut (*Arachis hypogea* L.) seeds with PEG to improve vigour and some biochemical activities. *Seed Science and Technology* 16, 197–212.

Gorecki, R.J., Ashino, H., Satoh, S. and Esashi, Y. (1991) Ethylene production in pea and cocklebur seeds of differing vigour. *Journal of Experimental Botany* 42, 407–414.

Habdas, H., Szafirowska, A. and Sokolowska, A. (2000) Cytological and physiological effects of matriconditioning on low viable cucumber seed germination. *Acta Horticulturae* 517, 113–120.

Hua, J. and Meyerowitz, E.M. (1998) Ethylene responses are negatively regulated by a receptor gene family in *Arabidopsis*. *Plant Cell* 10, 1321–1332.

Khan, A.A. (1994) ACC-derived ethylene production, a sensitive test for seed vigor. *Journal of the American Society for Horticultural Science* 119, 1083–1090.

Lalonde, S. and Saini, H.S. (1992) Comparative requirement for endogenous ethylene during seed germination. *Annals of Botany* 69, 423–428.

Lizada, M.C.C. and Yang, S.F. (1979) A simple and sensitive assay for 1-aminocyclopropane-1-carboxylic acid. *Analytical Biochemistry* 100, 140–145.

Matilla, A.J. (2000) Ethylene in seed formation and germination. *Seed Science Research* 10, 111–126.

McDonald, M.B. (1998) Seed quality assessment. *Seed Science Research* 8, 265–275.

McDonald, M.B. (2000) Seed priming. In: Black, M. and Bewley, J.D. (eds) *Seed Technology and its Biological Basis*. Sheffield Academic Press, London, pp. 257–286.

McKeon, T.A., Hoffman, N.E. and Yang, S.F. (1982) The effects of plant-hormone pretreatments on ethylene production and synthesis of 1-aminocyclopropane-1-carboxylic acid in water-stressed wheat leaves *Triticum aestivum*. *Planta* 155, 437–443.

Petruzzelli, L., Perrino, P., Harren, F. and Quagliotti, L. (1993) Ethylene and pea germination. *Acta Horticulturae* 363, 159–166.

Samimy, C. and Taylor, A.G. (1983) Influence of seed quality on ethylene production of germinating snap bean seeds. *Journal of the American Society for Horticultural Science* 108, 767–769.

Siriwitayawan, G. (2002) The involvement of ethylene in determining seed vigor and the time required for the completion of germination. Dissertation, University of Kentucky, Lexington, Kentucky.

49 Recent Accomplishments and New Opportunities in Seed Research

RALPH L. OBENDORF AND SILVIA ODORCIC

Seed Biology, Department of Crop and Soil Sciences, 617 Bradfield Hall, Cornell University Agricultural Experiment Station, Cornell University, Ithaca, NY 14853-1901, USA

Recent Accomplishments

The workshop

The Seventh International Workshop on Seed Biology in Salamanca, Spain, organized by Professor Gregorio Nicolás, was highly successful with 270 participants from 33 different countries and 290 presentations, including 66 oral presentations/invited lectures and 224 posters. In addition to ten invited lectures, sessions were organized into five categories of symposia. Presentations included seed development (15 oral presentations and 29 posters), seed germination and dormancy (14 oral presentations and 78 posters), desiccation and other stress tolerance and conservation (11 oral presentations and 49 posters), seed ecology (10 oral papers and 33 posters) and seed biotechnology (6 oral presentations and 35 posters). Approximately one-third of the attendees in Salamanca were first-time participants in the International Seed Research Workshops, a major accomplishment.

Salamanca

Salamanca is a very beautiful and historic city, offering workshop participants an enormous variety of superb eating and sightseeing adventures and hotel accommodation ranging from the historic and archaeological to the ultramodern, all within walking distance of the workshop venue. With sessions beginning at 8.00 a.m., lunch at 1.30 p.m. and dinner after 9.00 p.m., participants learned to operate on few hours of sleep nightly in order to enjoy the ambience and diversity of Salamanca.

Overview

Approaches to research in seed biology are changing very rapidly. Molecular biology tools are now used in nearly every aspect of seed biology. Genomics provides new opportunities to characterize genes and gene regulation translates results from one species to another. Proteomics is being used to identify enzymes, pathways and protein products involved in various processes. Metabolomics offers new possibilities to identify numerous metabolites, quickly pointing to the pathways involved. Multidisciplinary approaches and collaboration are required to utilize these new technologies in seed biology.

Arabidopsis remains an important model plant and system for the study of seed biology. However, *Arabidopsis* does not provide all the answers (e.g. endosperm development) and so there is still room for other species in seed research.

Many new terms and acronyms were introduced during the workshop, including PCD (programmed cell death), d*ek* (mutants), EF1α, MIK, MARK, H3, GADPH, Rab 28, HKX, Cat, SOD, GR, ESR, EST, RAPD, SlowDry, ColdTest, *HvPRT1*, *LeMSide1*, *LeMan1*, *PvGLP1*, *Vp1*, *SnRK1*, *SIP1*, *SNF4*, (Snuff and Sneeze) . . . but we are not sure about the last two!

Seed development

Since our last meeting in Merida, pathways for the biosynthesis of the raffinose family of oligosaccharides (RFOs) have been defined. Galactinol synthase, raffinose synthase and stachyose synthase have been cloned from multiple species and heterologously expressed for enzyme characterization and substrate specificities. Galactinol synthase is a key enzyme in stachyose biosynthesis. Galactinol synthase is controlled by cold, drying and abscisic acid (ABA). Stachyose synthase is not controlled by ABA, but stachyose accumulation probably is controlled by galactinol synthase. New information on galactosyl cyclitol biosynthesis has been discovered. Except for galactinol, which may be synthesized in leaves of some plants, the galactosyl cyclitols appear to be biosynthesized in seeds and more specifically in embryo tissues. Whereas the cyclitol *myo*-inositol appears to be present in all living cells, D-ononitol, D-pinitol, and D-*chiro*-inositol appear to be biosynthesized in vegetative tissues and transported to the seeds to form their respective galactosyl derivatives. Galactosyl derivatives of *O*-methylated cyclitols appear to be biosynthesized by stachyose synthase in embryo tissues. By contrast, current evidence indicates that galactosyl derivatives of D-*chiro*-inositol may be formed by galactinol synthase in embryo tissues.

Knowledge on health-related compounds in seeds has been enhanced. For example, while some wish to lower RFOs to reduce flatulence, others emphasize the health benefits of consuming RFOs. While some consider phytic acid to be an anti-metabolite, a chelator of essential heavy metals and a site source of phosphorus pollution from animal manures, others consider phytic acid to be a beneficial antioxidant and beneficial in supplying essential dietary nutrients. Fagopyritol A1, a galactosyl derivative of D-*chiro*-inositol found in buckwheat seeds, has a molecular structure isosteric with a putative insulin mediator believed to be deficient in subjects with non-insulin dependent diabetes mellitus and polycystic ovary syndrome. Two unique series of galactosyl oligomers of D-*chiro*-inositol have been characterized.

Applications of genomics have revealed the presence of many genes for which an identity or function remains to be elucidated. From analysis of maize expressed sequence tag (EST) clones, as many as 500 genes from developing endosperm remain unidentified and their function remains unknown. In legume embryos, a normal epidermal layer including transfer cell activities appears to be essential to normal embryogenesis. Storage tissues of both cereals and legumes undergo programmed cell death either during seed maturation, as in the central endosperm of cereals, or at sometime during seedling growth for aleurone, scutellum and cotyledon tissues.

Seed germination and dormancy

Seed priming, modelling and predicting were dominant themes in this workshop. Modelling thermotime, hydrotime and osmotime were recurring themes. Water relations (solute potential) thresholds of –1.7 MPa for seed maturation (cessation of dry matter accumulation) and –1.4 MPa for germination seem to be recurring values with predictive potential. Predicting genotypic cold test performance using proteomics to identify regulatory reactions and pathways, and predicting seedling emergence using three-dimensional arrays of soil particles and soil environments were emerging technologies with implications for breeding programmes and agronomic improvement. Rather than being an inhibitor, ABA is classified as a promoter of regulatory factors and events with strong interactions with gibberellin (GA) and other regulators. Clearly, different genes are expressed at different times and in different tissues during pre-germination and post-germination. Early metabolic activities require only minimal levels of hydration and are not correlated to pockets of greatest hydration. The role of Vp1 in dormancy and preharvest sprouting is still being debated. While Vp1 is a regulatory factor involved in controlling many processes normal to seed maturation and germination, regulation of Vp1 may be a consequence, rather than a cause, of dormancy status. In one unique study, weedy rices are being bred for map-based cloning of dormancy genes. QTL mapping is being used in genetic analysis of seed development, dormancy, germination and assessment of vigour.

Desiccation and other stress tolerance and conservation

Desiccation tolerance was reviewed with emphasis on the mechanisms and processes contributing to the observed responses. Under certain circumstances, rigid membranes are acceptable and not necessarily bad or catastrophic. The role of glasses and molecular mobility was reviewed in detail, with emphasis on the potential to predict seed lifespan from measurements of molecular mobility. Down-regulation of metabolism is essential for the development of desiccation tolerance. Oxidative stress, immobilization of the membrane surface and various signalling cascades, whether a hexokinase-dependent pathway or a hexokinase-independent pathway, are being studied actively for down-regulation of metabolism. Recent results suggest a regulatory role for the SnRK1 complex, the sucrose non-fermenting kinase, in signalling pathways leading to desiccation tolerance. cDNA microarrays, proteomics and genetic fingerprinting are being used extensively to predict up-regulated and down-regulated genes, as well as pathways related to desiccation tolerance and the down-regulation of metabolism essential for desiccation tolerance and preservation of germplasm. Recalcitrant systems may be dried to a lower moisture before damage is incurred if rapidly desiccated rather than being slowly desiccated.

Seed ecology

Communication is the key to seed ecology. If we do not communicate effectively with each other, then what we name things becomes irrelevant. However, when we do talk to one another, communication in a common language is essential. Learning to communicate can only be accomplished if we adopt and use each other's terms (note some listed above).

'Know your dormancy' was a dominant theme. Major classes of dormancy have been described and knowledge of the dormancy classification of your favourite seed may lead to predictable responses in some experiments, or to predictable failure in other experiments. It is important to distinguish between dormancy and persistence, especially when predicting a response to the soil environment. Phenotypic plasticity and germination deferment strategies were discussed as adaptation strategies in natural environments.

Sufficient data are now accumulated to support modelling of thermotime and hydrotime in an ecological setting.

Smoke, components of smoke and other factors were reviewed and updated.

Seed banks and preservation of germplasm

PCR-RAPDs are being used to study the effect of cryopreservation on genetic stability. Seed lifespans are being related to stability of genomic DNA. Refinements in the predictive modelling of seed longevity has renewed interest. Improved technologies for cryopreservation and mechanical methods for recording germination and vigour are being developed.

Seed biotechnology

Seeds deliver genes, including designer genes, encoding unique designer products that have storage stability within the seed. Molecular farming captures the potential of seeds as ideal repositories for recombinant proteins. Designer oil and lipid products are now widely produced using seeds as factories. Using oil seeds for protein production in which the unique properties of oil bodies and oleosins are exploited has led to a 'seed based manufacturing system for a wide range of therapeutic and industrial proteins'. Seeds have long been used as food and feed and the nutritional composition of seeds has been a matter of selection for thousands of years. A renewed emphasis on the use of seeds for health is fuelled by the recent identification of numerous naturally occurring compounds in seeds with potential for improved health. Seeds are also being exploited as factories for a variety of designer products (protein, oil, starch, sugar, cyclitols and galactosyl cyclitols, as well as non-natural products) with specific health-related benefits.

Improved techniques are being developed to detect adventitious genetically modified (GM) materials in non-GM seeds by using statistical sampling models, improved specificity in molecular detection, and quantification of the number of insertions present in the genome.

Molecular methods are being used in all areas of seed research.

New Opportunities in Seed Research

Seeds for the future

In the opening lecture to this workshop, the importance of seeds to the development of civilization, historical and modern agriculture, international trade and human sustenance was emphasized. Cereals alone account for 50% of the caloric and protein needs of humankind by direct human consumption and 75% if indirect consumption through meat, milk, eggs and alcoholic beverages is included. The modification of these species more than 8000 years ago for cultivation by our ancestors led to irreversible changes that made seeds of our staple food crops incapable of survival without human intervention. These historic changes to our staple food and feed seeds were far more drastic than those of modern GMO concerns. A major challenge and sobering responsibility for seed biologists and ecologists is the preservation of genetic resources for our staple food and feed crops upon which future generations will be dependent. We have a daunting responsibility to continue to understand and to improve the storability, stress tolerance, performance, nutritional and utilitarian composition, and healthfulness of seeds.

Communicate and collaborate with each other. Communicate and collaborate between public and private sectors. It was stated repeatedly during this workshop that the needs and objectives are different between public and private sectors. Academic institutions provide trained personnel needed by the private sector. Keep in mind the needs of the private sector when planning training programmes. With shrinking resources for seed biology training in many academic institutions, the private sector may benefit from supporting academic programmes. Research grants, fellowships, traineeships, internships, reciprocal exchanges of scientists and reciprocal non-disclosure agreements are but a few of the many ways for private and public sectors to support each other.

Tolerate the terms used by others. Work at learning a common language. Get together to conduct research. Debate is advantageous, but be tolerant of contrasting points of view and terminologies. Once you speak a common language, differences may be minimal.

Metabolomics is a new thrust. Microscale analytical procedures are being developed to analyse hundreds of metabolites for thousands of samples in a short time, thereby providing another tool for genetic evaluation of metabolic pathways. Proteomics is being practised. Genomics offers regulatory features and identification of gene families that may not be fully elicudated using other technologies. Remain watchful of other 'omics', many of which may yet be defined.

Cross the traditional discipline lines in searching for solutions to important problems. Keep an eye on the questions raised by our president, Professor Côme, at the beginning of this workshop. What is germination? What is dormancy? There is still much to be learned.

Young scientists represent a significant proportion of our participants (one-third attended this workshop for the first time) and a powerful resource for the future of seed biology and ecology, ready to adopt and expand on the many emerging technologies with energy, enthusiasm and unbiased perspectives.

Embrace diversity. This workshop represents a very diverse group of scientists from 33 different countries addressing an impressive array of subjects in seed biology and ecology. In addition, there are a large number of women amongst us who are willing and professionally capable of leadership roles.

Take a chance. Dare to be different. Try a risky experiment. And do something you love every day. Your happiness depends on it.

Index

Page numbers in **bold** refer to illustrations and tables